ASYMMETRICAL FUNCTION OF THE BRAIN

ASYMMETRICAL FUNCTION OF THE BRAIN

EDITED BY

MARCEL KINSBOURNE

Professor of Pediatrics and Psychology
The University of Toronto
Toronto

CAMBRIDGE UNIVERSITY PRESS

CAMBRIDGE

LONDON · NEW YORK · MELBOURNE

Published by the Syndics of the Cambridge University Press
The Pitt Building, Trumpington Street, Cambridge CB2 1RP
Bentley House, 200 Euston Road, London NW1 2DB
32 East 57th Street, New York, NY 10022, USA
296 Beaconsfield Parade, Middle Park, Melbourne 3206, Australia

© Cambridge University Press 1978

First published 1978
Reprinted 1979

Printed in the United States of America
Typeset by Automated Composition Service, Inc., Lancaster, Pa.
Printed by Hamilton Printing Co., Rensselaer, N.Y.
Bound by Payne Edition Bindery, Inc., Chester, N.Y.

Library of Congress Cataloging in Publication Data
Main entry under title:
Asymmetrical function of the brain.
Includes index
1. Cerebral dominance. 2. Laterality.
3. Brain - Localization of functions. 4. Cognition.
I. Kinsbourne, Marcel.
[DNLM: 1. Laterality. 2. Brain - Physiology.
3. Behavior. 4. Cognition. W1335 A861]
QP385.5.A85   612'.825   77-8633
ISBN 0 521 21481 5

Contents

Editorial preface vii   Contributors ix

I. Biological origins: preamble

1 Biological determinants of functional bisymmetry and asymmetry
  Marcel Kinsbourne   3

II. Studies of lateralized lesions

2 Functional cerebral hemispheric asymmetry
  Sidney Weinstein   17

3 Hemispheric asymmetry as evidenced by spatial disorders
  Ennio De Renzi   49

4 Spatial and temporal factors in visual perception of patients with unilateral cerebral lesions
  Amiram Carmon   86

5 Direct examination of cognitive function in the right and left hemispheres
  Robert D. Nebes   99

III. Studies of behavioral asymmetry

6 Lateral dominance as a determinant of temporal order of responding
  E. Rae Harcum   141

7 Mapping cerebral functional space: competition and collaboration in human performance
  Marcel Kinsbourne and Robert E. Hicks   267

8 Lateral asymmetries revealed by simple reaction time
  James Swanson, Alexa Ledlow, and Marcel Kinsbourne   274

9 Asymmetry of electrophysiological phenomena and its relation to behavior in humans
  Gail R. Marsh   292

10 Weber on sensory asymmetry
   J. D. Mollon   318

IV. Comparative studies

11 Manipulative strategies of baboons and origins of cerebral asymmetry
    Colwyn Trevarthen — 329

12 Dichotic listening and the development of linguistic processes
    M. P. Bryden and F. Allard — 392

13 Sex differences in spatial ability: possible environmental, genetic, and neurological factors
    Lauren Julius Harris — 405

14 Human handedness
    Robert E. Hicks and Marcel Kinsbourne — 523

V. Biological origins: perspectives

15 Evolution of language in relation to lateral action
    Marcel Kinsbourne — 553

Name index — 567
Subject index — 578

Editorial preface

In principle, the fact that some cognitive functions are asymmetrically represented in the human brain has been accepted by neuropsychologists for more than a century. But only recently has it become apparent how pervasive a property of higher mental functions is their asymmetrical representation at the cerebral and even the thalamic level. The left lateralization of verbal processes had long been regarded as a curious exception to the generally bisymmetrical organization of the vertebrate central nervous system. As an increasing number of different left-lateralized processes came to light, an increasingly strained effort was made to write them all off as verbal or at any rate verbally mediated. This conservatism finally succumbed to the clear demonstration of complementary specialization of the two hemispheres. Now that we know both hemispheres exhibit specialization, the question is the reverse: Are there any cognitive functions that are not asymmetrically represented? It cannot, as yet, be asserted that there are.

This book illustrates the types of investigations that have led to this upheaval in neuropsychology. In Part II, productive clinical neuropsychology programs are summarized that bear on double dissociation of cerebral representation of cognitive function. In Part III, the early discovery and modern interpretations of asymmetries in normal human behavior are considered. In Part IV, light cast by behavioral asymmetries on species-, age-, sex-, and handedness-related differences in cerebral organization are considered. Parts I and V place these issues in a biological framework.

The editor respectfully dedicates this book to Professor Oliver Zangwill, who pioneered the concept of complementary specialization of the cerebral hemispheres and has had fruitful influence on generations of neuropsychologists. We are all much in his debt.

Marcel Kinsbourne

Toronto, November 1977

Contributors

F. Allard, Ph.D., Assistant Professor, Departments of Psychology and Kinesiology
University of Waterloo
Waterloo, Ontario, Canada

M. P. Bryden, Ph.D., Professor of Psychology,
University of Waterloo
Waterloo, Ontario, Canada

Amiram Carmon, M.D., Ph.D., Professor and Director, Department of Behavioral Biology,
Medical School, Technion – Israel Institute of Technology
Haifa, Israel

Ennio De Renzi, M.D., Full Professor and Head, Neurological Department,
University of Modena
Modena, Italy

E. Rae Harcum, Ph.D., Heritage Professor of Psychology,
College of William and Mary
Williamsburg, Virginia, U.S.A.

Lauren Julius Harris, Ph.D., Associate Professor of Psychology
Michigan State University
East Lansing, Michigan, U.S.A.

Robert E. Hicks, Ph.D., Associate Professor of Psychology
State University of New York
Albany, New York, U.S.A.

Marcel Kinsbourne, D.M., M.R.C.P., Professor of Pediatrics and Psychology,
University of Toronto
Director, Neuropsychology Research Unit, Hospital for Sick Children
Toronto, Ontario, Canada

Alexa Ledlow, Ph.D., Postdoctoral Fellow
Neuropsychology Research Unit, Hospital for Sick Children
Toronto, Ontario, Canada

Gail R. Marsh, Ph.D., Associate Professor of Psychiatry
Duke University Medical Center
Durham, North Carolina, U.S.A.

J. D. Mollon, D. Phil., Lecturer in Experimental Psychology
Cambridge University
Cambridge, England

Robert D. Nebes, Ph.D., Assistant Professor of Psychiatry
Duke University Medical Center
Durham, North Carolina, U.S.A.

James Swanson, Ph.D., Research Associate
Neuropsychology Research Unit, Hospital for Sick Children
Toronto, Ontario, Canada

Colwyn Trevarthen, Ph.D., Reader in Psychology
University of Edinburgh
Edinburgh, Scotland

Sidney Weinstein, Ph.D., President, NeuroCommunication Research Laboratories, Inc.
Editor-in-Chief, International Journal of Neuroscience
Danbury, Connecticut, U.S.A.

PART I

BIOLOGICAL ORIGINS: PREAMBLE

1
Biological determinants of functional bisymmetry and asymmetry

MARCEL KINSBOURNE

At times progress in an area of research encounters technological limitations, and at times the limitations are conceptual. At present, progress in the area of brain organization as it relates to behavior is limited for the most part by the use of inadequate theoretical models. The *switchboard model* (Kimura, 1961; Geschwind, 1965) conceives of brain function as an exercise in information flow and ignores the possibility of dynamic interactions between areas of brain. It is concerned with the time required for a message to travel from one point in the brain to another and the extent to which its intelligibility is impaired by the journey. Although it has served a useful purpose, this model underestimates the capabilities of the brain and our ability to understand them. Switchboard theorists also exaggerate the range of behavioral phenomena that lend themselves to this passive type of modeling. After all, connections between any two points in the nervous system not only may transmit information, they also may mediate either excitation or inhibition (Sherrington, 1906; Kinsbourne, 1974a). The switchboard model is based on the notion of passive stimulus-response relationships. Rather than adopt this pretence that it is natural for animals to be passively exposed to input and to jerk lawfully in response, we acknowledge that normally an animal searches actively for information. Here follows an idealized minimal account of the behavior common to all active bisymmetrical organisms (Bilateria).

THE SEARCH CYCLE

Begin observing the individual while he is in an inert, vegetative state (parasympatheticotonus). After a time, under the influence of some drive, he enters an alert or sympatheticotonic state. This is a state of search (or of problem-solving thought, its internal equivalent). Both external search and problem-solving thought are characterized by specific mental sets that in themselves are fashioned from expectancies and their relative payoffs.

Orientation

The beginnings of search are anticipatory receptor orientations, exploratory in various directions and by means of various sense modalities. In the course of these orientations, if they are external, the individual may happen upon an event, which he then experiences. If they are internal, he may remember an event. That event then generates further (reactive) orientations for the purpose of data acquisition (the data being events or memorandums). Each orientation is followed by data analysis. Reactive orientation continues until there is enough information to satisfy a preset decision criterion. That decision, however elaborate it is, must incorporate a choice between approach and withdrawal (Schneirla, 1959). If the decision is to withdraw, then the search continues and subsequently resumes the above-mentioned sequence. If the decision is to approach, there follow locomotion (if the object is at a distance), grasp, and consummation. The proceeds of consummation restore the inert parasympatheticotonic state, which later is the point of departure for a repeat of the search-initiated cycle. With respect to this sequence, search through consummation, we can now consider which component steps must for adaptive purposes be bisymmetrically represented and which can plausibly depart from bisymmetry without detriment to their adaptive usefulness.

Bisymmetry is an improbable state of affairs. For any biological system to be exactly counterpoised could hardly occur by chance. Bisymmetry must be the outcome of natural selection based on adaptive advantage. It was first developed for purposes of efficient locomotion and swift turning to either side and is not necessarily perpetuated if the organism evolves to a sedentary mode. Even when it occurs, bisymmetry is presumably approximate rather than perfect, as is the rule with biological systems (in contrast to engineering systems). So there must be further fine controls to correct inequalities inherent in the primary balances (between right and left turning; Kinsbourne, 1974a).

Biological determinants

To illustrate these principles, consider turning behavior from the point of view of the most ancient bisymmetrical organism, a wormlike creature (Hyman, 1940). Being elongated and terrestrial, the worm must be able to turn right or left. As worms do not have special senses, we can relate the origin of bisymmetry primarily to this motoric requirement: to be able to turn with equal effectiveness and speed either to the left or to the right in relation to external events that with equal probability could occur on either side. So, right orienting and left orienting in terrestrial animals are the basic behaviors that make bisymmetry necessary, and whatever control center encodes such turning in a particular animal must therefore be represented in both halves of the nervous system. These opposing tendencies are in mutual inhibitory balance, and the organism travels along the vector that results from their interaction (Kinsbourne, 1974a). In such turning behavior we see directional biases only if we set up artificial conditions specially designed to make manifest any inherent lack of complete equality in the balance between these opposing tendencies. Suppose an animal has more tendency to turn to the right than to turn to the left, although adaptively it is most useful to be able to turn with equal facility to either side. Such an innate bias to one side can be corrected in two ways: by external stabilization using the spatial framework provided by sensory input to hold the animal on its course, or by "voluntarily" deciding to turn more to the weakly represented side than the other. For instance, a lower animal that normally travels in a straight line might, when deprived of its distance receptor information, travel in a trajectory with a systematic bias in one direction (Kinsbourne, 1974b). Analogously, if a human being is asked to divide his attention between an auditory message to the right ear and another to the left, without knowledge of results or preconceptions about how he is doing, he might reveal an otherwise compensated bias to one side. But, if he knew he had missed some trials on one side, he could readily compensate by shifting his attention to that side (at the expense of the other), thereby masking the basic imbalance. A special case in which a turning bias goes uncorrected is when the subject is preoccupied with some other task and the axis of attention is not relevant to that main task. This reveals a bias which, had the subject been aware of it, could have easily been overcome. This conceptualization applies to experiments on task-specific lateral direction of gaze while thinking. When right-handed people are asked to solve verbal problems, they usually look to the right while thinking about them; if they are asked to solve spatial problems, they look up and to the left while doing so (Kinsbourne, 1972, 1974c). However, they could easily think in these modes without deviating gaze in these ways and, indeed, do so once they become self-conscious about where they are looking while thinking.

Coding

When the orienting is completed and data analysis proceeds at a more abstract level without further need to refer to external stimuli, bisymmetry becomes less necessary. Abstracting or coding the information need not be done equally on the two sides. So the adaptive pressure toward bilateral symmetry relaxes with respect to that stage of information processing, and as bisymmetry occurs only when there is need for it, there is, in fact, asymmetry when that need is absent. Therefore, higher mental functions would be expected not to be symmetrically represented, there being no reason why they should be; and, in fact, they are not.

Locomotion

Orienting is the first step in the approach sequence. Given the decision to proceed, it acts as point of departure for locomotion to the target (unless an approach decision is made for a target that happens already to be within reach, in which case grasp supervenes at once). Basic to any decision to act is the choice between approach and withdrawal (Schneirla, 1959). That, too, must be bisymmetrical, because where one goes is determined by external circumstances not under his control that are distributed unpredictably to the right and left in space. So the locomotor apparatus is bisymmetrical; but even there minor wrinkles can be demonstrated. For example, in order to help them locomote, fish have fins that steer them in various ways like rudders, and these fins are not equally thick and muscular on both sides (Hubbs & Hubbs, 1944). In most species they are more muscular on the right. But for actively mobile species, the somatic asymmetry is not so severe as to impair the streamlining that makes for mechanical advantage in movement.

Consummation

Having approached its target, the animal exhibits consummatory behavior. For instance, it manipulates and chews. Here the need for symmetry depends on what is done, which in turn is limited by what the species is capable of. Most species act bimanually with the forelimbs in coordination, and for them bisymmetry is advantageous and therefore expected. So we would not expect paw preference to be a good indicator of brain organization in those species that do not normally use one paw in isolation for biologically important functions. Nor do laboratory tasks that call for single-limb response necessarily tap limb preferences even if they exist. With species that do practice unimanual activities in the natural environment, there is not necessarily any biological pressure to use each hand equally well, because once the object is apprehended one can, at leisure, so position himself that one and the same hand will do in virtually any situation. Because there is no adaptive need for bisymmetry, the scene is set for asymmetry. So at this behavioral stage one would expect asymmetry, and (in the human) it is there.

Thus, there are two functional locuses for major asymmetry in some species; at orienting and at consummating. One example of asymmetrical turning in humans is the spontaneous turning of newborns, four times more often to right than to the left (Turkewitz et al., 1965; Siqueland & Lipsitt, 1966). Another is the tendency to show a performance differential in hemifield presentation and dichotic listening in favor of stimuli emanating from one side of space rather than the other (e.g., Kimura, 1961, 1967). Both are basically elaborations of asymmetry: turning sideways, scanning sideways (Kinsbourne, 1970a, 1973, 1975). There is either physically obvious turning – of the eyes, the head, the body, singly or in combination – or a premotor attentional shift based on the same mechanism, though of shorter latency. These orienting operations are quite distinct from the consummatory acts, typically the grasp. Because hand preference as usually estimated relates to the use of the limb in manipulation, not in pointing, there is no particular reason why hand preference and side of attentional preference should be perfectly correlated. Nor are they (see Chapter 14). The asymmetries at the consummatory stage include paw preference, unequal dexterity between the two hands that usually favors the right (Provins, 1967; Annett, 1972;), and the observation that 3-month-old babies maintain grasp far longer with the right hand than with the left hand (Caplan & Kinsbourne, 1976).

In summary, asymmetry represents a relaxation of the need for symmetry. Asymmetry occurs in inverse degree to the biological advantage of bisymmetrical representation. That advantage depends on the particular stage in behavior catered to by the control mechanism under scrutiny. Advantage may, of course, also accrue from the asymmetry itself, but this is not a necessary condition for its appearance in phylogeny.

DEMONSTRATIONS OF BEHAVIORAL ASYMMETRIES

One can demonstrate behavioral asymmetries caused by innate or preprogrammed inequalities, which normally are functionally compensated for, by experimentally observing the effect of strategically located damage or by disposing of stabilizing factors.

Lateralized brain damage

Take the phenomenon of unilateral neglect of space. When people are damaged in certain parts of one or other cerebral hemisphere, they show a striking lack of orienting turns to the side opposite to the damage. They tend not to look over to that side, but instead unduly frequently look to the same side as the damage. Unilateral neglect is an imbalance in orienting tendencies made manifest by brain damage (Kinsbourne, 1970b, 1977). It is a pathological exaggeration of the imbalance we have drawn attention to in newborn babies. An interesting finding that has been in search of an explanation is that in humans lesions of the right hemisphere of the brain cause much unilateral neglect of space in many cases, whereas left hemisphere lesions cause a little neglect in only a few cases. Why? Right and left turning in the intact human being has to be equilibrated by corrective mechanisms; there is an innate propensity to turn to the right more than to the left. Right hemisphere damage releases the overpowering right-turning tendency programmed by the left hemisphere. The effect of damage to the left side on the distribution of attention across space is not as serious, because the left-turning tendency programmed by the right hemisphere is not as great. A supplementary explanation is a takeoff point in introducing the principles controlling the effect of functional proximity of control mechanisms. In the left half of the brain are located not only the facility that turns attention to the right (in all persons) but also the verbal processor (in some 96% of persons). When the verbal processor is activated because a person is anticipating speaking, listening to speech, or speaking, then that activation overlaps the adjacent right-turning control center and the verbal activity biases attention to the right. The left-turning center, in the right hemisphere, is far removed from the site of activation and is overpowered by its activated opponent. Consider the patient with brain damage and the clinician approaching him. The patient's right hemisphere is damaged, his left is in control, so he tends to turn to the right. The clinician addresses the patient. This activates the patient's left hemisphere and attention swings to the right. So the verbal interchange exacerbates the attentional imbalance, which thus becomes more observable clinically. Take the opposite case. The left hemisphere is damaged. The clinician approaches and makes verbal overtures as before. Again, the verbal processor in the patient's left hemisphere is activated. But this time it minimizes the imbalance. It is as if a "behavioral electrode" were applied to the left hemisphere to strengthen its opposition to the left-turning bias of the intact right hemisphere. This proposed mechanism is an instance of a more general principle with respect to the manner in which the *functional distance* between control centers affects their interaction. Any two control centers can be ranked on a continuum of how closely they approximate to each other. The degree of proximity is operationally defined as follows: The closer control centers are to each other, the more effectively they collaborate on concordant tasks and the less effectively they time share on discordant tasks. Conversely, the further apart they are functionally, the less effectively they collaborate on concordant tasks, but the more effectively they time share on discordant tasks.

Dual-task interaction studies

We have demonstrated the effect of functional distance in intact humans in two main ways: by showing how two processors in one hemisphere interact and by showing how a processor in one hemisphere interacts with the orientor in the same hemisphere. Incidentally, these considerations bring us closer to a mechanism for the perceptual asymmetry that has been much described in the literature.

Interaction of processor and orientor
If a subject who, like most people, is left-lateralized for language prepares for verbal activity, there is neuronal activation of his left hemisphere. That activation triggers a rightward attending in whatever is the relevant modality. If that rightward attentional bias suits the task, performance is unimpaired or even enhanced. If the verbal task is unaffected as regards its performance by the concurrent direction of gaze, there is no asymmetry. If the performance is enhanced by looking to the right, then it benefits from the left hemisphere activation. If it is enhanced by looking to the left, the reverse is true. So, if the subject expects a verbal stimulus and it happens to appear on the right, that is advantageous because the left hemisphere's verbal processing is consistent with right turning; they both are triggered within the same hemisphere, in areas close together. However, if in order to pick up verbal information the subject has to scan to the left, the two activities are discordant. Verbalizing activates the left hemisphere, and looking left inhibits it. On account of this conflict, performance is suboptimal. This explains asymmetries in perception (Kinsbourne, 1970a, 1973, 1975).

In contradistinction to what would follow from structural hypotheses explaining perceptual asymmetries, the orientational model does not lead one to expect the degree of asymmetry in dichotic listening or hemifield viewing directly to mirror the degree to which the relevant function is lateralized to the opposite side of the brain. In fact, we doubt the validity of the concept that the degree of lateralization of cognitive function varies between individuals as distinct from the side of lateralization. The extent to which attention swings contralateral to the more active hemisphere is determined not only by where the active processor is lateralized, but also by how hard it is working. How hard it is working is in turn determined by how difficult the task is and how willing and able the subject is to rise to the challenge of the task. It is even affected by what else he is thinking about at the time. And until these potentially relevant variables are disentangled, it is premature to infer, on the basis of perceptual asymmetries, that elderly people

Biological determinants

are more or less lateralized than young people, that males are more or less lateralized than females, or even that processors for particular stimulus materials within a modality are represented by mechanisms more or less lateralized than others.

Interaction of the processors

Subjects speak or remain silent while balancing a dowel rod on either the right or the left index finger (Kinsbourne & Cook, 1971). The subjects can balance the dowel longer on the right than on the left in the silent condition, but longer on the left than on the right in the speaking condition. It is as if speaking knocks the stick off the finger. This illustrates the concept that the verbal processor and the control mechanism for right-hand movement are functionally close together and therefore interfere when incongruent in their respective activities (see Chapter 7). We have been able to validate this principle with concurrent speaking and finger tapping in 5-year-old children and in a person with callosal section. A reduced brain capacity, whether because of immaturity (in a child; Kinsbourne & McMurray, 1975) or restriction of cerebral space (as following callosal section; Kreuter et al., 1972), amplifies these interference effects. A paradigm that gives like detectable asymmetry in mature adults could give greater asymmetry in a child. If properly tested for, asymmetries may actually be more detectable in children than in adults (Kinsbourne & Hiscock, 1977). It is not a question of finding out when in development a cognitive function lateralizes to one hemisphere. Function is lateralized where it is going to be from the start (Kinsbourne, 1976). The biasing effect of lateralization becomes of less consequence as the further sophistication of the brain corrects for these adaptively irrelevant biases.

SUMMARY

We arrive at the following statements. Bisymmetry is not a given; it is specifically programmed only if functional pressures make it necessary. Those functional pressures are particularly clear in relation to externally directed behaviors – turning right, turning left, walking right, walking left. For parts of the body whose functions are unrelated to considerations of symmetry, there is none.

REFERENCES

Annett, M. 1972. The distribution of manual asymmetry. *Br. J. Psychol. 63:* 343–358.

Caplan, P. J., & Kinsbourne, M. 1976. Baby drops the rattle: Asymmetry of duration of grasp by infants. *Child Dev. 47:* 532–534.

Geschwind, N. 1965. Disconnexion syndromes in animals and man. *Brain 88:* 237–294, 585–644.

Hubbs, C. L., & Hubbs, L. C. 1944. Bilateral asymmetry and bilateral variation in fishes. *Papers Michigan Acad. Sci. 30:* 229–311.

Hyman, L. H. 1940. *The Invertebrates.* New York: McGraw-Hill.

Kimura, D. 1961. Cerebral dominance and the perception of verbal stimuli. *Can. J. Psychol. 15:* 166–171.

  1967. Functional asymmetry of the brain in dichotic listening. *Cortex 3:* 163–178.

Kinsbourne, M. 1970a. The cerebral basis of lateral asymmetries in attention. *Acta Psychol. 33:* 193–201.

  1970b. A model for the mechanism of unilateral neglect of space. *Trans. Am. Neurol. Assoc. 95:* 143–145.

  1972. Eye and head turning indicate cerebral lateralization. *Science 176:* 539–541.

  1973. The control of attention by interaction between the cerebral hemispheres. In S. Kornblum (ed.). *Attention & Performance IV.* New York: Academic Press.

  1974a. Lateral interactions in the brain. In M. Kinsbourne & W. L. Smith, (eds.). *Hemispheric Disconnection and Cerebral Function.* Springfield, Ill.: Thomas.

  1974b. Mechanisms of hemispheric interaction in man. In M. Kinsbourne & W. L. Smith (eds.). *Hemispheric Disconnection and Cerebral Function.* Springfield, Ill.: Thomas.

  1974c. Direction of gaze and distribution of cerebral thought processes. *Neuropsychologia 12:* 279–281.

  1975. The mechanism of hemispheric control of the lateral gradient of attention. In P.M.A. Rabbitt & S. Dornic (eds.). *Attention & Performance V.* New York: Academic Press.

  1976. The ontogeny of cerebral dominance. In D. R. Aaronson & R. W. Rieber (eds.). *Developmental Psycholinguistics and Communication Disorders.* New York: N. Y. Academy of Sciences.

  1977. Hemi-neglect and hemisphere rivalry. In E. Weinstein (ed.). *Hemi-inattention and Hemisphere Specialization.* New York: Raven Press.

Kinsbourne, M., & Cook, J. 1971. Generalized and lateralized effect of concurrent verbalization on a unimanual skill. *Q. J. Exp. Psychol. 23:* 341–345.

Kinsbourne, M., & Hiscock, M. 1977. Does cerebral dominance develop? In S. Segalowitz & F. A. Gruber (eds.). *Language Development and Neurological Theory*. New York: Academic Press.

Kinsbourne, M., & McMurray, J. 1975. The effect of cerebral dominance on time sharing between speaking and tapping by preschool children. *Child Dev. 46:* 240–242.

Kreuter, C., Kinsbourne, M., & Trevarthen, C. 1972. Are deconnected cerebral hemispheres independent channels? A preliminary study of the effect of unilateral loading on bilateral finger tapping. *Neuropsychologia 10:* 453–461.

Provins, K. A. 1967. Motor skills, handedness, and behaviour. *Aust. J. Psychol. 19:* 137–150.

Schneirla, T. C. 1959. An evolutionary and developmental theory of biphasic processes underlying approach and withdrawal. In M. R. Jones (ed.). *Nebraska Symposium on Motivation*. Lincoln: University of Nebraska Press, 1–42.

Sherrington, C. S. 1906. *Integrative Action of the Nervous System*. New Haven: Yale University Press.

Siqueland, E. R., & Lipsitt, L. P. 1966. Conditioned head-turning in human newborns. *J. Exp. Child Psychol. 3:* 356–376.

Turkewitz, G., Gordon, E. W., & Birch, H. G. 1965. Head turning in the human neonate: Spontaneous patterns. *J. Genet. Psychol. 107:* 143.

PART II

STUDIES OF LATERALIZED LESIONS

2
Functional cerebral hemispheric asymmetry

SIDNEY WEINSTEIN

The concept of cerebral dominance has been based upon the frequent relationship of aphasia and injury to the left cerebral hemisphere and the concomitant fact that right-handedness occurs in some 95% of the population. However, after discovery of multiple functional asymmetries favoring the right as well as the left cerebral hemisphere, I have become convinced that even though the so-called minor hemisphere is less frequently concerned with speech than the so-called major hemisphere, its reputation as "minor" is otherwise undeserved, and therefore that the concept of cerebral dominance as classically expressed is incomplete.

EVIDENCE FOR HEMISPHERIC ASYMMETRY

Those concerned with cerebral hemispheric asymmetry soon become aware of the abundant literature dealing with handedness, the theories concerning its origins (genetic or environmental), the countless publications on cerebral dominance, and the vast literature on aphasia. Each of these topics has its own history, much of which goes back at least a century and some a good deal beyond that.

Hand preference

As evidence for functional differences between the hemispheres, let us first consider the most pervasive of all relevant data: hand preference. The earliest evidence of dextrality as a preferred mode of behavior dates back to Egyptian drawings made about 2500

B.C. (Dennis, 1958). A biblical reference dating from about 1400 B.C., in the Book of Judges (probably the earliest reference to sinistrality), speaks of a battalion of left-handed soldiers in the army of the children of Benjamin.

The major, still unsolved problem of hand preference concerns its genesis. Among the theories that attempt to account for specific preference for one hand over the other are those concerned with cultural pressures, pre- or postnatal damage to the nervous system, intrauterine position of the fetus, and genetic mechanisms. For complete comprehension of the mechanisms that result in symmetry or asymmetry of the cerebral hemispheres, the validity of a complementary theory of handedness must be demonstrated. That is, if cultural factors have predisposed an individual toward right- or left-handedness, then one may not be inclined to look for genetic factors that predispose the left hemisphere to speech representation. But if genetic factors are important in establishing hand preference, such factors may also play a role in establishing the speech representation of the left hemisphere. Although a genetic basis for handedness might suggest a genetic basis for speech representation, the possibility also exists that handedness and speech representation are both genetically determined but not necessarily interdependent.

Crossed aphasia

Examples of a possible independence of hand preference and laterality of speech representation are persons with crossed aphasia. *Crossed aphasia*, a phenomenon well known for many years, refers to an exception to the well-established, frequent relationship of aphasia to lesions of the left hemisphere. Individuals with crossed aphasia show aphasic symptoms after injury to the hemisphere ipsilateral to the preferred hand.

Some investigators, in an attempt to maintain the concept of cerebral dominance, explain the lack of relationship between hand preference and the speech-dominant hemisphere in crossed aphasia by assuming that the individual's "true" hand preference, whether genetically or congenitally established, has been overcome by subsequent environmental pressures. This view is quite plausible for dextral patients with crossed aphasia, because cultural pressures toward dextrality are common. Such an explanation for sinistral aphasics with left hemisphere lesions is less likely to be valid.

As far back as 1916, Kennedy explained crossed aphasia on the basis of *stock-brainedness,* that is, the independent occurrence of handedness and "brainedness." By *brainedness* was meant a genetic tendency for a specific hemisphere to be speech dominant. Kennedy viewed speech and hand preference as having separate genetic origins, which, despite their tendency to be related in the dextral individual – some 90% of the time – were based upon completely independent mechanisms. Much evidence has subsequently been acquired to dispute the popular view, based on Broca's observations of a century ago for dextrals and Hughlings Jackson's (1868) and Broca's (1888) for sinistrals, that there is a dependent relationship between handedness and speech upon a dominant cerebral hemisphere.

Handedness and speech representation

The fact that the relationship between handedness and speech representation is not perfect has been reported by a number of writers (Chesher, 1936; Roberts, 1951, 1956; Humphrey & Zangwill, 1952; Goodglass & Quadfasel, 1954; Brown & Simonson, 1957; Bingley, 1958; Zangwill, 1960).

This lack of identity between the hemisphere contralateral to the preferred hand and the hemisphere implicated in speech has been clearly demonstrated by Kimura (1961a, b), who found that if patients with epileptogenic focuses in various parts of the brain have competing verbal material simultaneously presented to the two ears, they report more of the material from the ear contralateral to the speech-dominant hemisphere, when speech dominance was determined by means of the Wada sodium amytal technique (Wada & Rasmussen, 1960). These findings were true of the group whose left hemisphere was dominant for speech, regardless of whether they were sinistral or dextral. Handedness was also irrelevant for the group whose right hemisphere was dominant for speech. Also pertinent in this regard are the numerous reports (Penfield & Roberts, 1959) demonstrating the plasticity of the young brain regarding speech representation. Children below the age of 4 years who sustain brain injury to the speech areas of the left hemisphere frequently recover their verbal ability within months after the initial impairment. By contrast, similar cerebral insult in the adult most frequently results in permanent dysphasia.

Bilateral asymmetry in anatomy

Despite the well-established doctrine of bilateral symmetry, it has frequently been pointed out, most recently by Von Bonin (1962), that "asymmetries of paired organs are the rule, rather than the exception in the human body." As evidence, Von Bonin pointed out that the right arm is a little longer than the left one, and, for the bones of the leg and the clavicle, the opposite is true. In 1931, Woo found that the right skull bone in the frontal and parietal region was consistently larger than that on the left.

With regard to studies of the gross anatomy of the cerebral hemispheres, a plethora of papers over the last century has demonstrated conflicting asymmetries. Thus, although Boyd (1861), Ogle (1871), and Broca (1875) all found the left hemisphere to weigh somewhat more than the right, Wagner (1864) and Thurnam (1866) found the opposite results. A gross anatomical difference, based upon non-neural cerebral tissue, which appears to be reliable, was found by Di Chiro (1962), who showed that the vein of Labbé predominates in the right hemisphere and the vein of Trolard in the left hemisphere. Cole and Glees (1951) have hypothesized, on the basis of Marchi degeneration studies of the spinal cord of monkeys and baboons, that paw preferences are caused by asymmetrical distribution of corticospinal fibers. Unfortunately, thus far there have been no subsequent studies to confirm this hypothesis.

The myriad studies with their resulting contradictory data have understandably made Bodian (1962) hold out only "a slight hope for the morphological approach."

Bilateral asymmetry in physiology

Studies that have employed a pharmacological or neurophysiological approach rather than a gross anatomical approach to explore the asymmetry between hemispheres, although more recent and much fewer in number, have been more successful. For example, the Wada technique (Wada & Rasmussen, 1960), which employs injection of sodium amytal into the internal carotid artery, consistently enables determination of the hemisphere most concerned with speech; and Feinberg et al. (1960) have demonstrated a high correlation between intellectual performance on a standard test of intelligence and oxygen uptake of the left hemisphere. This higher general metabolic activity of the left hemisphere may reflect the continuing verbal activities with which the left hemisphere is concerned. However, the finding is consistent with the report of intellectual impairment in the absence of aphasia after damage to the left parietotemporal area (Weinstein & Teuber, 1957b).

In a study of the conduction velocities of the ulnar and median nerves, Cress et al. (1963) have shown a significant relationship between this neurophysiological measure and handedness. This result lends support to and generalizes the hypothesis of Cole and Glees (1951) that fiber distribution in the peripheral nerves is asymmetrical.

Concerning the physiological bases of hand preference are the studies of Peterson and his collaborators. Peterson (1951) found that forced practice with the nonpreferred paw changed paw preference in rats. Lest this result be interpreted as indicating that paw preference is entirely dependent upon practice, Peterson and Devine (1963) have shown that transfer of paw preference in the rat follows production of small lesions, 0.5 to 1 mm^2 in size, in the frontal cortex contralateral to the preferred paw. Furthermore, Peterson and Rigney (1950) were able to demonstrate a shift in paw preference after application of acetylcholine to the hemisphere contralateral to the nonpreferred paw. Taken together, the studies indicate that fine neuroanatomical or chemical distinctions between the hemispheres, possibly projecting down the spinal cord to the peripheral nerves, may determine which hand or paw is preferred.

NORMAL LATERALITY DIFFERENCES

Let us now consider some of the normal functional differences found in various behavioral studies comparing the performance of the right and left sides of the body. To my knowledge, the earliest study demonstrating a sensory difference between the right and left hands of normal individuals was that performed by Semmes et al. (1960). This study determined absolute pressure sensitivity thresholds in a group of normal right-handed subjects and found a significant superiority for the left hand over the right. An apparent explanation for the left hand's greater sensitivity to pressure was that, because most of the subjects were right-handed, the greater thickening of the skin of that hand associated with its more frequent use lessened its sensitivity. If it is true that thickening of the skin of the preferred hand reduces its sensitivity, one might predict that the right hand would be more sensitive in left-handed individuals. Furthermore, one would predict that areas of the skin not subject to differential usage, for example, the forearm or sole, would not show laterality differences.

Weinstein and Sersen (1961) investigated these questions. Figure 2-1 demonstrates that a significantly greater percentage of dextrals than sinistrals were more sensitive individually on the left for palm, forearm, and sole, as well as for all three of these body parts. Five times as many dextral subjects as sinistral subjects were more sensitive on the left than on the right for all three body parts studied.

In a subsequent study (Sersen et al., 1966), superiority of the nonpreferred palm for pressure in dextrals was again demonstrated. In addition, that study found the first indications of a laterality difference more marked in sinistrals than in dextrals. Together with the finding of a significant relation between two-point discrimination and point localization in sinistrals but not in dextrals, the data suggest that these measures represent different functions in the two groups as well as hemispheric differences for sensory processes.

A study of the relative tactual sensitivity of the phalanges of both hands (Fig. 2-2) demonstrated that each phalanx of each finger of the left hand was more sensitive than the right hand homolog in a dextral male and a dextral female (Weinstein, 1962a). A normative study dealing with the relative tactual sensitivities of the breasts of normal sinistral and dextral women demonstrated that a majority of both groups had better pressure sensitivity on the left and better two-point discrimination on the right breast (Weinstein, 1963).

A subsequent, more extensive study (Weinstein, 1968) employed three measures of tactual sensitivity: pressure sensitivity, two-point discrimination, and point localization on 20 pairs of body parts in 24 dextral men and 24 dextral women. In general, the results support the findings of increased pressure sensitivity and two-point discrimination on the left side of the body in both men and women. Figures 2-3 and 2-4 show the superiority for pressure sensitivity of the left hand particularly the fingers in males and females. The study also demonstrated significant superiority of the left side for two-point discrimination of the sole, the calf, the thigh, the back, and the forearm in men and women.

Garfinkle (1965) tested the thumbs and forearms of 200 normal dextral and 200 normal sinistral children aged 5 through 12 for

Fig. 2-1. Percentage of right-handed (*white bars*) and left-handed (*black bars*) subjects whose palms, forearms, and soles were more sensitive on the left side and on the right side.

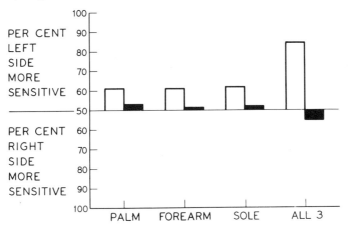

Fig. 2-2. Mean pressure sensitivity thresholds for all phalanges (*A, B, C*) of four fingers of both hands in a right-handed male (*broken line*) and a right-handed female (*solid line*). Overall means refer to all fingers for both subjects.

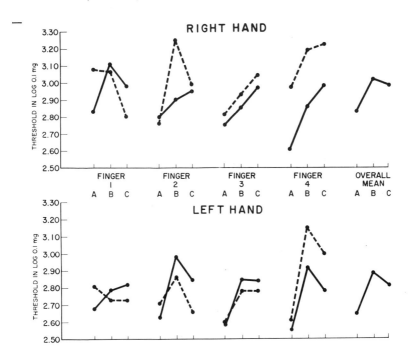

pressure sensitivity. He obtained the expected finding that a significant majority of the dextral subjects were more sensitive on the left thumb. However, the converse expectation, based on the usage or callus hypothesis – that the sinistrals would be more sensitive on the right thumb – was not confirmed. Instead, equal numbers of sinistrals were more sensitive on either thumb.

In a developmental study concerned with differences in sensitivity between the sides as a function of age and sex, Ghent (1961) also found a superiority of the left thumb and upper arm for pressure sensitivity thresholds.

That superiority of the left hemisphere is not restricted to classic tactual stimuli is demonstrated by the following studies. Green et al. (1961) found responsiveness for significantly lower electrical stimulation of the left compared with the right arm. Wolff and Jarvik (1964), studying pain sensitivity and pain tolerance in large

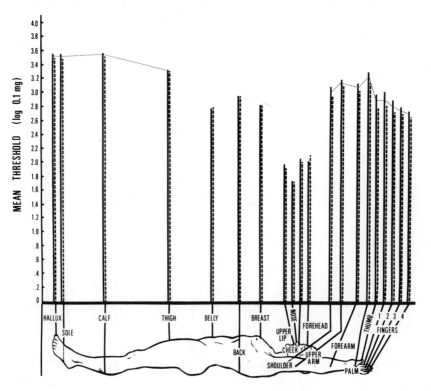

Fig. 2-3. Mean pressure sensitivity thresholds for the right (*solid line*) and left (*broken line*) sides of the body in 24 right-handed males.

groups, found that the intercorrelations between thresholds were consistently higher for the right than for the left hand. In another study, Wolff et al. (1965) found that the nondominant hand was consistently more sensitive to pain than the dominant hand. Unfortunately, a recent study (Weinstein et al., 1967) was unable to confirm this last finding, but its results seem to suggest that cerebral asymmetry also may play a significant role in pain perception.

LOSS OF COMPLEX ABILITIES AFTER UNILATERAL CEREBRAL DAMAGE

Let us now consider the effect of injury to either cerebral hemisphere on somewhat more complex abilities. All the work reported in this section was based on study of brain-injured male veterans of either World War II or the Korean campaign. The control subjects were identical in all respects to the brain-injured except that their

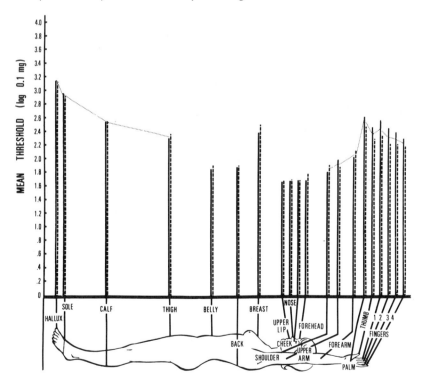

Fig. 2-4. Mean pressure sensitivity thresholds for the right (*solid line*) and left (*broken line*) sides of the body in 24 right-handed females.

wounds were of the peripheral nerves, not the brain. All measures employed were quantitative in nature, and all impairments were determined on the basis of normative statistics.

The most obvious impairment of complex ability after injury to one cerebral hemisphere is aphasia. The relationship of verbal loss to lesions of the left cerebral hemisphere has been well known since the work of Dax (1836). However, despite similarly prolonged interest in the effects of brain wounds on intellectual functions, there has been little agreement about the localization of those lesions that might produce deficits on test of intelligence. One hypothesis derivable from Lashley's (1929) results obtained with rats might specify that any brain injury of sufficient size would produce a deficit. In opposition to this view have been the recurrent claims that only lesions of the frontal lobes are followed by loss of intelligence (Halstead, 1947). Much of the confusion has derived from the testing of patients, often of advanced age, with neoplastic lesions or vascular accidents of the brain and patients subjected to lobotomy or other forms of psychosurgery because of psychosis, severe neurosis, or intractable pain (Hebb, 1939; Halstead, 1947).

Preinjury scores were obtained from the Army General Classification Test (AGCT), a standardized test of general intelligence administered to each man on induction into the armed services, for 62 men who subsequently sustained penetrating brain wounds and for 50 controls (Weinstein & Teuber, 1957a, b; Teuber & Weinstein, 1958; Weinstein, 1962b). Approximately 9 years after being wounded, the men were retested with the civilian edition of the AGCT. The 62 brain-injured subjects were subdivided into groups according to location of lesion (frontal, temporal, parietal, or occipital lobe, in left, right, or both hemispheres), and the data were subjected to a series of analyses of variance. The analyses indicated that lesions of the frontal or occipital lobes did not produce a significant decline in score and that only groups with wounds of the parietal or temporal lobe of the left hemisphere showed a significant decrease in score. The groups with lesions of the parietal, temporal, or parietotemporal area of the left hemisphere were each statistically significantly inferior to the control group, and the group with damage to the left parietotemporal area was significantly inferior to every other brain-injured subgroup except the one with injury to the left temporal lobe. When the analyses were restricted to nonaphasics, the group with damage to the left parietotemporal area still showed a mean loss in score and was significantly inferior to every other group. If the results are presented in terms of the number of subjects in each subgroup manifesting statistically significant losses in score, essentially the same results are seen. Only 1 of the 50 controls scored significantly lower on retest, and except in the group with lesions of the left parietotemporal area no more than 2 subjects in any group showed a significant deficit on retest. However, the scores of 9 of the 10 men with damage to the left parietotemporal area dropped significantly.

In 1954, Teuber and Weinstein demonstrated a striking effect of right hemisphere wounds on a spatial aspect of perception. A formboard was first presented in a given orientation to the blindfolded subject; after he had placed all the forms, the board was rotated 180° without his knowledge, and he was retested. Negative transfer (i.e., inability to reduce errors or time on retest) was found in 5 of the 6 subjects with lesions of the right temporal lobe, whereas only 3 of the 28 subjects with lesions elsewhere than in this region showed this lack of spatial transfer. The localization of the lesion suggests that some higher-order level of spatial impairment is involved, rather than a sensory or motor impairment.

Weinstein (1962b) studied the effects of brain injury on three-dimensional size discrimination in 58 brain-injured subjects and 20 controls without brain injury. The task consisted of palpating a wooden cube and then attempting to find, in a comparison array, the one that was identical in size. Surprisingly, the group with unilateral lesions of the right hemisphere made a significantly greater average number of errors than both the group with injuries to the left hemisphere and the control group. The control group and the group with left-sided lesions did not differ significantly from each other. It might be thought that the inferiority of the group with right hemisphere damage merely reflects a greater proportion of men with injury to the central region in this group. However, this was not the case: Approximately one-third of the men with right-sided and one-third of those with left-sided damage had central lesions.

In an attempt to elucidate some of the *sensory* factors that might contribute to the impairment, we considered the characteristics of those individuals whose size discrimination was significantly impaired with regard to both the locus of the lesion and the somatosensory status of the hand. By contrasting individuals with unilateral lesions of the right and left hemispheres, we obtained a significant χ^2 value: 11 of the 19 men with right-sided lesions showed defective size discrimination, whereas only 4 of the 23 men with left-sided lesions showed such a defect. Because we were interested in the relationship between lesions of the right hemisphere and the presence of defective two-point discrimination, we compared men with injury to the right hemisphere who had this sensory defect with men who had left hemisphere lesions and the same defect. The χ^2 value was significant: All 6 men with right-sided lesions *and* defective two-point discrimination showed significantly defective size judgment; only 1 of the 5 men with left-sided lesions *and* defective two-point discrimination showed such defective size judgment.

Finally, to investigate whether severe somatosensory impairments of any type would produce defective size judgment regardless of the side of the lesion, we compared the 7 men with right hemisphere lesions who had severe somatosensory impairment with the 7 men with left-sided lesions who had severe somatosensory impairment. We found that all 7 men with lesions of the right hemisphere were significantly defective in size judgment, whereas only 1 of the 7 men with left-sided lesions was so impaired.

Obviously, two factors are related to impairment of tactual, three-dimensional size discrimination after brain injury: lesion of the right hemisphere and defective two-point discrimination. Either factor alone may produce defects of size discrimination; however, a combination of both produces the greatest frequency of defects.

Right hemisphere lesions do not impair all aspects of somatic sensation. The following discussion supplies specific examples of measures that are not differentially affected by lesions of either hemisphere. Weinstein et al. (1958) employed a test of roughness discrimination analogous in design and procedure to the test of three-dimensional size discrimination. Brain-injured and control subjects were required to palpate and match patches of sandpaper. Although clear-cut impairment was related to injury in a region of the postcentral area, the subjects showing such impairment had lesions that were equally distributed between right and left hemispheres. Despite the similarity of the tests, size discrimination showed a striking laterality relation, but roughness discrimination did not. The intrinsic differences between tests within the same modality, which in the one case show and in the other case fail to show differential effects of lesions to either hemisphere, may reflect the distinction between an intensive and a spatial measure of sensitivity.

Roughness discrimination was not alone in failing to differentiate the hemispheres after injury. Studies of time-error for weight or size judgment (Weinstein, 1955a, b) and studies of size or weight discrimination (Weinstein, 1954; 1955c) also failed to yield laterality differences.

In previous reports (Semmes et al., 1955; Weinstein et al., 1956; Weinstein, 1958), we have considered the effects of brain injury spatial orientation and on the body schema. One subtest of the body schema test consisted of a diagram-reading task made up of diagrams of the male body with numbers superimposed over various body parts. The subject's task was to touch the appropriate part of his body, anterior or posterior, on the right or left side, in the order indicated by the numbers. The spatial orientation task consisted of visual and tactual sets of maps. The subject was to read each map and walk along a path on the floor, connecting by his journey dots arranged in three rows of three dots each.

To compare the effects of brain injury upon these analogous measures, we analyzed the data according to whether the injury was in the right or left hemisphere and whether in the anterior or posterior segment of the brain (Semmes et al., 1963). We found that the locuses of injury followed by a significant decrement were partially coincident and partially separate. *Both* personal orientation (i.e., body schema) and extrapersonal orientation (i.e., spatial orientation) were impaired by lesions of the posterior sector of the left hemisphere. Personal orientation was particularly impaired also by lesions of the anterior sector of the left hemisphere. Injury to the posterior sector of the right hemisphere produced defective extrapersonal orientation.

The final comparison of the hemispheres for complex abilities concerned tachistoscopic thresholds following right or left temporal lobectomy and following surgery of the right or left hemisphere, exclusive of the temporal lobes (Johnson). The threshold consisted of the time in milliseconds needed for 100% correct identification of numbers randomly presented in a clock-dial orientation. Although the differences failed to reach a statistically significant level, the thresholds of the group with right temporal lobectomy were double those of the group with nontemporal right hemisphere lesions and almost double those of the group with left temporal lobectomy. This finding of perceptual impairment after injury to the right temporal area thus adds generality to the previous data implicating lesions of this region of the right hemisphere in various spatial and perceptual impairments.

In comparing the right versus the left visual fields for short-term memory for Arabic numerals successively presented from 0 to 500 msec between the offset of the stimulus and onset of the comparison array, a significant superiority of the right field was demonstrated by the controls (71%). Both groups with lesions of the right hemisphere (temporal and nontemporal) showed superiority for the right field, expectedly greater than that of the controls (81%), whereas both groups with left hemisphere lesions showed a diminished superiority for the right field (65%). Thus, all groups (the normal and the groups with right or left hemisphere injury) demonstrated a relative superiority of the right field for short-term memory of visually presented digits, the degree of right-field superiority being somewhat enhanced or diminished by injury to the relevant hemisphere.

The studies described above demonstrate that a variety of complex behavioral deficits are selectively impaired after injury to either the right or the left hemisphere. These defects are in addition to the well-established ones, such as dysphasia, which have been associated with injury to the left hemisphere. Let us consider now some of the laterality differences noted after amputation of a body part or transection of the spinal cord.

LATERALITY DIFFERENCES AFTER AMPUTATION, APLASIA, OR PARAPLEGIA

Somatosensation and motor performance

Recent research has emphasized the superior tactual sensitivity of the stumps of amputees over the homologous contralateral areas of skin. Shapiro (1966) investigated some of the more complex somatosensory changes on the stumps of individuals with below-elbow amputations. Although he confirmed that the stump was more sensitive than the homologous region, he failed to find a difference between right and left stumps for "simpler" measures such as pressure sensitivity, two-point discrimination, or point localization. However, of greater interest was the highly significant finding that size discrimination was better on the left stump than on the right stump. This finding is all the more interesting in view of the previously discussed relationship of right hemisphere lesions to impaired three-dimensional size discrimination. Sensory reorganization after amputation not only concerns differences between the stump and the homologous region but also appears to reflect laterality of amputation.

In a series of studies dealing with the effects of congenital aplasia, or absence of limbs, we studied the strength and coordination of children born with only one upper extremity (Weinstein et al., 1964). The principal motivation for this study was to evaluate the hypothesis that hand preference has a genetic basis. We reasoned that because a child with only one upper extremity performs all manual activities with it alone, groups of children with unilateral aplasia of either hand should have the same degree of manual experience. The only difference between such groups should reside in genetic factors that differentiate the right from the left side of the body. Testing the assumption that a great majority (95%) of the normal population is right-handed because of genetic factors, we hypothesized a greater probability for children with right upper extremities alone to perform better on measures of strength and coordination than children with left upper extremities alone.

That the data failed to support this hypothesis can be accounted for on the basis of any of the following explanations: (1) genetic factors do not operate in the production of handedness, (2) environmental factors (e.g., practice) may override such genetic predispositions in that the left hand may differentially benefit more than the right, or (3) the factor that induced the aplasia may also have affected the genetic basis of handedness. Our data do not permit us to select among these alternatives.

Phantom perception

Other studies concerned with the effects of amputation or paraplegia deal with phantom perception, the report of the awareness of a body part absent because of aplasia or amputation, or which, although present, has been deafferentated. In a study of phantom

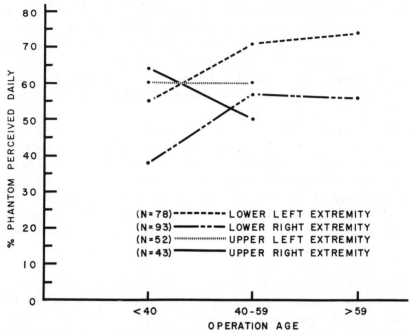

Fig. 2-5. Percentage of patients experiencing daily phantom perception, according to side and site of amputation and age at operation.

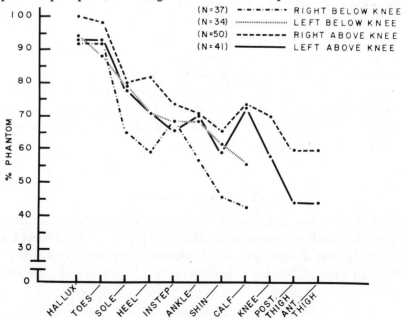

Fig. 2-6. Percentage of patients with lower extremity amputation experiencing phantom perception, according to site and side of amputation.

perception after amputation of various body parts, Sersen, Vetter, and Weinstein (unpublished) examined some 400 subjects with amputations above or below the major joints of the upper or lower extremities. In a preliminary analysis, we found unusual relationships concerning laterality and location of the amputation. For example, amputations above the right knee were associated with significantly *more* phantom perceptions than were corresponding left amputations; amputations below the right knee were associated with significantly *fewer* phantom perceptions than were left amputations (Fig. 2-5). We also found that amputation of the left leg (above or below the knee) produced more frequently occurring phantom perceptions than the corresponding right amputation (Fig. 2-6). Again, the reorganization of the cerebral systems that occurs after amputation seems to occur differentially for areas above and below major joints and for right and left sides of the body.

A report of the effect of a cerebral lesion upon a contralateral phantom limb (Head & Holmes, 1911) prompted us to study the effects on phantom perception of cerebral vascular accidents (CVA) that preceded or followed amputation (Weinstein et al., 1969). In 44 patients with CVA before amputation, it was found that a right CVA tended to delay the onset of phantom perception in both ipsilateral and contralateral limbs, and a left CVA apparently had no such effect (Fig. 2-7). As time of appearance of the phantom

Fig. 2-7. Percentage of patients with preamputation left (*solid line*) and right (*broken line*) cerebral vascular accident reporting onset of phantom perception at varying times after amputation.

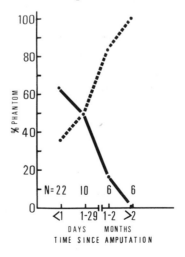

in patients with prior left CVA roughly parallels that for amputees without brain injury, it seems likely that a right CVA has an inhibitory influence on the onset of phantom perception rather than that a left CVA is facilitatory.

Let us consider one more study dealing with phantom perception after amputation (Fig. 2-8). In over 200 women who had undergone mastectomy, we found a significant tendency for phantom perception to decrease in incidence with age after left mastectomy, but not after right mastectomy (Weinstein et al., 1970). Furthermore, the phantom appeared significantly sooner after left mastectomy than after right mastectomy (Fig. 2-9). We have, as yet, no ready explanation for this discrepancy between left and right breasts for onset and frequency of phantom perception. It is interesting to recall, however, that the left breast was found to be significantly more sensitive to pressure than the right in a majority of normal women (Weinstein, 1963). It is possible that the same mechanisms responsible for the normally enhanced pressure sensitivity of the left breast is responsible for the early appearance of phantom perception after left mastectomy.

We also studied phantom perception after paraplegia in 150 patients (Weinstein, 1962c). We subdivided the group according to the presence of lesions from the third cervical (C3) to the third thoracic (T3) segment and compared this group with those with lesions from T4 to T9 and T10 to L5 (fifth lumbar). Of the phantoms reported by the patients with the lowest (T10 to L5) lesions, 67% were more intense on the left; of the phantoms reported by patients with midthoracic lesions (T4 to T9), 74% were more intense on the right. Although we cannot as yet account for this large disparity, it may be noted that the region of the cord associated with more intense left-sided phantom perception is that concerned with the sensory and motor innervation of the lower extremities. At the same time, it can be pointed out again that the majority of normal individuals are more sensitive on the left lower extremity than on the right. How these facts may ultimately interrelate can at present be little more than speculation.

To summarize the data dealing with the effects of amputation or deafferentation of body parts, left amputations yield better size discrimination and phantom perception that appears earlier but decreases in frequency with time; low spinal cord lesions produce more intense phantoms of the left side. Limb amputation on the

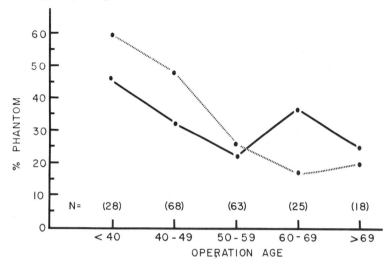

Fig. 2-8. Percentage of patients experiencing phantom perception following right (*solid line*, 98 women) and left (*broken line*, 104 women) mastectomy, according to age at operation.

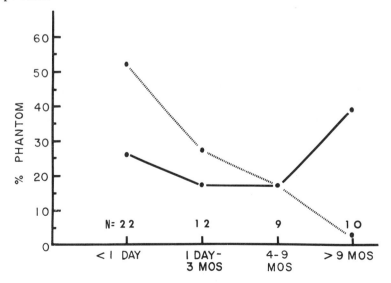

Fig. 2-9. Percentage of patients with right (*solid line*) and left (*broken line*) mastectomy reporting onset of phantom perception at varying times following operation.

left produces more frequently occurring and more complete phantom perception than similar right amputations. These data, we feel, may be related to the normal superiority of the left side of the body for both intensive and extensive measures of somatosensation.

LATERALITY DIFFERENCES AFTER PENETRATING BRAIN INJURY

Motor performance

The motor task employed a device that consisted of two mechanical counters with small buttons attached to a flat board. The task was to depress the button as rapidly as possible with the index finger of either or both hands in six 10-sec periods. A group of 72 brain-injured and 13 control subjects was tested (Weinstein, 1959). The

Fig. 2-10. Rate of index finger oscillation for right and left hands in brain-injured and control groups. *B-C*, bilateral central cerebral lesions; *B-NC*, bilateral noncentral lesions; *L-C*, left central lesion; *L-NC*, left noncentral lesion; *R-C*, right central lesion; *R-NC*, right noncentral lesion; *CON*, control group.

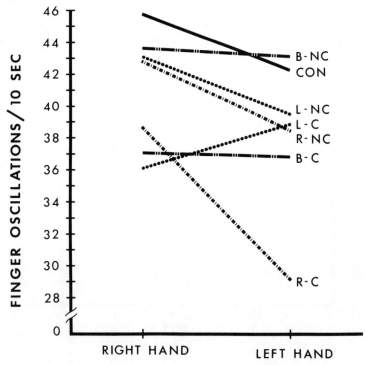

brain-injured group was subdivided into those with lesions of the right, left, or bilateral central (i.e., Rolandic) area, and those with lesions of the right, left, or bilateral noncentral area. Several expected results were obtained; for example, injury to the central region of either hemisphere produced a significantly greater slowing of the contralateral hand than injury to other parts of the brain (Fig. 2-10). But surprisingly, injury to the right central sector had a greater contralateral effect than injury to the left central sector. The impairment of the left hand after injury to the right central area exceeded that of the right hand after injury to the left central area. Furthermore, lesions of the right central sector also tended to produce bilateral slowing (i.e., both hands were affected), whereas lesions of the left central sector resulted only in contralateral impairment. Figure 2-11 shows the effects of these lesions on unimanual and on bimanual performance. Again, note the rela-

Fig. 2-11. Rate of index finger oscillation for one and both hands in brain-injured and control groups. *B-C,* bilateral central cerebral lesions; *B-NC,* bilateral noncentral lesions; *L-C,* left central lesion; *L-NC,* left noncentral lesion; *R-C,* right central lesion; *R-NC,* right noncentral lesion; *CON,* control group.

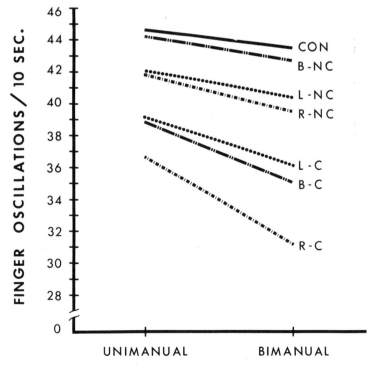

tively poorer performance of the groups with central lesions: The patients with damage to the right central area showed by far the poorest performance of all, and those with lesions of the right noncentral area showed poorer performance than the other groups with noncentral damage.

Somatosensory functioning

On relatively simple measures of somatosensation we found a somewhat different effect of brain injury than was obtained for motor functioning (Semmes et al., 1960). We employed four measures of somatic sensation in a group of 124 subjects with traumatic, penetrating brain wounds and a group of 33 controls with injury to the peripheral nerves of the leg. For each hand, we employed four measures of tactual sensitivity: pressure sensitivity threshold, two-point discrimination threshold, point localization threshold, and an assessment of the sense of passive movement of the index fingers.

We found that the left hand was somewhat more often affected by contralateral nonsensorimotor injuries and was twice as often affected by ipsilateral lesions than the right hand. Defects of the right hand occurred significantly more often following injury to each of the three subsectors of its contralateral sensorimotor region (precentral, central, and postcentral) than following injury outside this region but within the left hemisphere. In contrast, sensory defects of the left hand did not occur significantly more often following injury to the right sensorimotor region than following other lesions of the right hemisphere or, indeed, within the rest of the brain. It thus appears that defects of the right hand are more localizable than those of the left hand.

In addition, lesions of the left sensorimotor region were more commonly followed by defects that involved the ipsilateral hand (with or without concomitant defects of the contralateral hand) than by defects that were strictly limited to the contralateral hand. The reverse was the case following lesions of the right sensorimotor region. In other words, defects restricted to the contralateral hand tended to occur after injury to the right sensorimotor region, whereas defects involving the ipsilateral hand (with or without concomitant defect of the contralateral hand) predominated after injury to the left sensorimotor region.

This asymmetrical diffuseness and concentration of representation of the hands is indirectly confirmed by the studies of Goff et al. (1962) of cortical evoked potentials to somatosensory stimulation. Similar indirect confirmation for the motor sphere has been obtained by Gilden and Hariton (1967) in studies of antecedent cortical motor potentials to thumb flexion. The authors reported a tendency for greater bilaterality of motor representation for the right hand and greater contralaterality of such representation for the left hand.

At this point, it may be well to note the difference between sensory defects and motor defects with regard to representation in the ipsilateral or contralateral hemispheres. For the motor task, the left hand was maximally impaired by lesions of the right central area, whereas the right hand could be impaired by lesions involving either the left or the right central sector. By contrast, the right hand tended to sustain sensory impairment maximally after injury to the left sensorimotor region, whereas the left hand tended to have a more diffuse and bilateral representation. Simply put, the right hand tends to have a bilateral representation for motor functions, and the left hand tends to have only a contralateral representation. Conversely, the right hand tends to have a relatively contralateral representation for somatosensory functioning, whereas the left hand tends to have a bilateral and diffuse representation.

CONCLUSIONS

Young (1962) has expressed a theory to account for the fact that we have two cerebral hemispheres. His theory states that bilaterality was originally necessary for organisms with nervous systems that operate by means of a maplike analog system. Specifically, in the visual system there is an inversion of the image produced by the lens, which the optic chiasm corrects. Young believes that the appearance of the optic chiasm probably determined all other crossings in the brain. He believes that because the more complex computing function of the human brain makes it less dependent on topographical mapping, and therefore makes bilaterality less important, the isomorphism found in the brains of lower organisms has become less necessary in man. Because bilaterality is associated with a means of representing correct spatial relations, and because the coding system becomes more refined in man, a departure from isomorphism can be found. In Young's words: "The picture becomes more abstract, more impressionistic." Young has concluded that perhaps the "nondominant" (right) hemisphere is a vestige. Although the theory is interesting, the material presented in this chapter makes it highly unlikely that the right hemisphere can be merely a vestigial organ. In fact, my findings have led me to conclude that the hemispheres cannot be considered as dominant and nondominant (or vestigial); they are asymmetrical in function, but equivalent in importance, with neither one major or minor.

My associates and I have demonstrated that ability to perform a variety of tasks – relatively simple or complex, sensory, motor, perceptual, spatial, or cognitive in nature – may be impaired after injury to one or the other cerebral hemisphere. The classic concept that only the left hemisphere is concerned with complex functions thus appears indefensible. The fact that relatively simple functions, such as coordination and primary tactual sensitivity, are asymmetrically represented in the two hemispheres also argues against the designation of either hemisphere as major or minor. The fact that we have been able to demonstrate impairments of a relatively complex nature after lesions of either hemisphere, as well as impairments of a relatively simple nature after injury to either hemisphere, makes us reject the concept of cerebral dominance and resist labeling a hemisphere as "motor dominant," "sensory dominant," "spatially dominant," etc. Even "speech dominance" is not inextricably tied to handedness. Indeed, crossed aphasia, as we have pointed out, occurs in some 10% of dextral aphasics and 30% of sinistral aphasics (Humphrey & Zangwill, 1952; Goodglass & Quadfasel, 1954).

Taken together, these findings suggest that handedness and the various other motor and sensory asymmetries may have differing genetic bases. The studies of transfer of paw preference in rats (Peterson & Rigney, 1950; Peterson, 1951; Peterson & Devine, 1963) and the anatomical studies of asymmetrical corticospinal fiber concentrations (Cole & Glees, 1951) point to anatomical or biochemical asymmetry of specific regions of the brain, spinal cord, or nerve as the basis for behavioral differences of the right and left sides of the body.

Aphasia seems also to be related to these anatomical considerations of sensory and motor abilities. However, the various dissociations of handedness from sensory dominance (whether auditory or tactual), and of sensory dominance from speech dominance, suggest that all these functions may be anatomically independent, despite frequent coincident relationships.

If the findings of a differential role for the two hemispheres in sensation, motor functions, complex perception, and spatial and cognitive abilities are reliable, the origin of this asymmetry should be investigated.

> Bodian has stated:
> Comparisons of hemisphere size, like comparisons of whole brain size among individuals of a single species are largely unrewarding. The main contribution of evidence of gross asymmetry may be to suggest that differences at another order of magnitude may be present, namely, the microscopic level . . . Those who seek for anatomical correlates of unilateral hemisphere dominance may be up against the possibility that the anatomy of the hemispheres is entirely irrelevant to this problem. [Bodian, 1962]

I agree that the study of the gross anatomy of the hemispheres may be quite unrewarding in explaining the problems of cerebral dominance. However, our agreement is based upon different reasons. Bodian feels that the anatomical substrate for cerebral dominance can best be sought by means of the electron microscope. Because I reject the doctrine of general cerebral dominance and urge instead that we seek hemispheric asymmetries in all functions, I believe that any anatomical or biochemical approach, such as exploring neuron or fiber density, or acetylcholine or cholinesterase concentration, is doomed to failure if the investigator merely compares these phenomena in the total "dominant" and "nondominant" hemispheres, however these are defined. Such approaches, however, may well yield worthwhile data if the interhemispheric comparisons are restricted to functionally or behaviorally distinct systems.

I have proposed (Weinstein, 1967) that the normal development of perceptual ability depends on three factors: (1) a genetic, structural (anatomical, chemical, etc.) basis; (2) optimal sensory input during the critical period of growth; and (3) continuation of a level of stimulation appropriate (i.e., neither insufficient nor excessive) for maintenance of the sensory systems.

The paragraphs that follow present a theory based upon similar reasoning that attempts to account for the fact of cerebral asymmetry and its development.

Sufficient evidence exists in the animal and plant kingdoms to indicate the genetic bases for various lateral asymmetries in otherwise symmetrical organisms. Unless strong environmental factors intercede at critical periods, these genetic predispositions culminate in lateral asymmetries in the adult typical of the species.

The major postulate of this theory is that a genetic predisposition exists toward lateral asymmetry of a given, early developing system. It is not critical to the theory whether this initial asymmetry occurs first in the sensory or the motor system. For purposes of exposition, we shall assume priority of motor development.

If motor functions (and their asymmetrical lateralization) are the first to appear in the child, it is possible that a sequentially dependent compensatory lateralization of the sensory system may follow during development. This compensatory sensory lateralization may have asymmetrical cerebral anatomical or biochemical sequelae, laterally opposite to the motor system in, for example, magnitude of structures, number of fibers or cells, or concentration of chemical factors in a given hemisphere or side of the spinal cord.

This compensatory growth in the opposite hemisphere and hemicord may reflect (1) a greater availability of neurons on the side chronologically secondary in development (to the motor system), (2) availability of volume for cell or fiber growth, or (3) a chemical environment more accepting to the newly developing sensory systems.

If we assume an initial (genetic) motor asymmetry and a subsequent, laterally complementary, compensatory sensory asymmetry, the stage may be set for the establishment of other asymmetries of more complex functions, such as language, spatial ability, and general intellect. Thus, laterally asymmetrical establishment of sensory systems may result in induction of some consequent neural (or chemical) substrate concerned with complex forms of behavior, which, because of the existing sensorimotor asymmetry, are themselves also forced into asymmetry of representation in the hemispheres. This new asymmetrical establishment of a substrate involved in complex behavior is not immutable; unusual environmental forces, such as injury (e.g., from anoxia) or prolonged or intense laterally unequal stimulation, may attenuate, exaggerate, or even reverse the direction of the potential cerebral asymmetry of a given system.

Depending on the temporal sequence of still other developing neural systems for dealing with yet more complex forms of behavior, and on whether such unusual forces occur during critical periods of development, one might predict a progression of lateralizations consequent upon the prior degree and direction of the asymmetry.

The admittedly broad theoretical position expressed here has some heuristic value. Experiments based upon predictions so derivable can be designed within the various disciplines that must be involved if the problem of dominance or cerebral asymmetry is to be more fully resolved. Of considerable value would be cross-sectional studies of concentration of sensory and motor fibers in fetal, neonatal, and older developing and fully developed organisms. Such a study would compare, as a function of age (and hence degree of external environmental or internal induction forces), the degree of lateral asymmetry. Such studies should not be restricted merely to fiber or cell counts, but should also investigate differential lateral concentrations of various chemical agents such as neural transmitters, their precursors, and cytoarchitechtonics.

A second series of studies should involve attempts to enhance or attenuate any existing asymmetries, or to produce them in their absence, in the neonate (or even in the fetus) by laterally differential stimulation within various modalities at varying periods of neonatal (or fetal) development. For example, one might continually stimulate tactually the right paw (the left, in littermates) of a group of neonates and study the right and left sensory (and motor) systems for the consequences of introducing the stimulation at various ages and after varying the durations of such stimulation. Such a study might include, at first, neuroanatomical and histochemical techniques, and subsequently, neurophysiological and behavioral techniques.

A third strategy in the older organism might attempt to utilize laterally differential complex experiences to enhance, attenuate, or even reverse whatever asymmetries of structure (or function) may exist or to attempt to create nonexistent ones.

Finally, studies of simple sensory and motor or complex experiential deprivation (or overload) in the adult organism might be employed to determine their effects upon establishment, enhancement, or attenuation of asymmetry of structure or function.

Confirmatory experimental data would indeed be a rewarding culmination to the preliminary findings reported here, which have led to this theory.

ACKNOWLEDGMENTS

Much of the research reported here was supported by the following grants to the author: Vocational Rehabilitation Administration RD-1495-M; National Institute of Mental Health MH-10393-02; National Aeronautics and Space Administration NsG-489; and National Institute of Neurological Diseases and Blindness NB-07404-01. The author is also indebted to the Korein Foundation for supporting the preparation of this manuscript.

REFERENCES

Bingley, T. 1958. Mental symptoms in temporal lobe epilepsy and temporal lobe gliomas with special reference to laterality of lesion and the relationship between handedness and brainedness. *Acta Psychiatr. Scand. 33:* 151.

Bodian, D. 1962. Discussion. In V. B. Mountcastle (ed.). *Interhemispheric Relations and Cerebral Dominance.* Baltimore: Johns Hopkins University Press, pp. 25–26.

Boyd, R. 1861. Tables of the weights of the human body and internal organs in the sane and insane of both sexes at various ages arranged from 2,114 post-mortem examinations. *Philos. Trans. 151:* 241–262.

Broca, P. 1875. Instructions craniologiques et craniométriques de la Société d' Anthropologie de Paris. *Bull. Soc. Anthropol. (Paris) 6:* 534–536.
 1888. Mémoires sur le cerveau de l'homme. Paris: Reinwald.

Brown, J. R., & Simonson, J. 1957. A clinical study of 100 aphasic patients. 1. Observations on lateralization and localization of lesions. *Neurology 7:* 777–783.

Chesher, E. C. 1936. Some observations concerning the relation of handedness to the language mechanism. *Bull. Neurol. Inst. N.Y. 4:* 556–562.

Cole, J., & Glees, P. 1951. Handedness in monkeys. *Experientia 8:* 224–226.

Cress, R. H., Taylor, L. S., Allen, B. T., & Holden, R. W. 1963. Normal motor nerve conduction velocities in the upper extremities and their relation to handedness. *Arch. Phys. Med. 44:* 216–219.

Dax, M. 1836. Lésions de la moitié, gauche de l'encéphale coincident avec l'oubli des signes de la pensée. Read at Montpellier, 1836. Published in *Gaz. Hebdom. 11:* 259–260, 1865.

Dennis, W. 1958. Early graphic evidence of dextrality in man. *Percept. Mot. Skills 8:* 147–149.

Di Chiro, G. 1962. Angiographic patterns of cerebral convexity veins and superficial dural sinuses. *Am. J. Roentgenol. Radium Ther. Nucl. Med. 87:* 308–321.

Feinberg, I., Lane, M. H., & Lassen, N. A. 1960. Senile dementia and cerebral oxygen uptake measured on the right and left sides. *Nature 188:* 962–964.

Garfinkle, M. 1965. Tactual thresholds in children as a function of handedness, sex, and age. Doctoral dissertation, Yeshiva University.

Ghent, L. 1961. Developmental changes in tactual thresholds on dominant and nondominant sides. *J. Comp. Physiol. Psychol.* 64: 668–671.

Gilden, L., & Hariton, B. 1967. Motor potentials of contralateral and ipsilateral hemispheres with unimanual movement. Paper presented to Eastern Psychological Association, Boston.

Goff, W. R., Rosner, B. S., & Allison, T. 1962. Distribution of cerebral somatosensory evoked responses in normal man. *Electroencephalogr. Clin. Neurophysiol.* 14: 697–713.

Goodglass, H., & Quadfasel, F. A. 1954. Language laterality in left-handed aphasics. *Brain* 77: 521–548.

Green, J. B., Reese, C. L., Peques, J. J., & Elliott, F. A. 1961. Ability to distinguish two cutaneous stimuli separated by a brief time interval. *Neurology* 11: 1006–1010.

Halstead, W. C. 1947. *Brain and Intelligence.* Chicago: University of Chicago Press.

Head, H., & Holmes, G. 1911. Sensory disturbances from cerebral lesions. *Brain* 34: 102–254.

Hebb, D. O. 1939. Intelligence in man after large removals of cerebral tissue: Defects following right temporal lobectomy. *J. Gen. Psychol.* 21: 437–446.

Humphrey, M. E., & Zangwill, O. L. 1952. Dysphasia in lefthanded patients with unilateral brain lesions. *J. Neurol. Neurosurg. Psychiatry* 15: 184–193.

Jackson, J. H. 1868. Defect of intellectual expression (aphasia) with left hemiplegia. *Lancet* 1: 457.

Johnson, L. The effects of temporal lobe surgery on short-term visual memory in man. Doctoral dissertation in preparation, New York University.

Kennedy, F. 1916. Stock-brainedness, the causative factor in the so-called "crossed aphasics." *Am. J. Med. Sci.* 152: 849–859.

Kimura, D. 1961a. Some effects of temporal-lobe damage on auditory perception. *Can. J. Psychol.* 15: 156–165.

1961b. Cerebral dominance and the perception of verbal stimuli. *Can. J. Psychol.* 15: 166–171.

Lashley, K. S. 1929. *Brain Mechanisms and Intelligence.* Chicago: University of Chicago Press.

Ogle, S. W. 1871. On dextral pre-eminence. *Med. Chir. Trans.* 35: 279–301.

Penfield, W., & Roberts, L. 1959. *Speech and Brain Mechanisms.* Princeton: Princeton University Press.

Peterson, G. M. 1951. Transfers in handedness in the rat from forced practice. *J. Comp. Physiol. Psychol.* 44: 184–190.

Peterson, G. M., & Devine, J. V. 1963. Transfers in handedness in the rat resulting from small cortical lesions after limited forced practice. *J. Comp. Physiol. Psychol.* 56: 752–756.

Peterson, G. M., & Rigney, J. W. 1950. Influence on handedness of acetylcholine locally applied with other chemicals to the cerebral cortex of the rat. *J. Comp. Physiol. Psychol. 43:* 264-271.

Roberts, L. 1951. Localization of speech in the cerebral cortex. *Trans. Am. Neurol. Soc. 76:* 43-50.

——— 1956. Handedness and cerebral dominance. *Trans. Am. Neurol. Soc. 81:* 143.

Semmes, J., Weinstein, S., Ghent, L., & Teuber, H.-L. 1955. Spatial orientation in man after cerebral injury: Analyses by locus of lesion. *J. Psychol. 39:* 227-244.

——— 1960. *Somatosensory Changes after Penetrating Brain Wounds in Man.* Cambridge: Harvard University Press.

——— 1963. Correlates of impaired orientation in personal and extrapersonal space. *Brain 86:* 747-772.

Sersen, E. A., Weinstein, S., & Vetter, R. J. 1966. Laterality differences in tactile sensitivity as a function of handedness, familial background of handedness, and sex. Paper presented to Eastern Psychological Association, New York.

Shapiro, G. 1966. Somatosensory thresholds and discrimination in amputation stumps. Paper presented to Eastern Psychological Association, New York.

Teuber, H.-L., & Weinstein, S. 1954. Performance on a formboard task after penetrating brain injury. *J. Psychol. 38:* 177-190.

——— 1958. Equipotentiality versus cortical localization. *Science 127:* 241-242.

Thurnam, J. 1866. On the weight of the brain and the circumstances affecting it. *J. Ment. Sci. 57:* 1-43.

Von Bonin, G. 1962. Anatomical asymmetries of the cerebral hemispheres. In V. B. Mountcastle (ed.). *Interhemispheric Relations and Cerebral Dominance.* Baltimore: Johns Hopkins University Press, pp. 1-6.

Wada, J. A., & Rasmussen, T. 1960. Intracarotid injection of sodium amytal for the lateralization of cerebral speech dominance: Experimental and clinical observations. *J. Neurosurg. 17:* 266-282.

Wagner, H. 1864. Massbestimmungen der Oberfläche des grossen Gehirns. Wigand: Cassel & Göttingen.

Weinstein, S. 1954. Weight judgment in somesthesis after penetrating injury to the brain. *J. Comp. Physiol. Psychol. 47:* 31-35.

——— 1955a. Time-error in tactile size judgment after penetrating brain injury. *J. Comp. Physiol. Psychol. 48:* 320-323.

——— 1955b. Time-error in weight judgment after brain injury. *J. Comp. Physiol. Psychol. 48:* 203-207.

——— 1955c. Tactile size judgment after penetrating injury to the brain. *J. Comp. Physiol. Psychol. 48:* 106-109.

——— 1958. Body image: The psychophysiological approach. Paper read at Symposium on Body Image and Brain Damage: A Critical Evaluation, American Psychological Association, Washington, D.C.

1959. The effect of traumatic brain injury on speed of finger oscillation. Paper presented to Eastern Psychological Association, Washington, D.C.

1962a. Tactile sensitivity of the phalanges. *Percept. Mot. Skills 14:* 351–354.

1962b. Differences in effects of brain wounds implicating right or left hemispheres: Differential effects on certain intellectual and complex perceptual functions. In V. B. Mountcastle (ed.). *Interhemispheric Relations and Cerebral Dominance.* Baltimore: Johns Hopkins University Press, pp. 159–176.

1962c. Phantoms in paraplegia. In *Proceedings Eleventh Annual Clinical Spinal Cord Injury Conference.* Washington, D.C.: Veterans Administration.

1963. The relationship of laterality and cutaneous area to breast-sensitivity in sinistrals and dextrals. *Am. J. Psychol. 76:* 475–479.

1967. Sensory deprivation: Implications for visuoperceptual development. Paper presented at Thirteenth International Symposium of Neuropsychology, Quimper, France.

1968. Intensive and extensive aspects of tactile sensitivity as a function of body part, sex, and laterality. In P. R. Kenshalo (ed.). *The Skin Senses.* Springfield, Ill.: Thomas.

Weinstein, S., Richlin, M., Weisinger, M., & Fisher, L. 1967. *The Effects of Sensory Deprivation on Sensory, Perceptual, Motor, Cognitive, and Physiological Functions.* Washington, D.C.: National Aeronautics and Space Administration, CR-727.

Weinstein, S., Semmes, J., Ghent, L., & Teuber, H.-L. 1956. Spatial orientation in man after cerebral injury: Analyses according to concomitant defects. *J. Psychol. 42:* 249–263.

1958. Roughness discrimination after penetrating brain injury in man. *J. Comp. Physiol. Psychol. 51:* 269–275.

Weinstein, S., & Sersen, E. A. 1961. Tactual sensitivity as a function of handedness and laterality. *J. Comp. Physiol. Psychol. 54:* 665–669.

Weinstein, S., Sersen, E. A., & Vetter, R. J. 1964. Phantoms and somatic sensation in cases of congenital aplasia. *Cortex 1:* 276–290.

1970. Phantoms following mastectomy. *Neuropsychologia 8:* 185–197.

Weinstein, S., & Teuber, H.-L. 1957a. The role of preinjury education and intelligence level in intellectual loss after brain injury. *J. Comp. Physiol. Psychol. 60:* 535–539.

1957b. Effects of penetrating brain injury on intelligence test scores. *Science 125:* 1036–1037.

Weinstein, S., Vetter, R. J., Shapiro, G., & Sersen, E. A. 1969. The phantom limb in patient sustaining cerebral vascular accidents. *Cortex 5:* 91–103.

Wolff, B. B., & Jarvik, M. E. 1964. Relationship between superficial and deep somatic thresholds of pain with a note on handedness. *Am. J. Psychol. 77:* 589–599.

Wolff, B. B., Krasnegor, N. A., & Farr, R. S. 1965. Effect of suggestion upon experimental pain response parameters. *Percept. Mot. Skills 21:* 675-683.

Woo, T. L. 1931. On the asymmetry of the human skull. *Biometrika 22:* 324-352.

Young, J. Z. 1962. Why do we have two brains? In V. B. Mountcastle (ed.). *Interhemispheric Relations and Cerebral Dominance.* Baltimore: Johns Hopkins University Press, pp. 7-24.

Zangwill, O. L. 1960. *Cerebral Dominance and Its Relation to Psychological Function.* Edinburgh: Oliver & Boyd.

3
Hemispheric asymmetry as evidenced by spatial disorders

ENNIO DE RENZI

This paper reviews the results of some experimental investigations, carried out over the years 1964-1971 by my colleagues and myself in the Clinic for Nervous and Mental Disease of the University of Milan, with the aim of studying the relationship between disorders of spatial abilities and unilateral brain damage. The differential contribution of the two hemispheres to the processing of spatial information has been assessed on the basis of systematic collection of data, standardized procedure of evaluation, and definition of the nature of the abilities underlying each tested performance.

Disorders implying impaired cognition of extrapersonal space have long been known to neurologists, who have described them under a host of headings, customarily defined on the basis of the deranged performance. This performance was often important from a practical standpoint (e.g., reading, counting, or drawing), but did not necessarily reflect the impairment of a single basic ability. Benton (1969) summarized neurologists' case reports on this topic and drew up a list of seven main types of spatial disorders: (1) defective localization of stimuli in extrapersonal space, (2) defective short-term memory for spatial location, (3) defective route finding, (4) reading and counting disability, (5) defective topographical memory, (6) visuoconstructive disabilities, and (7) simultaneous agnosia. Benton's analysis shows clearly that each of these forms of spatial impairment may result from disruption of different mechanisms (e.g., defective route finding is produced by unilateral neglect as well as by a failure to build up the basic spatial scheme expressed in the route), but it is also evident that derangement of the same ability may be reflected in symptoms classified under different categories (e.g., defective scanning can underly defective localization of stimuli, reading and counting disabilities, and visuoconstructive apraxia).

A more convenient approach would certainly be to investigate spatial disorders with reference to the basic mechanisms that guide the orientation in space of normal subjects. To date, however, these mechanisms are only imperfectly understood; moreover, it is uncertain to what extent they are related to discrete neuronal networks liable to be independently damaged by brain lesions. The compromise solution we have chosen has been to distinguish different levels of disruption of spatial behavior on the basis of clues provided by clinical reports and to devise ad hoc tests in order to obtain a graded evaluation of performance. Because this work is still in progress and many relevant topics have not yet been investigated, any definite classification of the basic deficits produced by brain lesions would be premature; the work reported here is simply intended as an empirical attempt at clarifying this tangled issue. With this proviso in mind, we propose to summarize our findings on disorders of spatial behavior following unilateral brain damage under the following headings: (1) space exploration and localization, (2) space perception, (3) intelligent elaboration of spatial information, and (4) spatial memory. A final section is devoted to constructional apraxia, not so much because this symptom represents, in our opinion, a unitary, basic disorder, as because the spatial problems it involves deserve a separate discussion.

MATERIAL AND METHOD

Although the studies reported in this chapter were carried out on different samples of patients with damage restricted to one cerebral hemisphere, certain characteristics were shared by all the experimental groups investigated; these are briefly summarized here.

Our subjects were patients admitted to neurological wards and, therefore, showed the clinical features of the corresponding brain-damaged population. Approximately 60% to 70% suffered from cerebrovascular accident, 15% from neoplasm, and 15% from trauma; the proportion of patients whose disease lasted less than 3 months exceeded that of patients more chronically ill.

A patient qualified for inclusion in the sample with either left or right hemisphere damage when he was right-handed; presented clinical symptoms restricted to one hemisphere; had no past or present evidence of involvement of the other side of the brain; and when the available electroencephalographic, neuroradiological, and brain scanning findings did not contradict the assumption of unilateral damage made on a clinical basis. It can be rightly argued that the criteria followed in assigning patients to one or the other hemisphere-damaged group were not sufficiently reliable to prevent inclusion of patients actually suffering from bilateral disease. However, in view of the criteria followed in selecting the patients, there is no reason to believe that patients with bilateral damage prevailed in one more than in the other group, thus biasing interhemisphere comparisons. To avoid missampling, special care was taken to examine every testable patient who met the requirements for being assigned to one hemisphere-damaged group. That this attempt was successful is confirmed by data recorded in the course of one investigation (De Renzi, Scotti, & Spinnler, 1969). Over a definite period, note was taken of how many patients admitted to the wards could not be tested and of the reasons that prevented their examination. Out of a total of 197 patients with unilateral lesions, 29 were not examined, 16 with damage apparently confined to the right hemisphere and 13 with injury to the left hemisphere. Of the first group, 12 died before testing, and 4 were excluded because of severe disturbances of consciousness. Of 13 patients with left hemisphere damage, 3 died before testing, 6 were excluded because of disorders of consciousness, and 4 were excluded because severe aphasia prevented their comprehension of the test instructions. These data show that aphasia was not a major cause of exclusion, at least in studies entailing the administration of easily understandable tests, as was the case for most of our investigations.

In most of our researches, patients were subdivided not only for the side of the lesion but also for presence or absence of visual field defects (VFD). This symptom was ascertained by the confrontation method carried out separately for the superior and inferior quadrants and supplemented by double simultaneous stimulation in subjects giving apparently negative responses. The examiner sat in front of the patient, keeping his arms outstretched, and from each closed fist extended his forefinger, alternately on the left, on the right, or on both sides. The patient was asked to fix his gaze on the eyes of the examiner and was considered as suffering from VFD only if he failed to see or showed visual extinction consistently on the same side. More refined techniques, implying the use of perimeters, were not employed because they require sustained co-operation on the part of the subject and would have yielded dubious results or been inapplicable to some of our patients.

The rationale for classifying patients on the basis of presence or absence of VFD is that this symptom allows a rough distinction between patients with predominantly pre-Rolandic and patients with predominantly post-Rolandic damage. The central optic pathways run in the white matter underlying the border between the temporal and the parietal lobes and end in the calcarine fissure of the occipital lobe; they are, therefore, likely to be injured by lesions located behind, and spared by lesions located in front of, the Rolandic sulcus. There are, of course, exceptions to this rule, as exemplified by injuries to the superior parietal lobe, which do not involve optic pathways and, consequently, would be considered together with pre-Rolandic lesions. This kind of error would tend, however, to decrease rather than to enhance the difference in scores between patients with and patients without VFD and, therefore, does not detract from the value of the obtained differences. It must be added that available clinical and ancillary data would have allowed a more precise localization of the cerebral damage in a certain number of subjects, but we thought it preferable to use the same criterion for every patient, even at the expense of some loss in precision and accuracy.

A significant association of VFD with failure on a given task can be assumed to point to a critical role of the retro-Rolandic areas in the performance of the task, provided one can exclude the possibility that the impairment simply reflects the deficit of visual functions. We have considered this requirement to be complied with when either of the following results was obtained:

1 Presence of VFD affected the performance of only one hemisphere-damaged group. Since there is no known difference in the anatomy and physiology of central optic pathways of the two sides of the brain, the differential performance of patients with VFD according to which hemisphere is injured can be taken as evidence that the visual deficit is a symptom coincidental with damage to a definite cerebral area.
2 The impairment found in a visual test was similar to that found in a tactile test carried out without the aid of vision.

A final point is worth mentioning with respect to the anterior versus posterior subdivision of samples. Owing to the predominance of vascular causes of brain damage in our subjects, patients with damage to prefrontal areas are likely to be sparsely represented in groups with pre-Rolandic lesions, as this region is in general rarely involved by cerebrovascular accident. Therefore, definite inferences concerning the functioning of prefrontal areas in spatially oriented behavior are not justified on the basis of our data.

The cause and duration of illness were always controlled and found not to differentiate the groups to be compared. The possible influence of age and years of schooling on performance has been canceled out by introducing them as concomitant variables in the analysis of covariance or by pairing groups with respect to these factors when nonparametric methods had to be used because of the skewed distribution of scores.

Most of the tests used in the researches reviewed in this chapter were contrived ad hoc to evaluate one specific spatial ability. The main difficulty was to devise tasks that, though elementary, would yield well-distributed scores. These two requirements are not easy to fulfill. A simple test is likely to be performed perfectly by most normal subjects and by several brain-damaged patients; consequently, its scores cannot be analyzed by parametric methods, by which one may study interaction and combine several variables in the same analysis. On the other hand, if the test is made sufficiently complex to differentiate among normals, it risks being sensitive to more than one kind of cerebral disorder and thus loses specificity. We managed to cope with this issue by keeping the task as elementary as possible and refining the method of evaluating the performance. On some occasions, a graded assessment was allowed by the very nature of the measure adopted; on other occasions, we contrived to improve the sensitivity of the score by giving time credits for successful performances and canceling out the nonspecific speed factor by introducing simple visual reaction times into the analysis of covariance.

DEFECTIVE EXPLORATION AND LOCALIZATION OF STIMULI IN EXTRAPERSONAL SPACE

The ability to scan extrapersonal space and locate stimuli with respect to one another and in relation to the observer may be considered a prerequisite of every operation implying spatial perception and cognition. Severe disorders in this area have been described following bilateral posterior brain damage and are known as Balint's syndrome, which consists of fixation of gaze, inattention to whatever stimuli arise outside the fixation point, and optic ataxia (Hécaen & Ajuriaguerra, 1954). In the case of unilateral injury, visual hemiinattention for the contralateral field has been the most commonly reported symptom, although mislocation of stimuli within this field has also been observed.

Systematic investigations carried out on patients with unilateral damage have essentially been concerned with the question of whether omission in perceiving or reproducing stimuli located in the visual field contralateral to the injured side of the brain occurred with different frequency after left and right hemisphere lesions (Battersby et al., 1956; De Renzi & Faglioni, 1967; Gainotti, 1968; Costa et al., 1969). The results of these researches have provided clear-cut evidence for the prevalence of unilateral neglect in patients with right hemisphere damage, but have not permitted a graded evaluation of the performance, as only errors of omission were scored. Moreover, investigations have been restricted to the visual modality. Owing to these limitations, some relevant questions concerning exploration of space remain unsettled: for instance, to what extent left hemisphere damage impairs scanning of the right field, whether exploratory disorders occur when space is explored through nonvisual sensory channels, and whether hemiinattention is a merely sensorimotor phenomenon or also involves other mental functions.

To gain a better understanding of these issues, the ability to detect a stimulus on a display was investigated under two different testing conditions, one visually and the other tactually guided (De Renzi, Faglioni, & Scotti, 1970). In the *visual test*, the patient was given a card bearing a one-digit number and was requested to find its match on a cardboard displaying a random sequence of 100 numbers, from 0 to 99, of four different sizes. In the *tactile test* (Fig. 3-1), the patient's forefinger ipsilateral to the damaged hemisphere was guided to the center of a maze hidden by a curtain and had to move as quickly as possible along its alleys in search of a marble that had been placed at the end of one of the four lateral arms of the maze. Four different cardboards and two mazes were alternated over the eight trials of each test, and the item to be found was placed twice in each of the four quadrants, thus permitting a comparison of performance in the fields ipsilateral and contralateral to the damaged cerebral hemisphere. The score was the time required by the patient to find the number (or the marble) up to a maximum time of 90 sec. A graded evaluation of the performance was thus possible.

It is worth pointing out that the tactile test differs from the visual test not only in terms of the sensory channel through which information is conveyed but also because the exploration of the maze is guided essentially by its representation and not by its perception, as is the case for the visual display. In the tactual performance, the patient does not receive simultaneous information from a broad sensory area, comparable to the visual field, but only from the point where his forefinger lies, and, therefore, he must rely on the representation of the borders of the maze and on recall of the already explored areas for programming his searching strategy.

When data from the visual test were examined, it was found that the scoring procedure adopted entailed an unforeseen drawback that made it difficult to compare the performances of groups with right and left hemisphere damage. Normal subjects showed a

Fig. 3-1. One of the mazes used in the tactile searching test. (From De Renzi, Faglioni, & Scotti, 1970)

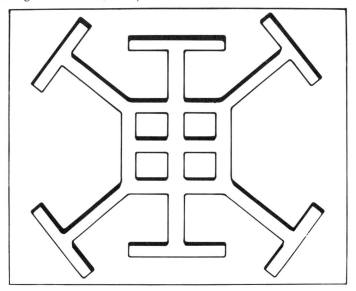

gradient between the searching times spent in the two halves of the display. They were significantly slower in finding the number in the right than in the left field ($p < 0.02$), probably as a consequence of reading habits, which require left-to-right scanning in Western languages. If one keeps in mind that the right field is contralateral to the side of the lesion in patients with left hemisphere damage and ipsilateral in those with right hemisphere damage, its "normal" disadvantage can be expected to remain or to increase in patients with left hemisphere lesions and to decrease or to reverse in those with right hemisphere lesions, especially those with VFD. Inspection of Figure 3-2 confirms this prediction: The left-right gradient is present in three out of four brain-damaged groups and is enhanced with respect to controls in the group with left hemisphere damage and VFD; on the other hand, it is reversed in patients with right hemisphere damage and VFD, whose searching time is slower in the left field. Although this result suggests that damage to the posterior region of one' hemisphere impairs scanning of the contralateral field, the occurrence of the gradient prevents a definitive comparison of the differential effect on per-

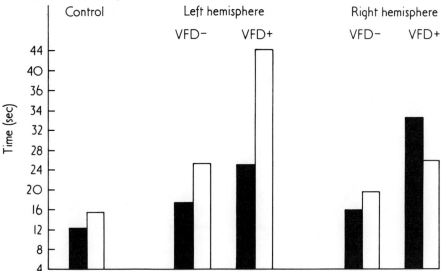

Fig. 3-2. Mean visual searching time of controls and of patients with right and left hemisphere damage, with and without visual field defects (*VFD*), in left (*black bars*) and right (*white bars*) visual fields. (From De Renzi, Faglioni, & Scotti, 1970)

formance of left versus right cerebral injury. This is likely to be an unavoidable limitation inherent in any visual searching task scored in terms of time.

More precise information was derived from the tactile test. Table 3-1 clearly shows that every brain-damaged subgroup took longer to find the marble in the contralateral than in the ipsilateral field, no matter whether this was the right or the left field. These interfield differences were significant both in patients with right ($p < 0.05$) and with left hemisphere damage ($p < 0.01$), although the mean scores of the controls (half of whom carried out the task with the right hand and half with the left hand) were nearly the same on both fields: 29.10 sec for the left field and 27.60 sec for the right field. It must be stressed that these scores are not better than those obtained by any brain-damaged group in the field ipsilateral to the side of the lesion; therefore, the view can be held that only the searching time for the contralateral field is affected by unilateral brain damage.

The finding that the slowing of searching time is not greater in the group with right hemisphere damage than in the group with left hemisphere damage suggests that there is no asymmetry in the proficiency of the two halves of the brain in the exploration of extrapersonal space. However, because the lack of hemispheric difference may be dependent on the ceiling effect produced by the time limit of 90 sec set for finding the marble, it was deemed worth analyzing how many patients failed to find the marble in the contralateral field within the time limit of 90 sec. Over the four trials carried out in each field, 5 controls missed the marble once, but none more than once. Accordingly, any patient who failed more than once on the contralateral field was considered to have per-

Table 3-1. *Mean scores of patients with left and right hemisphere damage, with and without visual field defects (VFD), on tactile maze in fields ipsilateral and contralateral to side of lesion*

	Patients with left hemisphere damage		Patients with right hemisphere damage	
	VFD − (N = 46)	VFD + (N = 25)	VFD − (N = 30)	VFD + (N = 20)
Ipsilateral field	23.25	28.25	25.25	27.25
Contralateral field	30.75	36.50	28.00	38.25

formed defectively on this task. There were 15 brain-damaged patients with defective performance, 10 of them had right hemisphere damage with VFD, 2 had right hemisphere damage without VFD, and 3 had left hemisphere damage without VFD. The great majority of failures were made by patients with right hemisphere damage and VFD, 50% of whom missed the marble in the contralateral field two or more times out of a maximum of four. The difference between this percentage and those found in the three other brain-damaged subgroups is significant beyond the 0.001 level of confidence and constitutes clear-cut evidence that unilateral neglect following retro-Rolandic right cerebral injury occurs also when space is tactually explored.

Based on this experiment, the following conclusions seem warranted:

1 Disorders of space exploration after unilateral brain damage are not restricted to the visual modality, as clinical reports suggest, but occur also in tactually guided tasks. Taken in conjunction with previous clinical observations indicating hemispatial disorientation for auditory stimuli (Denny Brown et al., 1952; Bender & Diamond, 1965), this finding supports the notion of a supramodal mechanism guiding the sensorimotor neuronal networks involved in scanning extrapersonal space through the various modalities.
2 Each hemisphere is exclusively involved in scanning the contralateral half of space, but damage to the right side of the brain impairs exploration of the left field more than damage to the left side impairs exploration of the right field.
3 Each half of the brain is concerned not only with the perception but also with the representation of contralateral space, and thus unilateral neglect may be conceived of as depending, at least in part, on a cognitive defect, understandable as mutilated representation of space.

Examining these same patients, Faglioni, Scotti, and Spinnler (1971) carried out a second investigation with the aim of studying the ability to locate exactly a point in bidimensional space. A visual test, previously described by De Renzi and Faglioni (1967), was used, and an equivalent tactile version was prepared. In the *visual test,* three model sheets presenting 6, 5, and 6 crosses, respectively, were given in succession together with sheets of identical size on which the patient was required to reproduce the crosses exactly in the same positions. Of the 17 model crosses, 9 were drawn in the right field of the sheet and 8 in the left. The score was the distance in millimeters separating the intersection point of the model cross from that of the nearest cross of the copy. Whenever a cross was omitted, an arbitrary score of 140 mm – corresponding to half of the length of the sheet – was given.

In the *tactile version* of the test, sheets were replaced by cork boards and crosses by drawing pins; a curtain prevented the subject from seeing both the model and the copy board, so that exploration and localization of the drawing pins were carried out only by feeling their heads. The examiner provided the patient with one drawing pin at a time up to the total number of drawing pins present in the model. From this standpoint, the tactile test was substantially different from the visual test, because in the former no drawing pin could be omitted, and in the latter the subject did not know the number of the crosses to be reproduced and was, therefore, likely to neglect one or more of them. Table 3-2 presents the means obtained by the five experimental groups and shows that the same pattern of performance was present in the two tests. The $F < 1$ value found in the interaction between group and test confirmed that there was no difference in the contribution of the visual and tactile test toward discriminating the groups. Intergroup comparisons, carried out by Roy and Bose's method (1953), disclosed that, when contrasted to the control group, patients with either left or right hemisphere damage and VFD scored worst, and the corresponding groups without VFD were not significantly impaired. On the other hand, when the four brain-damaged groups were compared two by two, only patients with

Table 3-2. *Mean scores of controls and of patients with left and right hemisphere damage, with and without visual field defects (VFD), on visual and tactile localization tests*

Test	Controls (N = 30)	Patients with left hemisphere damage		Patients with right hemisphere damage	
		VFD – (N = 46)	VFD + (N = 25)	VFD – (N = 30)	VFD + (N = 20)
Visual test	248	327	390	277	632
Tactile test	551	688	792	672	988

right hemisphere damage and VFD stood out from all the others because of their inferior performance; no other comparison turned out to be significant. These results indicate that the cerebral damage must encroach upon the retro-Rolandic regions to affect performance on this test, but they also show that there is a striking asymmetry in function between the posterior areas of the two hemispheres, as injury to the right side produces a much more clear-cut disruption of ability to locate stimuli in extrapersonal space.

Before accepting these results as evidence of a specific disorder of localization, one must cope with the objection that the copying tests used in this experiment unavoidably implied scanning of space and that disorders of exploration and not of localization could have been responsible for the impairment shown by patients with injury to the right posterior part of the brain. To distinguish the influence on performance of deficits of exploration as compared with those of location, the occurrence of unilateral neglect (i.e., of absent or incomplete exploration of the contralateral field) was assessed. In the visual test, unilateral neglect was assumed to occur when one or more crosses were omitted in the field contralateral to the side of the lesion (no control patient made such an error). This omission was made only by patients with VFD, and by far more patients with right-sided (9 out of 20, or 45%) than with left-sided (2 out of 25, or 9%) damage ($p < 0.0001$). The concentration of omission errors in the group that was more impaired on the cross-copying test prevents us from distinguishing the impact on performance of exploratory as compared with localization deficit. The tactile test provides a more definite answer because, as mentioned above, it was administered in such a way as to make omissions practically impossible. To verify whether defective exploration nevertheless played some role in the performance, the displacement of drawing pins from the contralateral to the homolateral field of the board was assessed as possibly indicating a tendency to unilateral neglect. One drawing pin was occasionally displaced by a control patient, and thus only displacement of more than one drawing pin was considered defective. Of 20 patients with right hemisphere damage and VFD, 3 showed defective performance, and this proportion was not different from that found in the other groups ($\chi^2 = 2.20$, $p < 0.50$, with 3 DF). It must be stressed that absence of evidence for contralateral neglect is strictly dependent on the testing technique used in this tactually guided task, because the findings previously reported on the tactile maze demonstrate that hemiinattention does exist in the tactile as well as in the visual modality. Nevertheless, the very fact that the present test was not sensitive to unilateral neglect is advantageous, because it permits one to ascribe the poor performance of patients with right posterior brain damage to the locating deficit only and supports the conclusion that damage to the retro-Rolandic region of the right hemisphere selectively deranges the ability to utilize the spatial cues that guide location of a point in bidimensional space.

DISORDERS OF SPATIAL PERCEPTION

Spatial perception is a loose term used by neurologists and psychologists to encompass under the same label a variety of performances ranging from elementary tasks (e.g., perception of depth or setting a line to the vertical) to deduction of the spatial relationship underlying a series of complex patterns (as in Raven's Progressive Matrices). We thought it advisable to keep spatial perception distinct from the intellectual analysis of spatial information and to investigate one of its more basic dimensions, orientation in space, at two different levels: the appreciation of direction (1) of one single segment and (2) of a set of segments making up a shape.

In the first task (De Renzi, Faglioni, & Scotti, 1971), the two pairs of rods shown in Figure 3-3 were presented to the patient. Each pair consisted of a vertical rod, movable along its axis, and a second rod, fixed to the first by a hinged joint that could be lifted and lowered in the sagittal plane. By appropriate rotation of the two rods, the second could be placed in any position in space that corresponded to the radiuses of an ideal sphere having its center in the joint between the two rods. One pair of rods was placed in a predetermined position by the examiner, and the patient was asked to set the other pair parallel to the standard, using only the hand ipsilateral to the side of the brain lesion. The task was presented both in the visual and in the tactile modality. In the *visual test*, the patient could see the two pairs of rods, but was not permitted to touch the model; in the *tactile test*, the apparatus was hidden from sight by a curtain, and the patient could estimate the position of the rods of both pairs only by feeling them.

There were five different positions of the standard pair to be reproduced, and they were presented to the patient in both test modalities, with the model first on the same side as the damaged hemisphere and then on the contralateral side. The order of presentation of the test modalities was alternated from subject to subject.

Each copying performance was scored by measuring the mean of the unsigned deviations from the model of the *vertical* angle made by the two rods and of the *horizontal* angle made by the second rod with respect to the sagittal plane. These two scores were treated separately for the left and the right positions of the model and for the visual and tactile tests in order to assess whether there was any first-or higher-order interaction between these factors and the performance. No interaction was found significant, and we shall compare only the overall means of the combined scores. As Figure 3-4 shows and the statistical analysis confirms, only the patients with right hemisphere damage and VFD performed poorly as compared with both controls and each of the other three brain-damaged groups; none of the other comparisons disclosed signif-

Fig. 3-3. Materials used in the rod orientation test. (From De Renzi, Faglioni, & Scotti. 1971. Judgment of spatial orientation in patients with local brain damage. *J. Neurol. Neuros. Psychiat.* Reproduced with permission.)

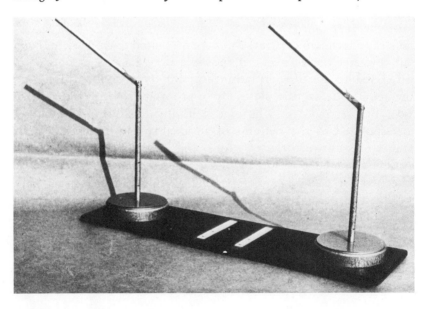

icant differences. Therefore, this is an example of pure right hemisphere dominance, not simply of asymmetry of function, as would be the case if damage to either posterior area affected the performance, though more so following injury to the right side.

The tactile shape-discrimination test used by De Renzi and Scotti (1969) may be assumed to evaluate the sense of spatial orientation at a somewhat more complex level, as the subject had to organize into a whole the spatial direction of successively perceived segments. The patient was asked to place the hand ipsilateral to the damaged hemisphere under a curtain and to follow the lateral sides of a block with his forefinger outstretched vertically downward. In the meantime, he watched a board bearing 6 alternatives and had to point to the match of the block he was exploring tactually as soon as he thought he had identified it. Because he was prevented from palpating simultaneously the lateral sides of the block, he had to rely entirely upon the successive changes of direction made by his exploring finger in order to build up the

Fig. 3-4. Mean error scores on rod orientation test of controls (C) and of patients with left (LH) and right (RH) hemisphere damage with and without visual field defects (VFD). (From De Renzi, Faglioni, & Scotti, 1971)

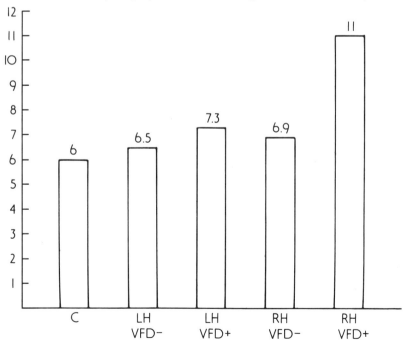

Fig. 3-5. The twelve blocks of the shape-discrimination test. (From De Renzi & Scotti, 1969)

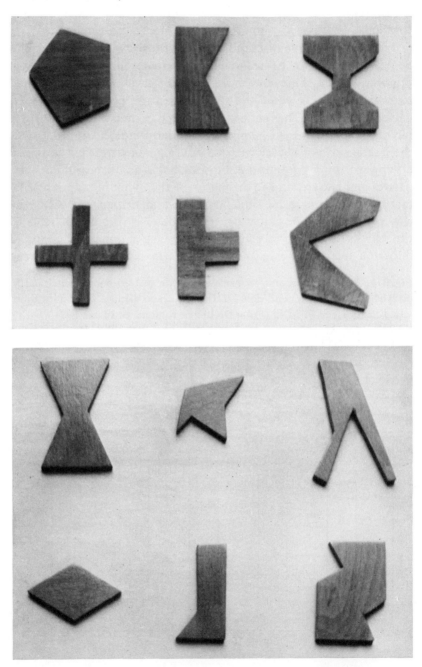

Table 3-3. *Mean scores of controls and of patients with left and right hemisphere damage, with and without visual field defects (VFD), on shape-discrimination test*

Controls		Patients with left hemisphere damage		Patients with right hemisphere damage	
Left hand (N = 15)	Right hand (N = 15)	VFD − (N = 46)	VFD + (N = 25)	VFD − (N = 30)	VFD + (N = 20)
24.76	25.06	21.97	20.45	19.37	15.33

shape. There were 12 shapes to identify (Fig. 3-5), and for each of them a score of 3 was given when made within 15 sec, a score of 2 when made within 30 sec, and a score of 1 when made within 60 sec. The maximum score was, therefore, 36. Half of the control patients carried out the test with the right hand and the other half with the left hand.

The mean scores of the five groups are presented in Table 3.3. The analysis of covariance – age and years of schooling being the concomitant variables – showed no hand effect in control patients and a significant impairment in patients with left and patients with right hemisphere damage with respect to the control subgroup that had used the same hand. Within the brain-damaged group, patients with right-sided damage performed significantly worse than patients with left-sided damage ($p < 0.0001$), a result that inspection of the means of Table 3-3 suggests to be mainly attributable to the low scores obtained by the subgroup with right hemisphere damage and VFD, although the interaction between hemisphere and VFD fell short of the 0.05 confidence level. Compared with the findings of the rod orientation test, these results confirm the major role played by the right hemisphere in analyzing direction in space, but point out also that increasing the complexity of the task makes the performance vulnerable to injury of the left hemisphere also and diminishes the strict association between impairment and damage to a definite area of the brain.

IMPAIRMENT ON TASKS OF SPACE PERCEPTION INVOLVING INTELLIGENCE

Most of the early studies that aimed at a systematic investigation of space perception in brain-damaged patients employed tasks derived from rather complex mental tests originally devised to assess the performance of normal subjects. This choice may have prevented the appreciation of the respective roles of the two hemispheres in mediating the spatial aspects of behavior, because the impairment of the many abilities involved in these tasks is likely to have clouded the interpretation of findings. Let us take as examples the Block Design and the Object Assembly subtests of the Wechsler Adult Intelligence Scale (WAIS). They have been considered by Wechsler as sensitive tools to evaluate general intelligence, but they have also been used by some investigators (Andersen, 1951; Heilbrun, 1956; Costa & Vaughan, 1962) to measure space perception and by others (Piercy & Smyth, 1962) to assess constructional abilities. There is nothing to object to regarding the assumptions underlying these researches, provided one is aware that all these aspects of behavior are reflected in the performance and that they may be differentially affected by lesions localized in separate areas of the brain. The same line of reasoning can be applied to Raven's Progressive Matrices; to Gottschaldt's test, in which the patient has to trace a geometrical figure embedded in a complex visual pattern; and to Elithorn's test (Elithorn, 1955), which requires the subject to discover a route on a two-dimensional lattice maze (Fig. 3-6). Studies from our laboratory have dealt with performance on these tests by patients with damage to different cerebral hemispheres.

Arrigoni and De Renzi (1964) investigated unilateral brain-damaged patients with the Block Design and the Object Assembly tests, Colonna and Faglioni (1966) with Raven's and Elithorn's tests, Russo and Vignolo (1967) with Gottschaldt's test. Table 3-4 reports the mean scores obtained by the two hemisphere-damaged groups on these tests, adjusted, except for Block Design and Object Assembly, for age, years of schooling, and reaction times. The general trend is rather consistent: Patients with right hemisphere damage scored somewhat lower than those with left hemisphere lesions, but the difference never reached the statistical level of confidence. In Raven's, Elithorn's, and Gottschaldt's tests, samples were subdivided also for the presence or absence of visual field defects: Patients with VFD and lesions of either hemisphere scored significantly lower than patients without VFD on Raven's and Elithorn's tests, but not on Gottschaldt's test. When, on the other hand, aphasics were compared with nonaphasic patients with left hemisphere damage, their performance was poorer on all three tests. This finding may be interpreted as evidence that disruption of verbal mediation constitutes a serious drawback to carrying out these apparently non-language-dependent tests, though it is difficult to conceive how silent verbalization can aid the patient in finding the hidden meaningless figures of Gottschaldt's test. The broader hypothesis can be advanced that whenever the processing of perceptual data goes beyond the level of "pure" detection of spatial orientation and involves intellectual analysis (such as deduction of relationships or discovering a hidden solution intermingled with irrelevant information), the contribution of the left hemisphere to performance increases and tends to attenuate the right hemisphere's superiority.

The same view can be applied to other mental operations dealing with spatial aspects of reality, for example, the ability to reverse perspectives in imagery. A study by De Renzi and Faglioni (1967) required patients to imagine how a pattern looked when rotated 180°. The subject was given a display card, showing nine scrawls in three rows, and a separate copy of one of these scrawls presented

Fig. 3-6. Items of Gottschaldt's test (*top*) and Elithorn's test (*bottom*).

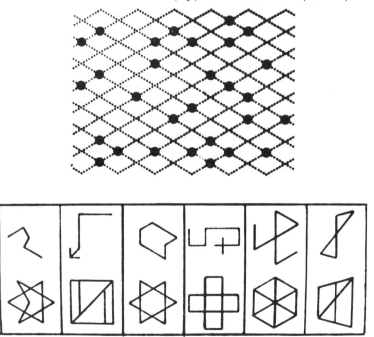

Table 3-4. Mean scores of patients with left (LH) and right (RH) hemisphere damage on five tests of space perception

Raven's test		Elithorn's test		Gottschaldt's test		Block Design		Object Assembly	
LH (N = 59)	RH (N = 53)	LH (N = 59)	RH (N = 53)	LH (N = 55)	RH (N = 40)	LH (N = 68)	RH (N = 55)	LH (N = 68)	RH (N = 55)
23.08	22.30	13.98	11.59	13.16	12.91	8.97	7.56	12.23	10.72

in an inverted position. He was not allowed to turn the item upside-down and was required to find the corresponding scrawl on the display within a time limit of 40 sec. The task was repeated with 10 different cards and individual scrawls. As indicated in Table 3-5, the mean score of patients with right-sided damage (5.9) was only slightly inferior to that of patients with left-sided lesions (6.3), and the difference fell far short of the 0.05 confidence level. The presence of VFD was highly significant in lowering the performance ($p < 0.001$), but the hemisphere-VFD interaction was not significant. This lack of interhemispheric difference has been confirmed by Butters and Barton (1970) and by Benson and Barton (1970), who examined the effect of unilateral parietal and nonparietal damage on the performance of three tasks requiring the patient to assume different perspectives in imagery. One is reminded that the task of finding routes through map reading, employed by Semmes et al. (1955), which also demands the rotation of a pattern in imagery, did not disclose right hemisphere superiority, although it showed the critical role of both parietal lobes.

DISORDERS OF SPATIAL MEMORY

Disorders of spatial memory manifest themselves in patients with localized brain lesions mainly as topographical amnesia, a failure in finding one's way about in familiar surroundings that is not explicable in terms of space perception deficit or of global amnesia. This disorder is difficult to evaluate systematically on a neurological ward because many patients are prevented from walking by paralysis of the lower limbs and verbal description of routes known before disease may be hindered by aphasia. Maze-learning tests have been used to assess spatial memory and have proved very sensitive both to right temporal (Milner, 1965; Corkin, 1965) and to right parietal

Table 3-5. *Mean scores of patients with left and right hemisphere damage, with and without visual field defects (VFD), on inverted scrawl test*

Patients with left hemisphere damage		Patients with right hemisphere damage	
VFD − (N = 55)	VFD + (N = 19)	VFD − (N = 40)	VFD + (N = 23)
7.6	5.1	7.2	4.7

(Newcombe & Russell, 1969) injury. However, the fact that the performance of these tasks is usually scored in terms of mean errors or of trials to criterion makes it difficult to distinguish to what extent failure is caused by orientation or by learning disorders, as errors resulting from these two different sources are pooled in the overall score.

Our effort at evaluating disorders of spatial memory in brain-damaged patients had been centered on a more elementary ability: recollection of the position an object occupies in space. This approach was prompted by the clinical observation of two patients, who, after the onset of illness, never learned to find their way about in familiar surroundings, though their performances were unimpaired on general memory tasks and not proportionally impaired in space perception. They also failed on such simple tasks as pointing with closed eyes to the layout of objects they had just before carefully watched and named, or locating the newspaper column they had been reading after the paper had been removed for a few seconds (De Renzi & Faglioni, 1962; Scotti, 1968). These clinical observations suggest that the memory trace of the position an object occupies in space can be selectively impaired in the

Fig. 3-7. The formboard test.

absence of general memory disorders and that this might be the basic deficit in topographical disorientation.

Up to now two studies have been concerned with brain-damaged patients' performance on memory-for-position tasks, and though the question of the relationship between impairment of this ability and failure in route learning has not been settled, the relative independence of spatial memory deficit from spatial perceptual disorders has been substantiated.

The first investigation (De Renzi, Faglioni, & Scotti, 1968) dealt with a tactually guided task. Blindfolded patients were handed six blocks, one at a time, and required to discover in which of the cutouts of the formboard, shown in Figure 3-7 each of them had to be placed. Five timed trials were given, the order of presentations of blocks in each trial being different, according to a listed schedule. The score for each trial corresponded to the time elapsed from the moment the first block was handed the patient to the moment the sixth block was inserted into its corresponding cutout. As indicated by the mean scores of Table 3-6, the overall performance on this task was particularly poor in patients with right hemisphere-damage and VFD, and statistical analysis confirmed this finding, yielding a significant hemisphere-VFD interaction ($p < 0.001$).

At the end of the fifth trial, the formboard was screened from sight, the blindfold was removed, and the patient was required to place the blocks on the table in the same arrangement he imagined them to be on the board. In this way it was possible to evaluate whether groups had a differential recall of the position occupied by each block on the formboard and to relate this performance to that on the first task. A significant inferiority ($p = 0.008$) of patients with VFD was found on this block-arrangement test (Table 3-7), regardless of which hemisphere was damaged. Thus, there is

Table 3-6. *Mean scores of patients with left and right hemisphere damage, with and without visual field defects (VFD), on tactile formboard test*

Patients with left hemisphere damage		Patients with right hemisphere damage	
VFD − (N = 30)	VFD + (N = 14)	VFD − (N = 22)	VFD + (N = 14)
94	112	101	177

a dissociation between performance on the tactile perceptual test, where the group with right posterior brain damage failed, and performance on the memory-for-position test, on which the lowest score was obtained by patients with left hemisphere damage and VFD. The failure of this group is all the more remarkable as it had scored on the formboard at approximately the same level as the two groups without VFD.

These results have been replicated in a second study in which memory for position was specifically investigated with two comparable tests, one employing the visual and the other the tactile modality (De Renzi, Faglioni, & Scotti, 1969). In the *visual test* the patient was first shown the position occupied in a two-row array by six geometrical patterns, which were subsequently hidden by a small sheet of paper. He was then given a copy of each item, one at a time, and asked to point to the hidden match on the cardboard. In the *tactile test* a wooden box composed of six cells, arranged in two rows, was placed behind a curtain and the patient's hand ipsilateral to the lesion was guided to palpate the items that had been put within each cell. These were pieces of material, easy to distinguish by touch from one another. Then, the items were removed from the cells and handed, one at a time, to the patient, who was requested to place them beneath the curtain in the same cell in which he had palpated them. The entire procedure (including the demonstration) was repeated twice for each test, which therefore called for a total of 18 responses. The mean scores on the two tests are shown in Table 3-8. Again the effect of VFD was highly significant ($p < 0.0001$), although neither the hemisphere in which the lesion was located nor the VFD-hemisphere interaction was. Furthermore, the relation between VFD and test score reached the statistical level of confidence ($p < 0.05$) and inspection of the

Table 3-7. *Mean scores of patients with left and right hemisphere damage, with and without visual field defects (VFD), on block-arrangement test*

Patients with left hemisphere damage		Patients with right hemisphere damage	
VFD − (N = 30)	VFD + (N = 14)	VFD − (N = 22)	VFD + (N = 14)
6.57	3.71	6.36	4.29

means indicated that the difference between patients with and patients without VFD was higher for the tactile (3.54) than for the visual test (1.98). This result suggests that tactile tasks may be more powerful than visual tasks in discriminating brain-damaged groups for certain types of spatial abilities, a finding in keeping with data reported in Corkin's (1965) and Milner's (1965) studies. In any event, the more pronounced impairment of patients with VFD on the tactile than on the visual test clearly indicates that hemianopic defects do not play a role per se, but point to damage to posterior areas of the brain.

Taken together, the results of the two studies support the view that memory for position is an ability at least in part independent of spatial perception and of general nonverbal memory. The first conclusion is based on the previously mentioned finding that patients with left retro-Rolandic lesions showed no specific impairment on the formboard test, but nevertheless failed even more than those with right retro-Rolandic damage when asked to remember the position of blocks. The second claim is implied by data from our laboratory, which show that the association between poor scores and VFD is characteristic of memory for position and is not found in other nonverbal memory tasks. The memory-for-position tests we used were based on the paired-associates technique, in that they required patients to associate drawings or objects with places (white sheets or cells). The same technique was used by Boller and De Renzi (1967) in investigating visual memory for drawings: Patients were given a series of paired associates, one pair at a time, made up in one test of figures of objects and in the second test of meaningless figures. At the end of the series the left figure of one

Table 3-8. *Mean scores of patients with left and right hemisphere damage, with and without visual field defects (VFD), on two memory-for-position tests*

Test	Patients with left hemisphere damage		Patients with right hemisphere damage	
	VFD − (N = 46)	VFD + (N = 25)	VFD − (N = 30)	VFD + (N = 20)
Visual test	11.22	9.96	12.42	9.75
Tactile test	12.08	8.33	13.36	9.92

of the pair was presented together with 10 alternatives, among which the patient had to find the associate. On both tests only the hemisphere effect proved significant, patients with left brain-damage performing worse. For instance, the mean scores obtained on the meaningful figure test, out of a maximum score of 30, were as follows: patients with left hemisphere damage without VFD, 16.28; those with left-sided damage with VFD, 13.45; those with right-sided damage without VFD, 19.71; and those with right hemisphere damage with VFD, 19.54. The group with right hemisphere lesions and VFD performed on this test as well as the group with right-sided lesions without VFD, a finding never observed in memory-for-position tests.

CONSTRUCTIONAL APRAXIA

Constructional apraxia (CA) is considered separately because this symptom, probably owing to its alleged localizing value, has been the subject of long debate and numerous investigations by neurologists. It is doubtful, however, that constructional tasks are suitable for eliciting unequivocal information about the nature of the abilities underlying the tested performance, because they entail both a perceptual and an executive aspect and eventually confound in the same performance the influence of a wide spectrum of deficits that could more profitably be evaluated independently. The original assumption made by Kleist (1934) that CA is an autonomous disorder that reflects disconnection between perceptual and motor processes has never been substantiated and does not account for the performance of the great majority of patients showing this disability, even if it applies to single cases (Dee & Benton, 1970). Notwithstanding these limitations, CA has acquired historical value because it prompted the first systematic investigations of hemispheric asymmetry of nonverbal functions. The two main questions that have been and still are under discussion can be summarized as follows:

1. Is there a difference in frequency and severity between constructional disorders following damage to the right and to the left hemisphere?
2. Does CA reflect the impairment of the same basic ability in lesions of either side of the brain or does it include under the same label disorders mediated by different mechanisms?

Our contribution to the first issue is based on two studies (Arrigoni & De Renzi, 1964; De Renzi & Faglioni, 1967). In the first, CA was assessed by three tests, requiring the patients to copy (1) drawings, (2) block constructions, and (3) scattered tokens. On each of these tests and on all three cumulatively, a significantly greater proportion of patients with right than with left hemisphere damage scored below the entire distribution of scores of the control group: this apparently supports the notion of the major role played by the minor hemisphere in this kind of activity. The cogency of this conclusion is, however, challenged by the results obtained on an elementary and apparently unrelated test, visual reaction time. Apraxic patients were slower than nonapraxic patients and the group with right hemisphere damage was slower than the group with left hemisphere damage. Because reaction time tests were performed by each subject with the hand ipsilateral to the side of the lesion (and, hence, with the dominant hand by patients with right hemisphere injury and with the nondominant hand by those with left-sided injury), the hemisphere-related difference could hardly be attributed to a specific derangement of the half of the brain ruling the performance. The question then arose whether the lengthening of reaction time reflected the general disorganizing effect of cerebral damage, possibly related to the extension of the lesion. In this perspective the greater occurrence of CA among patients with right hemisphere damage would be accounted for by an unintentional missampling of the two groups, the group with right sided lesions being more severely ill. To obviate this possible bias, the original samples were reduced to smaller groups of 44 pairs of patients with left- and right-sided brain damage, matched for reaction times. Apraxia was found in 17 patients with left and in 27 patients with right hemisphere damage. The corresponding χ^2 value fell short of the 0.05 confidence level. It has been rightly pointed out (Benton, 1967) that this failure to attain the statistical level of confidence cannot be adduced as evidence against the major contribution of the right hemisphere to CA, because the proportion of apraxic patients in the two reduced groups was approximately the same as in the larger original groups and only the smaller number of patients may have prevented statistical confirmation of the difference. Accordingly, the only legitimate inference to be drawn from these data is that the hemispheric asymmetry, if it exists, is tenuous and requires large series of patients to be demonstrated. A second study from our laboratory (De Renzi & Faglioni, 1967) provided more convincing support for the need to evaluate cautiously right-left differences associated with CA and highlighted the problems inherent in sampling hemisphere-damaged groups. The same drawing-copying test used in Arrigoni and De Renzi's research was administered to 70 controls, 74 patients with left hemisphere damage, and 63 patients with right hemisphere damage. Table 3-9 gives the number and proportion of patients in the two brain-injured groups who scored below the lowest score of controls in the two investigations; the χ^2 test yields a nonsignificant probability value ($p < 0.20$) for the De Renzi and Faglioni study.

This failure to replicate Arrigoni and De Renzi's results was surprising, since the testing procedures and the clinical population from which the brain-damaged subjects were drawn did not differ in the two studies and the number of tested patients was even greater in the second one. Therefore, we were led to suspect that sampling could have been the critical variable in determining the discrepancy between the two studies and that the bias may have been due to the fact that the first research involved also the administration of intelligence tests (Raven's Progressive Matrices and WAIS), while no such demanding test was included in the second research. It can be argued that the most impaired patients (those with left hemisphere lesions) were unintentionally excluded by Arrigoni and De Renzi because severe aphasia made them unsuitable for testing, but were included by De Renzi and Faglioni. If the discrepancy between the two investigations is really the result of a difference in selecting the group with left-sided brain damage, one would pre-

Table 3-9. *Number and percent of apraxic patients with left and right hemisphere damage who scored below lowest score of controls in two consecutive investigations of constructional apraxia*

Investigation	Patients with left hemisphere damage		Patients with right hemisphere damage	
	No.	%	No.	%
First investigation (Arrigoni & De Renzi, 1964)	13	18	21	38
Second investigation (De Renzi & Faglioni, 1967)	20	27	24	38

dict an increase in the percentage of apraxic patients with left-sided lesions from the first to the second study, but no change in the proportion of apraxic patients with right-sided lesions. This interpretation is supported by the finding, shown in Table 3-9, that the proportion of apraxics with right hemisphere damage was the same in both investigations, and that of apraxics with left hemisphere lesions rose from 18% to 27%. It must be added that neither in Arrigoni and De Renzi's nor in De Renzi and Faglioni's investigation did the severity of CA – as measured by the score obtained by each patient – discriminate the two brain-damaged groups. On the basis of these findings we are inclined to think that the difference between the proportion of apraxic patients with right and with left hemisphere damage is so weak that it can hardly be confirmed statistically, provided care is taken not to make a biased selection of samples. This assumption is confirmed by more recent investigations on this topic (Warrington et al., 1966; Benson & Barton, 1970; Dee, 1970), all of which failed to show the prevalence of CA in patients with right hemisphere damage that was claimed to exist early in the sixties.

In the course of the De Renzi and Faglioni study it was observed, however, that CA is significantly associated with the presence of VFD in patients with right hemisphere damage but not in those with left hemisphere damage. This differential relationship does suggest an asymmetry of representation of abilities underlying the constructional performance in the two hemispheres, as it points to a more restricted localization of lesions giving rise to CA in patients with right-sided than in patients with left-sided lesions, a finding already reported by Ajuriaguerra et al. (1960). Such an asymmetry leads us to consider the second question concerning the relationship between CA and the hemispheric locus of the lesion: Does the nature of the disorder differ according to which half of the brain is injured? There have been two ways to deal with this question: (1) to determine whether errors made by apraxics with right and left hemisphere damage are qualitatively different, and (2) to investigate whether CA is specifically associated with the impairment of different nonconstructional abilities, according to which hemisphere is involved. A wide spectrum of differential errors contrasting the performance of persons with right- and left-sided brain lesions has been reported by McFie and Zangwill (1960); Piercy et al. (1960); Benton (1962); Warrington et al. (1966), and Gainotti and

Tiacci (1970). We were able to demonstrate a greater number of omissions among subjects with right hemisphere damage and a tendency to oversimplification in reproducing complex drawings (e.g., a cube) among those with left hemisphere damage. On the other hand, one has to agree with Benton (1967) that the distinctive errors so far reported are produced only by a few of the apraxics with lesions of either hemisphere, so that the evidence derived from these findings, though suggestive, fails to show conclusively a difference in nature between the constructional performance of the two brain-damaged groups. We turn, therefore, to a search for the impairment of some basic ability differentially associated with dyspraxia in the two groups. Since Duensing (1953) the suggestion has been advanced that patients with right-sided brain damage fail because of a visuospatial disorder, whereas patients with left hemisphere damage fail because of an executive disorder (Piercy et al., 1960; Warrington, 1969). But the evidence so far presented in favor of a different basis for constructional disturbances in patients with right and left hemisphere damage is still unconvincing. The opposite instead seems to be the case, at least insofar as apraxic patients with lesions of either hemisphere have been found impaired in spatial perception. Piercy and Smyth (1962) reported that apraxics with right hemisphere and those with left hemisphere damage were inferior to nonapraxics with corresponding brain damage on Raven's Progressive Matrices. The finding that apraxics with right-sided brain lesions scored more poorly than apraxics with left-sided damage both on this and on construction tests was consequently interpreted as evidence that CA of right hemisphere origin differs from that of left hemisphere origin simply because the underlying visuospatial disorder is more severe. Even more negative have been the results of recent studies by Dee (1970) and Dee and Benton (1970), who found failure on construction tests to be closely associated with impairment on visual and tactile spatial perceptual tasks in patients with left as well as in those with right hemisphere lesions.

Our data, though somewhat more positive, also fail to give a definite answer to the question. De Renzi and Faglioni (1967) administered to their patients the cross-copying test and the inverted scrawl test (described earlier in this chapter) and analyzed the performance of apraxic and nonapraxic patients in relation to the presence or absence of VFD. As indicated in Table 3-10, a markedly poor score was common to the apraxic with damage to either hemi-

Table 3-10. *Mean scores of apraxic and nonapraxic patients with left and right hemisphere damage, with and without visual field defects (VFD), on spatial perception tests*[a]

	Patients with left hemisphere damage				Patients with right hemisphere damage			
Test	Nonapraxic VFD − (N = 41)	Nonapraxic VFD + (N = 13)	Apraxic VFD − (N = 14)	Apraxic VFD + (N = 6)	Nonapraxic VFD − (N = 30)	Nonapraxic VFD + (N = 9)	Apraxic VFD − (N = 10)	Apraxic VFD + (N = 14)
Inverted scrawl	8.1	6.9	7.1	3.3	8.6	5.9	5.7	3.5
Cross copying	227	348	254	653	200	425	455	741

[a]Higher scores correspond to better performance on inverted scrawl test and to worse performance on cross-copying test.

sphere and with VFD. But the pattern was different for apraxics without VFD: Those with right-sided lesions were impaired in comparison with nonapraxic patients without VFD, whereas those with left-sided lesions were not. Therefore, it can be claimed that visuospatial impairment is a common accompaniment of CA in persons with right hemisphere damage, whereas some apraxics with left-sided brain damage are free from clear-cut space perception disorders and their CA must be traced back to the disruption of some other still undetermined ability.

The first part of this statement must, however, be evaluated with caution, because our data can also be adduced to argue against a necessary relationship between space perception deficits and CA in patients with right hemisphere damage, as the spatial impairment shown by apraxics without VFD was not essentially different from that found in nonapraxics with VFD. Thus, the evidence derived from these results points to a widespread representation of space functions in the right hemisphere more than to their direct contribution in determining CA. Only apraxic patients with VFD stand out from the other groups with right-sided brain damage for their striking impairment of space perception, but in this respect the situation is comparable to that found in subjects with left hemisphere lesions, because apraxics with left-sided damage and VFD also show severe impairment in space perception. On the basis of the findings on space perception reported previously, it would be plausible to expect that the failure on construction tasks of patients with right hemisphere damage is caused by space disorders, and that of patients with left hemisphere damage results from a variety of reasons, such as defective analysis of the spatial relationship of the model, severe ideomotor apraxia, or possibly general mental impairment.

One must concede, however, that the evidence so far collected from our work as well as from other researches for a differential pattern of impaired abilities underlying CA in subjects with right- and left-sided lesions is far from compelling. It may be sensible to argue that the spatial tasks used up to now were not critical for assessing the kind of space disorder supposed to underlie CA. This is primarily a defect in evaluating orientation in space of the single elements (line, stick, block, etc.) constituting both the model and the copy and in appreciating their spatial relationship, and Raven's Matrices, the Minnesota Paper Form Board, and the inverted scrawl test, to cite only a few of the tests used, may not be suitable for these purposes. The design of space-perception tests relevant to the performance required by constructional tasks, together with the assessment of the executive impairment supposed to be responsible for the failure of at least some dyspraxics with left hemisphere damage, should be the main concern of future research.

CONCLUSIONS

The most common interpretation of the right hemisphere superiority in spatial performance is that it reflects a cerebral representation of the corresponding abilities that is bilateral, although preponderant on the right side (Piercy et al., 1960; Piercy & Smyth, 1962; Benton, 1965). From this viewpoint, the hemispheric asymmetry of spatial function cannot be considered the counterpart of the left hemisphere dominance for language, which concentrates all the related mechanisms in one-half of the brain. Semmes (1968) has pointed out that the two hemispheres differ essentially in the contrasting mode of their neural organization and that the diffuse representation of elementary functions in the right hemisphere may provide the basis for its specialization in spatial abilities, whereas the increased localization of functions in the left hemisphere would favor its specialization for language.

On the basis of the studies reviewed in this chapter, one is led to question the appropriateness of considering hemispheric asymmetry in spatial abilities as a unitary phenomenon and to suggest the need for further qualification according to the ability under discussion. In fact, both the position that the right hemisphere is exclusively specialized for space and the claim that both hemispheres are equipotential find objective support in our data, and the validity of the assertion depends entirely on the task and on the underlying ability or abilities investigated. To summarize the results in a comprehensive statement, one could say that the more elementary the space ability tested, the more it appears to be confined to a focal area of the right hemisphere. Thus, the neural substrate subserving the processing of cues defining the orientation in space of a single element is exclusively represented in the right half of the brain and focally organized in its posterior region. No role is played in this case by the left hemisphere, which, instead, becomes involved as soon as the integration of successive changes of orientation in space is required, as in the tactile test of shape discrimination. With this type of performance, however, the superiority of the right hemisphere is still clear-cut, but it subsides as the task becomes more complex and demands space abilities that go beyond the perceptual level (e.g., in conceptual or memory tests). Does this shift from a condition of pure right dominance to a condition of balance between the hemispheres depend on the effect on complex performances of the aid given by covert verbalization and on its disruption consequent to the disease of the left hemisphere? The finding of a positive correlation between the errors made in this type of test and those made in a sensitive test of verbal comprehension (the token test; De Renzi & Vignolo, 1962) provides some support for this hypothesis, but it is doubtful that the theory can account for all the participation of the left hemisphere in the performance. The issue clearly deserves further investigation.

Coming back to the most elementary level of space exploration, the results of the tactile maze test suggest a type of hemispheric asymmetry that may throw some light on the old question of hemiinattention. The finding that disorders of tactile exploration are common in persons with damage to either hemisphere and are confined to the contralateral field supports the view that each hemisphere is exclusively concerned with the opposite half of space. When, however, the most patent manifestation of deficit in scanning is considered, namely, unilateral neglect, failure is found almost exclusively among patients with right hemisphere injury and VFD. Because in the tactile test sensorimotor mechanisms of scanning play a negligible role and the instigation to explore derives essentially from the representation of the maze, a possible explanation of this imbalance in the occurrence of hemiinattention is that on the left side there is only a mental image of the contralateral (right) field, whereas on the right side both fields are represented, albeit the ipsilateral (right) in a weaker way. It follows that damage to the right hemisphere leaves the brain unaware of the existence of the left field, although damage to the left can be somewhat compensated for by the duplicate representation of the right field laid down in the right cortex. This hypothesis has some relevance for the clinical phenomenon of unilateral neglect, because it draws attention to the importance of the mental factor, of the capacity to imagine that virtual space extends in both directions well beyond the space actually perceived. This is the mechanism by which the normal subject is impelled intentionally to turn his head back or to the side when looking for something he knows must be present in the surrounding space. There are hints in the behavior of inattentive patients that they do not even think of the existence of the left space. I submit that the capacity to imagine space is dependent on a definite neural substrate and is asymmetrically represented in the two hemispheres.

ACKNOWLEDGMENT

The work reported here was supported by a research grant from the Consiglio Nazionale delle Ricerche.

REFERENCES

Ajuriaguerra, J., Hécaen, H., & Angelergues, R. 1960. Les apraxies: Variétés cliniques et latéralisation lésionelle. *Rev. Neurol. 102*:566–594.

Andersen, A. L. 1951. The effect of laterality localization of focal brain lesions on the Wechsler Bellevue subtests. *J. Clin. Psychol. 7*:149–153.

Arrigoni, G., & De Renzi, E. 1964. Constructional apraxia and hemispheric locus of lesion. *Cortex 1*:170–197.

Battersby, W. S., Bender, M. B., Pollack, M., & Kahn, R. L. 1956. Unilateral "spatial agnosia" ("inattention") in patients with cerebral lesion. *Brain 79*:68–93.

Bender, M. B., & Diamond, S. P. 1965. An analysis of auditory perceptual deficits with observations on the localization of dysfunction. *Brain 88*:675–686.

Benson, D. F., & Barton, M. I. 1970. Disturbances in constructional ability. *Cortex 6*:19–46.

Benton, A. L. 1962. The visual retention test as a constructional praxis task. *Confin. Neurol. 22*:141–155.

  1965. The problem of cerebral dominance. *Can. Psychol. 6*:332–348.

  1967. Constructional apraxia and the minor hemisphere. *Confin. Neurol. 29*:1–16.

  1969. Disorders of spatial orientation. In P. J. Vinken and G. W. Bruyn (eds.). *Handbook of Clinical Neurology,* Vol. 3. Amsterdam. North Holland.

Boller, F., & De Renzi, E. 1967. Relationship between visual memory defects and hemispheric locus of lesions. *Neurology 17*:1052–1058.

Butters, N., & Barton, M. 1970. Effect of parietal lobe damage on the performance of reversible operations in space. *Neuropsychologia 8*:205–214.

Colonna, A., & Faglioni, P. 1966. The performance of hemispheric damaged patients on spatial intelligence tests. *Cortex 2*:293–307.

Corkin, S. 1965. Tactual-guided maze learning in man: Effects of unilateral cortical excisions and bilateral hippocampal lesions. *Neuropsychologia* 3:339-351.

Costa, L. D., & Vaughan, H. G. 1962 Performance of patients with lateralized cerebral lesions. 1. Verbal and perceptual tests. *J. Nerv. Ment. Dis. 134*: 162-168.

Costa, L. D., Vaughan, H. G., Jr., Horwitz, M., & Ritter, W. 1969. Patterns of behavioral deficit associated with visual spatial neglect. *Cortex 5*:242-263.

Dee, H. L. 1970. Visuoconstructive and visuoperceptive deficit in patients with unilateral cerebral lesions. *Neuropsychologia 8*:305-314.

Dee, H. L., & Benton, A. L. 1970. A cross-modal investigation of spatial performances in patients with unilateral cerebral disease. *Cortex 6*:261-272.

Denny-Brown, D., Meyer, S. J., & Horenstein, S. 1952. The significance of perceptual rivalry resulting from parietal lesion. *Brain 75*:433-471.

De Renzi, E., & Faglioni, P. 1962. Il disorientamento spaziale da lesione cerebrale. *Sist. Nerv. 14*:409-436.

1967. The relationship between visuo-spatial impairment and constructional apraxia. *Cortex 3*:327-342.

De Renzi, E., Faglioni, P., & Scotti, G. 1968. Tactile spatial impairment and unilateral cerebral damage. *J. Nerv. Ment. Dis. 146*:468-475.

1969. Impairment of memory for position following brain damage. *Cortex 5*:274-284.

1970. Hemispheric contribution to exploration of space through the visual and tactile modality. *Cortex 6*:191-203.

1971. Judgment of spatial orientation in patients with focal brain damage. *J. Neurol. Neurosurg. Psychiatry 34*:489-495.

De Renzi, E., & Scotti, G. 1969. The influence of spatial disorders in impairing tactual discrimination of shapes. *Cortex 5*:53-62.

De Renzi, E., Scotti, G., & Spinnler, H. 1969. Perceptual and associative disorders of visual recognition. *Neurology 19*:634-642.

De Renzi, E., & Vignolo, L. A. 1962. The Token Test: A sensitive test to detect receptive disturbances in aphasics. *Brain 85*:665-678.

Duensing, F. 1953. Raumagnostische und ideatorisch-apraktische Störungen des gestaltenden Handelns. *Dtsch. Z. Nervenheilk. 170*:72-94.

Elithorn, A. 1955. A preliminary report on a perceptual maze test sensitive to brain damage. *J. Neurol. Neurosurg. Psychiatry 18*:287-292.

Faglioni, P., Scotti, G., & Spinnler, H. 1971. The performance of brain-damaged patients in spatial localization of visual and tactile stimuli. *Brain 94*: 443-454.

Gainotti, G. 1968. Les manifestations de négligence et d'inattention pour l'hémispace. *Cortex 4*: 64-91.

Gainotti, G., & Tiacci, C. 1970. Patterns of drawing disability in right and left hemispheric patients. *Neuropsychologia 8*:379-384.

Hécaen, H., & Ajuriaguerra, J. 1954. Balint's syndrome (psychic paralysis of visual fixation) and its minor forms. *Brain* 77:373-400.

Heilbrun, A. B. 1956. Psychological test performance as a function of lateral localization of cerebral lesion. *J. Comp. Physiol. Psychol.* 49: 10-14.

Kleist, K. 1934. *Gehirnpathologie vornehmlich auf Grund der Kriegserfahrungen.* Leipzig: Barth.

McFie, J., & Zangwill, O. L. 1960. Visual-constructive disabilities associated with lesions of the left cerebral hemisphere. *Brain* 83:243-260.

Milner, B. 1965. Visually guided maze learning in man: Effects of bilateral hippocampal, bilateral frontal, and unilateral cerebral lesions. *Neuropsychologia* 3:317-338.

Newcombe, F., & Russell, R. W. 1969. Dissociated visual perceptual and spatial deficits in focal lesions of the right hemisphere. *J. Neurol. Neurosurg. Psychiatry* 32:73-81.

Piercy, M., Hécaen, H., & Ajuriaguerra, J. 1960. Constructional apraxia associated with unilateral cerebral lesions: Left and right cases compared. *Brain* 83:225-242.

Piercy, M., & Smyth, V. O. 1962. Right hemisphere dominance for certain non-verbal intellectual skills. *Brain* 85:775-790.

Roy, S. N. & Bose, R. C. 1953. Simultaneous confidence interval estimation. *Ann. Math. Statist.* 24:513-536.

Russo, M., & Vignolo, L. A. 1967. Visual figure-ground discrimination in patients with unilateral cerebral disease. *Cortex* 3:113-127.

Scotti, G. 1968. La perdita della memoria topografica: Descrizione di un caso. *Sist. Nerv.* 5:352-361.

Semmes, J. 1968. Hemispheric specialization: A possible clue to mechanism. *Neuropsychologia* 6:11-26.

Semmes, J., Weinstein, S., Ghent, L., & Teuber, H.-L. 1955. Spatial orientation in man: Analysis by locus of lesion. *Am. J. Psychol.* 39:227-244.

Warrington, E. K. 1969. Constructional apraxia. In P. J. Vinken & G. W. Bruyn (eds.). *Handbook of Clinical Neurology*, Vol. 4. Amsterdam: North Holland.

Warrington, E. U., James, M., & Kinsbourne, M. 1966. Drawing disability in relation to laterality of cerebral lesion. *Brain* 89:53-82.

4
Spatial and temporal factors in visual perception of patients with unilateral cerebral lesions

AMIRAM CARMON

Broca's observation that lesions in the left frontal lobe cause aphasia gave rise to the term cerebral dominance, which implies that one cerebral hemisphere can perform a task, in this case speech, which the other hemisphere cannot do. The clinical observations that lesions of the right hemisphere impair recognition, identification, and execution of spatial configurations were an impetus to extend the conceptual frame of hemispheric functional asymmetry beyond dominance for verbal behavior. Even a verbal-spatial hemispheric dichotomy cannot be considered satisfactory anymore, in view of the recent findings concerning hemispheric function summarized in the following paragraphs. Moreover, a verbal-spatial dichotomy is descriptive only and does not relate much about the mechanisms that underlie functional hemispheric asymmetry. There is evidently a need to define the processes common to all behavioral phenomena that are associated with hemispheric differences, and those definitions not only should be parsimonious, but should also stand up before experimental manipulations.

An initial attempt to define a basic process in respect to left hemisphere dominance was made by Efron (1963). He suggested that the superiority of the left cerebral hemisphere in verbal performance results from its ability to recognize temporal sequences. Evidence that confirms his concept was obtained recently in clinical studies of patients with lesions confined to the left hemisphere, who were found to be impaired in recognition of nonverbal sequences. Carmon and Nachshon (1971) showed that normal subjects and patients with right cerebral damage could identify audiovisual nonverbal sequences when stimuli were presented at the relatively fast rate of 5-10 Hz, but patients with left hemisphere damage could do this only at the much slower rate of 0.5 Hz. It is clear that such deficit can be ascribed, not to an impaired verbal performance, but rather to a defect in processing sequences. Similarly, Albert (1972) described in patients with left-sided lesions an inability to point to, or name, sequentially a series of objects. Popen et al. (1969) stressed that the same process – impairment in sequential processing – underlies aphasia. Because speech is a serially organized behavior, impaired perception of the sequences of auditory signals constituting speech changes its perception and causes sensory aphasia, although a defect in generating sequential motor programs probably underlies motor aphasia.

A parallel development in our concepts of hemispheric differences can be seen in respect to the functions of the right cerebral hemisphere. This hemisphere was described initially as a processor of visuospatial information, but its perceptual roles were redefined recently. Bogen (1969) suggested, on the basis of clinical evidence, that the right hemisphere should be considered as a nonanalytic or holistic processor, implying that the right hemisphere operates in perception by treating patterns as unitary pieces of information. Lesions of the right hemisphere thus result in impaired recognition of total patterns. Various deficits in visuospatial processing caused by right hemisphere disease were reported. The patients performed poorly on tasks such as matching fragmented circles (Meier & French, 1965), recognition of overlapping figures (De Renzi et al., 1969; Gainotti & Tiacci, 1971), recognition of abstract figures (Rubino, 1970), and detection of depth effect in random stereograms (Carmon & Bechtoldt, 1969; Benton & Hécaen, 1970).

However, the results of the numerous studies did not disclose whether information processing by the right hemisphere is done by perceiving simultaneously the total pattern (i.e., in parallel mode) or by scanning sequentially discrete parts of the pattern in order to achieve its recognition. In the study described here this problem was circumvented by presenting spatial patterns sequentially and by requiring the subjects to identify the order of presentation of discrete patterns. By manipulating both the rate of the sequential input and the spatial configuration, we hoped to be able to differentiate between specific impairments associated with right and with left hemisphere lesions.

Two hypotheses were formulated:

1 Because the left hemisphere processes information sequentially, patients with lesions in this hemisphere will require a slower rate of presentation of the sequences in order to perceive the correct order. However, this rate should be constant, irrespective of the spatial complexity of the pattern.
2 Because the right hemisphere can process visual sequences as complete patterns it is only by increasing the spatial complexity of the patterns, that performance of patients with lesions in this hemisphere will be impaired.

EXPERIMENT 1

Method

Subjects were presented with spatial patterns composed of illuminated segments in the form of bars with a width of 10 mm and a maximum length of 60 mm. The segments were parts of an alpha-

Fig. 4-1. Examples of sequentially illuminated visual patterns composed of two, three, and four segments.

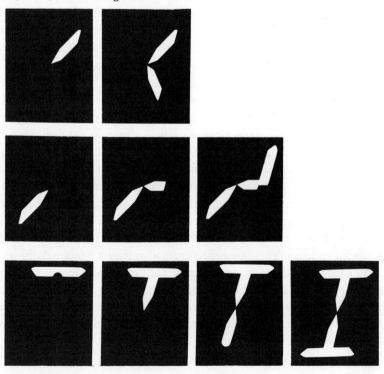

numeric electroluminescent panel (Sylvania). The order of illumination and the time intervals between appearance of individual segments were controlled by Massey-Dickinson behavioral programming modules. The subjects were seated about 0.5 meter from the panel, and thus even subjects with relatively low visual acuity were able to discriminate easily the various segments of the patterns. Illumination was set for 10 foot-lamberts brightness in a dark room. Rise and decay time of individual segments was not more than 1.0 msec. Three spatial conditions were used, each involving five different patterns. In the first condition patterns were composed of two, in the second of three, and in the third of four segments (Fig. 4-1). Individual segments within each pattern were illuminated in succession until the full pattern appeared on the screen. The delay between successive illuminations served as the temporal variable. This delay was increased in 10-msec steps from an initial 20 msec. In each time delay five trials were given. The subject was asked to identify the sequence in which the segments were illuminated by tracing it on the screen. Correct identification of the order of illumination was scored only if the subject identified the correct sequence in three consecutive delays; for example, identification was scored at 50 msec for correct responses at 50-, 60-, and 70-msec delays. Intertrial interval was 3 sec.

Subjects

Three groups, each composed of 20 right-handed patients, participated in this experiment. One group consisted of patients with lesions of the right hemisphere, the second group was made up of patients with lesions of the left hemisphere and the third, a control group, consisted of patients without any cerebral disease. The mean age was about 50 years in all groups and was not significantly different among the groups. Intrahemispheric locus of the lesions was evaluated by electroencephalography, brain scanning, angiography, or inspection at operation. The lesions were located anteriorly (in the frontal or frontotemporal regions) in 9 patients in each of the brain-damaged groups, and in the posterior parts of the hemisphere (temporoparietal or occipital regions) in the remaining 11 patients.

Examined separately were 8 left-handed patients, 4 with lesions in the left hemisphere and 4 with lesions in the right hemisphere.

Results

Three scores were derived for each patient. Each score was based on the mean temporal delay introduced between illuminations of successive segments in which order was correctly identified. A separate score was given for each spatial condition (two, three, or four segments).

The control patients performed better than the brain-damaged patients in all three spatial conditions. The mean delays in which recognition was achieved by this group were 40 msec for two-segment patterns, 40 msec for three-segment patterns, and 48 msec for four-segment patterns (Fig. 4-2). The respective mean scores for patients with right-sided lesions were 50, 56, and 105 msec; for those with left-sided lesions, 64, 58, and 71 msec. There was an interaction between the two brain-damaged groups: Patients with right hemisphere damage performed better than those with left hemisphere damage in the spatial conditions with two and three segments, and patients with left-sided lesions performed better than patients with right-sided lesions in the four-segment spatial condition.

Statistical evaluation of differences in performance among the three groups was done by a mixed design analysis of variance. All main effects – the overall differences among the three groups,

Fig. 4-2. Mean time required for correct identification of presentation order of segments in successively illuminated visual patterns by controls and by patients with right hemisphere lesions (*RHL*) and left hemisphere lesions (*LHL*).

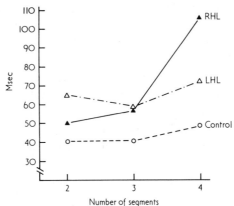

the overall differences among the three spatial conditions, and the interaction of the groups on these differences – were significant at the 0.01 level. The trends of the mean scores for the three spatial conditions were evaluated separately for each group. In the case of the control subjects, the trend did not show a significant deviation from zero ($F = 2.19$, $df\ 2/38$, $p > 0.10$). Thus, it could be inferred that the number of segments did not affect the performance of patients without cerebral disease. In the case of the patients with left hemisphere lesions the trend was significant, but only at the 0.05 level ($F = 3.84$, $df\ 2/38$). The mean scores of the group with right hemisphere lesions showed a markedly significant increase coinciding with the number of segments ($F = 14.92$, $df\ 2/38$, $p < 0.001$).

The experimental hypothesis was evaluated also by a series of statistical comparisons between the means. Under the first condition, identification of a sequence of two segments, the main factor was temporal resolution, that is the reciprocal of flicker fusion. Here the performance of the group with left-sided lesions, but not of the group with right-sided lesions, was worse than that of the controls (respective t-test values were 3.95 and 1.30). However, when the number of spatial components was increased to the four-segment condition, patients with right hemisphere damage performed significantly worse than the control group ($t = 3.90$), but the difference between patients with left hemisphere damage and the control subjects was not significant ($t = 1.61$).

The last step in the statistical analysis was the use of a criterion of impaired performance based on a probability of less than 1% that a score would fall within the distribution of the control group scores (Table 4-1). In the identification of two-segment patterns, 10 patients with left- and 2 patients with right-sided lesions exceeded this cutoff score. In the identification of four-segment patterns, 8 patients with right- and 4 with left-sided lesions exceeded the cutoff score. This finding is in agreement with the qualitative analysis because it shows a similar interaction between the groups. The effect of the intrahemispheric locus of lesion was not evaluated statistically because the small number of patients did not allow further subdivision of the groups. However, inspection of Table 4-1 shows that in most instances impaired performance was associated with lesions of the posterior part of the hemisphere, right or left.

Table 4-1. *Number of patients with left and right hemisphere lesions whose performance was impaired on test of identification of illumination sequence of visual patterns*

	Type of pattern	
Location of lesion	Two segments	Four segments
Left hemisphere		
Anterior (N = 9)	3	2
Posterior (N = 11)	7	2
Total	10 (50%)	4 (20%)
Right hemisphere		
Anterior (N = 9)	0	2
Posterior (N = 11)	2	6
Total	2 (10%)	8 (80%)

In the accessory group of left-handed patients, performance of all patients with right-sided lesions was impaired in comparison with that of the control group in all three spatial conditions. The performance of all 4 patients with left-sided lesions was not impaired in any condition.

Comment

The results of this experiment support the hypothesis that the effect of right and left cerebral lesions on visual perception of sequential patterns is different. Lesions of the left hemisphere were found to increase generally the time needed for perception of nonverbal sequences, and lesions of the right hemisphere were found to impair perception only in relation to the spatial complexity of the patterns. However, because the total pattern appeared on the screen for a very short time, the patients did not have enough time to point to all segments when illuminated, especially during trials with short delays. Therefore, the defective performance associated with lesions of the right hemisphere could result from an impairment of short-term spatial memory and not of spatial perception (Kimura, 1963; De Renzi, 1968). Because all patterns were continuous (i.e., segments were adjacent), they could be sequentially resolved by identifying only the first and the last segment and deducing logically the temporal location of intervening segments. Thus, patients with left hemisphere lesions could benefit by using the spatial configuration as a clue to the correct order. Another methodological problem was that 3 patients with right hemisphere lesions and 2 with left hemisphere lesions had visual field defects, and all 5 showed impaired performance.

In a second experiment, therefore, noncontinuous patterns that remained on the screen long enough to allow direct identification were utilized. In this experiment no patient with a visual field defect was included.

EXPERIMENT 2

Method

The same alphanumeric electroluminescent panel and the Massey-Dickinson behavioral programming modules were used. However, the panel was masked this time by an opaque heavy cardboard in which were cut six holes, 9 mm in diameter. The holes were arranged in two horizontal lines, each with three holes, separated by about 2 cm. Each hole corresponded to a different segment of the electroluminescent panel. In this way, an array of up to six disks could be illuminated. The disks could be programmed to be illuminated successively in all possible combinations.

Four different spatial combinations were used. In two, only one out of three or one out of six disks was illuminated later than all other disks. In the other two combinations, two out of five or two out of six disks were illuminated later. The subject's task was to identify the disks that were illuminated after a delay. He was informed before each trial how many disks would appear and how many of those would appear later than the other disks. Each of the four combinations was based on four different arrays of disks. The arrays were counterbalanced for the location of the delay in illumination. Thus, disks that were illuminated after a delay appeared on either of the two sides of the screen and in the two rows in equal chance. After the sequence of illumination was completed, the disks remained illuminated on the screen for 10 sec to allow identification by pointing. Five trials were given for each array, and scores were derived in the same manner as in the first experiment; that is, the mean temporal delay in which detection of the disks was achieved for each spatial combination.

Subjects

A group of 20 patients with right- and 20 with left-sided lesions participated in this experiment. A control group of 40 patients without any cerebral disease was examined as well. All patients were right-handed. Localization of the lesions was evaluated in the same manner as in the first experiment. The intrahemispheric locus of the lesion was posterior in 8 patients with right hemisphere damage and in 9 patients with left hemisphere injury. No patient who participated in this experiment had a visual field defect. Mean age was around 50 years in all three groups.

Results

As in the first experiment, control patients performed better than the two groups of brain-damaged patients and could identify the correct order in shorter delays. The two brain-damaged groups interacted on the delay needed for correct identification of order. In the condition where identification was of one out of three disks, patients with right-sided lesions performed better than those with left-sided lesions. The mean scores in this condition were 38 msec for the control subjects, 51 msec for patients with right hemisphere damage, and 56 msec for patients with left hemisphere damage (Fig. 4-3). In all other conditions patients with right-sided lesions performed worse than those with left-sided lesions. The maximal

Fig. 4-3. Mean time required for correct identification of order of illumination of arrays of disks by controls and by patients with right hemisphere lesions (*RHL*) and left hemisphere lesions (*LHL*).

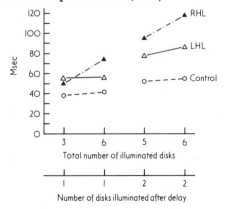

difference between the group was attained in the most complicated spatial condition (identification of two out of six disks). Under this condition the mean scores were 55 msec for the controls, 112 msec for patients with right hemisphere damage, and 83 msec for patients with left hemisphere damage.

Analysis of variance revealed all major effects as well as the interactions among the groups to be significant at the 0.01 level. Separate t-tests showed that under all conditions both brain-damaged groups performed significantly worse than the control group. The differences between the groups with right- and left-sided lesions were not significant except in the last condition (identification of two out of six disks). Analysis of the performance according to an "impaired performance score" was done, as in the first experiment. The number of patients with impaired performance in each brain-damaged group for the four spatial conditions is presented in Table 4-2. Among the patients with right hemisphere damage, the number with impaired performance increased with the spatial complexity of the task. However, in the least complex spatial condition, employing three disks, more patients with left- than with right-sided lesions showed defective performance. As in the first experiment, impaired performance was observed mainly in patients with lesions of the posterior parts of the hemisphere.

Table 4-2. *Number of patients with left and right hemisphere lesions whose performance was impaired on test of identification of illumination sequence of arrays of disks*

| | Type of pattern | | | |
Location of lesion	1/3	1/6	2/5	2/6
Left hemisphere				
Anterior (N = 11)	5	2	1	1
Posterior (N = 9)	5	5	4	3
Total	10 (50%)	7 (35%)	5 (25%)	4 (20%)
Right hemisphere				
Anterior (N = 12)	2	2	4	3
Posterior (N = 8)	5	7	7	7
Total	7 (35%)	9 (45%)	11 (55%)	10 (50%)

FINAL COMMENT

The results of both experiments demonstrate that lesions of the right and of the left hemisphere are qualitatively as well as quantitatively different in impairing visual perception. Damage to the left hemisphere resulted in an overall impairment of recognition of a visual sequence irrespective of its spatial complexity. Damage to the right hemisphere did not affect perception of sequences unless the number of spatial components was increased. When the spatial configuration was complex, a very marked impairment was observed in these patients' performance. This functional dichotomy in visual perception seems to be associated mainly with damage to the posterior parts of the cerebral hemispheres.

The validity of this conclusion applies only to right-handed subjects. In left-handed subjects damage to the right hemisphere compromised both the temporal and the spatial factors of perception, but damage to the left hemisphere did not affect performance to a significant degree. Thus, in left-handers no evidence was found for a complete reversal of the effect observed in right-handed patients.

The theoretical considerations that led to the formulation of the experimental hypotheses were, however, laid within a wide framework and not limited to visual functions. The hypothesis of a spatial versus sequential hemispheric functional dichotomy in visual perception was derived from the idea that speech perception is a complex form of frequency analysis (Neff, 1964). In support of Neff's concept stands evidence not only from studies of visual perception in patients with left hemisphere damage, but also from studies of auditory perception in normal subjects. Thus, in dichotic listening a right-ear preference, and therefore an implied left hemisphere superiority, was found for perception of nonverbal sequences (Krashen, 1972; Halperin et al., 1973; Robinson & Solomon, 1974).

Other auditory sequences, however, such as melodies and emotional nonverbal human sounds, are perceived better by the left ear (Kimura, 1964; Shankweiler, 1966; Carmon & Nachshon, 1973). It is possible that these patterns are perceived in a holistic manner and therefore are associated with right hemisphere superiority.

It can be concluded that, by manipulation of spatial and temporal variables associated with perception of visual patterns, it is possible to single out the specific role each cerebral hemisphere has in processing the information. Analogous situations should be investigated in audition and in somesthesis in order to generalize these findings beyond the visual modality.

REFERENCES

Albert, M. L. 1972. Auditory sequencing and left cerebral dominance for language. *Neuropsychologia 10:*245-248.

Benton, A. L., & Hécaen, H. 1970. Stereoscopic vision in patients with unilateral cerebral disease. *Neurology 20:*1084-1088.

Bogen, J. E. 1969. The other side of the brain: An appositional approach. *Bull. Los Angeles Neurol. Soc. 34:*135-163.

Carmon, A., & Bechtoldt, H. P. 1969. Dominance of the right cerebral hemisphere for stereopsis. *Neuropsychologia 7:*29-39.

Carmon, A., & Nachshon, I. 1971. Effect of unilateral brain damage on perception of temporal order. *Cortex 7:*410-418.

1973. Ear differences in perception of human nonverbal emotional sounds. *Acta Psychol. 37:*351-357.

De Renzi, E. 1968. Nonverbal memory and hemispheric side of lesion. *Neuropsychologia 6:*181.

De Renzi, E., Scotti, G., & Spinnler, H. 1969. Perceptual and associative disorders of visual recognition. *Neurology 19:*634-642.

Efron, R. 1963. Temporal perception, aphasia and déjà vu. *Brain 86:*403-424.

Gainotti, G., & Tiacci, G. 1971. The relationship between disorders of visual perception and unilateral spatial neglect. *Neuropsychologia 9:*451-458.

Halperin, Y., Nachshon, I., & Carmon A. 1973. Shift of ear superiority in dichotic listening to temporally patterned nonverbal stimuli. *J. Acoust. Soc. Am. 53:*46-50.

Kimura, D. 1963. Right temporal lobe damage: Perception of unfamiliar stimuli after damage. *Arch. Neurol. 8:*264.

1964. Left-right differences in the perception of melodies. *Q. J. Exp. Psychol. 16:*355-358.

Krashen, S. D. 1972. Language and the left hemisphere. Doctoral dissertation, University of California, Los Angeles.

Meier, M. J., & French, L. A. 1965. Lateralized deficits in complex visual discrimination and bilateral transfer of reminiscence following unilateral temporal lobectomy. *Neuropsychologia 3:* 261-272.

Neff, W. D. 1964. Temporal pattern discrimination in lower animals and its relation to language perception in man. In A. V. S. de Reuck & M. O'Connor, (eds.). *Disorders of Language.* London: Churchill, p. 192.

Popen, R., Stark, J., Eisenson, J., Forrest, T., & Wertheim, G. 1969. *J. Speech Hear. Res. 12:*288-300.

Robinson, G. M., & Solomon, D. J. 1974. Rhythm is processed by the speech hemisphere. *J. Exp. Psychol. 102*:508–554.

Rubino, C. A. 1970. Hemispheric lateralization of visual function. *Cortex 6*:102–120.

Shankweiler, D. 1966. Effect of temporal lobe damage on the perception of dichotically presented melodies. *J. Comp. Physiol. Psychol. 62*:115–119.

5
Direct examination of cognitive function in the right and left hemispheres

ROBERT D. NEBES

Since Broca's time (1861), the main method for determining whether various cognitive functions are asymmetrically represented in the two cerebral hemispheres has been to compare the performances on the same task of patients in whom either only the right or only the left hemisphere has been injured. The rationale is that if individuals with damage restricted to one hemisphere do worse on a given task than individuals with damage restricted to the other, then mental abilities required by that task must be predominantly organized in the first hemisphere. Obviously, in order for such a comparison to be meaningful, the two groups of unilaterally brain-damaged patients must be carefully matched for age, sex, education, premorbid intelligence, and most important of all, size and locus of the lesion. An even greater difficulty with this approach is that the examiner can never be sure whether the defective performance observed after unilateral injury originates from the damaged hemisphere, doing its best in the face of injury, or from the other hemisphere, doing its best in the face of an inherent lack of the needed abilities. For example, in left hemisphere damage, does the disturbed language expression and comprehension we see represent a subnormal performance by the injured "dominant" hemisphere or a normal performance by the "minor" hemisphere? If we are observing the disrupted output of an injured hemisphere, it becomes very dangerous to make inferences about hemispheric competence; Semmes (1968) has pointed out that, if both hemispheres are equally proficient in performing a task, but the neural substrate of the ability involved is more focally organized in one hemisphere than in the other, then limited damage to that side of the brain is more apt to produce a severe deficit, thus making that hemisphere appear to be the primary locus of the capacity being tested. These questions of pseudodominance and of which hemisphere is the origin of the observed performance are not issues if the functional abilities of the two hemispheres are studied directly rather than inferred from the disabilities that follow their injury.

There are three types of patients in whom direct examination of a single intact hemisphere is possible. In one, the crossing neural fibers connecting the two cerebral hemispheres have been severed (commissurotomy); in the other two, one hemisphere has been inactivated, either permanently through surgical removal (hemispherectomy) or temporarily through anesthetization of one side of the brain (sodium amytal test).

Section of the interhemispheric tracts to control epilepsy has been found to eliminate much of the normal integration of sensory information between the two sides of the brain, leaving each hemisphere cognizant only of contralateral sensory input (Sperry et al., 1969). The left hemisphere thus receives detailed information about visual stimuli only if they fall in the right visual half-field (Gazzaniga et al., 1965) and about somesthetic stimuli only if they contact the right side of the body (Gazzaniga et al., 1963). This division of the subject's peripheral sensory world makes it possible to restrict a stimulus to just one hemisphere, either by allowing only one hand to palpate the object or by flashing the stimulus tachistoscopically in one visual half-field for a duration too short to permit eye movements. The right and left hemispheres in these individuals can thus be examined independently on exactly the same task. The presence of both hemispheres also allows investigation of interhemispheric interactions, as well as of hemispheric hierarchies of response in various types of tasks. The main disadvantages of commissurotomized patients are the small number of subjects available, a large intersubject variability, and the need for special techniques, such as tachistoscopic presentation, to restrict information to one hemisphere. There is also some question about just how intact the two hemispheres are in these patients in view of their history of severe epilepsy.

Hemispherectomy involves the surgical removal of the entire cerebral cortex of one hemisphere, as well as a variable amount of subcortical tissue. The operation is performed either to excise an extremely large and invasive tumor or to remove a hemisphere which, damaged at birth and nonfunctional, is the source of epileptic seizures. Only individuals who have undergone hemispherectomy for tumor as adults are discussed here, as functional plasticity in the presence of early brain damage greatly alters the normal pattern of hemispheric lateralization of function. Hemispherectomy for tumor in adults, and especially removal of the left hemisphere, is relatively rare, and extensive neuropsychological testing of such patients is even more rare. These individuals are, however, very important, for in them there is no doubt about which hemisphere is performing a task; nor is there any possibility that one hemisphere may be suppressing or inhibiting abilities present in the other, as is the case in unilaterally brain-damaged or commissurotomized patients. It is also possible to follow the readjustments and reorganizations the remaining hemisphere makes as it copes with the loss of its counterpart. This permits us to investigate the limits of that hemisphere's capabilities, not only for its own specialized functions, but also for those functions normally carried out in the missing hemisphere.

Injection of sodium amytal into the internal carotid artery on one side temporarily anesthetizes the ipsilateral hemisphere, a sort of reversible hemispherectomy. Originally devised by Wada and Rasmussen (1960) to determine which hemisphere is dominant for language, the injections produce a flattening of the electroencephalogram on the affected side of the brain, along with hemiparesis, hemianesthesia, and hemianopsia (Terzian, 1964). This unilateral anesthesia lasts from 5 to 10 minutes, during which time tests of higher mental functions present in the other hemisphere may be carried out. The complexity and length of such tests are, of course, very limited, and there is little time for the conscious hemisphere to readjust to the situation. This technique is still useful, however, as both hemispheres can be anesthetized in turn, allowing intrasubject comparison of hemispheric abilities. Also, because the procedure is relatively benign, it is done fairly routinely in some situations, producing a large number of subjects, albeit subjects in whom some brain disease is suspected.

LANGUAGE

Language was the first higher cognitive function found to be asymmetrically represented in man's two cerebral hemispheres, and it remains the best-documented case of hemispheric specialization, with the results of many different approaches all confirming the dominance of the left hemisphere for language skills in right-handed individuals. The value of direct testing of intact hemisphere lies not in verifying this well-established dominance, but rather in investigating the language capacities that, resident in the right hemisphere, are normally overshadowed by the left hemisphere.

Language in the left hemisphere

Injection of sodium amytal to anesthetize the right hemisphere of right-handers only rarely produces any deficit in verbal skills. Usually, as far as can be tested, comprehension and expression are unaffected, outside of some slurring resulting from left-sided facial weakness (Branch et al., 1964). Removal of the right hemisphere also does not usually lead to any aphasia or gross deficit in language abilities as measured on standardized tests. In fact, verbal IQ usually excedes performance IQ by 20 to 26 points (Smith, 1969; Gott, 1973a). Vocabulary may be normal or even superior (Mensh et al. 1952; Gott, 1973a). As expected, the left hemisphere in commissurotomized patients is proficient in all language skills (Gazzaniga & Sperry, 1967; Sperry & Gazzaniga, 1967). Stimuli, such as words or objects, when presented to the left hemisphere either visually in the right visual half-field or haptically in the right hand, are readily identified both verbally and in writing by the right hand. In fact, in most patients for several weeks following surgery, only the right hand is capable of writing (Bogen, 1969a). Later, the major hemisphere apparently gains some motor control also over the left hand, allowing written answers by the left hand to stimuli received by the left hemisphere. The major hemisphere's comprehension of language can be demonstrated by the correct tactual retrieval with the right hand of items named, described, or defined by the examiner and by the correct performance of printed commands flashed to the right visual half-field. Similarly, in the realm of calculation, the major hemisphere can add, subtract, multiply, and divide. Thus, using input restricted to the left hemisphere, the verbal capacities demonstrated are roughly equivalent to those elicited from the whole subject under more conventional testing.

Language in the right hemisphere

The limits of right hemisphere language skills have been a matter of controversy even since Broca (1861) first suggested that speech is primarily organized in the left hemisphere. The amount and nature of the minor hemisphere's comprehension and expression are of considerable importance, both in designing therapy and in understanding the basic psychological makeup of the two hemispheres. In terms of speech output, studies using sodium amytal produce a very negative view of the minor hemisphere. After anesthetization of the left hemisphere, right-handed subjects are usually completely incapable of speech or any sort of phonation, even though they do show indications of language comprehension (Wada & Rasmussen, 1960; Terzian, 1964; Branch et al., 1964). The only evidence for right hemisphere speech comes from a study on right-handed stroke patients with aphasia (Kinsbourne, 1971), in which the patients' aphasic speech was not changed after anesthetization of the left hemisphere, but was abolished after anesthetization of the right hemisphere. Kinsbourne concluded that in these subjects the aphasic speech was originating in the normally minor hemisphere. This would suggest that given time – that is, the days between the stroke and the amytal injection – some language functions had become active in the minor hemisphere. It is unlikely that in these few days there had been any significant amount of language acquisition, but rather that capacities normally suppressed by the left hemisphere had been released.

If the right hemisphere is capable of some expressive language when given sufficient time to reorganize in the absence of left hemisphere interference, then patients subjected to left hemispherectomy should yield the clearest data on the inherent limitation of language organized in the minor hemisphere. The results from the few left hemispherectomized patients studied in detail emphasize the poverty of right hemisphere expression. A 47-year-old man with complete left hemisphere removal was incapable of any voluntary or propositional language for several months after surgery; his speech consisted of stereotyped phrases and expletives, without even successful repetition of single words. Although occasional propositional utterances were heard in later months, his ability to communicate in words never recovered to any degree. Writing was, if anything, worse, the patient showing only a few instances of anything beyond copying. Hillier (1954) obtained somewhat better results with a boy who had developed a tumor at age 14. After left hemispherectomy, it was 16 days before he began to speak even isolated words. His vocabulary was said to be improving steadily, but unfortunately, no long term follow-up was reported to establish the limits of this improvement. Gott (1973b) and Zaidel (1973) studied a case of left hemispherectomy for tumor in an even younger child, aged 8 years at the onset of symptoms. Here again, the right hemisphere was able to produce only isolated words and phrases; she spoke automatic series (days of the week, etc.) and repeated words very well, but had great difficulty initiating words herself in a proper context. Her speech had good intonation and melody, but was nonfluent, with no paraphasic errors or neologisms. She was relatively poor at naming objects or actions, resorting to circumlocutions. She did, however, speak well enough to obtain a dull normal vocabulary score on the Wechsler Intelligence Scale for Children (WISC). Her writing abilities did not extend much above copying. The fairly low level of this patient's expressive language is impressive when you consider that her brain injury occurred at an age when extensive reorganization of the nervous system is thought by many to be still possible following brain damage (Basser, 1962).

Early reports on the first commissurotomized patients emphasized the lack of any evidence for a role of the minor hemisphere in expressive language (Gazzaniga et al., 1962; Gazzaniga & Sperry, 1967). The subjects were unable to name, verbally or in writing, words or objects seen in the left visual half-field or felt in the hand, although the results of nonverbal testing clearly showed that the minor hemisphere had perceived and comprehended the nature of these items. Expressive language in these patients appeared to be purely a left hemisphere skill.

As new subjects were tested and the old ones retested over the years, however, instances of limited right hemisphere expression were reported. When, instead of flashing visual stimuli at 0.1 sec or less, as was done in previous testing, the stimuli were exposed for longer periods, up to 20 sec, with appropriate controls over eye movements, Butler and Norrsell (1968) reported that one of the patients could say short words such as *cup,* identifying pictures seen in his left visual field. The lack of minor hemisphere speech in previous work could thus be viewed as a consequence of not allowing the right hemisphere sufficent time to produce the proper name. This implies that the persisting visual stimulus itself, and not just the memory of it, is necessary to evoke the name. It does not, however, explain why the subjects cannot also name items held in their left hand for long periods.

Another possible explanation for these results is that the longer stimulus durations used in the last experiment allowed the subject time to cross-cue the answer from the right to the left hemisphere. Gazzaniga and Hillyard (1971) found two of the younger patients to be capable of reporting verbally whether a *1* or a *0* had been flashed in the left visual field, although their verbal reaction time was 200 msec longer than if the stimulus was in the right visual field. When, however, other numbers were substituted, the left-field reaction time increased dramatically, a rise proportional to the size of the number. Thus, the reaction time to the number *2* was 2500 msec, and to the number *8* it was 5000 msec. There was no such proportional rise with numbers in the right half-field. One of the subjects actually said that the way he could tell which number had been flashed in his left field was to count silently from zero to nine until he came to a number that "stuck out." When Gazzaniga required immediate verbal reply to the left-field numbers, the two subjects' scores fell to chance. Interestingly enough, a similar mechanism was employed by the left hemispherectomized patient of Gott (1973b) who, in order to say a number or letter, had to start at the beginning and run through the entire sequence of numbers or letters until she came to the proper item. Teng and Sperry (1973) also found that commissurotomized subjects often correctly verbalized single digits (35% correct) and letters (20% correct) presented in the left visual field. They, however, interpreted this as evidence for right hemisphere speech, seeing no reason for digits to be more easily relayed between the hemispheres than letters. However, as the subjects were given 19 letters, but only 10 digits, to choose among, one might expect cross-cueing of digits to be more efficient. This is reflected in the reaction times found in this experiment, which ranged from 8 to 20 sec for letters and from 5 to 10 sec for digits, certainly sufficient time in view of Gazzaniga and Hillyard's results (1971) for cross-cueing.

Thus, at least some commissurotomized patients can produce a correct verbalization to a left-field visual stimulus if two conditions are met: (1) sufficient time is allowed and (2) the number of stimuli is limited and their identity known to the subject ahead of time. A possible mechanism would involve silent recitation by the left hemisphere of the list of possible choices, while the right hemisphere monitored either the peripheral mouth and throat movements or the covert internal speech and signaled the left hemisphere when it came to the proper name. This would explain the correct verbalizations given during the long periods of stimulation in Butler and Norrsell's (1968) experiment. A similar explanation can be given for Milner and Taylor's (1970) experiment in which commissurotomized patients were able to verbalize whether they had been touched on one or two places on the left arm. Rather than invoking minor hemisphere speech, it could be argued that the small number of verbal response choices allowed rapid cross-cueing of the answer from the right to the left hemisphere.

Evidence for minor hemisphere writing abilities comes from two young commissurotomized patients (AA and LB), both of whom were able to produce words blindly with the left hand, which they could not then verbalize (Levy et al., 1970; Nebes & Sperry, 1971). LB could, upon blindly feeling letters with his left hand, arrange them to create a word. Similarly, after feeling letters already ordered so as to make a word, he could then often write it with his left hand. In both situations, he was unable to verbalize correctly the word thus formed. On similar tests, AA correctly copied into script with his left hand, and then failed to name, printed nouns or verbs flashed in his left visual field (Fig. 5-1A). This inability to name words executed by the left hand is taken to indicate that the left (speaking) hemisphere did not participate in the production.

Fig. 5-1. Illustrations of writing by the left hand of commissurotomized patients after presentation of stimuli in the left visual field.

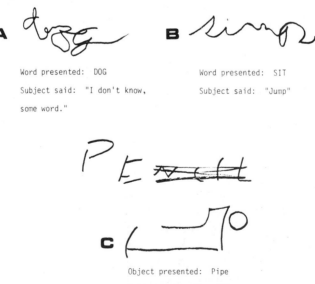

A

Word presented: DOG
Subject said: "I don't know, some word."

B

Word presented: SIT
Subject said: "Jump"

C

Object presented: Pipe
Subject said: "Pencil"

When, instead of copying words, the task required producing the names associated with objects, the right hemispheres of both patients had a great deal of difficulty. LB could at times get the first few letters written, but that was all. Similarly, AA only once wrote a correct word to a picture, and then failed to name it.

One very important factor in the limited nature of this minor hemisphere writing appears to be interference by the major hemisphere. On a majority of the trials in which a word was flashed to AA's left field, he wrote an incorrect answer with his left hand, a word that he could subsequently name, apparently representing a guess by his major hemisphere. This interference was even more striking on those trials in which the left hand had already begun to write the proper word, then stopped, and finished with letters appropriate to an incorrect guess that the subject then verbalized. Two good examples of this are shown in Figure 5-1. In the first (Fig. 5-1B), the word shown in AA's left field was *sit*. He wrote *si*, hesitated, added *mp*, and said "*jump*." In the second case (Fig. 5-1C), LB felt a pipe with his left hand and then with the left hand wrote *pi*, stopped, and using a lighter pressure finished *ncil*, stopped again, and crossed out the added letters. When asked to draw the object with his left hand, he drew a pipe.

Evidently, the left hemisphere interferes with the ability of the right hemisphere to respond in writing, but this is certainly not the sole reason for the limited nature of this writing. Rather, the language skills of the right hemisphere appear to be basically inadequate to the task of producing the name of an object. The greater left hemisphere interference seen when the right hemisphere had to generate the names of objects rather than just copy words is probably a consequence of the right hemisphere's failure rather than its cause; the left hemisphere fills in with a guess in the face of the right hemisphere's inability to program the correct answer. The minor hemisphere thus appears to have access to the motor patterning necessary to write words, but like a patient with nominal aphasia, it is unable to generate the name itself.

One further point to be remembered in generalizing from this work to a total picture of the minor hemisphere's capacity for expressive language is that the three subjects who gave the best evidence for minor hemisphere expressive language all had apparent cerebral birth injuries, and all had undergone operations at age 13 or 14. Extensive testing since that time, combined with the plasticity of the young brain, may have produced greater expressive abilities than might be expected from the right hemisphere of a normal adult.

The relatively greater success of the right hemisphere of commissurotomized patients in expressing itself in writing as compared with speech probably results because the speech musculature is a unitary system, apparently under sole control of the left hemisphere. It would thus be very difficult for the right hemisphere to gain access to the motor control even if it possessed the motor patterning necessary to say a word. In contrast, when the left hand is writing, the right hemisphere has the main control over that hand and may therefore more successfully resist major hemisphere intervention.

The possible results of competition between the two hemispheres for control over the vocal apparatus has been suggested by Jones (1966), who found bilateral representation of speech in four severe stutterers. Bilateral speech representation here means that anesthetization of either hemisphere with amytal produced cessation of speech, suggesting that both hemispheres were involved in, and were necessary for, speech. All these patients subsequently underwent surgery removing cortex from the speech areas in one or the other hemisphere. After surgery the subjects no longer stuttered, and amytal studies showed speech to be organized only in the unoperated hemisphere. Jones postulated that stuttering in these patients was a result of competition between the language centers in the two hemispheres, which was abolished by the surgery. Unfortunately, this cannot explain all stuttering: Up to 15% of the patients studied by Branch et al. (1964) had bilateral speech representation, but these investigators reported no association with stuttering. Also, two recent amytal experiments on a total of six normal stutterers found all of them to have solely left hemisphere speech representation (Andrews et al., 1972; Luessenhop et al., 1973). The factor distinguishing stutterers who have bilateral language representation from those who do not is unclear. Jones's subjects started stuttering at an early age, but their brain damage occurred when they were adults, so the brain damage itself was not the cause of their stuttering. The results to date certainly warrant further studies, both on the cerebral dominance of stutterers and on the types of language dysfunctions found in individuals in whom speech is organized in both hemispheres.

Comprehension of language by the right hemisphere has been a very fruitful field for investigation, with all three of the approaches discussed here demonstrating that the minor hemisphere's capacity to understand language greatly exceeds its capacity to produce it. Tests of right hemisphere language comprehension during left-sided anesthetization with amytal have been fairly superficial owing to the time limitations of the method and consist mainly of commands to perform gross left-sided body movements. The major aim is to make sure that any expressive language disturbances are not due merely to a loss of consciousness. Even so, it is obvious that during the period of left hemisphere anesthetization subjects, although totally incapable of speech, can understand and carry out simple verbal commands (Wada & Rasmussen, 1960; Terzian, 1964). Early reports of the clinical sequelae of left hemispherectomy also emphasized the relative intactness of verbal comprehension compared with expression (Hillier, 1954; Smith & Burklund, 1966). A later study by Smith (1969) showed his left hemispherectomized patient to score practically as well as three right hemispherectomized patients on the Peabody Picture Vocabulary Test, which examines comprehension of single orally presented words. Zaidel (1973) has done the most complete investigation to date on a left hemispherectomized patient (same patient reported by Gott, 1973b), using a series of standardized aphasia batteries. He found no deficit in phonemic discrimination. The patient was also fairly good at picking out pictures of objects and actions named by the examiner, and in understanding prepositions of place (except for left and right), as well as various syntactical constructions, including the possessive. When, however, the sentences became long, or there were several elements to be related within a sentence, the patient's understanding decreased drastically. For example, she did badly in making sense of short spoken paragraphs or sentences with several clauses. Her ability to read was still poorer, as she was unable to read even short words. A great deal of this patient's difficulty with aural understanding may be due to a poor verbal short-term memory, which would make any attempt to understand complex or multiple-item statements very difficult.

In view of the scarcity of minor hemisphere capacity for expressive language, tests of right hemisphere comprehension all rely on nonverbal readouts (Gazzaniga & Sperry, 1967; Sperry & Gazzaniga, 1967). To discover whether the right half-brain could read and understand nouns, printed object names were flashed in the left half-field, and the subject asked to retrieve blindly the named item with his left hand. This sort of task was accomplished fairly easily. As in the expressive language experiments, the patient's inability to name the objects successfully recovered by his own left hand demonstrates that this intermodal matching of objects to their names was accomplished by the right hemisphere. Similar results were obtained when the procedure was reversed and the patient asked to point out, from among 10 or more words lying in free view, the one associated with an object seen in his left field or felt in his left hand. These results demonstrate that the minor hemisphere can associate a written noun with an object. When the names were given orally and the tactile examination of the choices was restricted to the left hand, although the auditory stimulus was perceived by both hemispheres, only the right hemisphere had the necessary haptic information to make the correct choice. This it easily did.

Right hemisphere comprehension of language goes beyond simple association of objects to their names. The minor hemisphere can also tactually retrieve an item upon hearing an abstract description or a definition of its use. When AA was asked to find from among 15 stimuli not shown to him before the test the one that "makes things look bigger," he correctly picked out with his left hand a magnifying glass (Nebes & Sperry, 1971). To a command to find a round, thin object made of metal, he chose a coin. Again, the subject's verbalizations revealed that the left hemisphere was ignorant of what the items were, calling the first a telescope and the second a spoon. The minor hemisphere in these and other instances demonstrates that it comprehends not only which object is identified by a word but also the significance of a series of words in a definition. Thus, as stated before, the minor hemisphere is much like a patient with nominal aphasia, who, though knowing an object's use and the things and qualities with which it is associated, cannot produce its name.

When we turn from nouns and adjectives to verbs, the picture is less clear. From the earliest testing it was evident that the right hemisphere could understand fairly complex instructions and carry out the required tasks successfully – for example, copying words viewed in the left field. It was therefore quite surprising that when printed verbs were flashed in the left visual field, the right hemisphere could not carry out the actions named (Gazzaniga et al.,

1967). It made no difference whether facial motions (e.g., smile, chew) or hand motions (e.g., tap, wave) were requested; they were not performed. This was not the result of a primary motor disability, for the subjects could not even pick out a picture of the action named. When, however, the picture of the motion was flashed in the left field, it was successfully reproduced. The difficulty thus appeared to lie in comprehending the verbs.

Several lines of evidence contradict this apparent minor hemisphere incomprehension of verbs. Levy (1970) noted that the experiments with verbs had a different output from those with nouns, in that instead of tactually retrieving an object, the subject had to make a specific motion. She felt that with tactual retrieval only the minor hemisphere received the input and the left hemisphere was therefore unlikely to interfere. When, however, a specific act was required, the initiation of the motion was easily noticed by the left hemisphere and could be blocked, very much as the major hemisphere obstructed minor hemisphere writing. Levy therefore tested minor hemisphere comprehension of verbs flashed in the left field, using three different readouts: (1) performance of the named act, (2) pointing to a picture of the act, and (3) tactually retrieving an object associated with the verb – for example, a toy chair for the verb *sit*. Her results confirmed the inability of the subjects to perform the motion called for. Similarly the best score obtained in pointing out the correct pictorial representation of the activity was only 50% correct. In tactually choosing the associated object, however, the subjects were highly successful. Levy therefore attributed the minor hemisphere's failure to carry out motor acts upon seeing printed verbs to interference by the major hemisphere.

A second experiment that seemingly supports this interference model used dichotic listening (Gordon, 1973). Two differing commands were given simultaneously, one to each ear. Gordon found that often the left hand would carry out the command addressed to the left ear, while the right hand performed the command given the right ear. Afterward the subject could verbally report only the command to the right ear. It had previously been shown with dichotic testing that in these patients each hemisphere apparently pays attention only to the contralateral ear (Milner et al., 1968). As the subject could not say what act his left hand had performed, that act must have been initiated by the minor hemisphere. This successful execution of a command took place while the left hemisphere was occupied by its own instruction and thus unable to obstruct minor hemisphere output.

The major hemisphere interference model has one major problem. If the left hemisphere is inhibiting motions initiated by the right hemisphere, why does it not do so when the command is given by means of a picture? Although the actual motion is the same whether communicated by a printed verb or by a picture, it was only to the verbs that the minor hemisphere showed apraxia.

Gazzaniga and Hillyard (1971) also reported results difficult to explain on the basis of motor interference. In an attempt to define the upper limits of minor hemisphere comprehension, they flashed to the patient's left visual field pictures of a scene depicting some activity. The experimenter then read aloud two statements describing the action, and the subject had to indicate the correct one by a head nod. They tested the right hemisphere's ability to distinguish the active from the passive voice, the present from the future tense, and the singular from the plural. In none of these did the right hemisphere score above chance. It could only distinguish a positive from a negative statement. The left hemisphere apparently performed normally. Gazzaniga concluded that the right hemisphere in commissurotomized patients cannot recognize the relationship between subject, object, and verb. Because the subjects could accurately choose the name of an object shown in the left field, Gazzaniga did not believe this difficulty with action pictures could be caused by a perceptual deficit. However, as action scenes would be more complex than a single item, a perceptual problem cannot be ruled out.

Recent work by Zaidel (1973) has greatly clarified some of these issues. He devised an ingenious technique that permits presentation of visual material to a single hemisphere of a commissurotomized patient for minutes at a time. One eye is covered, and the other is fitted with a scleral contact lens on which is mounted a small screen. Half of this screen is opaque, blocking out vision in one half-field, and on the other half of the screen an image of the surface before the subject is projected. This allows the patient to scan freely visual material at will, and even monitor his own hand movements, and still have the visual information available only to one hemisphere. In one experiment using this technique, Zaidel examined the ability of the right hemisphere to comprehend different parts of speech.

He gave the subjects two vocabulary tests in which they had to choose from among several pictures the one that best described the meaning of a word said by the examiner. The overall result was that, although not doing as well as the left hemisphere, the right hemisphere did attain the vocabulary level of a 10- to 16-year-old child. Comparison of the relative scores on the various parts of speech showed both hemispheres to have a similar pattern: Both did better on nouns than on verbs, and better on verbs than on adjectives. In fact, the right hemisphere did almost as well on verbs (approximately 50% correct) as on nouns (approximately 60% correct). Both hemispheres also showed a similar effect of word frequency: The less frequently a word appeared in general usage, the less likely they were to get it correct. These results clearly demonstrate that the right hemisphere can understand orally presented verbs when asked to pick out pictures depicting actions and that comprehension of verbs compared with nouns is no worse in the right hemisphere than in the left. In a further series of studies using half-field-occluding contact lenses, Zaidel looked at the minor hemisphere's ability to understand syntax. He administered four standardized tests similar to that used by Gazzaniga and Hillyard (1971) except that, instead of indicating which sentence correctly described a picture, the subject had to choose from among several pictures, differing from one another in only one dimension, the picture corresponding to a word, phrase, or sentence said by the examiner. Zaidel's findings were totally different from Gazzaniga's. He found the right hemisphere to do significantly better than chance in understanding almost all types of syntactical constructions, including both active and passive sentences. Interrogatives were somewhat harder for the right hemisphere, as were possessive statements. Again, there was little difference between the comprehension of nouns and verbs. The most important factors limiting minor hemisphere performance seemed to be the length of the construction and the word order. Overall, the right hemisphere certainly could recognize the relationship between subject, verb, and object. Zaidel feels that Gazzaniga's results were an artifact of presenting the pictures tachistoscopically and that the minor hemisphere, though not approaching the level of the left hemisphere, can understand a great deal of spoken language.

If the right hemisphere can comprehend the nature of verbs, as these last studies indicate, then the question still remains of why it fails to execute the proper action to a command flashed in the left visual field (Gazzaniga et al., 1967). Although this may indeed represent left hemisphere interference, there is a confounding factor: The experiments that have shown the right hemisphere to comprehend verbs, by following commands (Gordon, 1973) or by picking out an object associated with a verb (Levy, 1970) or by choosing a picture associated with a verb (Zaidel, 1973), have all presented the verbs orally, but Gazzaniga et al. (1967) presented them in the form of printed words. Although it seems unlikely that the modality through which the verb is presented is a critical factor, the possibility cannot be dismissed without further testing, preferably using Zaidel's apparatus, to see (1) whether the right hemisphere can match a printed verb to a picture of an appropriate action, and (2) whether it can carry out printed commands presented in the left visual field for long periods. This would test whether the tachistoscopic nature of the task or the modality of the input is crucial in causing this difference in results.

NONLANGUAGE FUNCTIONS

Until fairly recently, investigations of the asymmetrical representation of cognitive functions in the human brain concerned primarily language abilities and the left hemisphere. Development of multifactor theories of intelligence emphasizing the diversity of intellectual skills and the creation of tests of nonverbal abilities have led to the discovery that these higher mental skills are also often lateralized, many in the right hemisphere. The types of mental functions for which the minor hemisphere is specialized, and the psychological factors that distinguish the modes of operation of the right and left hemispheres, are still matters for active investigation, and thus information obtained from direct testing of the two hemispheres is of great interest.

Spatial relations

One of the most dramatic deficits a patient can experience following cortical injury involves his capacity to perceive and manipulate the spatial relationships of objects to one another and to the subject himself (Benton, 1969). The patient may find it difficult to orient himself to his surroundings, becoming easily lost; or he may show a deficit in reproducing structures organized in space, such as in copying drawings, stick constructions, or Kohs block patterns. The right hemisphere, more often than the left, is the site of injury in these individuals. In such cases, however, the precise distribution of spatial abilities between the two sides of the brain is still not clear.

Three experiments on commissurotomized patients have tested the spatial perceptual abilities of the right compared with the left hemisphere. It has recently been found that if two arrowheads or clock hands of conflicting orientation were simultaneously flashed, one in each visual half-field, the patients, when asked to pick out from an array the one they had seen, chose the direction of line that had been shown in the left visual field (C. Trevarthen, J. Levy, & R. W. Sperry, unpublished data, 1971). Thus, in a free-choice situation, it was the orientation seen by the right hemisphere to which a response was made. Similarly, when two dots, one on either side of the visual midline, were made to appear to move simultaneously in different directions, the subjects again responded to the direction of movement perceived by the right hemisphere. These results suggest that at the basic level of line orientation and direction of movement the right hemisphere percept predominates. This result is consistent with the findings of Warrington and Rabin (1970) on line orientation in patients with unilateral brain damage.

A study of the commissurotomized patients' capacity to perceive a three-dimensional spatial structure from a two-dimensional representation was carried out by Levy-Agresti and Sperry (1968) using a modification of the Space Relations Test (Bennett et al., 1947) that allowed the subject's two hemispheres to be independently examined. The test consisted of a series of solid forms, each with a unique shape and pattern of textural surfaces. The subject blindly felt one of these stimuli with his right or left hand while simultaneously viewing three two-dimensional representations of solid forms that had been opened up (Fig. 5-2). The patient had to choose the two-dimensional layout that would, when folded up, produce the shape he was holding. The results showed the patients to be significantly more accurate with the stimuli felt in the left hand than with those felt in the right, demonstrating a minor hemisphere superiority for the task. The low correlation found to exist between the right and left hands' patterns of success on the various stimuli led Levy to propose that the two hemispheres were using different perceptual strategies to solve the task; that is, the left hemisphere analyzed the shapes in terms of the relationship of their details, and the minor hemisphere mentally folded up the two-dimensional layouts and visualized the solid forms.

Zaidel and Sperry (1973) also observed a difference in the approach used by the two hemispheres to solve another type of spatial problem. They employed a specially adapted form of Raven's Coloured Progressive Matrices in which the subject was shown in free view a pattern with a piece missing, while behind a screen he simultaneously felt three plates each of which had on it a design formed by raised lines. The subject's task was to select the tactile design that would complete the visual pattern. On this task, not only was the right-hemisphere–left-hand system far superior to the left-hemisphere–right-hand system, but it appeared to

Fig. 5-2. Two-dimensional layouts used to test ability to perceive three-dimensional structure from two-dimensional representation.

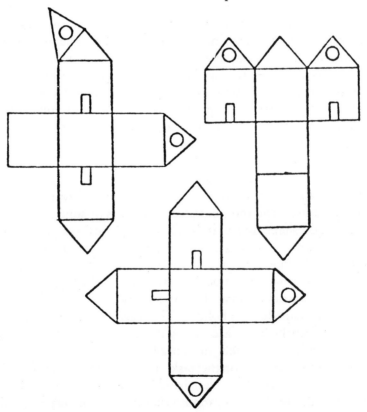

examine the choices in a different way, being more concerned with spatially related wholes. The left hemisphere, by contrast, put a higher priority on details such as number of elements, length of lines, and size of angles.

The actual construction of visuospatial arrangements also appears to be vitally dependent on minor hemisphere capacities. Both Smith (1966) and Gott (1973a) have shown the copying of geometrical shapes to be very poor following right hemisphere removal. Investigations into the abilities of the two hemispheres of commissurotomized patients to perform visuoconstructive tasks can be conducted only in the fairly short period following surgery when each hand is controlled solely by the contralateral hemisphere. Later, each hemisphere has some degree of motor control over both hands (Gazzaniga et al., 1967), making it difficult to determine which side of the brain is performing a given task. Tests run by Bogen (1969a) soon after surgery in these patients showed that, when asked to copy Greek crosses or two-dimensional representations of three-dimensional cubes, the patients performed far better with the left hand than with the right hand. This occurred at a time when they could write only with the right hand. This shows that the failure was not caused by a general right-hand motor deficit, but rather was an apraxia specific to the figure-copying task. The right-hand drawings showed many of the deficits of spatial articulation seen in copies by patients with parietal lobe disease of the nondominant hemisphere (Warrington et al., 1966). A similar right-hand apraxia was found with the Kohs blocks; that is, the left hand's responses were swift and accurate, while the right hand's responses were slow and spatially disorganized. In fact, the left hand would often spontaneously reach out and try to correct the right hand's mistakes. The same pattern of results is seen with the Block Design subtests of the Wechsler intelligence scales, on which both Smith (1969) and Gott (1973a) have reported that right hemispherectomized patients do considerably less well than do those with left hemispherectomy.

These experiments show that the right hemisphere generally surpasses the left in perceiving and manipulating spatial relationships. The full range of spatial abilities, however, has certainly not been explored in these studies, and the right and left hemispheres may prove equally competent on some spatial tasks, though perhaps differing in the strategies they use to solve them.

Closure factors

Two different closure factors were identified by Thurstone in his factor analysis of perceptual abilities. The first factor (Closure Speed) is described as the ability to perceive an apparently disorganized or unrelated group of parts as a meaningful whole, that is, the capacity to construct a whole picture from incomplete or limited material (Thurstone, 1944). Evidence that this factor is asymmetrically represented in the two hemispheres comes from studies showing recognition of objects portrayed in line drawings, most of the contour of which is missing, to be more disturbed by right than by left hemisphere damage (De Renzi & Spinnler, 1966; Lansdell, 1968). Three experiments have been conducted on commissurotomized patients to test the hypothesis that the right hemisphere in man is more efficient in perceiving the relationship between the part or parts of a stimulus and the overall configuration. The first study (Nebes, 1971) reduced the part-whole operation to its most basic level by requiring the patient to choose from among three different sizes of complete circles the one from which a given arc had come. To test each hemisphere separately, either the arcs or the circles or both were examined haptically by the subject's right or left hand. The test was given in three different forms. In the first, the subject blindly felt an arc hidden behind a screen while visually selecting from among three different sizes of circles the one it would make if it were completed (Fig. 5-3A). In the second procedure, the arcs were shown visually while the patient felt the three circles and indicated his choice by tapping it. In the third procedure, both the arcs and the circles were behind the screen for haptic examination.

Results showed the commissurotomized patients to be significantly more accurate in matching arcs to the appropriate-sized circles when using the left hand than when using the right, regardless of the inter- or intramodal nature of the task. The right hand generally performed at a chance level. On two control tasks where the subjects matched circles to circles or arcs to arcs, using the same three procedures as in the experimental test, there was no difference in accuracy between the two hands. Since the stimuli and the procedures were identical in the experimental and control tasks, it is obviously the part-whole matching required by the experimental test that led to the differential performance by the two hands. Thus, in matching a partial to a complete stimulus, the

Direct examination of cognitive function 119

right-hemisphere–left-hand system is far superior to the left-hemisphere–right-hand system.

The second experiment (Nebes, 1972) used more complex stimuli than the first but still required the subject to conceive the whole from its parts. The patient was shown a series of line drawings, each depicting a geometrical shape that had been cut up and the pieces moved apart, maintaining, however, their original orientation and relative positions (Fig. 5-3B). For each visual stimulus the subject had to reach behind the screen, feel three solid forms, and tap the one the fragmented figure would make if it was reunited. Here, as in the first experiment, the patients were very accurate with the left hand, but with the right hand performed at the chance level. On a control test where the visual stimuli were presented as complete forms, neither hand had any difficulty in selecting the correct tactile shape. Again, it was the mental operation of forming a concept of the complete stimulus from its pieces in which the right hemisphere surpassed the left.

The third experiment (Nebes, 1973) used as stimuli square arrays of dots organized so that the dots were more closely spaced along one dimension of the square than along the other (Fig. 5-3C). It

Fig. 5-3. Tests of capacity for part-whole perception. *A.* Somesthetic-visual matching of an arc to a complete circle. *B.* Stimuli for the matching of fragmented visual shapes to solid forms. *C.* Stimulus pattern composed of vertical columns of dots.

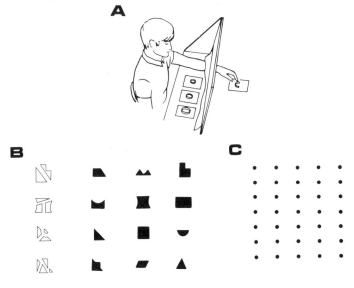

has been known for some time that people tend to see dots thus spaced as forming lines running parallel to the axis on which there is the greater concentration of points (Wertheimer, 1958). Perception of these lines of dots thus relies on the subject's noting the spatial relationship of the dots to one another and to the aggregate they appear to form. To discover whether the right hemisphere was more competent than the left on this type of part-whole task, dot arrays were produced in which the lines appeared to run either vertically or horizontally. The patient sat before a translucent projection screen upon which the stimuli were tachistoscopically flashed in either the right or the left visual field. He was told to lift the index finger of one hand if the lines ran vertically and not to move it if they ran horizontally. The data showed commissurotomized patients to be significantly more accurate in responding to left-field arrays than to those in the right visual field, regardless of which hand they used to signal. The minor hemisphere thus proved more efficient than the left in discerning the pattern inherent in the array due to the differential spacing of its parts.

All three experiments demonstrate that when required to perform a spatial transformation on sensory input (i.e., to generate a concept of the complete stimulus configuration from partial or fragmentary data), the minor hemisphere is far superior to the major. This suggests that in man the right cerebral cortex is responsible for forming from the incomplete information provided by our senses the spatial and cognitive map of our surroundings in which the planning and organization of behavior take place.

The second closure factor (Closure Flexibility) is described as the capacity to hold a configuration in mind despite distraction, to see a given configuration when it is hidden or embedded in a larger, more complex figure (Thurstone, 1944). Disturbances in this ability have been associated with unilateral left hemisphere lesions (Teuber & Weinstein, 1956). Zaidel (1973) using half-field-occluding lenses, investigated the hemispheric laterality of this factor in commissurotomized patients by comparing their performance on embedded figure tasks in their right and left visual fields. The subjects were given the Visual Closure Subtest of the Illinois Test of Psycholinguistic Abilities (ITPA) in which they were shown a line drawing of a common object and then had to pick it out in a complex scene. Commissurotomized patients were significantly better in the right visual field (left hemisphere) than in the left. This was also true when simple geometrical shapes were used and the subjects had to trace them out in the background of a more complex shape. These results, along with the work on unilaterally brain-damaged subjects, indicate that although the ability to form a gestalt from incomplete information may reside in the right hemisphere, the ability to discover a shape within an irrelevant background gestalt resides in the left.

Nonverbal stimuli

Another type of operation in which the right hemisphere has been said to outperform the left is the accurate perception and memory of stimuli that have no verbal label or are too complex or too similar to specify in words (Kimura, 1966; De Renzi, 1968). Subjects with minor hemisphere damage have difficulty in recognizing and remembering faces, nonsense figures, complex patterns, and music. Recent work directly testing the two hemispheres has confirmed and extended these findings.

Musical ability, in the form of singing, was shown to be present in patients who had undergone complete left hemispherectomy (Smith, 1966; Zaidel, 1973). However, when the right hemisphere was selectively inactivated by an amytal injection, speech was preserved, but ability to produce a tune was grossly disturbed (Bogen & Gordon, 1971; Gordon & Bogen, 1974). Subjects tended to sing in a monotone, and although they often changed pitch at the appropriate place, the change was very inaccurate. Rhythm seems much less affected. Following left-sided injections in the same patients, singing recovered long before any significant amount of speech, although the subjects did not begin to sing until at least some speech had returned. These findings show an obvious dissociation between speech and at least some aspects of musical production in the two hemispheres.

Rather than compare the successive performance of the two hemispheres of commissurotomized patients on the same task, Levy et al. (1972) presented stimuli simultaneously to both hemispheres, creating a conflict situation about which hemisphere would succeed in expressing itself. The stimuli were visual chimeras, composites formed of the right half of one stimulus and the left half of another. The chimeras were tachistoscopically flashed so that the junction of the two halves of the stimulus fell at the subject's point of fixation. Each hemisphere in the commissurotomized individual had been previously shown to perceive as complete and bilaterally symmetrical any stimulus that extended across the visual midline into the half-field from which it received no visual input (C. Trevarthen & M. Kinsbourne, unpublished data, 1971; Trevarthen, 1974). This phenomenon of completion meant that with the chimerical stimuli each hemisphere saw a different bilaterally symmetrical and whole image. Because the two hemispheres were receiving conflicting stimuli, the type of behavioral readout used became a significant variable. The subjects were required to report in one of two ways: (1) pointing with the right or left hand to the correct item among an array of complete stimuli, or (2) naming or verbally describing the stimulus.

The first experiment used faces for stimuli (Fig. 5-4A). The results showed that when the subjects pointed to their choice, regardless of which hand they used, they picked out the face shown in the left visual field and thus seen by the right hemisphere. If, however, they described the stimulus verbally or said the name

Fig. 5-4. Chimerical stimuli used to test simultaneous performance of both cerebral hemispheres in commissurotomized patients. *A*, faces; *B*, antlers; *C*, objects.

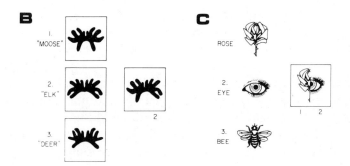

they had been taught to associate with that face, it was the right-field face that was identified. There were practically no double responses (i.e., picking out both stimuli shown on a particular trial); rather, either one field or the other was exclusively reported, depending on the mode of the response. The subjects did not report anything unusual about the stimuli, but said they had seen a complete face. To discover whether both hemispheres were simultaneously perceiving their own stimulus, or whether only one hemisphere was active at a time, depending on the nature of the readout, the type of response was switched without warning on selected trials. For example, if the subject had been pointing to his choice for a series of trials, on a certain trial after the stimulus had been flashed, he was asked instead to report verbally. In these instances, subjects were found to say the name of the right-field stimulus and then, when asked to point, to pick out the left-field face, despite any reservations they had over the consistency of their choices.

The generality of these findings was tested by using chimeras composed of other complex visual stimuli, such as ambiguously shaped solids resembling antlers (Fig. 5-4B), vertical strings of crosses and squares, drawings of common objects (Fig. 5-4C), or even words. All these produced the same pattern of results: The subjects pointed to the stimulus seen in the left field and named the one seen in the right. They gave no indication that they had seen the stimuli as anything but complete and regular.

Levy et al. (1972) concluded that the minor hemisphere's superiority demonstrated in these experiments involved not just the level of the spatial-perceptual transformations required by previous tests, but rather the direct apprehension of shape. They suggested that the primary tendency is for all visual stimuli to be perceived by the minor hemisphere, even words, as long as only visual recognition is required and not a verbal transformation. It is especially significant that in the type of conflict situation employed by these tests, the minor hemisphere controlled the motor output (i.e., pointing) not only of the left hand but also of the right. This is in striking contrast to the experiments involving writing, where the major hemisphere dominated. Thus, in the competition for the motor channels, the hemisphere that is most competent for the function involved assumes control over the motor system. This point is emphasized by an experiment in which the subjects were asked to point to a picture of an object whose name rhymed with the name of the object they had seen. Although the subject was pointing, the right side of the chimera predominated. Here, because a verbal transformation rather than a visual recognition was required (i.e., matching of auditory images), the left hemisphere now controlled the output (Levy, 1972).

An additional point brought out by these experiments is the great difficulty commissurotomized patients had in learning to associate names with specific faces or antlers. Rather than forming a relationship between the stimulus as a whole and its name, they were reduced to forming associations between the name and stimulus peculiarities, such as the face called Bob had glasses. Levy feels this represents a disconnection effect in which the verbal naming faculties in the left hemisphere have been cut off from the facial recognition faculties in the right hemisphere.

This series of experiments using the chimerical technique demonstrates the superiority of the minor hemisphere in commissurotomized patients for the processing of visual images of both nameable and nonnameable objects and even of words themselves. As long as a verbal readout or transformation was not required, visual image recognition appeared to be a right hemisphere function.

Another experiment in this series investigated the way in which the two hemispheres differ in the conceptual bias with which they approach a problem. Here, the chimerical stimuli were composed of line drawings of common objects, and the subject was asked to pick out from a stimulus array the one that was "similar" to the item he had seen flashed on the screen. Similarity could be achieved in this test by choosing either a visually similar object or one that was conceptually similar, that is, related in its use to the test stimulus. Matches made by the right hemisphere were found to be structurally similar; those made by the left were conceptually similar (Sperry & Levy, 1970). Kumar (1973) found a similar pattern in the types of concept formations employed by the right and left hemispheres of commissurotomized patients when he asked them to examine and sort stimuli blindly with the right or left hand, attempting by trial and error to discover which dimension was the proper one to differentiate the items. When the stimuli were blocks varying in a number of spatial dimensions (height, size, shape, etc.), subjects did much better with the left hand than with the right, suggesting superior right hemisphere performance; when, however, the stimuli were everyday objects related in their use, the patients were better at sorting with the right hand.

Smith (1969) and Gott (1973a) both reported that on the Similarities Subscale of the Wechsler Adult Intelligence Scale (WAIS), a test of verbal abstract reasoning, right hemispherectomized patients do quite well; left hemispherectomized patients, extremely poorly. Just the opposite pattern was found with a nonverbal reasoning task involving geometrical figures (Raven's Coloured Progressive Matrices), where a left hemispherectomized patient scored normally, and right hemispherectomized patients were definitely impaired (Smith, 1969). All these results suggest strongly that although both cerebral hemispheres are capable of reasoning and forming concepts, they do so in quite different ways. Further characterization of this difference may bring us close to discovering the essential nature of the psychological variables underlying hemispheric specialization.

The above-described work showing a minor hemisphere preference for structural similarity does not mean that the right hemisphere is incapable of understanding or forming concepts based on semantic similarity. Zaidel (1973) administered the Visual Reception and Visual Association subtests of the ITPA to the right and left hemispheres of commissurotomized patients by using half-field-occluding lenses. In the Visual Reception Subtest the subject is shown a picture of an object, and then has to choose from among four objects the one that is conceptually similar; for example, if shown an hourglass, he would pick out a watch. The correct choice is perceptually different, but is a member of the same overall concept based on usage. In the Visual Association Subtest, the subject chooses from among four object pictures the one that is most meaningfully related to a given object; for example, if shown a stove, he would choose a frying pan. In both tests, the right hemisphere of commissurotomized patients scored above the level of a 5-year-old child, not much worse than the left hemisphere's score in the same subjects. Thus, the minor hemisphere can elicit semantic information from pictures and can form complex semantic associations, perhaps in the absence of the verbal symbols that seem to mediate this type of function in the left hemisphere.

MEMORY

Although many of the experiments already described have involved memory to some degree, as when commissurotomized patients had to retrieve tactually an object whose name had been briefly flashed, there have been relatively few studies specifically designed to test directly the abilities of the right and left hemispheres to remember different stimulus materials. Fedio and Weinberg (1971) presented patients with a string of object pictures, requiring that as they saw each one they name it and then recall the name of the preceding picture. They found recall after a left-sided injection of sodium amytal to be significantly reduced from preinjection levels for up to 10 minutes after the injection. This memory deficit outlasted by many minutes any aphasia or anomia. Right-sided injections produced much less effect. The nature of this memory deficit is not clear; it could stem from a difficulty either in storage or in retrieval. Serafetinides (1966) asked subjects to study words, numbers, and Bender Gestalt figures until they had them memorized. Left-sided amytal injections produced a decrease in recall of the verbal material that was evident after the aphasia had cleared. No such loss was found after minor hemisphere anesthetization. Results of these verbal memory tests seem to show a defect in the retrieval mechanism, as the subject could accurately recall the material just prior to injection. There was no relationship between the side of the injection and recognition of the Bender Gestalt figures, but as most subjects had perfect scores, this task was obviously not sufficiently difficult to serve as a test of nonverbal memory.

Zaidel (1973) found the left hemisphere to be superior to the right on the Visual Sequential Memory Subtest of the ITPA (McCarthy & Kirk, 1963). The patient was shown for 5 seconds a line of nonsense figures in the right or left visual field while wearing half-field-occluding contact lenses. He then had to reproduce this sequence by arranging chips bearing the corresponding nonsense figures, using the hand ipsilateral to the visual field in which the stimuli had appeared. Despite the nonverbal nature of the shapes, the sequential memory aspect of the task apparently was a more critical factor, leading to superior left hemisphere performance. When the patterns were given tactually, this hemispheric difference disappeared, as left hemisphere scores decreased from the visual condition much more than did those of the right hemisphere. Apparently, this was not the result of any gross difficulty in perceiving the patterns, as the subjects did very well with either hand in tactually matching single shapes. In another experiment with these patients, however (Milner & Taylor, 1972), the right hemisphere did surpass the left in tactile memory for single nonsense shapes. Five of seven commissurotomized patients failed to exceed a chance level in blindly retrieving from among four choices a nonsense shape felt by the right hand, even if there was no delay before they started the search. Two other patients failed with a 15-second delay. Using the left hand, these same subjects successfully retrieved a shape up to 2 minutes after they had felt it. When reproduction rather than recognition of geometric shapes was required, Kumar (1973) also found the right hemisphere of commissurotomized patients to be superior to the left. He had subjects blindly palpate with the right or left hand a pattern from a tactile version of the Memory for Designs Test, and then blindly draw with that same hand the design they had felt. The left hand (right hemisphere) was considerably better at this than was the right.

There have been two studies in which standardized memory tests have been administered to hemispherectomized patients. Smith (1969) found the digit span to be decreased in a left hemispherectomized patient, but fairly normal in three patients with right hemispherectomy. The reverse was true with a nonverbal short-term memory task (Benton Visual Retention Test), on which the right hemispherectomized patients did very poorly. Of course, in view of the language deficit in the patient with a left hemispherectomy, the low digit span score cannot be attributed to a problem specific to short-term or immediate verbal memory. The left hemispherectomized patient reported by Gott (1973a) did much worse than two patients with right hemispherectomy on both verbal (Wechsler Memory Scale) and nonverbal (Memory for Design) memory tests. These results seem to be very much influenced by the age at which the patient first suffered brain damage.

Overall, the work on hemisphere specialization in memory processes using direct hemisphere testing has not added greatly to our knowledge and has not produced any new insights into the neural mechanisms underlying memory. Certainly in view of the fairly clear results following temporal lobectomy (Milner, 1974), a great deal more should be done in this area.

EMOTION

Although the ultimate control over the expression of emotion is mediated predominantly by subcortical centers, cortical events in the minor as well as the major hemisphere play an important role in initiating emotional responses. The left hemispherectomized patient reported by Smith (1966) showed obvious frustration and anger at his inability to speak. Commissurotomized patients reacted strongly with blushing and giggling to the presentation of pictures of nudes in the left visual field (right hemisphere), even while the major hemisphere showed by its verbalization that it had no idea why this was happening (Gazzaniga, 1965; Sperry & Gazzaniga, 1967). Similarly, unpleasant odors presented solely to the right hemisphere elicited grimaces and exclamations of disgust for which the left hemisphere had no explanation (Gordon & Sperry, 1969).

The only evidence that the two hemispheres differ in their emotional makeup comes from amytal studies (Perria et al., 1961; Terzian, 1964), where catastrophic reactions and feelings of guilt and depression were common after left hemisphere anesthetization, while feelings of euphoria often followed right-sided injections. The feeling of depression was reported to begin toward the end of the aphasic episode, 4 to 6 minutes after injection, and to last from 1 to 10 minutes (Perria et al., 1961). This makes it difficult to decide whether this represents a release of a right hemisphere emotional bias or a rebound effect as the left hemisphere regains awareness. The former explanation appears more likely, as a similar pattern of emotional reactions has been found with unilateral hemisphere damage (Gainotti, 1972). Along these lines it is interesting that Hommes and Panhuysen (1971) have reported that anesthetization of either hemisphere disrupts language in individuals suffering from depression. The depth of the preinjection depression was also negatively correlated with the strength of the patient's cerebral dominance as measured by the length of the postinjection aphasia. This suggests that, in these individuals, depression was linked to an abnormal state of cerebral dominance; that is, both hemispheres were involved in speech production. Hommes interprets this as both hemispheres acting as nondominant hemispheres, because amytal injected into either side produced a euphoria that, in normal individuals, usually occurs only after injection into the nondominant hemisphere. This, of course, is not how bilateral language representation would normally be interpreted, but if one considers it as evidence that these subjects have essentially two dominant hemispheres, it becomes hard to reconcile this finding with previous results showing *inactivation* of the dominant hemisphere to produce a depressive state. Obviously, if some balance of emotional tendencies exists between the two hemispheres, it is not as simple as depression being associated with language and the side of the brain in which it is organized. Investigations along these lines in commissurotomized patients would be valuable, especially if it proved possible to link different emotional biases with what we already know of the contrasting styles of information processing in the right and left hemispheres.

CONSCIOUSNESS

The term *consciousness* has been applied to concepts as diverse as simple awareness of external events, awareness of self as an independent entity, and even awareness of the cosmic interrelationship of all things. On each of these levels, there have been suggestions that the right and left cerebral hemispheres are not equally involved in the phenomenon of consciousness. Serafetinides et al. (1965) found subjects to become unreactive to external stimuli (i.e., unconscious) after anesthetization by amytal of the hemisphere dominant for language, but not after anesthetization of the nondominant hemisphere. They felt that this reflected an asymmetry in the relative influence exerted by the two hemispheres on the reticular activating system, so that the hemisphere dominant for speech was also dominant for the maintenance of consciousness. Rosadini and Rossi (1967), however, using a more sensitive measure of external awareness – a vigilance task requiring reaction to a light or tone – found no such lateralization of consciousness. Rather, the only time a subject became totally unresponsive was when electroencephalography showed inactivation of both hemispheres.

Lately, the question of hemispheric lateralization of consciousness has again been raised because tests on commissurotomized patients have showed both the right and left hemispheres to be capable of independently carrying out many of the higher cognitive activities normally associated with consciousness. Indeed, Sperry (1968) spoke of two separate streams of consciousness existing in commissurotomized patients, one in each hemisphere. Other investigators (MacKay, 1966; Lishman, 1971; Eccles, 1973) have emphatically rejected this interpretation, taking the position that consciousness is an individual subjective state we perceive in ourselves and infer in others from their verbal reports of their introspective insights. Such an inference, therefore, requires more than just evidence of the existence of higher mental abilities; it requires verbal communication, and thus, by definition, this view of consciousness rules out any possibility of ascertaining self-awareness in the minor hemisphere. Eccles (1973) has gone even further, equating consciousness with the operations of the language hemisphere and seeing the minor hemisphere as nothing more than an automaton or a superior animal brain. When he talks of actions carried out by the right hemisphere as being "all unbeknown to the *subject*" (by which he obviously means the left, speaking, hemisphere), it is quite evident he believes that actions or mental events not available to the speech centers do not give rise to conscious experiences. Thus, when a commissurotomized subject feels an object in his left hand and cannot verbally identify it, his demonstrated ability to use it correctly, to match it across modalities, to pick out its name, to assign it to its proper conceptual category, and to remember it, is all unconscious, as are any apparently purposeful movements whether spontaneous or in response to command. Eccles thus casts into the limbo of unconsciousness all global aphasics, left hemispherectomized patients, and deaf mutes. This view of consciousness as a property of the left hemisphere is, of course, irrefutable, as Eccles has characterized consciousness in terms of major hemisphere functions, turning it into a matter of definition.

A radically different conclusion about right hemisphere consciousness has been reached by Ornstein (1972), who suggested that the different styles of information processing used by the two hemispheres reflect an underlying difference in their "modes of consciousness." He has taken the dualities present in many philosophies, most prominently those of the Orient, and has identified them with the anatomical and functional duality of the human brain. Thus, the left hemisphere form of consciousness is rational, verbal, and active, in contrast to that of the right, which is intuitive, spatial, and receptive. The evidence for an intuitive mode of consciousness comes not so much from scientific experimentation (which is a left hemisphere approach in Ornstein's model anyway), as from experiential and introspective self-examination. He sees intuitive consciousness as being heavily involved in creation – the visualization of new combinations and interrelationships. For example, Einstein's capacity to conceive the relationships of time, matter, and energy, coupled with his great difficulty in expressing these concepts in symbolic form, represents a problem in changing right hemisphere intuitions into a left hemisphere form. The association of this type of consciousness with the right hemisphere is, of course, based predominantly on the apparent use by the right hemisphere of a holistic form of information processing.

At the present time, discussions of the neural bases of consciousness often become mired in arguments over what constitutes an adequate definition of the phenomenon (Kinsbourne, 1974). Without a clearer understanding of the nature of consciousness, it is all too easy to define it in terms of properties possessed by the brain area in which the investigator wishes it to reside. Just as attempts to discover "centers" in the brain for various mental functions proved futile until the basic components of the activities were more thoroughly analyzed, assigning consciousness to part of the brain seems an unprofitable task without a great deal more knowledge than we presently possess.

DISCUSSION

Several major contributions to our knowledge of brain organization have resulted from direct investigation of hemisphere function. The first of these concerns language abilities present in the minor hemisphere. It has been shown that, although relatively incapable of producing either written or spoken language, the right hemisphere can understand even fairly complex syntactical constructions. Although apparently biased in favor of using structural or visual concepts, the right hemisphere can also form and understand semantic concepts.

These studies have also produced insights into the basic dissimilarity between the right and the left hemispheres' methods of processing sensory data and have exposed the interaction and hierarchy of dominance present between the two hemispheres in processing input and, more importantly, in competing for control over motor output.

Two general types of mental operations were shown by these studies to be more competently performed by the right than by the left hemisphere. The first is the ability to generate from fragmentary information a percept of the whole. This can be viewed as a spatial function in which, from limited data, we infer the structure and organization of our environment without having to submit the whole sensory array to a detailed analysis. The chimerical studies also clearly demonstrated that it is not just the type of perceptual stimulus or the mode of readout used that determines which hemisphere is dominant, but rather the type of information processing required to solve the given problem. If only visual recognition is called for, even if the material is verbal, the right hemisphere acts. If, however, a verbal transformation is demanded, even if the material is nonverbal, it is handled by the left hemisphere.

This dichotomizing of the two hemispheres according to the functions they perform rather than by their preferred input or output has led several investigators (Levy-Agresti & Sperry, 1968; Bogen, 1969b) to propose a model of hemispheric action in which the minor hemisphere is seen to organize and treat data in terms of complex wholes, being in effect a synthesizer with a predisposition for viewing the total rather than the parts. The left hemisphere in this model sequentially analyzes input, abstracting out the relevant details to which it associates verbal symbols in order to manipulate and store the data more efficiently. At present, this model can be best viewed as an approach to hemispheric differences rather than a comprehensive statement of the organization of the cerebral functions in man. Future experiments will reveal whether this is an accurate picture of how man's two hemispheres differ in their specialization.

The competition between the left and right hemispheres of commissurotomized patients for control of the motor output and the response hierarchy this entails were clearly revealed in two different studies. In the experiments with right hemisphere writing, the frequent seizure of control of the left hand by the left hemisphere demonstrated that, even though the main motor control for the left hand is in the right hemisphere, when the response is a verbal one, the left hemisphere can override and dominate the action. This is in striking contrast to the experiments using chimeras where, on tests of right hemisphere function, this hemisphere controlled the right as well as the left hand. These results point to a response hierarchy in terms of competing responses programmed by the two hemispheres, with the hemisphere whose specialty the task demands gaining access to the motor system. This decision between the two hemispheres' claims to the response apparatus may well be a result of cortical influences balancing out at the brain stem level.

ACKNOWLEDGMENTS

The work reported here was supported by NINDS grant NS 06233. A shorter version of this chapter appeared in *Psychological Bulletin 81*:1-14, copyright (1974) by the American Psychological Association. Reprinted by permission.

REFERENCES

Andrews, G., Quinn, P. T., & Sorly, W. A. 1972. Stuttering: An investigation into cerebral dominance for speech. *J. Neurol. Neurosurg. Psychiatry* 35: 414-418.

Basser, L. S. 1962. Hemiplegia of early onset and the faculty of speech with special reference to the effects of hemispherectomy. *Brain* 85:427-460.

Bennett, G. K., Seashore, H. G., & Wesman, A. G. 1947. *Differential Aptitude Tests (Space Relations, Form A)*. New York: Psychological Corporation.

Benton, A. L. 1969. Disorders of spatial orientation. In P. J. Vinken & G. W. Bruyn (eds.). *Symposium on Fatigue*, Vol. 3, pp. 212-228. Amsterdam: North Holland.

Bogen, J. E. 1969a. The other side of the brain: Dysgraphia and dyscopia following cerebral commissurotomy. *Bull. Los Angeles Neurol. Soc.* 34: 73-105.

1969b. The other side of the brain. 2. An appositional mind. *Bull. Los Angeles Neurol. Soc.* 34:135-162.

Bogen, J. E., & Gordon, H. W. 1971. Musical tests for functional lateralization with intracarotid amobarbital. *Nature* 230:524-525.

Branch, C., Milner, B., & Rasmussen, T. 1964. Intracarotid sodium amytal for the lateralization of cerebral speech dominance. *J. Neurosurg.* 21:399-405.

Broca, P. 1861. Remarks on the seat of the faculty of articulate language, followed by an observation of aphemia. In G. Von Bonin (ed.). *Some Papers on the Cerebral Cortex*. Springfield, Ill.: Thomas, 1960.

Butler, S., & Norrsell, U. 1968. Vocalization possibly initiated by the minor hemisphere. *Nature* 220:793-794.

De Renzi, E. 1968. Nonverbal memory and hemispheric side of lesion. *Neuropsychologia* 6:181-189.

De Renzi, E., & Spinnler, H. 1966. Visual recognition in patients with unilateral cerebral disease. *J. Nerv. Ment. Dis.* 142:515-525.

Eccles, J. C. 1973. Brain, speech and consciousness. *Naturwissenschaften* 60: 167-176.

Fedio, P., & Weinberg, L. K. 1971. Dysnomia and impairment of verbal memory following intracarotid injection of sodium amytal. *Brain Res.* 31: 159-168.

Gainotti, G. 1972. Emotional behavior and hemispheric side of the lesion. *Cortex* 8:41-55.

Gazzaniga, M. S. 1965. Psychological properties of the disconnected hemispheres in man. *Science* 150:372.

Gazzaniga, M. S., Bogen, J. E., & Sperry, R. W. 1962. Some functional effects of sectioning the cerebral commissures in man. *Proc. Natl. Acad. Sci. U.S.A.* 48:1765-1769.

1963. Laterality effects in somesthesis following cerebral commissurotomy in man. *Neuropsychologia* 1:209-215.

1965. Observations on visual perception after disconnection of the cerebral hemispheres in man. *Brain* 88:221-236.

1967. Dyspraxia following division of the cerebral commissures. *Arch. Neurol.* 16:606-612.

Gazzaniga, M. S., & Hillyard, S. A. 1971. A language and speech capacity of the right hemisphere. *Neuropsychologia* 9:273-280.

Gazzaniga, M. S., & Sperry, R. W. 1967. Language after section of the cerebral commissures. *Brain* 90:131-148.

Gordon, H. 1973. Verbal and nonverbal cerebral processing in man for audition. Doctoral dissertation, California Institute of Technology.

Gordon, H. W., & Bogen, J. E. 1974. Hemispheric lateralization of singing after intracarotid sodium amylobarbitone. *J. Neurol. Neurosurg. Psychiatry* 37:727-738.

Gordon, H. W., & Sperry, R. W. 1969. Lateralization of olfactory perception in the surgically separated hemispheres of man. *Neuropsychologia* 7:111-120.

Gott, P. S. 1973a. Cognitive abilities following right and left hemispherectomy. *Cortex* 9:266-274.

1973b. Language after dominant hemispherectomy. *J. Neurol. Neurosurg. Psychiatry* 36:1082-1088.

Hillier, W. F. 1954. Total left cerebral hemispherectomy for malignant glioma. *Neurology* 4:718-721.

Hommes, O. R., & Panhuysen, L. H. H. M. 1971. Depression and cerebral dominance. *Psychiatr. Neurol. Neurochir.* 74:259-270.

Jones, R. K. 1966. Observations on stammering after localized cerebral injury. *J. Neurol. Neurosurg. Psychiatry* 29:192-195.

Kimura, D. 1966. Dual functional asymmetry of the brain in visual perception. *Neuropsychologia* 4:275-285.

Kinsbourne, M. 1971. The minor cerebral hemisphere as a source of aphasic speech. *Arch. Neurol.* 25:302-306.

1974. Cerebral control and mental evolution. In M. Kinsbourne & W. L. Smith (eds.). *Hemispheric Disconnection and Cerebral Function*. Springfield, Ill.: Thomas.

Kumar, S. 1973. The right and left of being internally different. *Impact Sci.* 23:53-64.

Lansdell, H. 1968. Effect of extent of temporal lobe ablations on two lateralized deficits. *Physiol. Behav.* 3:271-273.

Levy, J. 1970. Information processing and higher psychological functions in the disconnected hemispheres of human commissurotomy patients. Doctoral dissertation, California Institute of Technology.

1972. Lateral specialization of the brain: Behavioral manifestations and possible evolutionary basis. In J. A. Kiger, Jr. (ed.). *Proceedings of the 32nd Annual Biology Colloquium on the Biology of Behaviors*. Corvallis: Oregon State University Press.

Levy, J., Nebes, R. D., & Sperry, R. W. 1970. Expressive language in the surgically separated minor hemisphere. *Cortex* 7:49-58.
Levy, J., Trevarthen, C. B., & Sperry, R. W. 1972. Perception of bilateral chimeric figures following hemispheric deconnection. *Brain* 95:61-78.
Levy-Agresti, J., & Sperry, R. W. 1968. Differential perceptual capacities in major and minor hemispheres. *Proc. Natl. Acad. Sci. U.S.A. 61*:1151.
Lishman, W. A. 1971. Emotion, consciousness and will after brain bisection in man. *Cortex* 7:181-192.
Luessenhop, A. J., Boggs, J. S., LaBorwit, L. J., & Walle, E. L. 1973. Cerebral dominance in stutterers determined by Wada testing. *Neurology 23*: 1190-1192.
MacKay, D. M. 1966. In J. C. C. Eccles (ed.). *Brain and Conscious Experience*, p. 312. New York: Springer-Verlag.
McCarthy, J. J., & Kirk, S. A. 1963. *The Illinois Test of Psycholinguistic Abilities*. Urbana: University of Illinois Press.
Mensh, J. N., Schwartz, H. G., Matarazzo, R. G., & Matarazzo, J. D. 1952. Psychological functioning following cerebral hemispherectomy in man. *Arch. Neurol. Psychiatry* 67:787-796.
Milner, B. 1974. Hemispheric specialization: Scope and limits. In F. G. Worden & F. O. Schmitt (eds.). *The Neurosciences: Third Study Program*, pp. 75-89. Cambridge: MIT Press.
Milner, B., & Taylor, L. B. 1970. Somesthetic thresholds after commissural section in man. *Neurology 20*:37.
   1972. Right hemisphere superiority in tactile pattern recognition after cerebral commissurotomy: Evidence for nonverbal memory. *Neuropsychologia 10*:1-16.
Milner, B., Taylor, L., & Sperry, R. W. 1968. Lateralized suppression of dichotically presented digits after commissural section in man. *Science 161*: 184-186.
Nebes, R. D. 1971. Superiority of the minor hemisphere in commissurotomized man for the perception of part-whole relations. *Cortex* 7:333-349.
   1972. Dominance of the minor hemisphere in commissurotomized man on a test of figural unification. *Brain 95*:633-638.
   1973. Perception of dot patterns by the disconnected right and left hemisphere in man. *Neuropsychologia 11*:285-290.
Nebes, R. D., & Sperry, R. W. 1971. Hemispheric deconnection syndrome with cerebral birth injury in the dominant arm area. *Neuropsychologia* 9:247-259.
Ornstein, R. E. 1972. *The Psychology of Consciousness*. San Francisco: Freeman.
Perria, L., Rosadini, G., & Rossi, G. F. 1961. Determination of side of cerebral dominance with amobarbital. *Arch. Neurol. 4*:173-179.
Rosadini, G., & Rossi, G. F. 1967. On the suggested cerebral dominance for consciousness. *Brain 90*:101-112.

Semmes, J. 1968. Hemispheric specialization: A possible clue to mechanism. *Neuropsychologia* 6:11-26.
Serafetinides, E. A. 1966. Auditory recall and visual recognition following intracarotid sodium amytal injections. *Cortex* 2:367-372.
Serafetinides, E. A., Hoard, R. D., & Driver, M. V. 1965. Intracarotid sodium amylobarbitone and cerebral dominance for speech and consciousness. *Brain* 88:107-130.
Smith, A. 1966. Speech and other functions after left (dominant) hemispherectomy. *J. Neurol. Neurosurg. Psychiatry* 29:167-171.
1969. Nondominant hemispherectomy. *Neurology* 19:442-445.
Smith, A., & Burklund, C. W. 1966. Dominant hemispherectomy: Preliminary report on neuropsychological sequelae. *Science* 19:1280-1282.
Sperry, R. W. 1968. Mental unity following surgical disconnection of the cerebral hemispheres. *Harvey Lect.* 62:293-323.
Sperry, R. W., & Gazzaniga, M. S. 1967. Language following surgical disconnection of the commissures. In F. L. Darley (ed.). *Brain Mechanisms Underlying Speech and Language.* New York: Grume & Stratton.
Sperry, R. W., Gazzaniga, M. S., & Bogen, J. E. 1969. Interhemispheric relationships: The neocortical commissures; syndromes of hemisphere disconnection. In P. J. Vinken & G. W. Bruyn (eds.). *Handbook of Clinical Neurology*, Vol. 4, pp. 273-290. Amsterdam: North Holland.
Sperry, R. W., & Levy, J. 1970. Mental capacities of the disconnected minor hemisphere following commissurotomy. Paper presented to American Psychological Association, Miami.
Teng, E. L., & Sperry, R. W. 1973. Interhemispheric interaction during simultaneous bilateral presentation of letters or digits in commissurotomized patients. *Neuropsychologia* 11:131-140.
Terzian, H. 1964. Behavioral and EEG effects of intracarotid sodium amytal injection. *Acta Neurochir. (Wien)* 12:230-239.
Teuber, H.-L., & Weinstein, S. 1956. Ability to discover hidden figures after cerebral lesions. *Arch. Neurol. Psychiatry* 76:369-379.
Thurstone, L. L. 1944. *A Factorial Study of Perception.* Psychometric Monographs, No. 4. Chicago: University of Chicago Press.
Trevarthen, C. B. 1974. Functional relations of disconnected hemispheres with the brain stem, and with each other: Monkey and man. In M. Kinsbourne and W. L. Smith (eds.). *Hemispheric Disconnection and Cerebral Function.* Springfield, Ill.: Thomas.
Wada, J., & Rasmussen, T. 1960. Intracarotid injection of sodium amytal for the lateralization of cerebral speech dominance: Experimental and clinical observations. *J. Neurosurg.* 17:266-282.
Warrington, E. K., James, M., & Kinsbourne, M. 1966. Drawing disability in relation to laterality of cerebral lesions. *Brain* 89:53-82.
Warrington, E. K., & Rabin, P. 1970. Perceptual matching in patients with cerebral lesions. *Neuropsychologia* 8:475-487.

Wertheimer, M. 1958. Principles of perceptual organization. In D. C. Beardslee & M. Wertheimer (eds). *Readings in Perception*, pp. 115-135. New York: Van Nostrand.

Zaidel, D., & Sperry, R. W. 1973. Performance on the Raven's Coloured Progressive Matrices by subjects with cerebral commissurotomy. *Cortex 9*: 34-39.

Zaidel, E. 1973. Linguistic competence and related functions in the right cerebral hemisphere of man following commissurotomy and hemispherectomy. Doctoral dissertation, California Institute of Technology.

PART III

STUDIES OF BEHAVIORAL ASYMMETRY

6
Lateral dominance as a determinant of temporal order of responding[1]

E. RAE HARCUM

The title of this chapter focuses attention on the tautological problems of interpreting data in terms of lateral dominance. This wording implies that the dominant side of the brain determines which element among multiple stimuli will enjoy primacy in the perceptual and reproductive processes. However, the wording could with justification be rearranged to imply that the temporal order of responding evokes the manifest lateral dominance. The temporal primacy and the accuracy of perception are, in fact, correlated, but the causal relationship is not known. Therefore, we must not only discover those variables that elicit these lateral asymmetries, but also provide a compendium that can lead to an understanding of the specific mechanisms underlying them.

The present approach toward such an understanding differs from the neuropsychological orientation found in other chapters. The neuropsychological approach holds that the different cerebral hemispheres are structurally more capable of handling different classes of stimuli: the left, verbal; the right, nonverbal. In contrast, the thesis presented here is that the key to understanding these lateral asymmetries lies in the information-processing dynamics of the subjects. These dynamics of perceptual acquisition often vary according to stimulus characteristics and according to many subject and response variables. The obtained plasticity of the asymmetrical performance is more in harmony with the information-processing interpretation than with the argument for innate superiority of a particular hemisphere.

In place of the structural-functional dichotomy implied above, a possibly helpful distinction is that between sensory and perceptual mechanisms (Hebb, 1963). For example, the greater sensory impact of the projections to one hemisphere may be critical in controlling hemifield differences, but such a so-called structural variable may be important only because it influences a functional (perceptual) mechanism. It may cause one stimulus to be perceived earlier or more vividly than others because it attracts "focal attention" (Neisser, 1967).

Instead of the traditional contrast between structural mechanisms and learned mechanisms, this chapter employs two rubrics: (1) *subject-specific processes*, in which the relative performance of visual hemifields is not determined by the operations of the experimenter; and (2) *task-specific processes*, in which the experimenter does exert such control. Even here, however, the distinction parallels the degree of stability for the asymmetrical performance, with the most stable effects clearly indicating subject-specific mechanisms, and the most labile effects reflecting task-specific processes. In between are semistable effects, such as those caused by persisting attentional sets.

Substantive content

This chapter is primarily concerned with the ways in which perceptual accuracy for stimuli in right and left visual hemifields is affected by changes in the conditions of testing. The findings imply that the obtained hemifield differences are not completely determined by the relative structural potency or integrity of the two cerebral hemispheres. According to this view, the numerous factors subsumed under the general term *dominance* are important because they foster primacy in a perceptual process. The directionality of the perceptual process with respect to the various parts of the visual field is a natural consequence of this initial selection of starting point, as is the observer's preferred order for overt reproduction of the stimulus components. Because this perceptual process is selective, so that selective set variables are not easily distinguished from sensory acuity or response variables, it is necessary to search for stable results that retain validity under various testing conditions.

Definition of dominance

This chapter applies the maxims of operational measurement (Garner et al., 1956) to research on lateral differences in perceptual performance. Specifically, when correlational procedures are used to establish individual differences in types or traits, such as would be represented by cerebral dominance or specialization, allowance must be made for correlations spuriously generated by the technique of measurement. If a real entity rather than a procedural artifact is being measured, the results of two different ways of measuring that entity must show a higher correlation than the results for two different entities measured by the same technique (Campbell & Fiske, 1959).

Cerebral hemisphere dominance. Unfortunately, with cerebral hemisphere dominance, increasing the variety of laterality tests for a given individual increases the probability of obtaining some one result that is inconsistent with the others (Penfield & Roberts, 1959). Moreover, Zangwill (1960) concluded that so-called left-handed persons were more likely than right-handers to have mixed laterality of the brain. He also observed left-handed subjects whose speech function was apparently localized in the left cerebral hemisphere. Rossi and Rosadini (1967) found a high relationship between handedness and lateral dominance for speech as determined by injection of sodium amorbarbital, but with inconsistencies, particularly for left-handed subjects. Bryden (1965), also finding left-handers less consistent in their lateral asymmetries on tachistoscopic and dichotic listening tests, concluded that cerebral dominance was not a unitary process. Zurif and Bryden (1969) were forced to distinguish between familial and nonfamilial left-handers, for whom they found perceptual differences. Therefore, cerebral dominance does seem to vary with specific tasks, particularly for left-handed persons.

Eye dominance. With respect to eye dominance, Gilinsky (1952) similarly doubted whether there was a unitary entity. Again, the concept of ocular dominance tends to lose explanatory value when the number of behavioral tests is increased and the results of these tests fail to correlate consistently.

Conclusion. The above discussion indicates that one must be cautious about using the concept of dominance for explanatory purposes. Although there is sufficient stability of performance to indicate that the operations are converging on some entity, there is enough inconsistency to discount the possibility that this entity can be conceived simply. The problem is to discover on which tasks the relative performance of the left and right members is determined by subject-specific variables (i.e., lateral dominance) and on which it is controlled by stimulus, response, and methodological variables.

Plan of the chapter

This chapter considers first possible artifacts in procedures and measurement that could produce some lateral differences, now erroneously attributed to given experimental variables. Next, empirical differences in performance for stimulus elements presented to the two visual hemifields are considered. The condition in which stimuli appear completely on one or the other side of fixation, which Heron (1957) called successive exposure, is here called *lateral presentation*. The condition in which stimuli appear simultaneously to the right and left of fixation, called simultaneous exposure by Heron, is here referred to as *bilateral presentation*. Next, a special case of bilateral presentation, called *continuous presentation* because the pattern extends so far to the left and right that its ends are not perceptible, is examined. The Discussion section attempts to provide a general theoretical framework at a molar level for the results described.

POSSIBLE ARTIFACTS

An artifact occurs in an experiment when a result is produced by some uncontrolled factor instead of the independent variable to which it is ascribed. Because experiments are rarely replicated exactly, an artifact and an interaction between experimental variables can be easily confused. The existence of an artifact is operationally demonstrated when manipulation of a previously uncontrolled factor produces a given effect in the absence of the alleged independent variable. An artifact is also demonstrated when a further experiment with improved procedures eliminates the effect. The latter method is more difficult to employ successfully because the researcher must show that the new method does, in fact, entail improvement. An axiom of research science is that the less precise procedure is less likely to provide statistical support for an empirical difference. In the present context, this places a heavy onus on those attempting to prove the absence of a lateral difference under a given experimental condition, because at least some empirical difference is usually obtained.

Three categories of artifacts that would generate apparent left-right differences in perception are discussed:
1. Fixation error by the observer, causing the stimuli in one hemifield to be projected onto a more sensitive region of the retina.
2. Response bias, including both the possibilities that the requirements of a given experimental technique permit cognitive deduction – instead of a direct perception – of which stimulus was presented and that some elements of a multiple-element display can be favored temporally because the elements are reported sequentially.
3. Factors related to measurement biases. Performance differences between hemifields can vary according to the characteristics of the indicator response and according to whether the subject knows about specific aspects of the experimental procedure and conditions. Another type of measurement bias is more likely to affect the size of differences than to produce differences. It is most important in interpreting the meaning of differences.

Fixation error

Fixation error is most likely to influence results with near-threshold stimuli, when the subject's task is to detect them or to discriminate their critical detail. On tasks in which individual stimulus elements are otherwise easily detectable and the subject must apprehend and remember many of them at once in a brief exposure, the absolute retinal locus of an element may be relatively unimportant. Even in such a case, however, a fixation error may cause the observer to look at a position within a pattern, such as at one end, which would shift the reference point for his perceptual organization.

Evidence about the importance of fixation locus to hemifield differences comes from three kinds of studies: (1) inferential studies, in which the evidence must be deduced from the empirical data; (2) studies using controlled eye position, in which attempts are made to restrict possible errors in eye position before the stimulus is exposed; and (3) studies in which eye position is recorded, so that the locus of fixation is actually documented.

Inferential studies

In some studies the location of the visual pattern is systematically varied with respect to the fixation point, and effects of the different relative locations are examined. If the resulting fixation "errors" are larger than what would be reasonably expected to occur naturally, and no effect of fixation locus on relative performance is obtained, then the natural errors can be dismissed as a major cause of asymmetries in performance. A related line of inferential evidence establishes whether the distributions of errors parallel the gradients of retinal sensitivity, with fewest errors about the foveal regions.

Thirty-six eccentric positions. Harcum and Rabe (1958b) simulated gross fixation errors by presenting some of the linear stimulus patterns so eccentrically that all elements lay in the same direction from fixation. The centers of different linear arrangements of four open and four blackened circles sometimes registered with the fixation cross at the center of the visual field, and at other times they were eccentric along eight cardinal radiuses of the visual field. Regardless of the location of the target center, the array was inclined 0°, 45°, 90°, or 135° clockwise from vertical. Nine visual sectors were defined by the eight radially eccentric locations of the rotational center (36 eccentric patterns), plus that location which coincided with the fixation point (four central patterns). The presentation sequence for patterns and orientations was random. The subject attempted to reproduce both the inclination of the array and the binary pattern of elements.

Fig. 6-1 shows curves of errors per half-target that are nearly symmetrical in the north and south sectors, with maximums at about 0° and 180° orientation from the center of the sector. The greater maximum is at 0° for the north sector and at 180° for the south sector. The curves for the west and east sectors are virtually horizontal mirror images of one another, with maximums at 270° to the west, but at about 90° to the east. Therefore, for patterns along radiuses of the visual field, the elements away from fixation are less accurately reproduced. Because the curves in the west and east sectors cannot be made to correspond to those in the north and south sectors by allowing for differences in the loca-

Lateral dominance and responding order 147

tion of fixation, these curves are not explicable simply on the basis of relative distance from fixation.

Figure 6-2 represents distributions of errors among element positions for laterally presented patterns along the horizontal meridian plus the results for the horizontal orientation of bilateral targets. The curves for lateral arrays show minimums near fixation and decreases in errors at the most eccentric positions. These data imply that the observer was fixating properly. However, the means of errors for elements with central fixation are generally flat curves,

Fig. 6-1. Means of errors in reproducing four-element half-targets specified according to rotation angle from above center for each eccentric sector of the visual field. N, north; S, south; E, east; W, west. (After Harcum & Rabe, 1958b. Used with permission of the Division of Research Development and Administration, the University of Michigan.)

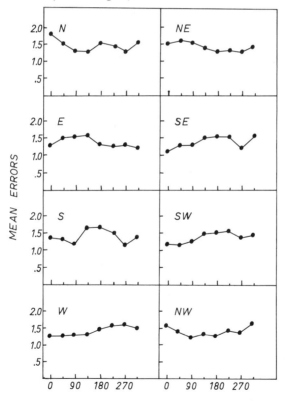

suggesting a broad span of attention. As the subjects did not know the location of the pattern before exposure, the difference between the lateral and bilateral exposures for elements at identical retinal locations had to be a function of stimulus differences and responding differences, rather than of differences with respect to fixation locus. The absence of lateral differences was caused by experimental conditions that presumably did not permit a consistent directional set.

Systematic horizontal displacement. Camp and Harcum (1964) also studied the reproduction of horizontal binary patterns when the patterns appeared at 11 different horizontal positions with respect to fixation. The results were rather insensitive to changes in the location of the pattern, except in one experiment in which the subject was forced to identify elements relative to the fixation point. Consequently, in the one case in which the fixation locus seemed crucial, the evidence suggested that the effect was mediated by a change in subject strategy, rather than by fixation locus per se.

Fig. 6-2. Percentage of errors at various element positions for binary patterns in lateral and bilateral exposure selected from 36 possible orientations. (Data from Harcum & Rabe, 1958b)

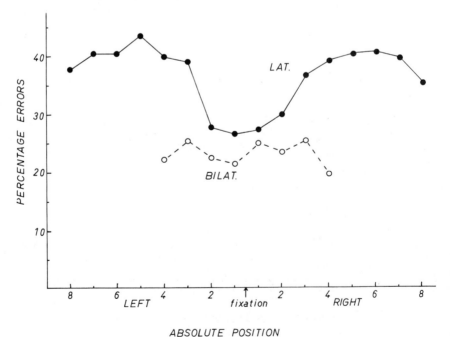

Retinal "sensitivity" curves. Other evidence indicates that in tachistoscopic pattern perception the distribution of errors among elements is determined by relative location of the elements within the pattern, rather than by absolute retinal locus. Such evidence has been obtained by Harcum (1964c,d, 1966a) and Harcum and Skrzypek (1965) for binary patterns, and by Crovitz and Schiffman (1965) for letter patterns.

Studies using eye-position control

Direct monitoring. LaGrone (1942) watched the observer's eyes by means of a mirror to ensure fixation; he still discovered hemifield differences in both lateral and bilateral exposures.

Forcing devices. Harcum, Hartman, and Smith (1963) employed a device to prevent the subject from activating the exposure mechanism of the tachistoscope until he was properly fixating. An exposure could be initiated only when a small faint light appeared for 1 sec at the fixation cross. In studies with both binary and letter patterns (Harcum & Smith, 1963, 1966; Harcum, 1964c), no difference in hemifield effects was found between such a forced-fixation condition and fixation maintained only by instructions.

A reduction in the hemifield difference with binary patterns using a fixation-forcing device was obtained by Ayres (1966). He argued that hemifield differences in binary pattern reproduction were produced by the combination of fixation error, responding order bias, and preexposure set, as attempts to control these factors eliminated the statistical evidence for lateral asymmetry. Harcum (1967a,c) challenged this conclusion with respect to fixation error on two grounds: (1) the subjects in Ayres's study were probably less reliable, being students in introductory psychology, so that there was less chance of demonstrating the significance of the empirical difference; and (2) a reduction of hemifield differences with a potential decrease in fixation errors does not mean that complete elimination of fixation errors would completely eliminate hemifield differences.

Strong instructions. Bryden and Rainey (1963) found superior perception to the left of fixation with single letters and outline objects presented bilaterally, but only when the subject was cautioned to fixate accurately before each exposure. If this left-field superiority was caused by fixation error, then the strong admonition to maintain accurate fixation should have reduced, rather than enhanced, the hemifield difference.

Studies with eye position recorded
In the course of investigating post exposure movements of the eyes, several studies have recorded eye position during bilateral tachistoscopic exposures (Bryden, 1960; Crovitz & Daves, 1962; Mandes, 1966). These studies yielded hemifield differences when the subject was accurately fixated at the initiation of exposure.

Correlation of fixation and performance. Further evidence that hemifield differences are not dependent on fixation errors has been obtained in an unpublished study from this laboratory. The purpose of the experiment was to deter the subject from shifting his fixation locus in anticipation of the exposure and to document the actual fixation locus. Each of 20 stimulus patterns consisted of eight typewritten versions of the capital letter O, of which four were blackened to make binary patterns. The visual angle between the centers of elements was approximately 1°, and the exposure duration was 0.1 sec. The eye movements were recorded on film by an ophthalmograph onto which the tachistoscope had been mounted. Trials to be filmed were selected haphazardly, with six to nine records made per subject. The subject's task was to reproduce the pattern on the answer sheets by blackening the circles that were filled. He was also told:
> When your head has been positioned so that the experimenter can correctly film your eye movement *and* so that you can comfortably view the field, you will concentrate your gaze on the X fixation cross before you. When you are in this position and are maintaining your fixation, you will say "Ready." Approximately 2 sec afterward, the pattern of circles will be flashed across the screen. You are to remain fixating until you hear a buzzer. Then mark your answer on the sheets provided.

The buzzer was delayed 5 sec. Figure 6-3 shows the results for six subjects. The ordinate gives the percentage of errors for 60 observations, and the abscissa represents element position from the left end. The arrow on the baseline identifies the location of the fixa-

Lateral dominance and responding order 151

tion marker, and the point of the triangle indicates the average measured locus of ocular fixation at initiation of exposure for that subject. Within each panel the two numbers to the left and right of the data curve represent the means of errors to left and right of fixation on only those exposures for which the film record was obtained. These numbers are consistent with the direction of the difference between hemifields for the data from all observations. The six curves for pattern reproduction illustrate the usual variations that occur on this task. The data in the two top panels show a small constant error of fixation to the left, but more errors on

Fig. 6-3. Percentage of errors at eight element positions of binary patterns with recorded locus of fixation for six subjects. See text.

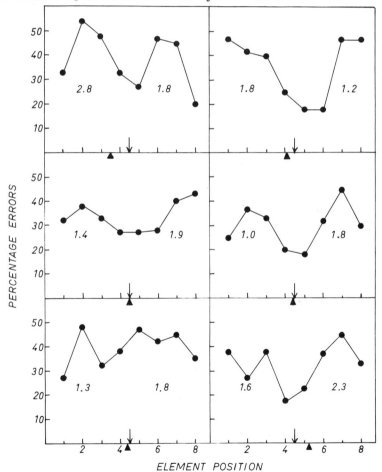

the left. The bottom right panel shows a small fixation bias to the right, and more errors on the right. Therefore, the results from the three subjects who show some fixation error indicate more errors on the potentially biased side. Moreover, performance is better for the element closer to the fixation marker than for the element closest to the actual locus of ocular fixation. Therefore, the advantage for the central elements accrues from their proximity to an instruction-emphasized fixation marker rather than nearness to ocular fixation. The remaining three subjects show little or no fixation error, but still show a bias in reproduction accuracy favoring the elements on the left.

The emphasis on accurate fixation undoubtedly increased the tendency for the subject to use the fixation marker as a reference point for the perception. Most of these subjects reveal effects of both end segregation and fixation saliency (Harcum, 1957a), inferred from dips in errors near the ends of the pattern and at fixation, respectively. Although there are gross individual differences in the shapes of the curves, these differences appear to be produced by subject-specific factors not closely related to fixation locus.

Response bias

Response biases are discussed under two general headings: order of report and stimulus reconstruction. The first topic considers whether a stimulus element reported later in a sequence of several elements is reproduced with less accuracy entirely because of the temporal delay. The second category refers to a possible cognitive operation whereby the subject infers the identity of unperceived elements from knowledge of what he has perceived because of redundancy within the stimulus pattern.

Order of report

Stimulus dimensions. A persistent problem in discussions of selective perception is whether the perception itself, or merely the report of it, is selective. No one doubts that responding order ordinarily affects performance. The question is whether the order of response is solely responsible for selectivity in perception. Lawrence and Coles (1954) have argued that response ordering does cause the apparent effect, because information to the subject about the order of report is equally effective before and after exposure, regardless of the discriminability of the response categories. Harris and Haber (1963), on the other hand, have demonstrated that perception itself can be selective. Two reviews of the literature on this subject (Haber, 1966; Egeth, 1967) have supported this conclusion, as has Neisser (1967) in his book.

Spatial arrays. The ordering-of-report accuracy for the various elements within multiple-element patterns does correspond to the observer's order of reporting the elements (Glanville & Dallenbach, 1929). Anderson and Crosland's (1933) subjects were instructed after exposure to report arrays of letters from left to right, from right to left, or from the center out alternating letters on the left and right. They found with bilateral exposure that not all the hemifield differences were eliminated by this counterbalancing of report orders and concluded that the perceptual process itself favored the letters on the left.

The Anderson and Crosland technique was used with tachistoscopic binary patterns by Ayres and Harcum (1962) and by Harcum, Hartman, and Smith (1963), with the same conclusion. Ayres (1966) subsequently challenged this conclusion, but his argument was rebutted in two papers by Harcum (1967c, 1968a). Freeburne and Goldman (1969) have recently redone the experiment with letter patterns and have concluded that perception of tachistoscopically presented patterns is selective, but is not consistently biased in direction. The latter conclusion is discussed later in this chapter.

Bryden (1960) also employed right-to-left and left-to-right orders of report for bilaterally exposed patterns of forms or letters. Although the results for the forms followed the order of report, the accuracy of reproduction was superior in the left hemifield for letters regardless of reporting order. Bryden concluded that there was a polarization of the perceptual process with the verbal material, which was consequently perceived from left to right and later reordered from right to left if that was required by the experimenter.

Partial report. Another technique to control for order of report, in which the subject is told after exposure to reproduce only part of the pattern, or some aspect of it, was devised by Sperling (1960).

He obtained evidence for a positional bias in letter perception, favoring elements to the left and above. Klemmer (1963) also obtained superior performance to the left using postexposure cues for the subject to recall the binary status of individual lights within a horizontal array.

Crovitz et al. (1966) investigated tachistoscopic perception of seven-letter arrays fixated at the first, last, or middle letter. In one experiment they required the subjects to report letters only, without regard to their location, and found results similar to those of total report – the typical serial-position curve. In another experiment, the subject was asked to locate a given letter after exposure. The effect of partial report was an increase in accuracy of performance about fixation, indicating the increased importance of fixation referenced localization. When the report required information about element location, the hemifield differences were eliminated with lateral exposure, but were still present with bilateral exposure.

Winnick and Bruder (1968) flashed groups of four letters simultaneously to the left and right of fixation and required the subjects to indicate whether a given letter had appeared at a given location. Element-position curves in each field showed superior performance for the leftmost element and maximum errors for the third element from the left. The direction of difference favored the right field, but the difference between pairs of letter positions was significant for the fourth eccentricity only. Probably Winnick and Bruder (1968) failed to find large lateral differences in bilateral exposure because they employed an interval between the two central letters twice as great as that between the remaining pairs of letters. Harcum and Skrzypek (1965) and Harcum (1968a) have concluded that such differential spacing between pairs increases the probability that the subject will treat the stimuli in left and right hemifield as different targets. Therefore, the results become more like those with lateral exposure, that is, improved relative performance for the stimuli on the right. Although Winnick and Bruder concluded that their results supported Ayres's (1966) conclusion that the evidence for the scanning mechanism is due to experimental artifacts, their data do in fact support the scanning interpretation. Even with a control for the order of report in their experiment, there was an empirical hemifield difference, which under the conditions of their study favored the right visual field – although it was statistically significant for only one element position. This is hardly negative evidence.

Recording of optional order. Harcum, Hartman, and Smith (1963), Harcum and Friedman (1963b), and Harcum (1968c) permitted subjects to reproduce binary patterns in any order and recorded the order actually used by each subject on each exposure. There was as usual a high correlation between accuracy of reproduction and responding order, but the correlation was far from perfect. It is, therefore, apparent that accuracy was not determined by order of report, although ordinarily both responses followed the observer's cognitive operation on the stimulus traces.

Stimulus reconstruction

A nonperceptual source of variance in tachistoscopic perception of meaningful words is the possibility that the subject can reconstruct the stimulus cognitively from partial perception, using his knowledge of redundancy in the language (Miller et al., 1954) and of relative word frequencies (Havens & Foote, 1963). The constraints of the language produce maximum uncertainty for the middle letters and corresponding minimums at the ends (Garner, 1962), but tachistoscopic accuracy is usually superior for the beginning letters, probably because the subject habitually reads from the left.

Broerse and Swaan (1966) required Dutch subjects to predict seven-letter Dutch words when two, three, or four of the initial or final letters were given them. The beginning letters were the more helpful to the subject in guessing the word. Part of the explanation for this was that the initial letters conveyed more information about the word (i.e., provided fewer alternative solutions). Broerse and Swaan argued also that the subject was able to use this greater information at the beginning more effectively, but their evidence is not convincing.

Horowitz and his coworkers (1968, 1969) have shown that a meaningful word is more easily recalled if the subject is subsequently shown fragments of it selected from, respectively, the beginning, end, and center. Because a greater importance of the initial letters in judging similarity is present even with preschool children (Marchbanks & Levin, 1965), Horowitz et al. (1968) doubted that it is conditioned by experience in learning to read. Possibly it results from an "initial reproducing tendency" (Müller & Pilzecker, 1900), which has been proposed as a general law of behavior.

To test the notion that the first part of a word yields the most cues to the identity of the word, Mishkin and Forgays (1952) presented eight-letter words bisected by fixation, with the four letters on the left or right partly blurred. The results revealed no difference for the case in which first (left) or second (right) halves of the word were obscured. Although this result indicated no effect of such a bias due to perception of initial letters, it is not convincing in view of later evidence from bilateral exposure, which typically shows superiority to the left of fixation. Such a tendency for left-field superiority may have compensated for a possible constructional bias in predicting the part of the word on the right from knowledge of the letters on the left.

Other studies of Mewhort (1966) and Harcum (1968a) with redundant letter patterns support the conclusion that the superior reproduction of these patterns is not wholly due to the possibility for cognitive reconstruction. Neisser (1967) concluded that the effect is due partly to a construction of a total percept from perceived fragments in a process he called "figural synthesis." This seems to be similar to the "redintegration" process of Horowitz et al. (1968).

Measurement biases

This section considers measurement biases caused by experimental procedure and handling of results. It does not include differential effects resulting from differences in task or from the particular aspect of performance the experimenter chooses to study. Also, this section does not discuss the problems of defining thresholds by the classic psychophysical techniques, as these problems are well known from analyses of subliminal perception (e.g., Pierce, 1963; Price, 1966; Neisser, 1967).

Absolute versus relative measurement

A general problem in assessing specific patterns of performance differences between groups is whether the data should be weighted according to the overall level of performance for the individual subject. This could occur when, for example, one is comparing the results of young and aged subjects, easy and difficult tasks, normal and abnormal subjects. Harcum (1968b, 1970a), discussing this problem in the context of differences between visual hemifields, pointed out that perceptual performance for two different experimental conditions may appear unequal when there is a gross difference in overall accuracy, whereas the same data are identical after performance is weighted according to the overall level of performance. The decision about whether there is a difference between conditions depends upon how one conceptualizes the data. For example, Glanzer (1966) found, with bilateral exposure of tachistoscopic patterns, apparently flatter element-position curves when the task was made easier in certain ways. Harcum (1968a, 1970a) explained this as merely smaller differences when the totals of errors were smaller.

Three measures of the degree of asymmetry in perceptual performance have been used in addition to the absolute differences (Harcum, 1970a): (1) the percentage of the total errors that occurred for the stimuli on one side of fixation, or equivalently a conversion of all raw data into logarithms; (2) the frequency with which subjects show superior perception for stimuli in a given hemifield; and (3) the rank order of the individual elements in terms of the relative accuracy of reproduction, a measure applicable to multiple-element patterns. When the location of the elements that are least perceptible becomes closer to one end of the total array, the distribution becomes more asymmetrical.

Harcum (1970a) concluded that the relative concept of differential performance was more fruitful than the absolute measure, although all ways of looking at data can potentially provide insights into the underlying mechanisms.

Measurement procedures

Thresholds. The most frequently used procedure for determining differential perception in the two visual hemifields is to present a given experimental condition and record the accuracy of report. This procedure avoids the problem of interpreting data in terms of threshold. A study by Wyke and Ettlinger (1961), for example, used a modified method of limits to obtain thresholds and found results that were the reverse of those usually obtained: lower thresholds in the right field for outline forms in bilateral exposure. Since the method of limits is particularly susceptible to effects of set and expectancy, the result may be due to the interaction of a selective set toward stimuli in the right field and the method of measurement. For example, Haber (1966) and Harcum (1964a) found that subjects who expect to see a stimulus again tend to withhold correct perceptual information. On the other hand, the subjects also gain partial information in prerecognition exposures that foster certain hypotheses about the stimulus, possibly aiding or even hindering later perceptual accuracy (Wyatt & Campbell, 1951; Havens & Foote, 1963; Harcum, 1964a).

Signal-detection methods. To avoid the above problems and obtain further evidence about the perceptual mechanism, some researchers now employ the methods of signal-detection theory (Tanner & Swets, 1954; Price, 1966). Differences in performance of selected clinical groups can be attributed to a difference in the criterion for positive reports used by the different groups, rather than to differences in sensitivity (Clark, 1966). Winnick and Bruder (1968) have shown, for stimuli presented bilaterally at various positions, that superior recognition in terms of probabilities of correct responses is paralleled by measures of both d' (detectability) and β (confidence). Derks et al. (1969) found that the subjects' tendency to make overt errors in reproducing binary patterns increases when the event proportion is greater than 50% and decreases when the event proportion is less than 50%. This result holds when the subject is required to mark open binary circles, as well as when he is told to mark the filled elements.

Introspective reports. Dallenbach (1920) employed introspective reporting to get at different levels of perceptual clarity for visual stimuli at various locations. Although cognitive clarity, which was reflected in reproduction accuracy, tended to follow attributive clarity, which was the sensory impact or attensity of the stimuli, these two levels of awareness did not always coincide, suggesting different spatial distributions of perceptual attention.

Haber and Hershenson (1965) have modified the method with a similar goal in mind. According to Hershenson (1969a,b), appropriate methodology could produce "perceptual reports" (Natsoulas, 1967, 1968), which are responses unambiguously related to given stimulus events. The perceptual report procedure attempts to eliminate postperceptual factors such as guessing (Haber & Hershenson, 1965). The observer examines the particular stimulus array before the experimental exposure and is instructed to report only those letters he actually saw. Hershenson (1969a) obtained perceptibility gradients for such known arrays that generally corresponded to visual acuity gradients and concluded that he had obtained perceptual reports of visual images. Harcum (1970b) doubted the conclusiveness of this evidence, since there were viable alternative explanations; for example, Harcum (1964c, 1965a,b, 1966a, 1967b) had attributed gradients of errors in reproduction primarily to the effects of an internal scanning mechanism. Any stimulus characteristic, or any state of the organism, that differentially draws the focus of attention to certain elements favors those elements by primacy in the perceptual sequence. According to this reasoning, an emphasis on fixation could cause the subjects to use the fixation marker as a frame of reference for mnemonic organization (Camp & Harcum, 1964; Harcum, 1964c, 1967b).

To answer the question of whether the perceptivity gradient varies with specific stimulus configurations, Camp (1960) obtained distributions of errors for three subjects in the reproduction of lateral tachistoscopic patterns of 10 binary elements. For each subject, these gradients appeared very similar to the results with lateral exposure given in Figure 2, that is, symmetrically increasing errors from fixation. Camp then attempted in two experiments to eliminate error gradients by adjusting the size and brightness of the individual elements for differences in "sensitivity," defined in terms of the relative absence of errors. He could not do so; although his failure does not prove that it is impossible to abolish such gradients, this was implied by the fact that the three subjects did not respond consistently to the experimental manipulations. The large dip in errors about fixation was in fact attenuated, but the gradients for different subjects were not similar as before. Therefore, the distributions of errors could not be related to a stable retinal sensitivity, but to some other factor, which Camp called *attensity*. In sum, when detectability of elements is not a problem, a perceptibility curve is determined not by the absolute retinal position of elements, but by the relative saliency of elements in an attentional process that responds to instructions, experience with the task, and certain physical factors that can "tune" the perceptual system.

The perceptual report procedure raises the question of validity of verbal reports. Results reported by Harcum (1970b) strongly suggest that the subject's expectations determined the distributions of errors in Hershenson's (1969b) experimental situation. Therefore, the delineation of the proper training and sophistication of the subjects remains a continuing problem for those who would use the perceptual report technique. This is particularly serious in view of the large individual differences on the tasks that are of most relevance here (Dallenbach, 1920; Harcum & Dyer, 1962; Harcum, 1964c; Kirssin & Harcum, 1967).

Complex tasks. In many psychological tasks, such as reading, speaking, or copying drawings, the subject can fail to achieve a given level of proficiency for several reasons. For example, Warrington, James, and Kinsbourne (1966) have shown that patients with cerebral lesions who fail to copy a given drawing correctly may have any one of several diagnostically different deficits. Therefore, it is important to establish measures for the various critical aspects of the total task. In addition, it is more fruitful to look for differences in mechanisms, rather than merely to establish the significance of differences between selected groups (Werner, 1937).

Method-specific correlations. The above point is important in the context of establishing the reality of clinical groups or types of subjects. As mentioned previously, Campbell and Fiske (1959) have pointed out the danger of concluding that a consistency within subjects has been found when in fact the correlation among scores is due to the commonality of method. Results of two different ways of measuring a given trait should correlate more highly than results of similar ways of measuring two different traits. This approach is particularly germane to efforts toward establishing either eye or cerebral hemisphere dominance from a battery of tests.

Negative data
In studies of variables possibly affecting hemifield differences it is often as important to establish that no difference exists as to establish the significance of an empirical difference. It is, of course, not possible to prove a negative, but it is possible by various means to provide convincing evidence for no effect (e.g., Harcum, 1969b). The problem is not solved on the basis of a nonsignificant statistic, contrary to the apparent assumption so frequently noted in this area of research. Often a negative conclusion is documented solely by the presentation of the nonsignificant statistic, with no argument or data addressed to the question of the variances within conditions. Particularly in the area of hemifield differences, the number of variables that can influence performance is enormous, resulting in such marked individual differences that many have proposed perceptual types (e.g., Ayres & Harcum, 1962; Harcum & Dyer, 1962).

Negative data can easily be generated by failing to control the important variables in order to produce a sufficiently small error variance. The use of a small number of subjects has the same effect. Although this fact is trite, it is often ignored in practice, so that some researchers attempt to refute theory and contradict other experimental results on the basis of positive data and nonsignificant probability values.

Loss of sensitivity in the measure has the same effect. For example, Fudin (1969b) concluded that hemifield differences in lateral exposure of letter arrays were not significant for most extreme eccentric locations of the stimuli, indicating that superior right-field performance was critically dependent on close proximity of the initial letters to the fixation point for the arrays in the right hemifield. This so-called negative result was used (Fudin, 1969a,b) in an attempt to discredit previous theories (Heron, 1957; Harcum & Finkel, 1963), but it probably represents no more than a failure of sensitivity in the measure. As the subject's task was to identify the correct position of six known letters, guessing if he was unsure, the probability of his being correct by chance should be on the order of one in six. For the six different test stimuli, the subject should get six letters correct by chance – three for each half-target. The subjects actually localized 3.86 letters correctly for each half-target at the most eccentric placement of the stimulus array, so the total accuracy was only about 15% above chance. Moreover, the ordering of the data points for this most eccentric stimulus location duplicated exactly the results for two other less eccentric locations and coincided with a third displacement except for a tiny inversion at one point. This situation illustrates and emphasizes the danger of attempting to draw definitive conclusions from so-called negative results, defined only in terms of failure to reject a null hypothesis.

To be able to conclude that an effect is not significant, it is important to examine the actual data for uncontrolled variance, for internal consistency, and for coherence with other results. In addition, there are measures of the degree of statistical association, such as ω^2 (Hays, 1965) and r_m (Friedman, 1968), which can help the researcher reach a conclusion about the importance of his results. For example, Freeburne and Goldman (1969) reached a negative conclusion about differences in intrinsic directionality of a "cogitive scanning" process on the basis of a nonsignificant F-test, despite many contrary prior conclusions (e.g., Anderson & Crosland,

1933; Bryden, 1960) and their own empirical result for practiced subjects, which was in the usual direction. Moreover, with the size of sample used in their study (Experiment II), according to the table provided by Friedman (1968), the difference between the means would have had to be 1.01 times the standard deviation of the scores within each group in order to achieve significance at the 5% level.

The above considerations apply also to guaranteeing positive results merely by testing a sufficiently large number of subjects. As Hays (1965) put it: "This kind of testmanship flourishes best when people pay too much attention to the significance test and too little to the degree of statistical association the finding represents" (p. 326).

LATERAL PRESENTATION

The lateral and bilateral methods for presenting tachistoscopic patterns usually produce different relative performance for the visual hemifields. Because in lateral presentation the primary projection of the stimuli is to one cerebral hemisphere, and in bilateral exposure stimuli are projected simultaneously to both hemispheres, the mechanisms in the former are probably less complicated. This relative complexity may be reversed, of course, if simple stimuli are presented bilaterally, and complex, multiple-element stimuli are used in the lateral exposure. The mutual relevance of stimuli to the right and left of fixation in a single exposure is also important. A totally dissimilar stimulus in the opposite hemifield does not affect performance (Heron, 1957), whereas a formally similar, but meaningfully unrelated, stimulus exerts a strong effect (Harcum, 1968a).

Sometimes experimenters have used both lateral and bilateral conditions in the same experiment, controlling for possible differences in preexposure set. Sometimes the limits of the difference between lateral and bilateral exposures have been deliberately tested by systematic variation from exposure to exposure of the relative numbers of stimulus components in each hemifield.

Task-specific variables

Task-specific variables in lateral exposure discussed here are stimulus type, number of elements in the stimulus configuration, orientation of stimulus elements, stimulus redundancy, hemiretinal differences, and type of response.

Type of stimulus

Verbal materials. LaGrone (1942) tachistoscopically presented, in lateral exposure, two capital letters, simple three-letter words, two words in a meaningful sequence, or simple three-word sentences. For the letters and single words the degree of eccentricity was varied in different exposures. The pairs of capital letters to the left were better recognized. The words and sentences were recognized with greater accuracy on the right, consistent with the usual result for this type of stimulus (e.g., Crosland, 1939; Heron, 1957).

Mishkin and Forgays (1952) found that English words in lateral exposure were reproduced more accurately on the right. For four-letter words the right-field superiority was not significantly greater at small eccentricities of the word center (about 1°) or for large eccentricities (about 5°), but it was significantly different for the intermediate eccentricities, in which the dynamic range of the task was obviously greater. As words presented above fixation in different exposures were reproduced more accurately than those below, some factor other than different degrees of training was responsible for the differential lateral effects. The appearance of the reverse lateral effect with Hebrew words for bilingual subjects further indicated that the effect was not due to a factor such as selective attention before exposure or to cerebral dominance. The results were attributed to differential training of the lateral visual systems specific to the normal direction of reading.

Later results of Orbach (1952) indicated that when bilingual subjects did not know in which language or on which side of fixation the stimulus would be, Hebrew words were recognized better on the left of fixation only when Hebrew had been the first-learned language. English words were always recognized more accurately to the right of fixation. More recent work (Orbach, 1967) casts doubt on the earlier findings with respect to Hebrew words, however. In his second study Orbach (1967) presented Hebrew and English words laterally to subjects whose first language was Hebrew, but who were also proficient in English. English words were recognized more accurately in the right field, as usual, and the differential was greater for right-handed subjects. The Hebrew words showed a slight right-field superiority for the right-handed subjects, but left-field superiority for the left-handed subjects. Thus, the lateral asymmetry was determined by directional scanning, cerebral dominance, and language structure of the stimulus words. To account for his failure to find completely reversed directions of lateral difference for the two languages, which are read in opposite directions, Orbach concluded that the English language is more "directional" than the Hebrew because its directionality is more influenced by context and less by intrinsic structure. Moreover, Hebrew readers frequently experience the left-to-right direction of reading with longer Arabic numerals and with written music. Therefore, their right-to-left directionality in the perceptual process is probably weaker than the English readers' left-to-right directionality.

The directional character of verbal stimuli appears to be enhanced by redundancy in the letter sequence. Dornbush and Winnick (1965) found significant right-field superiority with 0.15 sec lateral exposures of meaningful eight-letter words, but not with anagrams, when the subject did not know which class of stimuli to expect. The right-field superiority increased as the order of approximation to the English language increased, supporting a familiarity as opposed to a neurological basis for right-left differences in tachistoscopic perception. Crovitz and Schiffman (1965) also failed to find significant hemifield differences in lateral exposure of eight disconnected letters exposed for 0.1 sec. Bryden (1970) performed a similar study and obtained right-field superiority for first, as well as fourth, orders of approximation to English of four-letter arrays exposed for 0.03 sec. Although these results for nonwords seem inconsistent, possibly indicating critical differences in exposure duration or stimulus length, actually the difference is only a matter of conclusions based on statistical tests. Dornbush and Winnick (1965) and Crovitz and Schiffman (1965) did, in fact, obtain an empirial difference for nonsense letter arrays in the right field, but the differences were not statistically significant. As these researchers each employed only 20 subjects, compared with the 32 subjects used by Bryden (1970), the latter's significant effects would seem to carry more weight, particularly with the added support of other positive results (e.g., Heron, 1957).

Right-field superiority with lateral exposure of English words was also confirmed by Harcum and Jones (1962) and by Harcum and Finkel (1963). Not only was performance unequal for the two hemifields, but the shapes of the curves to the right and left were different also (Harcum and Jones, 1962; Harcum, 1966b). For normally oriented letters and sequences of letters, the curves on the right were more asymmetrical, with more errors away from fixation. This specific result, which was predicted from the argument for a more consistent direction for perceptual processing of those stimuli, is strong support for the explanation of lateral asymmetry as a temporospatial scanning process. The results of Crovitz and Schiffman (1965), using arrays of unrelated letters, duplicated these results in terms of the different shapes of the curves left and right of fixation.

Hirata and Osaka (1967) found superior recognition in the right visual field with lateral exposure of two-letter Japanese words. The letters were printed vertically, and the results were consistent with predictions based on the normal direction of reading from top to bottom. As a right-to-left reading habit is also relevant for Japanese writing, these results present problems for an interpretation in terms of a functional mechanism of horizontal perceptual scanning. To make such an argument hold, it seems necessary to assume in this context a functional correspondence between vertical scanning and horizontal scanning – with the appropriate transformation, of course. Although there is some empirical justification for the argument of equivalence between left-to-right and above-to-below scanning for English readers (e.g., Massa, 1968), it is not necessary to pursue it at this point, as not all instances of lateral asymmetry must be attributed to perceptual scanning (Harcum, 1967c).

Form stimuli. Heron (1957), using nonsense forms in one experment and familiar forms in another, tachistoscopically presented them singly in the left or right hemifield. The hemifield difference was not significant for either class of stimuli. Hirata and Osaka (1967) also found no significant hemifield differences with lateral exposure of single Japanese letters, random shapes, or Landolt rings.

Bisiach (1965) flashed one of eight meaningful figures to the right or left of fixation and asked the subject to report which figure had been presented. Before the exposure, the observers were told in which hemifield the stimulus would appear. Perceptual accuracy was greater for the stimuli to the left of fixation. This result is contrary to the findings of Wyke and Ettlinger (1961) and Bryden and Rainey (1963). In a repetition of the experiment Bisiach

(1966) used nonsense forms, but otherwise the same procedure, and found no lateral difference. This result is consistent with those of Heron (1957) and Bryden and Rainey (1963).

Binary patterns. Camp (1960), Camp and Harcum (1964), and Harcum (1958b) found no difference between hemifields in lateral exposure for binary patterns when the subject did not know beforehand the stimulus location with respect to fixation.

The results for horizontal patterns lateral and bilateral to fixation, from Harcum and Rabe's (1958b) study of tachistoscopically presented binary patterns binocularly at 36 orientations, are shown in Figure 6-2. The bilateral patterns were much more accurately reproduced than the corresponding eight elements from the two lateral targets. The functions for lateral targets ascended symmetrically from fixation, indicating no hemifield difference. These data are much the same as those obtained when the subject knew that only lateral targets would appear, but not on which side of fixation (Camp, 1960).

Harcum (1958c) presented eight-element binary patterns laterally along the eight cardinal radiuses of the visual field, or bilaterally across the four cardinal meridians, with four elements in each hemifield. Length of total pattern was 3°. The percentages of correct responses for horizontal patterns exposed for 0.2 sec are replotted in Figure 6-4, with the abscissa representing absolute position. The two similar curves for each location represent two groups of eight subjects, observing different counterbalanced sequences of 0.2-sec and 0.075-sec exposures. The lateral stimuli show a decided advantage for elements near fixation, similar to a sensitivity effect, but the most eccentric elements reveal obvious end-segregation effects – contrary to visual acuity gradients. Performance about fixation is not greatly different between central and lateral patterns, but the central fixation curve is again generally flatter, indicating that absolute retinal sensitivity is not responsible for the distribution of errors.

The curves in Figure 6-4 for bilateral patterns are virtually identical to those obtained by Hershenson (1969a) for letters unknown to the subject before exposure. The lateral exposures merely show an exaggerated effect at the fixation point for basically the same shape of curve. These results, including the bias toward superior performance on the left, are very similar to those Klemmer (1963) obtained with cued (partial) recall of patterns of lights. As his results did not change with varied angular separation of the lights, Klemmer concluded that a memory, rather than a perceptual, process was responsible. Harcum (1964b, 1967b) accounted for these results in terms of directional tendencies in a perceptual scanning process, in which the scans with binary patterns are more often from left to right for bilateral exposure and from fixation to periphery for lateral exposure. As the subject did not know which condition was to appear and could detect the end elements of even the eccentric patterns, the results for the bilateral patterns reflect a greater conflict of the tendencies for fixation primacy and left-field primacy.

Number of elements

Heron (1957) found that groups of letters in lateral exposure were more often correctly reported when they appeared to the right of

Fig. 6-4. Percentage of errors at various element positions for binary patterns in lateral and bilateral exposure, tested first and second. (Data from Harcum, 1958c)

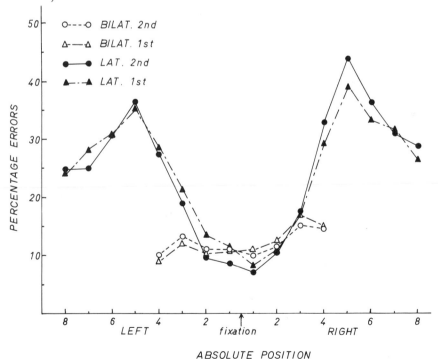

fixation. Single letters also sometimes exhibited a significant difference in accuracy of report favoring the right. Single forms in lateral exposure did not exhibit hemifield differences, a result supported by both Terrace (1959) and Bryden (1960).

To test the hypothesis that the usual right-field superiority was due to the greater proximity of the beginning (left) elements to the fixation point, Melville (1957) flashed three-letter and seven-letter words laterally. The beginning of a seven-letter word appearing to the left was further from the fixation point than the beginning of a three-letter word appearing to the left. Exposure duration was varied for different subjects and conditions, with the net result that a few more seven-letter than three-letter words were recognized. The degree of right-field superiority was greater for the seven-letter words, and only for them was the difference significant.

Bryden and Rainey (1963) obtained right-field superiority with lateral exposure of a single letter or familiar object. Geometrical forms were slightly more accurately reproduced on the right, but the difference was not significant. As these stimuli were rather large (about 5° of visual angle), the parts may have required a temporospatial sequence of perceptual analysis, producing the same results as for multiple-component patterns.

Orientation of elements
Harcum and Finkel (1963) and Harcum (1966b) succeeded, with lateral exposure of meaningful English words, in reversing the direction of hemifield differences by reversing certain directional characteristics of the letters. The results of Harcum's (1966b) study are shown in Figure 6-5. The subject did not know the stimulus condition or hemifield prior to exposure. Whereas the normal words were reproduced more accurately when they appeared in the right hemifield, the superior perception for mirrored words occurred in the left field. The results for sequence-reversed words were about the same as for normal words, except for the greater number of errors. Reversed letters tended to produce a greater so-called primacy effect at the right within the words, regardless of exposure field. Harcum concluded that for reversed letters the scan in the right-to-left direction for the stimuli in the left field conflicts with the scan in the meaningful direction – which would otherwise produce right-field superiority – eliminating hemifield differences.

Goodglass and Barton (1963), to test the notion of hemispheric effects on word perception, presented three-letter words aligned vertically in lateral exposure. Both left- and right-handed subjects recognized the words to the right of fixation more accurately, countering the argument for cerebral dominance as a determinant of the hemifield of greater accuracy. In a later study, however, Barton et al. (1965) found that right-handed subjects who were native Israelis perceived both Hebrew and English in the right field more accurately, lending support to the theory of hemisphere dominance.

In Bryden's (1970) study of four-letter words, of either first or fourth order of approximation to English, aligned vertically in left or right hemifields, all subjects were right-handed. Arrays of both approximations to English were more accurately perceived in the right field. The superior perception of more familiar sequences was not as marked as when the same stimuli were oriented horizontally.

Fig. 6-5. Means of errors per exposure in reproducing English words representing four conditions of stimulus directionality in lateral exposure. Curves refer to directional sequence and orientation of letters: BC, both correct; BR, both reversed; SR, only sequence reversed; LR, only letters reversed. (After Harcum, "Visual hemifield differences as conflicts in direction of reading," *Journal of Experimental Psychology*, 1966b, 72, 479–480. Copyright 1966 by the American Psychological Association. Reprinted by permission.)

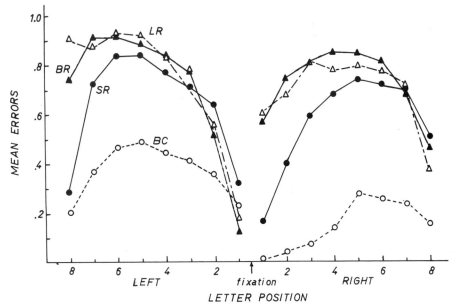

This indicated a true perceptual effect due to a more effective horizontal scanning, rather than a reconstruction artifact.

Nasal-temporal differences
Although Hirata and Osaka (1967) found consistent right-field superiority with lateral presentation of Japanese words, the asymmetry was greater with the left eye. The observers did not know to which eye the stimulus was exposed. The discrimination was always superior for Landolt C stimuli projected to the nasal·hemiretina, however. This contradicts the results of Markowitz and Weitzman (1969), who found consistently superior Landolt C perception by the temporal hemiretina. The only apparent procedural difference is that Markowitz and Weitzman told their subjects which eye was being used.

Markowitz and Weitzman (1969) found superior recognition in the right visual field for three-letter nonwords, but only for viewing with the left eye. This result with letter arrays supported the data of Overton and Wiener (1966), who horizontally presented single five-letter words monocularly at four different relations to fixation. The four positions incorporated lateral locations left and right of fixation, with one of the arrays in each field extending from fixation out about 1°, and the other extending distally from that point. Exposure duration was adjusted according to the observer's sensitivity. The subjects were required to guess the word after each exposure, aided by a list of stimulus words. Recognition was superior on the right, but this effect was restricted primarily to the left eye for the more eccentric stimuli.

Type of response

Estimation of number. In three experiments, Kimura (1966) found superior enumeration of items in the left hemifield with lateral exposure, unless the items were English letters; for letters she obtained the usual right-field superiority. She concluded that the right hemisphere of the brain mediates the enumeration of nonverbal stimuli, but "the left posterior part of the brain contributes critically to the identification of verbal-conceptual forms" (p. 283). There was no hemifield difference when the subjects were required to report the number of different forms. Bryden and Kinsbourne (personal communications) in several studies have failed to find hemifield differences in number estimation. Other data from this laboratory are generally negative, but inconclusive.

Reaction time. Jones and Jones (1947) tested reaction time in the correct naming of single letters presented laterally. Of four left-handed and three right-handed subjects, all but one produced slightly faster responses to the stimuli in the right field.

Van der Meer's (1959) subjects fixated a red point in a darkened room and were instructed to press keys with both hands as soon as they perceived a peripheral flash of light. The binocular reaction time was shorter for a light to the right of fixation. Van der Meer concluded that a right-directed tension in the dynamic visual field caused the subjective center of the field to shift to the right. Thus, the stimulus in the right field was presumably closer to the locus of attention. In monocular perception, the reaction time was faster in the nasal hemiretina. The difference in reaction time between nasal and temporal hemiretinas was greater for the right eye than for the left eye, presumably because the shorter nasal-area reaction time and the directional tension toward the right supplemented one another. For the left eye these tendencies apparently worked in opposite directions, with the nasal hemiretina facilitating reaction time but the right-directional tendency favoring the temporal hemiretina. Handedness and eyedness seemed to have little influence on the difference in reaction time between stimuli on the left and right.

Van der Meer (1959) predicted that with only one hand reacting the subjective center of the field would tend to shift in the direction of the performing hand. This assumption was supported by his data showing a facilitation of reaction time on the side of the hand with which the subject made his response.

Rutschmann (1966), using the right eye, obtained differences in the judged temporal appearance of stimuli at different locations on the retina. An eccentric stimulus had to appear before a foveal stimulus for the two to be judged as simultaneous, but this interval was smaller for the nasal than the temporal hemiretina. Thus, a stimulus in one hemifield may appear to the observer to arrive in the central nervous system before a simultaneous stimulus in the opposite hemifield, implying for these subjects possible dominance of the left side of the brain or the nasal hemiretina. Efron (1963) similarly concluded that the dominant cerebral hemisphere for language functions judges temporal asynchrony, that is, contains the "simultaneity center."

Hirata and Osaka (1967) also recorded the reaction time for perception of laterally exposed Japanese words, nonsense syllables, and Landolt rings. The measured response was a key press when the subject believed he made the correct perception. To increase reliability, Hirata and Osaka used trained subjects; the subject responded orally as soon as he could, coincident with the key press. Viewing was monocular, and the subject did not know which eye was used. There were no significant differences between hemifields or between viewing eyes, although a slight difference favored the word stimuli in the right hemifield for both eyes. A significant interaction between hemifield and viewing eye reflected faster reaction times for Landolt Cs in the temporal visual field (nasal hemiretina). For each type of stimulus there was a significant observer-eye interaction, but the triple interaction with hemifield was not significant.

Order of response. Heron (1957) reported that groups of letters arranged in a square, exposed to the left or right of fixation, were consistently reported in a sequence of upper left, upper right, lower left, and lower right, regardless of the hemifield in which they were exposed. The accuracy of the responses exactly followed the order of report. These square configurations appearing in the right visual hemifield were consistently reproduced more accurately than those in the left hemifield, however.

Bryden (1960) found, for groups of geometrical forms presented to the left or right of fixation, that the subjects consistently reported the elements from left to right. As performance was not different in the two hemifields, he concluded that the interaction of visual acuity and the tendency to report letters from left to right did not produce the right-field superiority for lateral presentation of letters.

Partial report. Crovitz et al. (1966) investigated errors of mislocalization and of omission in the tachistoscopic perception of seven-letter arrays fixated at the first, last, or middle letter, when the subject was required to report letters in the exact position within the array. For the lateral exposures, omission errors were less frequent for the end letters and fewer to left than right within the array regardless of location. Nevertheless, there was greater asymmetry for the curve on the right of fixation, with fewer errors on the right, overall. Mislocalization errors were less frequent and tended to show typical serial-position curves in each field. When Crovitz et al. merely required the subjects to report letters, without regard to location, results were similar to total report except that the letters close to fixation showed relative improvement. When another partial report method was used, in which the subject was asked after exposure to locate a given letter, there was greatest accuracy about fixation for the lateral arrays, with decreases at the most eccentric positions.

Subject-specific variables

Subject variables that affect performance with lateral presentation are discussed under the headings cerebral dominance, hemiretinal differences, instructional set, and cognitive style.

Cerebral dominance

Bryden (1964) reported a relation between the handedness of the subjects and the degree of right-field superiority with lateral exposure, but only for single letters and not for groups of letters. This is consistent with the data of Heron (1957), showing sometimes a semistable set toward the right for single letters, and a consistent right-field superiority for letter groups. Therefore, the effects of hemisphere dominance can be masked by effects of other stronger variables.

Bryden (1966b) investigated whether hemifield differences in lateral tachistoscopic exposure of single letters or three-letter patterns were produced by a directional scanning mechanism or by cerebral dominance. Both the single- and multiple-element stimuli were reproduced more accurately by right-handed subjects when presented in the right hemifield, but the differences were much greater for the multiple-element stimuli. Moreover, the correlation between the left-right differences on the two tasks was extremely low ($r = -0.04$ for the greater correlation). Thus, different processes seemed to underlie the single- and multiple-element tasks. Bryden (1966b) argued further that, if the scanning hypothesis was correct, there should be a negative correlation between hemifield differences for normal orientation and mirror images of the two types of targets. By the same logic, if the results were the same for normal and reversed orientations, then the cerebral dominance interpretation would be correct. Bryden flashed stimuli to the left or right of fixation for three groups of subjects. Each group saw one type of stimulus: left-right mirror images of single asymmetrical letters, normally oriented letters, or a mixed sequence of normal and reversed letters. Each group of subjects revealed superior accuracy for the letters on the right, but the differences tended not to be significant because of the large individual differences. For those subjects viewing the mixed stimuli, the results for normal and reversed stimuli correlated significantly ($r = +0.51$), indicating a stable effect such as would be consistent with cerebral hemisphere dominance. Therefore, these results support Bryden's (1966b) earlier conclusion that the processes with single letters are more stable in effect (i.e., subject-specific) than those with multiple-element stimuli, which can be interpreted as a result of directional scanning.

Zurif and Bryden (1969) presented single letters laterally in right and left visual fields to right- and left-handed subjects. Each left-handed observer was further identified as familial or nonfamilial (i.e., whether or not he had a close relative who was also left-handed). The stimuli on the right were more accurately reproduced by the right-handers and nonfamilial left-handers, whereas the familial left-handers were not consistent in laterality. These results underline the difficulty of specifying dominance in order to predict hemifield differences on different tasks.

Hemiretinal differences

Superior capability of one eye could produce hemifield differences because of differential performance by nasal and temporal hemiretinas. The results for various tachistoscopic tests of perceptual accuracy have been mixed. Overton and Wiener (1966) found greater accuracy on the right for left-eye viewing of more eccentric stimuli. Orbach (1967), on the other hand, did not find an effect of eyedness, measured by a sighting test, on the perception of English and Hebrew words laterally exposed to bilingual observers.

Data supporting the superiority of the nasal hemiretina for perceptual accuracy and for reaction time were reported by Hirata and Osaka (1967) for right-handed subjects with lateral exposure. For single Japanese letters and random shapes there was no significant lateral difference. The discrimination was more accurate for Landolt rings when they were presented to the nasal hemiretina. For groups of Japanese letters, however, perception was more accurate on the right of fixation. Overall, the scores were higher with the right eye, and the hemifield differences were greater with the left eye.

Hayashi and Bryden (1967) investigated lateral perception of single letters at two eccentric locations for subjects who were classified as left-eyed or right-eyed by a sighting test, but all of whom were right-handed. Binocular perception was superior on the right for both eye-dominance groups, the difference being greater for the more eccentric stimuli. For both groups of subjects the differences in hemifield recognition for single letters with binocular viewing were correlated with differences between the eyes in acuity tested with the Snellen chart. The hemifields were not different in performance for the subjects with superior left-eye acuity, but these differences were significant for subjects with superior right-eye acuity.

Instructional set

Mishkin and Forgays (1952) found that the superior perception of tachistoscopically presented English words to the right of fixation was greater when the subject was informed of the stimulus hemifield before exposure. This indicates a general tendency to attend to the left when the hemifield locus of the stimulus is unknown, which tends to counteract the effects of the scanning mechanism.

In several experiments, Heron (1957) presented simple forms or letters tachistoscopically to the left and right of fixation. Both nonsense and familiar forms were recognized about equally well to the right and left of fixation with bilateral exposure. However, four letters presented in the form of a square to the left or right of fixation yielded better perception on the right. For targets presented to the right of fixation, no effect resulted from giving the subjects prior knowledge of the side of fixation on which the target would appear. For targets on the left of fixation, recognition was better when the subjects were informed before presentation of the lateral location. This is also evidence for an attentional mechanism, but the direction of effect is opposite to that found by Mishkin and Forgays (1952).

Terrace (1959) and Bryden (1960) obtained the same direction of differential results as did Heron for letters and forms with lateral exposure, even though their subjects did not know the class of stimulus before exposure. Therefore, prior information about the type of stimulus is not critical for differential results from those stimuli. With lateral exposure of either four unrelated letters or four digits, White (1969a) did not find significant hemifield differences. As he pointed out, however, the random use of both lateral and bilateral arrays may have caused the subjects to alter their directional scanning.

Cognitive style

Cognitive style designates any relatively stable strategy of perception by the subject. The mechanism is similar to an attentional set except that it is less easily manipulated by the experimenter. The causal mechanism may be a lowered criterion for detection of elements at certain locations or a predisposition toward certain temporospatial sequences of information analysis. The more stable the inferred strategy, the greater the likelihood that it has a structural base.

Reading skill. Many studies have related reading ability and lateral differences in perceptual performance. Using elementary school children classified as superior or inferior readers, Crosland (1939) tachistoscopically presented nine-letter arrays to the left or right of fixation or bisected by fixation. For each location of the array, the inferior readers reproduced the letters on the right within the array more accurately than did the superior readers. This result suggests that the poorer readers had greater tendencies to scan the stimuli from right to left. The results were not affected by a change to monocular viewing with either eye. On a similar task, Crosland found that right-eyed college students performed better than left-eyed students on target elements to the left of fixation and vice versa. Thus, the right-eyed college students differed from the left-eyed college students the same way the superior elementary school readers differed from the inferior readers. Crosland's (1939) data showed little difference between left and right hemifields in lateral tachistoscopic exposure of letter arrays for inferior readers, although there was a slight right-field superiority. These readers tended to be left-eye-dominant. The superior readers yielded a larger difference favoring the letter arrays on the right.

LaGrone (1942) flashed capital letters, words, or short sentences in lateral exposure to college students. His findings generally corroborated those of Crosland (1939) in that superior readers (top quartile) tended to show superior perception relative to inferior readers for stimuli to the left within the array, whereas the inferior readers (bottom quartile) perceived better than superior readers on the right within the array. This discrepancy of results for best readers and other readers creates problems for hypotheses that attempt to explain hemifield differences in terms of eye movements in reading. In fact, LaGrone found no substantial relationship for any of his tachistoscopic tests, with bilateral as well as lateral exposure, to the number of regressive eye movements in reading. Therefore, as Tinker (1946) has concluded, a perceptual sequence is more important than the sequence of eye movements.

In an experiment by Forgays (1953), significantly superior perception for words to the right of fixation in lateral exposure did not appear until the children had reached the seventh grade. He concluded that the appearance of the differential effect is due to selective training through experience with reading, rather than to maturation.

Discussion

The operations described in the section on lateral exposure do not seem to be converging toward a single perceptual process, task-specific or subject-specific. If there is such a unitary process, it must be several steps removed from the operations thus far employed.

A typical situation occurs in the comparison of the results of Overton and Wiener (1966) and Bryden (1964). Although Overton and Wiener (1966) found a greater differential favoring recognition of laterally exposed words in the right field for observations with the left eye, for the more eccentric stimuli only, there were no differential effects of the subject's handedness or eyedness. Bryden (1964), however, found a relation between the handedness of the subject and the degree of right-field superiority with lateral exposure. He concluded, however, on the basis of the different results for the different types of stimuli, that sometimes "highly learned directional reading habits (cf. Heron, 1957) seem to override the effect of cerebral dominance" (p. 686). This argument is the same as that proposed by Harcum and Dyer (1962) for binary patterns in bisected exposure to apply to both cerebral dominance and eye dominance.

The best theory to account for the asymmetry in performance with lateral exposure of multiple-element stimuli is the one suggested by Heron (1957). This theory proposes a postexposure process after tachistoscopic exposure, which in effect reads the traces of stimulation in a perceptual sequence corresponding to the one that would be taken overtly by the eyes if the exposure were longer. Particularly if the visual stimuli are verbal material, the eye movements are inclined both toward the left, to pick up the beginning of the sequence, and toward the right, to execute the usual directionality of reading. For stimuli to the left of fixation, these directional tendencies conflict, degrading performance. For stimuli on the right, there is no such conflict between directional tendencies for reading, so perception on the right is superior to that on the left.

Although Fudin (1969a, b) challenged Heron's (1957) theory, as mentioned previously, his data do not support his argument. Fudin argued that with lateral exposure of letter arrays in the left visual field the attentional mechanism skips to the leftmost letter before the left-to-right scanning is initiated. Therefore, according to Fudin (1969a, b), information about the stimulus is gained only in the left-to-right scanning, eliminating the possibility of the conflict between two scans of different directionality. This alternative theory predicts that the stimuli in the left half of an array in the left field would be reproduced with greater accuracy than those in the right half of that array, probably more so than the usual results with letter recognition have shown (Harcum & Jones, 1962; Harcum & Finkel, 1963; Harcum, 1966b). Fudin's (1969a) data reveal the reverse of this result for each of four degrees of eccentricity of the arrays, probably because his task involves letter localization rather than letter recognition, as his subjects knew which letters would appear in each exposure, but not their location. An added problem in the interpretation of his data is that he introduced the task to the subjects as a test of visual acuity. Harcum (1970b) found evidence that subjects shown arrays of letters that are known before exposure expect to see those letters about fixation. This tendency is greater when more letters appear in one visual hemifield than the other, with one end of the array consequently closer to the fixation point.

BILATERAL PRESENTATION

With bilateral presentation the stimulus elements appear simultaneously to the left and right of fixation in a given exposure, and the task ordinarily incorporates two stimulus elements, or a single stimulus with separable components. Consequently, order and sequence effects in the perception and in the response should be more important than with lateral presentation.

Task-specific variables

The classes of task-specific variables discussed are type of stimulus element, stimulus directionality, stimulus configuration, number of elements, determinants of difficulty, mutual relevance of stimulus elements, stimulus locations, and response factors. In bilateral exposure, as in lateral exposure, several of these variables have often been manipulated in the same study.

Type of stimulus element

Letters. For tachistoscopically exposed bilateral groups of letters, Wagner (1918) obtained most errors about the middle of the pattern and more errors on the right than on the left. This distribution of errors cannot have been caused by differential retinal sensitivity, because greatest sensitivity occurs about the central fovea (e.g., Wertheim, 1894). The asymmetry in the function was attributed by Wagner to the subjects' scanning habits from experience with the left-to-right direction of reading. Anderson and Crosland (1933) confirmed these results and supported this conclusion.

Glanville and Dallenbach (1929) found that the subjects were most likely to perceive accurately the letters to the left in the upper of two rows of letters. In several experiments, Heron (1957) tachistoscopically presented simple forms or letters in bilateral exposure, finding no difference in recognition accuracy for nonsense or familiar forms to the right or left of fixation, but more accurate perception of letter groups on the left. Heron's (1957) explanation for these results was consistent with his treatment of the results for lateral exposure. The subjects were scanning persisting traces of the letter stimuli in a postexposure process that paralleled the usual sequence of reading. Therefore, for groups of letters in bilateral exposure, the perceptual process would include first a scan toward the traces of the left end of the pattern and then a scan toward the right visual field. Consequently, a primacy effect would favor the letters in the left field. As the forms presumably do not evoke this consistent directional scanning from left to right, primacy effects in perception would not consistently favor one or the other hemifield.

Typewritten binary patterns. Harcum (1957b) used tachistoscopically exposed binary patterns because they produce easily quantifiable results and probably require a perception of spatial relationships similar to that for conventional letter or digit spans. Such patterns do not, however, involve the meaningfulness and directionality of stimulus elements that are intrinsic to letters or digits. Because the stimuli are composed of multiple elements, the subject must reproduce them in some order, making the choice of order an indication of his information-processing strategy. Because reading is fastest for material in accustomed orientations (Chen and Carr, 1926; Aulhorn, 1948; Tinker, 1955), the prediction for the first experiments (Harcum, 1957b) was that the accuracy in reproducing linear binary patterns would be greatest for stimuli along the horizontal visual meridian.

The stimuli were nine zeros typewritten in straight rows, in which different binary patterns were created by variously filling in four of the zeros with the typewritten number sign. In one experiment, nine naive subjects were informed of the target orientation prior to 0.2-sec stimulus exposures. In a second experiment, the exposure duration was 0.25 sec, and four different observers were not informed of the stimulus orientation before presentation.

The differences in errors among the four meridians were small because the frequency of errors on radiuses to the right of fixation systematically exceeded the number of errors on the opposite radiuses. The particular radiuses on which most and fewest errors occurred varied among the subjects. Means of errors per element position for the targets of known orientation are shown in Figure 6-6. The marked reduction in errors at the most eccentric element positions was called the end-segregation effect because these end elements separate figure (i.e., stimulus pattern) from ground. This effect differs from the loss in accuracy that would be expected in the peripheral retina on the basis of retinal sensitivity as conventionally measured. It is apparently the same phenomenon as the masking effect discussed by Woodworth (1938, p. 720), attributed to a mutual interference of the stimulus traces of individual letters, with traces of the end letters lacking interference from one side. Such lateral inhibition effects are well known, behaviorally as well as neurologically. For example, Berger (1944) found that the horizontal space between letters had to be increased by 10% to make the discriminability of letters in five-letter arrays equal to those in two-letter combinations. The end-segregation effect also appears for laterally exposed stimuli, but for the elements adjacent to fixation it is confounded with a possible retinal sensitivity gradient.

These results for binary patterns duplicated those for letters (e.g., Wagner, 1918). The results with varied orientations of the arrays support the argument for a perceptual scanning process related to the subjects' habitual direction of reading. There was always some orientation of an array that failed to demonstrate a difference in reproduction accuracy between element groups on opposite sides of fixation, indicating an ambiguity in the inferred directional tendencies. Therefore, the degree of hemifield asym-

Fig. 6-6. Percentage of errors at nine element positions for binary patterns at four known orientations. N, north; S, south; E, east; W, west. (After Harcum, 1957b. Used with permission of the Division of Research, Development and Administration, the University of Michigan.)

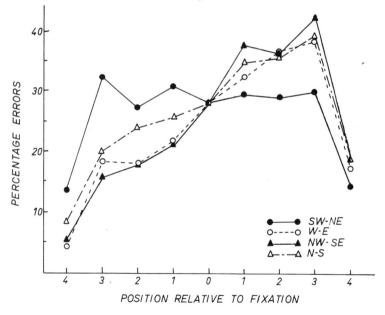

metry was attributed to the consistency with which the subject scanned in a given direction across the array. The so-called null meridian was the one that failed to show, for a subject under a given condition, a consistent primacy effect for one side, indicating no consistent directional scanning along that meridian.

Harcum (1958a) later repeated this comparison of known and unknown orientations for typewritten targets, using both naive and practiced subjects. As a consequence, the earlier results were corroborated and found to be more general. As all subjects were readers of English, the results confirmed expectations derived from an analysis of the directionality of reading.

Open and blackened circles. Because the subjective directionality of the stimuli seems to be influenced by whether they are cognized as verbal or nonverbal, possibly the typewritten binary patterns, which are actually nonverbal, are perceived as verbal by the subject because of an intrinsic directional aura about typewritten stimuli. Therefore, Harcum (1964d) performed the experiment with binary patterns, using various types of binary elements, as well as many different experimental conditions. The bilateral exposure often revealed superior reproduction of the binary elements to the left, which is the same outcome as that obtained with letter stimuli. On the other hand, there were also variations of results that seemed to depend upon rather subtle differences in experimental conditions. These conditions seemed to exert their greatest effects through cognitive changes in the subject. Individual differences were large enough to lead Harcum (1957b, 1964b, d) to suggest perceptual types. An example from a study by Harcum and Rabe (1958a) is shown in Fig. 6-7. The stimuli in this study were tachistoscopically exposed arrays of eight circles, some of which were blackened to form distinctive patterns. Eight subjects reproduced the pattern of blackened circles. Three subjects exhibited more errors on those radiuses of the visual field to the east of fixation, whereas three others made more errors in the western hemifield, and two produced symmetrical distributions of errors east and west of fixation. These general classifications of subjects appear frequently in such studies, and they seem to represent something more than normal variability.

Mandes (1966) also investigated lateral differences in perception of binary patterns of eight open and filled circles, using as subjects both children and adults. The patterns were oriented either horizontally or vertically. Adults made significantly fewer errors on binary patterns to the north and west relative to south and east. The children also showed slightly better performance to the north than to the south and better performance to the east than west, but neither difference was significant. Direction of eye movement after the exposure was positively correlated with the hemifield of greater accuracy. The result for the horizontal arrays with children as subjects was in the same direction as the significant difference obtained by Dyer and Harcum (1961) with six-element targets. Two important differences between the experiments could account for Mandes's failure to find significant differences for the children: first, he used eight-element patterns, and second – and probably critical – he employed a differentially larger space between the two central elements than between the remaining adjacent pairs of elements. Harcum (1968a) and Harcum and Skrzypek (1965) have shown that hemifield differences are reduced when the observer has reason to conceptualize the stimuli in opposite hemifields as not part of the same pattern.

Fig. 6-7. Means of errors for four-element half-targets on various radiuses of the visual field for three selected groups of subjects. N, north; S, south; E, east; W, west. (After Harcum & Rabe, 1958a. Used with permission of the Division of Research, Development and Administration, the University of Michigan.)

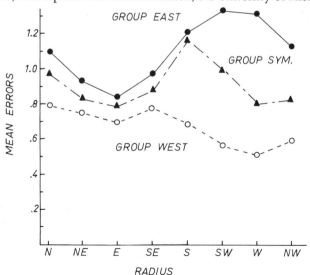

Cells in matrixes. Corroborating results with similar binary patterns were obtained by Massa (1968), who tachistoscopically exposed 1 X 6 matrixes containing selected patterns of blackened cells. These arrays were oriented horizontally or vertically in the visual field, with fixation always at the center of the line between the two central cells. When the subjects knew the orientation before exposure, equivalent performance was determined for the cells at the west and at the north. The accuracy was greater at both ends of the array, but superior at north and west, consistent with previous results with other binary patterns (Harcum, 1957b).

To show generality, Massa compared the results from the visual task and a similar kinesthetic tactile task. He moved six fingers of the subject simultaneously in a special apparatus, using the same pattern of stimulation as the blackened squares in the 1 X 6 visual matrixes. The patterns of errors for the visual and the kinesthetic tasks were generally similar, but there were differences also. For the tactile stimulation, there were about equivalent numbers of errors for each hand and fewer errors for the two outside fingers on each hand (the end-segregation effect). This facilitative effect at the center of the stimulus pattern was not seen in the visual task, apparently because of a smaller degree of cognitive separation between the visual hemifields than the discriminability between hands. Other visual studies that have used a differentially larger space between the elements about fixation in bilateral exposure (e.g., Harcum & Skrzypek, 1965) have shown a dip in errors for the central elements (i.e., an isolation effect). In Massa's study, the kinesthetic errors were primarily due to mislocalizations, as indicated by spatial generalization gradients around the correct finger.

Binary letters. If the differential in errors favoring primacy for the elements in the left visual field is due to the reading experience of the subjects, then a verbal target would be more likely to produce such a difference. To test this possibility, Harcum (1958e) employed binary patterns identical to those used in an earlier study (Harcum & Rabe, 1958a), except that the letters H and O replaced blackened and open circles, respectively. The subject was required to reproduce the pattern of Hs. The targets were exposed binocularly across fixation for 0.075 sec, at horizontal, vertical, or diagonal orientations, without the subject's knowledge of orientation.

Means of errors for the various elements again showed decreases in errors near fixation and at both ends of the target. Differences between opposite hemifields were minimal, however. Apparently the task for the subjects was not to discriminate H and O as letters, but to discriminate between two binary forms. Their subjective reports were of perceiving a "target" and a "null" form, rather than "letters." The fact that all subjects had received practice on similar tasks probably also contributed to the symmetry of performance (Harcum & Rabe, 1958a).

Therefore, the use of letter patterns does not guarantee that left-field superiority will be obtained. Harcum (1958e) concluded that the subject's cognition of the stimuli as letters is an important determinant of whether the stimuli will be perceived as a pattern having directionality.

Distinguishing features. In his study of lateral differences in children and adults, Mandes (1966) also used forms with distinguishing features. The distinguishing-features test was composed of outline visual forms identical except for one extremity, which provided critical details for form recognition. The critical details could appear at various locations in the visual field, depending on the orientation of the complete form. This test did not produce significant differences among up, down, right, or left orientations.

Two-dimensional display. Massa (1968) presented English letters in a 9 X 12 matrix, and the subject attempted to reproduce the letters without respect to location. Accuracy was greater for the letters to the west and north of fixation. Accuracy of performance was independent of whether 5 or 10 letters were presented, and the differences between so-called random and structured (i.e., spatially organized) arrays were small.

Chaiken et al. (1962) investigated binocular tachistoscopic perception of one different form in a 9 X 9 matrix, subtending about 5° and fixated at its center. For exposure durations of 10 to 200 msec, the form field was generally ovoid horizontally, with perception to the north superior to that to the south. The form fields were more extensive with longer exposures.

Crovitz and Friedman (1967) flashed eight letters binocularly on each of the eight cardinal radiuses of the visual field, equidistant from fixation in a circular pattern or in a smooth elliptical pattern with the horizontal eccentricities greater or less than the vertical eccentricities. Errors of reporting letters were fewest to the north and northwest, and greatest toward the south and southeast, regardless of the stimulus eccentricity pattern. The absence of differences between stimulus patterns indicates that absolute retinal sensitivity is not the determining factor in anisotrophy of the visual fields. The first letter reported tended to be to the west and north, with the second and third letters reported tending to be in a clockwise direction from the previously reported letter.

Mixed stimuli. Warrington, Kinsbourne, and James (1966) found that the span of apprehension for six-element arrays was significantly decreased when letters and digits were presented in mixed sequences – particularly when the number of changes within the series was greatest. For a single change of stimulus class, the span was greatest when the change was near the center of the array.

Dick and Mewhort (1967) mixed sequences of four letters and four numbers in tachistoscopic exposure, with the letters showing zero-order or fourth-order approximations to English. The letters were grouped together, alternated in pairs with numbers, or alternated singly with numbers; the initial item (i.e., the left one) was either a number or a letter. The subjects were asked to report letters or numbers first. Although the letters were strongly affected by the order of report, the numbers showed little such effect. The superior accuracy for fourth-order English was greater when letters were reported first. Grouping size for similar units did not affect reproduction accuracy for numbers, but the larger grouping produced greater accuracy for letters. The advantage of higher approximation to English was greater for the larger groupings and disappeared for the single alternation of items. The results were explained in terms of a greater requirement of spatial cues in the reproduction of letters. Numbers were not affected by delay in reporting. The differences between letters and numbers in relation to the order of report rule out the possibility that the usual relation between report order and accuracy is due to interaction of report sequence and image decay. The critical factor is interaction of perceptual scanning with image decay.

Stimulus directionality
Stimuli such as the letters of a language have intrinsic directionality because of the normal direction for reading that language. There-

fore, they have a correct orientation and a reversed orientation. Also, a directionality in a pattern of letters can be induced by their relative position in a series (i.e., approximation to the normal order in the languages).

Anderson (1946) randomly interspersed exposures of English letters or Hebrew characters forming nonwords bisected by fixation. He found that the bilingual subjects recognized more English letters to the left of fixation and more Hebrew letters to the right. Thus, the results seem to be task-specific. Anderson accounted for the superior perception on one side by the subjects' initial attention to those elements because of the general predisposition toward the usual direction of reading.

Harcum and Filion (1963) modified the usual superior tachistoscopic reproduction of letters to the left of fixation in bilateral exposure by reversing the right-left orientation of individual letters in English words, independently or combined with reversal of the letter sequence. The subjects did not know the condition prior to exposure. Harcum and Filion argued that if the orientation and sequence of letters in briefly exposed words affected relative reproduction for letters appearing to the right or left of fixation, then these right-left differences could be attributed to a functional concept, such as set or distribution of attention, rather than to a structural concept such as lateral dominance. Reversal of letter sequence alone had little effect on the usual superiority of the left hemifield, but reversal of the letters strongly affected these differences. For 7 of the 12 subjects in one experiment, there was in fact right-field superiority for the words with reversed letters.

In a later experiment Harcum (1964b) obtained results from the comparison of tachistoscopic asymmetrical and symmetrical letters that supported the implication of a directionality imposed on the perceptual scanning by directional characteristics of verbal stimuli. Figure 6-8 presents percentages of errors separately for the subjects who exhibited right-field superiority or left-field superiority – the subject-specific results. The curves for asymmetrical and symmetrical letters, respectively, show that the size of the hemifield difference was influenced by letter asymmetry – a task-specific variable. Presumably, the asymmetry of the letters contributed to the consistency of scanning direction for the individual subject. Bryden (1968) reproduced Harcum's result, but for only one critical spacing of the stimuli, and therefore rejected the scanning interpretation.

Stimulus configuration

Spacing of elements. Another variable that can affect the distribution of errors in the perceptual task is a change in the spacing between pairs of elements. Increasing the eccentricity of the elements at the ends of the patterns can change the relative performance for opposite hemifields by forcing the subject to attend selectively to certain segments of the target.

Harcum (1958c) investigated possible effects of the angular length of the total target upon relative numbers of errors in the right or left visual field. In one experiment, the target template was a linear array of eight open circles in which 38 individual patterns were formed by blackening certain circles. These patterns were identical to those used in an earlier study (Harcum & Rabe, 1958a), except that the space between elements was about three times larger, increasing the total subtense from about 5° to over 10°. The targets were presented binocularly for 0.075 sec, along unknown meridians, to both naive and practiced subjects. The re-

Fig. 6-8. Means of errors per exposure for asymmetrical (*A*) and symmetrical (*S*) letter patterns for subjects with left and right visual field superiority. (After Harcum, *American Journal of Psychology*, 1964b. Copyright 1964 by the Board of Trustees of the University of Illinois. Reprinted with permission.)

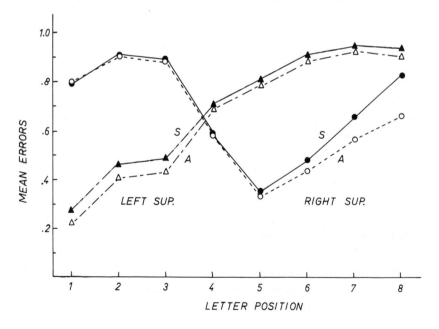

sults corroborated those of the previous studies in which the smaller interelement spacing was used.

Harcum (1958c) also studied the effects of decreased angular length of the pattern. The hypothesis was that shorter patterns were more likely to produce equality of errors to the left and right of fixation. Sometimes the fixation point registered with the center of the target (i.e., bilateral exposure), and sometimes all eight elements were eccentric along one radius of the visual field (i.e., lateral exposure). The four elements closest to fixation in an eccentric target registered with the four elements on that side of fixation in a centrally presented pattern. The results for 0.2-sec exposures are shown in Figure 6-4. These results with bilateral presentation for both exposure durations showed a gradual increase in errors starting at the west and north ends of the target and proceeding to the opposite ends (i.e., a primacy effect). There was no marked dip in errors near fixation and little end-segregation effect.

Harcum and Skrzypek (1965) found that increased spacing between the items of binary patterns did not greatly influence the distributions of errors, presumably because of the subjects' strong habits of organizing patterns in terms of the total stimulus configuration, with a left-to-right spatial ordering of the elements. The overall distributions of errors showed perceptual serial-position curves (i.e., fewer errors for the outer ends of all patterns and more errors on the right than on the left).

Consistent with Harcum and Skrzypek's (1965) work with binary patterns, Crovitz and Schiffman (1965) found relative position of items to be more important for perceptual accuracy than absolute retinal position of elements. Crovitz and Schiffman (1965) and Bryden (1966a) found decreased degrees of left-field superiority for tachistoscopic perception of eight-letter patterns with uniform increases in the spacing between letters. This result is not consistent with Harcum's (1964c) results for binary patterns. However, Kimura (1959) obtained left-field superiority in perception of square configurations of letters and forms at one eccentricity, but right-field superiority when element eccentricity was increased. This result certainly suggests a functional, rather than structural, basis for the hemifield differences.

Mewhort (1966) varied interelement spacing of letter patterns having different degrees of internal redundancy. He concluded that the main effect of uniform spacing between letters was not significant. The direction of an effect in his experiment was, however, consistent with that of other researchers for letter patterns (e.g., Bryden, 1966b). The increase in spacing hindered overall performance for the redundant stimuli, but not for the nonredundant stimuli. The subjects tended to report the leftmost letter first under all conditions, although less often for the more widely spaced words. Mewhort (1966) also found relatively improved performance on the right with the greater spacing. Thus, the larger spacing may decrease the probability that a subject will always begin the scan from the extreme left (Harcum, 1967b). The finding of both Mewhort (1966) and Bryden (1966a) that the frequency of left-first reporting was less with the wider spacing is consistent with this interpretation. Spacing did not influence reporting sequence, contrary to the result of Bryden (1966a). Bryden (1966a) suggested that the difference between his results and those of Harcum (1964c) may be attributed to the shorter exposure duration in his study, which may have hindered detection of the more eccentric letters with greater spacing.

Two rows of five letters were exposed by Massa (1968) at about the center of a matrix, with the letters in contiguous cells or with one-cell spacing between them. Results were equivalent for the two dispersions of the stimulus elements. For exposure durations of about 0.12 sec or more, there was no longer an advantage of knowledge of the specific locations at which the letters of the array would appear.

Michon (1964) presented to Dutch-speaking subjects a group of 12 letters of zero, one, or two degrees of approximation to the Dutch language for a total of 1.5 sec, either simultaneously or in temporospatial groups of two, three, four, or six letters. Performance was superior for the six-letter units, and the advantage of approximation to Dutch was greatest for the longer units. A consistent serial-position effect was evident in that the middle (embedded) units of the three-, four-, and six-unit arrays were less accurately reproduced.

Bryden et al. (1968) found with bilateral tachistoscopic exposure of six digits at two amounts of dispersion that the spacing did not influence overall performance, although the smaller spacing did produce greater left-field superiority. The consistency of left-to-right reporting was not affected by the spacing. When the subjects were told to report the letters from left to right or from right to left, lateral asymmetry followed the order of report when the reporting instructions were given before exposure. Instructions after exposure, however, revealed almost symmetrical functions of correct responses. Overall performance was favored by left-to-right report, particularly for subjects receiving preexposure instructions. Bryden et al. concluded that the perception of numbers is not as polarized left to right as is letter perception, but also it is not as responsive to order of report as the perception of forms.

Heron (1957) found that the left-field superiority, which resulted when horizontal groups of four letters were presented simultaneously to the left and right of fixation, was greater when the letters within the group of four were spaced farther apart. He suggested that the two groups of letters on the left and right of fixation were probably not as distinctly separated from one another with the greater interletter spacing; they were more likely to be perceived as a single array.

Monty (1968) showed that subjects used spatial cues as aids for retention in short-term memory. Retention was superior for meaningful forms when the spatial cues were provided, but superior for letters when less favorable spatial presentation methods were employed.

Massa (1968) investigated binocular tachistoscopic perception of patterns of blackened squares in a 9 × 12 matrix, horizontally subtending about 10°. The patterns consisted of 5 or 10 blackened squares, chosen to produce random or structured groups. The subject reproduced either the total pattern or only a part of it. Structured patterns were reproduced much more accurately. Moreover, performance improved if results were scored with respect to internal relations within a pattern, instead of in terms of absolute registration between response and location of stimulus element.

The effect of spacing is probably produced by changes in the subjects' strategy of observing as well as by changes in interelement inhibitory effects (masking). Changes in spacing alter the intrinsic organization of the array and, potentially, the tendency of the subjects to use certain elements as reference points. This inference is supported by the results of studies in which the perceptual saliency of elements is changed without changing the spacing of elements.

Isolation effects. Perceptual isolation of one item, as in the field of verbal learning (Harcum, 1969a), may decrease errors for that item, but with compensating increases for other items. This isolation effect in the tachistoscopic perception of binary patterns seems to be like that in serial learning (Harcum, 1965a,b, 1968c).

Harcum (1965a) tested the hypothesis that those mechanisms of serial learning which produce the characteristic bowed curve of errors also determine the distribution of errors among element positions in the perception of tachistoscopic patterns. As predicted, there was a relative decrease in errors for the isolated element.

A second study (Harcum, 1965b) tested the hypothesis that prior knowledge of isolation is critical for an isolation effect in tachistoscopic perception. Three groups of subjects reproduced 10-element binary patterns. Subjects in the "isolated" group saw vertical lines above and below the seventh element from the left, but subjects in the "unisolated" group observed no such lines. Isolated and unisolated patterns were mixed randomly for the "mixed" subjects, who were not informed before exposure which type of pattern was to be exposed. The results supported the conclusion that different distributions of errors are caused by different per-

Fig. 6-9. Means of errors per exposure at each element position for patterns isolated at the seventh position and for unisolated patterns. (After Harcum, 1965b)

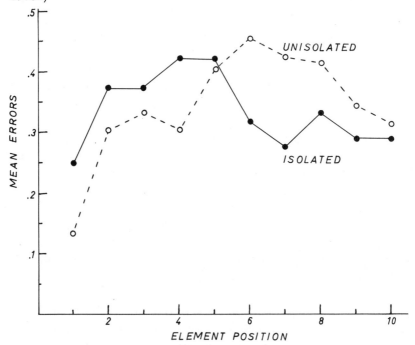

ceptual strategies, which are generally established by the subject prior to the exposure of the stimulus. Figure 6-9, an adaptation of the data from Harcum (1965b), illustrates the errors per exposure for each element position of a binary pattern isolated at the seventh position and one not so isolated. Although the totals of errors for the two conditions are not different, the distributions of errors differ markedly. An isolation effect was found for only those subjects who viewed isolated patterns exclusively and, therefore, knew before exposure that the seventh element would be isolated.

A later experiment (Harcum, 1968c) supplemented the earlier ones in which an isolation effect was obtained in perceptual serial-position curves when the subject had advance knowledge of the isolation. Contrary to the previous findings, however, the subjects seeing a mixed sequence of isolated and unisolated patterns also showed an effect of the isolation. Nevertheless, for these subjects the reproductions of unisolated and isolated patterns were not significantly different, supporting the earlier conclusion that a difference in perceptual strategy accounts for the differential result.

Harcum and Smith (1966) failed to find perceptual isolation effects when the subject had no prior knowledge of the isolation. Reference markers were placed within binary visual patterns in some exposures, and, for particular blocks of exposures, reference lines were reproduced on the responding templates. The perceptual curve never varied despite the manipulation of these situational variables. The absence of an isolation effect was attributed to the subject's critical lack of foreknowledge about the locus of isolation.

Bower (1965) obtained isolation effects by varying the utility (i.e., relative frequency of appearance of a target letter) at various locations within six-letter patterns. He interpreted the results in terms of selective filtering (Broadbent, 1958), however, rather than internal scanning (Harcum, 1965a,b).

Number of elements
Crosland (1931) concluded that left-field superiority in bilateral exposure was enhanced for longer series of letters. Heron (1957) also obtained this effect; single letters did not always produce a significant difference between hemifields, but groups of letters yielded fewer errors on the left.

Harcum (1958b) tested the hypothesis that hemifield differences are related to the length of the target in terms of the number of elements; that is, the more elements in the target, the more likely it is to extend beyond the subject's perceptual span. Assuming a set favoring the target elements to the left of fixation, the elements on the right of fixation will usually not be perceived. The test used 10-element targets and recording templates made by adding one circle on each end of the 8-element templates used in a previous study (Harcum & Rabe, 1958a). One of the added circles was also blackened. All the university student subjects had observed previously in experiments employing similar apparatus and procedure. They were not informed of the target meridian before exposure. For both 8- and 10-element targets the hemifield differences were significant. The absolute difference between hemifields was greater for the 10-element targets, but the problem of differences in totals of errors for 8- and 10-element targets makes one cautious about interpreting this result. All meridians except the vertical yielded, for both target lengths, progressive increases in errors from the western end of the target toward the east. The data on element position illustrate the fact that the sensitivity for target elements on the various meridians of the visual field is modified by psychological factors not ordinarily thought of as related to sensitivity. For example, the 8 central elements in the 10-element targets appeared at exactly the same location on the retina as the 8-element targets, but the numbers of errors were about double. Moreover, the decrease in errors for the former end elements, with 8-element targets, at the fourth positions from fixation almost disappeared for the 10-element targets. This end-segregation effect appeared at the fifth positions from fixation in the 10-element targets. Also, a relative decrease in errors near the fixation point was more pronounced for the 10-element targets. Therefore, the accuracy of reproducing an element at a given retinal position is not solely dependent upon the detection sensitivity or acuity at the retinal projection of that point, but rather depends upon complex information-assimilation factors determined by other visual stimuli within the visual field.

Harcum (1964c) tried to eliminate the usual end-to-end perceptual scanning of tachistoscopic patterns by using 17-element stimuli, expecting that the end elements would not be perceptible. As there was still generally poor accuracy of reproduction for the leftmost element, apparently the primacy effect still favored a perceptible left end of the visual target. On the other hand, 5-element patterns did not show left-right differences, regardless of the overall visual angle subtended. Either they were so short (i.e., easy to perceive) that sequential scanning was not necessary, or the scanning was not consistent in direction.

Acuity for single elements. Harcum (1958f) presented binocularly a single black or outline circle at one of four eccentric locations on one of the eight cardinal radiuses in different tachistoscopic exposures. An error was scored when the subject described the radius containing the target as other than the one on which the target in fact appeared. The 32 target positions and stimulus dimensions were almost identical to those in a previous study investigating binary pattern recognition (Harcum & Rabe, 1958a).

The results shown in Figure 6-10 for the outline circles support the hypothesis that detection localization accuracy for single elements is best, and equally good, along both the east and west radi-

Fig. 6-10. Percentage of errors in localizing the radius of single open circles at four eccentricities and eight radiuses of the visual field. N, north; S, south; E, east; W, west. (After Harcum, 1958f. Used with permission of the Division of Research, Development and Administration, the University of Michigan.)

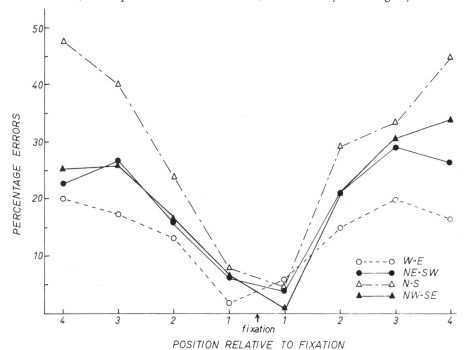

uses, and worst along the north and south radiuses. The filled targets, for which the data are not shown, apparently did not produce any problems of detectability, but nevertheless yielded slightly more errors for diagonal than for orthogonal radiuses (radial localization errors).

Determinants of difficulty

Delay of response. Glanzer (1966) reported results of tachistoscopic pattern perception in which the subjects responded with or without a distraction task interpolated between the exposure and the reproduction. When the subjects had to interpolate recitation of irrelevant material, the raw data revealed differences in total errors, but essentially parallel curves, showing more errors to the right of fixation. Harcum (1968b) explained the relatively less articulated curve after recitation in terms of a general interfering effect of the distractor task, for all elements equally.

Verbalization lengths. Glanzer (1966) also found that visual patterns of longer verbalization length, which were more difficult to reproduce, produced a larger absolute increase in errors for elements on the right. When these results were replotted in terms of the percentages of total errors, they did not show major changes in the shape of the curves (Harcum, 1968b). The perceptual difficulty at each element position increased by the same multiplicative factor – the same for the leftmost element as for the element that originally revealed the maximum of errors.

Exposure duration. Glanzer (1966) also varied the difficulty of the perceptual span task by manipulating exposure duration. Again the absolute difference on the right increased. Again the percentage plots failed to show a consistent change (Harcum, 1968b).

Harcum (1958c) found about the same hemifield difference for bilateral binary patterns exposed for 0.2 sec as for those exposed for 0.075 sec. This differs from Glanzer's (1966) result of a larger difference for shorter exposures. Presumably, as Ayres (1966) and Bryden (1966a) have suggested, the scanning mechanism has more opportunity to pick up the leftmost stimulus elements with the longer exposure, thus favoring the left in the perception and balancing the effect of overall task difficulty.

Mathewson et al. (1968) presented eight letters arranged either in a circle about the fixation point, or evenly spaced horizontally in bilateral exposure, at exposure durations of 0.02 and 0.12 sec. The purpose was to investigate Bryden's (1966a) suggestive finding that the distribution of errors may vary as a function of exposure interval. In the case of the linear arrays, the curves for both exposure durations revealed fewer errors about fixation and at the ends of the arrays, with fewer errors on the left than on the right of fixation. The superiority of performance with the longer exposures was greater for the letters to the left of fixation. Mathewson et al. (1968) concluded that the position-exposure interaction is not significant for the linear arrays, although they pointed out this trend in the data. As White (1969a) noted, the statistical evidence they report is, in fact, indicative of a significant effect. In the case of the circular arrays, the letters to the north were more accurately reported with the short exposures; the accuracy tended to improve more for these elements, too, but not significantly.

White (1969a) used four- and eight-element arrays, respectively, to test lateral and bilateral presentation in the same experiment. Some arrays were composed of capital letters and others consisted of digits. The stimuli were presented in blocks using different exposure durations, in the sequence of 25, 50, 75, and 100 msec, or the reverse of that sequence. Performance was assessed in terms of the number of elements reported correctly (identification) and the number of elements reported correctly in the proper relative position within the array (localization). The most interesting finding was, for the bilateral presentations, a significant interaction between exposure duration and element position in the localization measure. For the 25-msec exposures, performance was most accurate about fixation, monotonically decreasing, generally symmetrically, for more peripheral elements. As the exposure duration increased, however, accuracy progressively improved differentially on the left, with the elements on the right gaining relatively little from the increase in exposure time. Also, an end-segregation effect appeared. White interpreted the results as evidence for a left-to-right scanning that can start farther to the left with the increased time. The mean number of left-to-right orders of report also increased progressively with the longer exposures, consistent with the results of Bryden (1966a). Therefore, the perceptual strategy apparently changed as the exposure duration increased.

The above discussion implies that the effect of a change in exposure duration must be critically dependent upon other details of the experimental situation. If the subject can still scan from the left with the shorter exposure, the performance for the elements on the right will be further degraded, because the traces of those elements do not persist long enough to be processed by the perceiving mechanism. If, however, the shorter exposure hinders the initial scanning to the left, the usual bilateral scanning mechanism fails, and the prevailing scan to the right tends to facilitate the processing of stimuli in the right field.

Mutual relevance of stimuli
Of the many task-specific variables relevant to the information-translation process, one that might be especially important is redundancy of the information in the two hemifields. Specifically, if the stimulation from the two hemifields in bilateral exposure provides redundant information, then the subject may scan the stimulus configuration as he scans lateral exposures. The question is: Does redundant information in a cerebral hemisphere, presented by means of bilateral exposure, produce the same result as the absence of a primary projection to that hemisphere, as in the case of lateral exposure?

In a control experiment, Heron (1957) presented a horizontal black line on one side of fixation simultaneously with letters on the opposite side of fixation. The results of letter reproduction were the same as with lateral exposure of the letters alone. Thus, bilateral exposure of quite different types of stimuli did not invoke the "usual" mechanisms of bilateral exposure. Heron concluded there must be some initial perception of the types of stimuli, and then the stimuli are scanned as appropriate to their type.

Practice with binary patterns. One way to create redundancy in binary patterns is to employ a finite number of possible patterns and then permit the subject to become familiar with them. If the subject learns the population of stimulus patterns, his opportunity to predict the unperceived portion of any given pattern is thereby improved. This would be like perceiving some letters of a familiar word and predicting from them the rest of the word because of the redundancy in the language (Miller et al., 1954). Moreover, redundant patterns are apparently easier to encode in memory (Glanzer & Clark, 1963). Therefore, for redundant patterns, the accuracy of performance on both sides of fixation might be determined entirely by the capability of the superior side, producing equal accuracy to the right and left of fixation.

Harcum and Rabe (1958a) approached redundancy through familiarity with the stimuli. In one experiment, they selected randomly, from the 256 possible target patterns with an eight-element template, the particular pattern to be presented on any one exposure. Eight unpracticed subjects saw targets presented in random order across the four main meridians. Harcum and Rabe (1958a) also selected for use in a companion experiment 38 patterns from those containing only four filled elements, again with the eight-element template. These patterns were presented in random order along the four cardinal meridians of the visual field to four practiced and four unpracticed subjects.

For the unrestricted patterns (i.e., all 256 patterns as possible stimuli), the means of errors per radius of the visual field, when target inclination was unknown, showed maximums at north and south and minimums at east and at some point in the western half of the field. Corresponding results to the mean error data were found for errorless reproductions. A direct comparison of the results for four subjects who observed without knowledge of target orientation in both experiments – with the 38 selected patterns and with the unrestricted patterns – corroborates the comparisons of results with all subjects. For the selected targets, the results revealed fewer errors on the left for the early observations, but little difference between hemifields thereafter. Thus, as the overall perceptual accuracy improved, the hemifield difference disappeared, producing results like those for the unrestricted patterns. The task with the unrestricted patterns was probably easier because the 38 patterns included only those with four filled elements, which were intrinsically more difficult than patterns with more or fewer filled elements. It was not possible to conclude from these studies whether the effect of practice on lateral asymmetry with the selected targets operated through emerging redundancy or merely through a decrease in task difficulty.

Harcum, Filion, and Dyer (1962) did not find a gross change in the left-right differential when the binary pattern for each observation was chosen randomly from all possible patterns. This supports the explanation of Harcum and Rabe's (1958a) results in terms of the possibility that the subjects viewing selected targets were able to memorize particular patterns. Toward the end of the experiment, the subjects could have been able to perceive a few elements at one end of the learned target and thus predict the unseen elements more accurately.

Redundant binary patterns. In a preliminary experiment, for which only an abstract is published, Miller and Harcum (1963) used physically redundant binary patterns to test the hypothesis that the perceptual process in bilateral exposure employs redundancy to reduce errors for unperceived portions of the target. It was predicted that there would be left-field superiority with these patterns, but that the redundancy would reduce the size of the differences; that is, results would resemble those obtained with lateral exposure, which fail to show a left-right differential with such binary patterns.

The stimuli were horizontal binary patterns of open and filled circles, exposed for 0.05 sec. For 10-element bilateral patterns, the elements to the right and left of fixation were identical, permitting error-free reproduction of the total pattern if the elements *either* to the right or to the left of fixation were perceived accurately. The 5-element targets in lateral exposure were identical with the halves of the 10-element stimuli. The subjects were told that only one 5-element target would be presented in each exposure, but that in some of the exposures it would appear on both sides of fixation. They were not told before the exposure which stimulus condition would appear. The eight subjects serving under Condition 1 were instructed to mark on the recording template only the 5 elements on one side of fixation. For the 10-element patterns, each subject was to mark only one side, being permitted his option about which side he marked. The eight subjects under Condition 2 were told to mark the blackened circles on both sides of fixation after each exposure of a stimulus, even though only one 5-element target appeared in lateral exposure. The first 240 exposures for each subject included in haphazard order 160 observations of lateral exposure – 80 left and 80 right – and 80 bilateral exposures. In a final 60 observations in the experiment, unknown to the subjects, nonredundant 10-element patterns were substituted for the redundant patterns. If the subjects were making use of the redundancy, this scrambling of patterns should produce negative transfer, interfering with accurate perception, particularly on the side of fixation that had been less accurately perceived.

Both Conditions 1 and 2 exhibit in Fig. 6-11 fewer errors at the left with bilateral exposure, as usual. The lateral exposures with Condition 1 also show left-field superiority, contrary to the usual result. The results with bilateral exposure for Condition 1 cannot be safely interpreted because the subjects usually made the reproduction on the left of the responding template. Comparison of lateral and bilateral exposure for Condition 2 does not suggest differences in totals of errors, although left and right hemifields are more nearly equal for the lateral exposures. All subjects usually marked responses from left to right. More errors overall were made under Condition 2, probably because of the extra responding requirement.

The results for the last 20 observations under each condition before scrambling are shown in Fig. 6-12. Although these results

Fig. 6-11. Means of errors per exposure for redundant binary patterns in bilateral and lateral exposure with two different response requirements. (Data from Miller & Harcum, 1963)

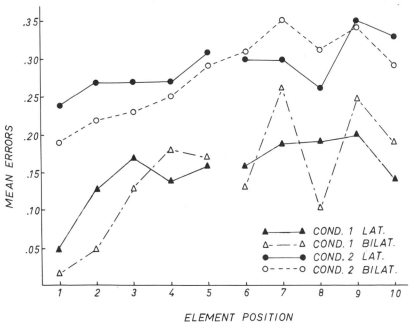

are included in the averages of Fig. 6-11, there are some apparent changes in the results due to practice. Left-field superiority continues only for Condition 2 with bilateral exposure. The apparent right-field superiority with bilateral exposure for Condition 1 is possibly an artifact of the subjects' more frequent choice to reproduce the elements to the left of fixation when they had the option of which side to reproduce, increasing the number of commission errors. Lateral exposure for Condition 1 reveals equality to left and right – the usual result, although the curve on the left ordinarily shows fewer errors for elements nearer fixation.

Fig. 6-13 shows the results after the bilateral patterns were scrambled. Only 1 subject out of the 16 detected the scrambling of the last exposures, and she did not respond differently from the others. Condition 1 shows a general increase in errors with bilateral exposure, as would be expected.[2] With lateral exposure for Condition 1, there is now right-field superiority; Condition 2 shows an increase in accuracy on the right with lateral exposure, but poorer perception on the right with bilateral exposure.

Fig. 6-12. Means of errors per exposure with redundant binary patterns after practice. (Data from Miller & Harcum, 1963)

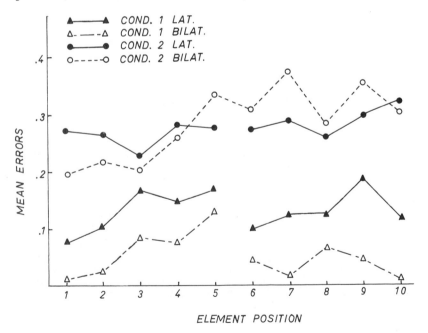

These data do not suggest simple interpretations. The finding of differences in errors on left and right with lateral exposure is contrary to results of previous studies employing nonredundant patterns (Camp & Harcum, 1964). Perhaps this is because there were fewer elements in these patterns than were used previously. It is not possible to determine whether the change to right-field superiority for both responding conditions after scrambling is due to a continuation of the trend as a result of practice, which was noted earlier, or to effects of the scrambling. The increase of errors to the right of fixation after scrambling for Condition 2 with bilateral exposure does indicate an effect of the previous redundancy. Thus, redundant bilateral exposures of binary patterns do not produce the same results as lateral exposures.

Although this study suggests new variables and problems for investigation, it presents further complications for the understanding of mechanisms. Not all the results were predictable from previous results, suggesting again task-specific variables probably interacting with subject-specific variables.

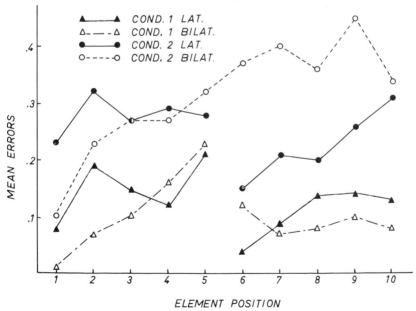

Fig. 6-13. Means of errors per exposure after scrambling of the redundant patterns. (Data from Miller & Harcum, 1963)

Compound words. Another unpublished experiment from this laboratory employed tachistoscopic displays of eight letters in bilateral exposure. Again one prediction was that for bilateral patterns redundancy in left and right visual hemifields would reduce the left-field superiority. A second prediction was that inclusion of a relatively large space between the central stimulus elements would decrease effects of hemifield redundancy because it would cause the observer to perceive the two sets of stimuli as unrelated.

Six stimulus conditions were produced by different meaningful relationships among four letters exposed in opposite hemifields for 0.05 sec. Four conditions employed words composed of two four-letter words, as follows: CW, compound words; WW, prefix words interchanged in order to form meaningless compound words; WN, compound words with the four letters on the right rearranged to form pronounceable nonwords; and NW, four letters on the left rearranged to form pronounceable nonwords. Conditions NN and 1W consisted of eight-letter nonwords and meaningful words, respectively, in which no word had a meaningful word of more than three letters included within it. For the eight subjects under the closed condition all spaces between letters were equal. For eight different subjects under the spaced condition, the space between fourth and fifth letters was increased, with all other letters becoming correspondingly more eccentric. The subjects did not know this stimulus condition before exposure.

Fig. 6-14 shows the means of errors per exposure under the various conditions. The most relevant comparisons are between Conditions WW and NN, in which there is no redundancy between the paired stimuli, and Conditions CW and 1W, which does include redundancy. In spite of the general tendency toward fewer errors for the letters to the left, for closed patterns, both Conditions CW and 1W tend to exhibit more nearly equal numbers of errors in left and right hemifields than Conditions WW and NN. This may be a spurious result, however, because fewer errors were produced under Conditions CW and 1W.

For all conditions the differential spacing seems to increase the relative efficiency of performance for stimuli in the right field, consistent with what would be expected from lateral exposure. Each spaced condition shows substantially more errors than the corresponding closed condition, probably because of the greater

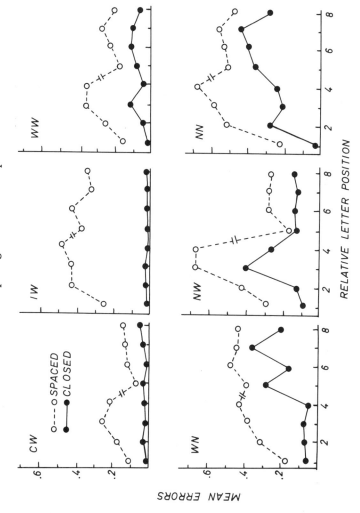

Fig. 6-14. Means of errors per exposure for letter patterns with different hemifield redundancies and two conditions of internal spacing. See text for explanation.

difficulty for the subject in adopting a perceptual strategy, not the greater retinal eccentricity of the elements (Crovitz & Schiffman, 1965; Harcum & Skrzypek, 1965).

The more meaningful stimulus of Conditions WN and NW produces more accurate perception, regardless of position. Again the spacing produces a greater disadvantage for the stimuli on the left.

As Heron (1957) and Harcum and Skrzypek (1965) concluded, the spacing of elements is important insofar as it contributes to the subjects' concept of a single stimulus configuration. Thus, both the differential spacing and the lack of redundancy, which presumably contribute to the perceived independence of the stimuli in the two hemifields, cause the mechanisms in bilateral exposure to approximate the mechanisms of lateral exposure. These results support the conclusion of Mewhort (1967) that part of the difficulty of reproducing nonwords lies in the difficulty of encoding them in memory. Also, greater spacing apparently makes encoding (chunking) more difficult (Mewhort, 1966).

This study with redundant letter patterns has been subsequently repeated in almost exact detail (Harcum, 1968b). The results were confirmed in all respects.

Approximations to English. With meaningful words, the constraints in both forward and backward directions would produce a symmetrical distribution of errors in tachistoscopic perception, with the maximum about the middle, although there usually is some asymmetry in the empirical results favoring the left-hand elements (Haslerud & Clark, 1957; Garner, 1962). This asymmetry is presumably caused by the left-to-right reading process (Garner, 1962). For nonwords, the distributions of errors are markedly asymmetrical (Harcum & Filion, 1963), presumably because the reading habit contributes relatively more to the final distribution of errors.

Mewhort (1966) varied redundancy and spacing of tachistoscopically exposed eight-letter arrays. He argued that increased spacing makes the sequential scanning of the words more difficult, decreasing the probability that the stimuli on the right can be coded into memory during the tachistoscopic exposure. Redundancy should facilitate scanning, perhaps because it facilitates the organization (chunking) of letter groups. Frequency of reports of the leftmost letter were not affected by word redundancy, but the subjects reported the redundant words from left to right more consistently. The increase in spacing hindered perception of the familiar words, but not the nonredundant words. Thus, the lateral differences are more similar for redundant and nonredundant stimuli at the greatest degree of spacing used. Actually, the performance gradients were essentially symmetrical from fixation with the largest spacing. The left-right differential is smaller for the redundant stimuli, but again the problem of large overall differences in errors confounds interpretation of this fact.

Stimulus locations
Harcum and Rabe (1958b) tachistoscopically presented eight-element binary patterns bilateral to fixation at four inclinations and at four inclinations through eight eccentric focuses. The subjects were unaware of stimulus position prior to exposure. Some of the results of the experiment are shown in Fig. 6-2. The results for the lateral targets generally corroborated the usual result when the subject knew that only lateral targets would appear, but not in which hemifield; the curves of errors were essentially symmetrical, rising equally from fixation in each hemifield. Although there were more errors for the far eccentric than for the near eccentric elements, the near eccentric elements produced substantially more errors than the central elements that occupied identical retinal positions. Thus, performance for stimuli at a given retinal area was affected by the coincident stimulation received by other areas, that is, by the relative location of the elements within the total stimulus configuration.

Harcum (1958c) presented eight-element binary patterns tachistoscopically along eight cardinal radiuses of the visual field, and also bilaterally with four elements on opposite radiuses. Stimulus orientation was not known before exposure. The patterns were exposed for 0.2 sec and for 0.075 sec, in counterbalanced design. The results for both exposure durations showed, as usual, fewer errors on the left with the bilateral patterns.

Camp and Harcum (1964) found that the element-position curve for 10 elements is asymmetrical even when the relative numbers of elements in each hemifield is systematically varied. The curve becomes symmetrical only when most of the elements appear in the right field.

Response factors

Reporting number of dots. LaGrone (1942) presented to college student subjects arrays of dots simultaneously in the left and right visual hemifields. There were two to five dots in either hemifield, with the total in both hemifields never more than six. The degree of eccentricity of the stimuli was also varied. The task for the subjects was to report the number of dots in each field. The stimuli in the left field were always more accurately reported.

Reporting number of filled circles. Harcum and Blackwell (1958) flashed linear binary configurations along four meridians of the visual field. Targets were arrays of eight open circles, with individual elements formed by blackening from zero to eight circles. Two such targets always appeared across a meridian such that each target fell entirely on one of the two opposite radiuses. A black rectangle separated the two targets, with the long axis at right angles to the target array. Each of the target configurations, which were otherwise presented in random order, appeared binocularly once along each meridian of the visual field in each of three experimental sessions, unknown to the subject prior to exposure. The subject was required merely to report the number of blackened elements appearing on each side of fixation. Each subject had served previously in experiments using similar apparatus, but in which he had been required to reproduce patterns rather than number.

The means of errors of number estimation for meridians averaged for all subjects are shown in Fig. 6-15. The upper curve in this figure represents errors averaged without regard to whether they are overestimations or underestimations. Worst performance occurred for vertical targets. There was little difference among the other three meridians, but again with a slight advantage for the horizontal. The stimuli to the east of fixation were generally more accurately described. The lower curve represents errors of number estimation averaged with regard to whether they are under- or overestimations. Here there was little lateral symmetry, but an underestimation in all regions except those below fixation.

The low number of errorless reproductions made in the experiment indicates that the subjects were in fact estimating the number of filled circles rather than counting or subitizing (Kaufman et al., 1949). It is impossible to estimate accurately what chance responses should be in the experiment because the number of filled elements in a total configuration was never more than 11 or fewer than 4. The number of cases in which errors per radius cancel to produce errorless meridians does not seem sufficiently large to implicate a visual mechanism for producing confusion about which of the two radiuses in a target configuration contained a given blackened element. Also, the no-error errorless reproductions could have been predicted from the empirical results for radiuses. Therefore, the meridional data presumably involve only the factors producing the errorless radiuses. Of course, the half-meridian responses were deliberately designed to be independent both perceptually and in the motor response.

As was true for pattern reproduction (e.g., Harcum & Rabe, 1958a), the subjects appeared to fall into three general categories. The proportions in each classification are somewhat different, but

Fig. 6-15. Means of errors of number estimation for filled binary elements on various radiuses of the visual field computed with and without regard to sign. N, north; S, south; E, east; W, west. (After Harcum & Blackwell, 1958)

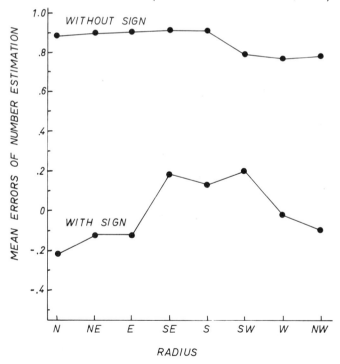

some subjects were difficult to classify. Most subjects made more errors east than west of fixation. Because only one subject clearly made more errors at west, the overall results reflect the responses of subjects who made more errors east of fixation. Thus, changing the relevant aspect of a visual pattern and changing the movements required to indicate the perception of that aspect apparently do not materially change the results.

Number reproduction for filled circles. Because Harcum and Rabe (1958a) asked their subjects to reproduce eight-element binary patterns, selected from all 256 possible patterns, the responses could also be scored for accuracy in the number of filled elements reproduced. The interpretation of the number-reproduction data is questionable because the subjects were asked to reproduce patterns and not to reproduce numbers directly. Fig. 6-16 summarizes

Fig. 6-16. Means of errors of number reproduction for filled binary elements on various radiuses of the visual field computed with and without regard to sign. N, north; S, south; E, east; W, west; (After Harcum & Rabe, 1958a. Used with permission of the Division of Research, Development and Administration, the University of Michigan.)

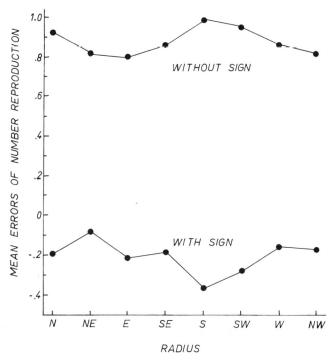

results. The upper curve represents the data without regard to sign, and the lower curve illustrates the results with sign. The errors without sign are maximal at north and south and minimal at east and northwest. The errors with sign show similar results, with consistent errors of underestimation, particularly for the vertical radiuses.

Bisection of horizontal distance. Takala (1951) found that a point of light adjusted to bisect a horizontal distance marked off by two other lights was generally set too far to the left, particularly under conditions of reduced information about the stimulus situation. Thus, the left half of the horizontal extent was overestimated. This result held for both viewing eyes. When a light was flashed in the dark and then flashed again after an interval, the second flash tended to be located to the left of the first one. Takala attributed the lateral asymmetry to the tendency for left-to-right scanning across the visual object.

Figure orientation. Other results of Takala (1951) indicated that perceptual organization of a figure more frequently involved an orientation toward the left. Also, the left half of an ambiguous drawing was more often perceived as a figure. Free-hand drawings of a ship approaching a pier in 74% of the cases showed the ship heading to the left, whereas a ship drawn to illustrate leaving a pier was pointed to the left only 37% of the time – a significant difference. Again, Takala (1951) concluded that the perceptual organization progresses "as if the attention were progressing from the left to the right and downwards from above" (p. 132).

Transposition errors. Koestler and Jenkins (1965) observed so-called transposition errors in tachistoscopic reproduction of sequences of five, six, or seven digits. In these errors, digits were correctly identified but interchanged in position. Most frequently the error consisted of interchanging the item at the most difficult position in the list with the one preceding it (i.e., to its left). The relative difficulty of perception for different positions, measured by the frequency of single omission errors, followed the typical serial-position curve for each length of series.

Derks et al. (1969) also found that mislocation errors in the reproduction of eight-element binary patterns followed the typical serial-position function, with fewer errors on the left.

Order of report. In Bryden's (1960) study, bilateral exposure of six letters yielded more accurate report for the letters in the left visual hemifield, regardless of the order of report used by the subject. The hemifield differences in the perception of six forms exposed across fixation were, on the other hand, determined by the sequence of reporting the stimuli; stimuli reported first were reproduced more accurately.

Harcum (1962) used a recognition procedure for eight-element binary patterns to reduce the importance of differential response delay for elements on the right. However, when the subject chose the exposed pattern from a group of four alternatives, the results were the same as when he reproduced the pattern without aid.

Partial report. Crovitz et al. (1966) tachistoscopically presented arrays of seven letters fixated at the first, fourth, or last letter and required the subject to report the exact position of the letters. Omission errors for bilateral exposure were less common at the ends, particularly on the left. Mislocalization errors showed the typical serial-position effect. When the subject reported only letters, without regard to position, perception improved for the letters close to fixation. When the subject had merely to locate a given letter after the exposure, for the bilateral stimulus, the end letters were most accurately reproduced; there was a marked dip in errors about fixation, and performance was slightly better on the right. These effects of response measure indicate that retinal sensitivity does not control the distribution of errors.

The data of White (1969a) for bilateral patterns of either letters or digits, interposed with similar lateral patterns, indicated best performance about fixation for both identification and localization of the elements when the exposure duration was less than 0.1 sec. For the localization measure subjects were required to reproduce not only the identity of the element, but also the location of that element within the pattern. Although the result was apparently not significant, the left-field superiority seemed to be somewhat greater for the localization task.

Fitzgerald and Marshall (1967) employed tachistoscopic presentation of letter arrays bisected by fixation. In one condition eight letters were evenly spaced, and in another condition a gap was inserted between the two central letters. In one of these conditions the subject was required to report only those stimuli on one side of fixation, as instructed after exposure. In a third condition, a group of four letters was bisected by fixation, and the subject attempted to report all letters. There were effects of both stimulus and response factors on hemifield differences. With full report the result, as usual, was superior perception to the left. With partial report, however, the stimuli to the right were more accurately reproduced, with the greater differential for the gapped arrays.

A partial report technique was employed by Massa (1968), using as stimuli blackened squares in a matrix. After exposure of a pattern, a pair of blackened squares was flashed, and the subject attempted to report whether those squares had been blackened in the pattern. With so-called structured patterns the performance improved somewhat with greater delay between stimulus pattern and comparison pairs, but with random patterns the performance dropped sharply after delay. Massa concluded that "the short-term memory contains an eidetic component which results in more similar performance for all patterns for very short delays" (p. 81). The random patterns decay faster than the structured patterns with interpolated delay of the mnemonic aid. Actually, for the structured patterns, the partial report remained higher than total report for at least 1 sec.

Signal detection. Winnick et al. (1967) used a signal-detection measure of letter recognizability within a six-letter tachistoscopic array cued after exposure. There was no significant effect of letter position, possibly because there was no serial-position cue in the response or because the response was partial rather than sequential.

Winnick and Bruder (1968) flashed four-letter arrays simultaneously to the left and right of fixation and required the subjects to indicate if a given letter had appeared at a given location. Numbers of errors in left and right hemifields were not significantly different, and element-position curves in each field showed superior performance for the leftmost element and maximum errors for the third element from the left. Performance was superior in the right field, but the difference between pairs of letter positions was significant for the fourth position only. These results are similar to those usually obtained with lateral presentation of stimuli. Harcum and Skrzypek (1965) and Harcum (1968a) have shown that a spatial separation between the pairs of elements next to fixation that is greater than the other interelement spaces, as was used by

Winnick and Bruder (1968), usually improves the relative performance for the stimuli on the right.

The effects of responding method on the distributions of different types of errors in perceiving binary patterns were studied by Derks et al. (1969). Regardless of the number of filled circles in an eight-circle template, the distribution of errors was asymmetrically bowed with more errors on the right, whether the subject was required to reproduce filled or unfilled elements only. Derks et al. concluded that both errors of perceived numerosity and errors in localizing elements were due to inaccuracies of stimulus registration and particular strategies of perceptual processing.

Immediate memory. Hirata and Osaka (1967) used an immediate memory technique, like that frequently used in dichotic listening, to investigate hemifield perception of numerals in monocular viewing. Three pairs of numbers were exposed in series, one pair at a time, for 80 msec with 80-msec intervals between slides. One member of each pair appeared in each hemifield. Superior reproduction occurred for numbers in the temporal visual field, but the result was significant only for the left eye. The modal recall sequence for the subjects began with a stimulus on the left and alternated left and right until the last number was reported. This sequence corresponds to the usual sequence in which one reads numbers.

Hines et al. (1969) investigated the asymmetry of the visual field with binocular viewing, also using an immediate memory task. Three pairs of digits were flashed, with one digit in each hemifield, and the subject was instructed to recall all items in any order. Both left-handed and right-handed subjects revealed superior perception for the left hemifield, but not significantly so, although significantly more subjects did reveal left-field superiority. The experiment was repeated with four pairs of digits and with monocular as well as binocular viewing. This time the left-field superiority was significant for both monocular and binocular viewing. In a third experiment three pairs of numerals were again presented, but this time one number appeared at fixation and the other in a given hemifield, and the subject was required to report all numbers at fixation before reporting the eccentric numbers. The first-reported eccentric numbers revealed a significant superiority in reproduction for the numbers in the right field. The numbers reported second showed a smaller right-field superiority, and those reported third did not yield a hemifield difference. Hines et al. argued that the procedure in their third experiment probably eliminated perceptual scanning, which had produced left-field superiority in bilateral exposure and with free recall (virtually left-to-right, as usual, with optional order). Thus, they concluded, the right-field superiority was due to the primary projection of the stimuli to the left side of the brain, which was presumably dominant for the right-handed subjects used in that experiment. This right-field superiority is consistent with dominance set, as contrasted to scanning set. The absence of an effect for the last (third) numbers is consistent with results for single elements. A serial-position effect implies an effect of memory (i.e., delayed recall).

Sampson and Spong (1961) presented pairs of digits simultaneously, with the right-hand digits viewed by the nasal hemiretina of the right eye and the left stimulus by the nasal hemiretina of the left eye. In each trial, four pairs of numerals were shown in sequence, but the subject could report the digits in any sequence. Unconventional digits were more accurately reproduced when they were paired with conventional digits, but conventional digits were not affected by the pairing. The typical response was to report the stimuli in temporal pairs, with the left stimulus first, indicating an effect of learned sequential processing.

Subject-specific variables

Subject-specific variables in bilateral exposure are discussed under the headings dominance effects, experience, set, and processing strategy.

Dominance effects
The variables relating to dominance were frequently manipulated together in the same experiments. Often the aim was to distinguish among possible effects of eye dominance, cerebral dominance, and hemiretinal differences.

Eye differences. With bilateral presentation, Crosland (1939) found generally more accurate tachistoscopic perception of letters near the left end of an array. Although children who were inferior readers were less accurate overall than superior readers, they were more accurate for the letters that were farthest to the right within the display. The poorer readers apparently had less consistent tendencies to scan the stimuli from left to right. Results were independent of viewing eye and of binocular versus monocular viewing. On the other hand, the results for right-eyed college students differed from those of the left-eyed students in the same way that the results of superior readers were different from the results of inferior readers among schoolchildren.

Knehr (1941) observed that eight subjects revealed better Landolt C acuity to the left of fixation, suggesting dominance of the right cerebral hemisphere. Moreover, the overall acuity was better in the right visual field for the left eye and in the left visual field for the right eye (i.e., superiority for the temporal hemiretina).

LaGrone and Holland (1943) found, with bilateral tachistoscopic presentation of single numerals or pairs of numerals and pairs of capital letters, that right-eyed subjects perceived more accurately to the left of fixation. Right-handedness, which correlated with right-eyedness, also produced left-field superiority in perception. Reading skill was correlated with left-field superiority; the authors concluded that reading disability resulted from the atypical field differences, rather than being the cause of them.

Harcum (1958d) found evidence that the difference between means of errors to the west and east of fixation, in recognition of binary patterns of solid circles and squares, is not basically dependent upon viewing eye. In three experimental sessions, viewing was with left eye, right eye, and binocularly, for each of 16 practiced subjects. Exposure duration was 0.1 sec. The means of errors for the radiuses under each viewing condition generally showed minimums of errors at east and at west. The symmetry of the data does not indicate equal capability in both halves of the visual field for all subjects, however. Some subjects did, in fact, show a differential in means of errors between the right and left halves of the visual field, but the numbers of subjects favoring the different halves were about equal.

Crovitz et al. (1959) presented a square grouping of three Xs and an O with the subject required to report the varied position of the O. The square grouping appeared about fixation in some exposures, or eccentrically in each of the four visual quadrants in others. Right-handed and right-eyed subjects tended to report the O as appearing in the left positions within the grouping, and the reverse was true for left-handed, left-eyed subjects. As a control, another group of right-handed subjects was asked to imagine the task and showed a greater frequency of right position reports. Therefore, the apparent effect of handedness was not due to biases in report.

Bryden (1964) presented evidence that cerebral dominance controlled hemifield differences for single-element stimuli in bilateral tachistoscopic exposure, but this effect of dominance was masked by scanning and organizational processes for multiple-element stimuli.

Wyke and Chorover (1965) compared visuospatial discrimination in the monocular hemifields. The subject's task was to report whether a comparison spot of light, presented for 1 sec, appeared centrally or peripherally to a standard spot previously presented for 1 sec. The subjects using the left eye performed more accurately than those using the right eye.

With bilateral exposure of Japanese words, Hirata and Osaka (1967) found that the words to the right of fixation were reported first regardless of viewing eye. However, only for the left eye were the words on the right more accurately reported. With geometrical forms, the stimulus projected to the nasal hemiretina was reported first, but there was no significant hemifield difference in perceptual accuracy for either eye.

Hemiretinal differences. Kephart and Revesman (1953) flashed pairs of trigrams to the nasal hemiretina of each eye and required the subjects to report which letters were seen. The subjects were screened to eliminate phoria or large differences in visual acuity between the eyes. The dominance score was the number of letters out of 14 exposures seen by the left eye minus the number seen by the right eye. Thus, maximum asymmetrical score was ±42. The distribution of scores appeared to be normal (mean = − 0.5), with a range of individual scores from − 39 (right dominance) to +35 (left dominance).

Crovitz and Lipscomb (1963) discovered, in a binocular color-rivalry experiment, that the color flashed to the nasal hemiretina was most frequently reported. When one color was presented to the left eye and another to the right eye, the color presented to the left eye was frequently localized to the left of the color presented to the right eye. There were large individual differences, however.

McKinney (1967) investigated the relative rates of disappearance of luminous figures in the dark, presented both binocularly and monocularly to the left and right of fixation. Overall, the stimuli in the right hemifield were more stable than those in the left, but the effect was critically dependent upon viewing eye. For binocular viewing there was no effect of the subject's handedness, but the right-field superiority was greater for right-eyed subjects. For monocular viewing, only the left eye revealed significant differences, indicating greater stability of images in the right visual field (temporal hemiretina). Further analyses indicated that there were more fragmentations of the visual image in the temporal hemiretina of the left eye and more fragmentations in the temporal hemiretina of the dominant eye.

Cerebral dominance. Jones and Jones (1947) found shorter reaction time in the right visual field for bilaterally exposed single letters. There was, however, no systematic difference between left- and right-handed subjects.

Efron (1963) applied shocks to fingers on the left and right hands and presented flashes of light binocularly to the left and right of fixation. For both stimulus modalities the stimulus that projected to the dominant hemisphere was perceived as temporally first. On the basis of detailed analysis of the laterality of the so-called left-handed subjects, he concluded that the judgment of temporal asynchrony is executed by "the hemisphere which is dominant for language functions" (p. 283).

Eyedness and handedness. Takala (1951) discovered generally more frequent preferences for "left halves" and "direction to the left" in various experiments for right-handed subjects and especially right-eyed subjects of this group. The only task showing more frequent left-field preference by left-handed subjects was bisection of a horizontal extent in the dark with long exposure duration of the stimuli.

Sampson and Spong (1961) tested 20 subjects designated according to handedness and eyedness. Pairs of digits were presented simultaneously in a binocular viewer that divided the visual field, causing the right-hand digit to be viewed by the nasal hemiretina of the right eye and the left-hand stimulus to be seen by the nasal hemiretina of the left eye. The digits were either conventional or unconventional and were paired with like or unlike mates. For each trial, four pairs of numerals were shown in sequence, and the subject was required to report as many numbers as he could in any sequence. The right-eyed, right-handed subjects were superior in both speed and accuracy to subjects with left-eye dominance. Poorest perceivers were the left-handed, left-eyed subjects. When the stimulus to each eye was of the same type (conventional or unconventional), the speed and accuracy were best for right-handed, right-eyed subjects viewing with the right eye. Accuracy was poorest for left-handed, left-eyed subjects viewing with the right eye. When the same type of digit was projected to each eye, the relation of handedness and eye dominance to both speed and accuracy of responding was dependent upon the viewing eye. When a different type of digit was projected to each eye, the right-handed, right-eyed subjects were most accurate for both stimuli, and the left-handed, left-eyed were least accurate. Regardless of stimulus type or like versus unlike pairing, the speed of response for each group of subjects was faster for right-eyed viewing. The first stimulus reported on any trial almost invariably had been presented to the left eye, and thus it was independent of handedness or eyedness. For subsequent digits the report was influenced by handedness and eyedness.

Corballis (1964) tachistoscopically presented four letters in different quadrants of the visual field in a device that permitted separate exposure of a complete field to each eye without the subject's knowledge of which eye was stimulated by material in either field. In one experiment, four conditions were used: (1) all stimuli exposed to the left eye; (2) all stimuli exposed to right eye; (3) two letters exposed to the nasal hemiretina of each eye; and (4) two letters exposed to the temporal hemiretina of each eye. There was no fixation point, but the subject was told to fixate on the center of the field. Regardless of exposure condition, the right-handed and right-eyed subjects reported the letters as they appeared in the visual field in the sequence northwest, northeast, southwest, and southeast. There was no difference in perceptual accuracy for the two eyes, but the performance was superior for the temporal hemiretinas. Moreover, there was no overall difference between left and right hemifields. In a second and a third experiment, four different letters were similarly presented to each eye without a fixation point and with a fixation point, respectively. Therefore, the sets of four letters appeared superimposed. The superimposed images presented to the left eye were reported slightly earlier than those presented to the right eye.

Immediate memory. Zurif and Bryden (1969) investigated, with right-handed subjects, dichotic recognition and a tachistoscopic visual analog of the dichotic listening task. The bilateral viewing and dichotic listening tasks employed both free recall and recall of the letters presented to one hemifield or to one ear first. In each task the material projected to the left hemisphere was more accurately perceived. Whereas nonfamilial left-handers showed right-field superiority in all tasks, familial left-handers were inconsistent in direction of lateral differences and showed no significant differences between ears or visual hemifields. In free recall after dichotic listening, familial left-handers tended to report a stimulus to the left ear first, whereas the remaining subjects reported a stimulus to the right ear first. In bilateral viewing all subjects tended to start reports with a letter presented to the left visual field. Left-right difference scores in the various tasks tended to correlate significantly and positively within a sensory modality, but to be very low for the tests from different modalities. Zurif and Bryden (1969) concluded that "hemispheric dominance is not a unitary function: perhaps the hemisphere specialized for the processing of auditory verbal material is not necessarily as dominant for visual material" (p. 186).

Experience
Hemifield differences with bilateral exposure also have been related to the directionality of reading. The child develops a predisposition for attending to certain features of a stimulus field first, and he tends to organize his perception around this differentiation of the field.

Gottschalk et al. (1964) required children from ages of about 3 to 6 years to name 20 pictures of familiar objects arranged in a 4 × 5 cell matrix. The more mature and experienced the children, the more organized was the sequence in which the pictures were named. Such organization of sequence was defined as naming adjacent items successively and naming successively pictures at the beginning and end of two adjacent rows or columns.

Regan and Cropley (1964) have shown that children who use

the English language develop the tendency to orient stimuli on a horizontal plane, and then to use a left-to-right direction for ordering multiple stimuli. Studies of visual search (Brandt, 1940; Anderson & Ross, 1955) show that eye movements of English-reading subjects favor stimuli to the left, which are in turn more accurately reproduced.

Haslerud and Motoyoshi (1961) found, in an incidental learning test, that sixth-grade Japanese students tended to recognize the irrelevant stimulus (Japanese letters) when it appeared north of the critical stimulus. American students exhibited greater incidental learning when the irrelevant stimulus was to the west of the critical stimulus.

Tachistoscopic linear patterns of open and filled circles are also usually reproduced from left to right by English-reading subjects, with greater accuracy on the left (Harcum, 1957b, 1958a, b, c; Harcum & Rabe, 1958a). The perceptual scanning follows the subject's expectation about the probable order of responding (Ayres & Harcum, 1962; Harcum, Hartman, & Smith, 1963). Results of Harcum, Hartman, and Smith's (1963) American observers are shown in Figure 6-17, as are Harcum and Friedman's (1963b) results under similar conditions for Israeli subjects. The Israeli subjects, whose native language reads from right to left, tended to reproduce the patterns from right to left and to show greater perceptual accuracy on the right, as was predicted from the hypothesis relating directionality of reading and perceptual scanning.

Preschool children tested by Dyer and Harcum (1961) did not show a significant hemifield differential in the perception of tachistoscopic binary patterns. Children in the first and second grades did, however, exhibit greater accuracy for the elements at the right. This was attributed to the learning of a left-to-right direction in reading, plus a perceptual recency effect which, for the children, was greater than the primacy effect. There was no effect of eyedness or handedness.

Practice level. Harcum, Filion, and Dyer (1962) predicted from the hypothesis of internal scanning that the magnitude of the difference between left and right halves of the visual patterns would be reduced as the subjects became more practiced, because attentional selectivity would be reduced owing to an increase in the perceptual span. They concluded that the perceptual mechanisms did not seem to be changed in terms of the predictions for the original dependent variable, namely, a change (flattening) of the shape of the perceptual serial-position curve. They concluded that merely the overall efficiency of performance had improved, but noted two aspects of their data that weakened this conclusion. The first was that the actual element-position curves before and after practice were not strictly parallel. This would imply canceling effects of changes in the perceptual mechanisms on the overall left-right hemifield differences. The second problem was that the left-right differences for performance before and after practice would lead to a quite different conclusion if the element-position data were replotted in relative terms, because of the decreases in the total number of errors. According to the present interpretation, there are two main bases for predicting improvement in performance with practice: (1) the left-to-right scanning becomes more consistent and, therefore, more efficient (Harcum & Jones, 1962; Harcum & Finkel, 1963; Harcum, 1965a); and (2) the encoding

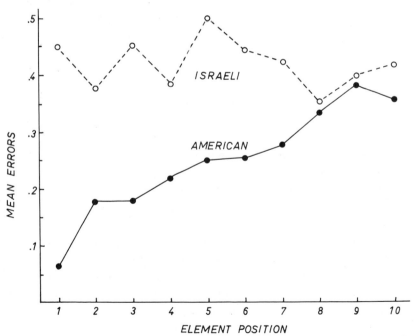

Fig. 6-17. Means of errors per exposure in reproducing binary patterns for Israeli and for American subjects. (Israeli data from Harcum & Friedman, 1963b; American data from Harcum, Hartman, & Smith, 1963)

process becomes more effective (Harcum, 1968b, 1970a). The increase in the left-to-right consistency of scanning tends to improve relative superiority for elements to the left of fixation, and the increase in encoding capacity results in a relatively greater improvement for the items on the right. These two mechanisms are not mutually exclusive. They may improve at about the same rate, leaving no change in the shape of the curve, as defined in relative terms, or they may change at different rates, changing the shape of the curve. The actual shape of the element-position curves before practice implied a difficulty for the observer in perceiving the end elements, as the curves were similar to the so-called sensitivity curves drawn by Harcum (1958f, 1966a) for such binary patterns under some conditions. The end-segregation phenomenon appeared at the end of practice, indicating relative improvement in the perception of the end elements. The proportional improvement of perception of the elements at the left implied increased consistency of the left-to-right scanning.

Braine (1968) tachistoscopically presented figures with distinguishing features at various orientations, and also horizontal binary patterns, to Israeli subjects for whom Hebrew was the first or only language. Developmental stage of subjects ranged from third grade to college level. On both tasks the younger children perceived stimuli on the right more accurately, whereas the reverse effect was found for the college students. This change occurred at about the seventh grade. The order of reproducing elements of binary patterns was more frequently from right to left for the youngest group (fifth grade), but reversed to more frequent left-to-right reporting for college students. To discover if the hemifield of superior accuracy was caused by order of report, Braine analyzed results separately for left-to-right and for right-to-left orders of report on each trial. For seventh-grade and college students the performance was, in fact, superior in the hemifield that was reported first, but for the fifth-grade students the stimuli in the right field were more accurately reported, regardless of reporting order. The seventh-grade students showed a bimodal distribution in the frequencies of left and right directions of responding.

Set

The term *set* can have many meanings, depending upon the degree of stability intended. The type of mechanism implied would likely be different for a transient set, such as might be implied in selective perception; for a stereotyped approach toward solving a problem, such as would be implied in the Einstellung experiment; or for a fixed method of viewing the world, such as would be implied from results of a projective test of personality.

Harcum (1967c) has proposed two mechanisms of set to account for lateral differences in performance: *dominance set* and *scanning set*. The former refers to a stable set toward superior perception for stimuli projected to one cerebral hemisphere, which for that particular task at least could be identified as the dominant hemisphere. The latter mechanism, the scanning set, refers to a less stable predisposition to adopt a particular sequential analysis of multiple-element stimuli. White (1969b), in his review of the literature on laterality, came to a similar conclusion. He concluded that a postexposure scanning mechanism can account for lateral differences for suprathreshold multiple-element stimuli, but that higher-order structural components seem to be implicated in the perception of single elements at threshold levels of stimulation.

Dominance set. Perhaps the most relevant of the so-called sensory tasks would be the relative saliency of simple stimuli simultaneously exposed to the right and left of fixation. Dallenbach and his co-workers (Dallenbach, 1923; Burke & Dallenbach, 1924; White & Dallenbach, 1932) have evaluated relative attensity for stimuli appearing simultaneously on opposite radiuses of the visual field. *Attensity* refers to the attention-attracting power of a stimulus, the degree to which it stands out in consciousness. Right-handed subjects reported greater attensity for stimuli to the left of fixation, but left-handed subjects exhibited the reverse effect. The side of greater attensity agreed with the eyedness of the subject (White & Dallenbach, 1932). Dallenbach concluded that the visual processes projected to the nondominant hemisphere of the brain were more attense because they were less subject to interference by other excitations.

Kirssin and Harcum (1967) repeated White and Dallenbach's (1932) experiment, but with different results. The results for each of 13 subjects are shown individually in Figure 6-18. The abscissa represents the physical difference between the variable and the standard stimulus patches – ranging through much weaker, weaker, equal, stronger, and much stronger intensity of the variable, respectively. The ordinate represents the percentage of observations in which the subject indicated that the variable stimulus was the more attense. As the two curves in each panel represent identical conditions except for the side on which the variable appeared, differences between the curves show the hemifield differences directly. Of the 13 subjects, 5 reported greater attensity more frequently on the same side as the dominant eye, and 9 subjects indicated greater attensity on the side opposite the dominant hand; 7 subjects selected the stimulus on the right as more attense; 3 of them were right-handed and 4 left-handed. Thus, these measures of the laterality characteristics of the subject and his reports of relative attensity in the left and right hemifields are apparently not correlated. These data suggest a general tendency for left hemisphere dominance, consistent with Heron's (1957) notion that there is a general tendency for the subjects to be set to attend to the right field prior to stimulus exposure. It also supports Bryden's (1964) proposition of left hemisphere dominance in the perception of single stimuli.

Fig. 6-18. Percentage of times each of 13 subjects judged the variable to be of greater attensity to the right (*solid line*) and left (*broken line*) of fixation for five degrees of relative intensity. Letters R and L indicate the subject's handedness (first letter) and eyedness. (After Kirssin & Harcum, 1967)

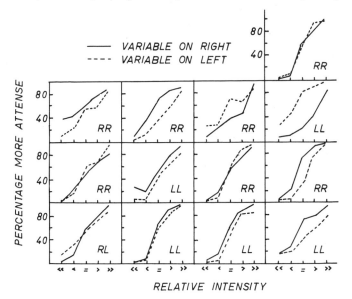

Kirssin and Harcum (1967) also found that the relative attensity of hemifields was not correlated with accuracy in pattern perception. With tachistoscopic binary patterns exposed across fixation, all subjects exhibited greater accuracy to the left of fixation.

Dallenbach employed a method using alphanumerics that is similar to the one proposed by Hershenson (1969a, b) for perceptual report (e.g., Dallenbach, 1920; Glanville & Dallenbach, 1929). He found that introspectively the sensory impact (attensity) did not always correspond to cognitive clarity (identification of items), although it usually did. Moreover, maximum attensity did not always correspond to the locus of ocular fixation. There were important individual differences in the locus of greatest attensity, as subsequently also found by Kirssin and Harcum (1967). The judgment of attensity would seem to be even more pristine than a perceptual report of what the subject said he saw.

Winters and Gerjuoy (1966) have shown that a stimulus form to the right of fixation is perceived as larger than an identical form presented to the left of fixation. The shorter the intertrial interval, the smaller this difference. Gerjuoy and Winters (1966) found right-side enhancement of size with educable retarded subjects, but only when exposure duration was unlimited.

Scanning set. When Pillsbury (1897) flashed words containing one or more disfigured letters, the subjects noticed the misprinted letters more often when they appeared at the beginning of the word. He concluded: "This seems to indicate a general tendency of the subject to *read through the word from left to right,* and thus to give the first letters of the word a more prominent part in the recognition of the word as a whole" (p. 350). For the first letter, the subject had little expectation of the word, but for later letters he did have this expectancy.

Evidence for a habitual directional sequence in the mechanisms of tachistoscopic presentation, independent of stimulus directionality, has also been derived from the use of linear binary patterns (Harcum, 1957b, 1958a, b, c; Harcum & Rabe, 1958a, b). Not only are these patterns meaningless, but they lack intrinsic directionality. The parts of the patterns in the left hemifield are usually more accurately reproduced (Harcum, 1957b), apparently because they are also scanned from left to right. They are, as verbal patterns, reproduced from left to right by English-reading subjects. Presumably, both the direction of scanning and the direction of reproduction are due to the subjects' habit of reading English from left to right (Anderson & Crosland, 1933; Sperling, 1960; Ayres & Harcum, 1962; Harcum, Hartman, and Smith, 1963). When Ayres and Harcum (1962) and Harcum, Hartman, and Smith (1963) allowed their subjects to mark their reproduction of tachistoscopic binary patterns of opened and blackened circles in any sequence, most subjects consistently marked the elements from left to right across the response template and consistently exhibited greater accuracy for the elements on the left. This correlation between responding sequence and perceptual accuracy suggests that the fewer errors are due to primacy rather than recency effects. The subject's knowledge about the required order of recalling the patterns influenced the distribution of errors. Thus, the perceptual scanning seemed to follow the subject's expectation about the probable order of responding.

The conclusions of Harcum, Hartman, and Smith (1963), with respect to the effects of knowledge of reporting order before and after exposure, were partially supported by Winnick and Dornbush (1965). However, those investigators employed 16-element patterns of letters and numbers in bilateral presentation. Although there was a significant interaction between time of instruction (before and after exposure) and hemifield differences, when the subject was to report the total pattern, there was no overall right-left difference. This lack of right-left differential may be attributable to the overall difficulty of the task. Perhaps the end elements were not perceived, producing the condition called continuous presentation (Harcum, 1969b) – discussed later – that seems to modify the perceptual process.

In the study by Chaiken et al. (1962), in which they obtained superior accuracy to west and north for perception of an odd stimulus in a two-dimensional display of similar forms, three of five subjects saw more target stimuli to the west, and two saw more to the east. These differences were large for only two subjects, however. Massa (1968) also found that recall was more likely to be superior to the west and north in a matrix for letter stimuli than for blackened squares. Hirata and Osaka (1967) failed to find significant hemifield differences for single Japanese letters and geometrical forms in bilateral exposure, but for two-letter Japanese words there was significant right-field superiority for the left eye.

L'Abate (1960) simultaneously flashed unrelated trigrams of varying association value to the left and right of fixation. Although trigrams of higher association value were generally more accurately reported, the stimuli to the left were consistently more accurately reported, even when the association value on the left was 0% and that on the right was 100%.

Harcum and Filion (1963) found that reversal of letter orientation and sequence caused some subjects to reverse the hemifield difference. Although the results for individual subjects were not presented, one example is shown in Figure 6-19. The curves represent the various conditions of simulus directionality. In the LRW condition the letters alone are reversed, and in the DRW the arrays are doubly reversed, with both letters and sequences of letters reversed from normal English. SRW are sequence-reversed words, and BCW are words with both letters and sequences of letters in correct orientation. LRN are nonwords with letters reversed, and BCN are nonwords with the letters correctly oriented. This subject showed consistently fewer errors in the left visual field, although the differences were reduced with letter reversal. In Figure 6-20 the results of another subject show a reversal to right-field superiority

Fig. 6-19. Means of errors per exposure for one subject for letter patterns representing six conditions of stimulus directionality. See text for explanation. (Data from Harcum & Filion, 1963)

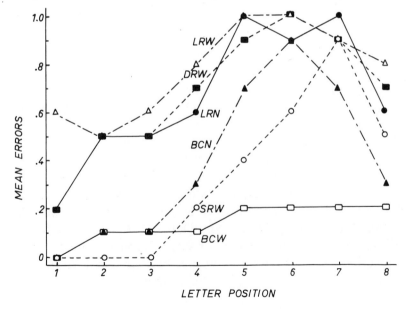

Lateral dominance and responding order 229

when the individual letters were reversed. Harcum and Filion (1963) inferred a sequential perception of the letters that was affected by stimulus characteristics. What appears to be lateral dominance could be any factor causing the sequential perception to begin at one particular end of the visual pattern. It could be responsible for what Harcum (1957a) has called the directional "consistency of scan."

To maintain a constant set under each condition, and to reduce the possibility of selective fixational errors, Harcum and Filion (1963) did not inform the subjects prior to each exposure of the specific stimulus condition to be represented. Harcum and Smith (1963) repeated the experiment, telling the subject before each exposure which stimulus condition would appear. If the differences between left and right visual hemifields were produced by a temporospatial process whose sequence is determined by the orientational characteristics of the stimulus, then the advance information about the stimulus conditions could enhance these effects. The preknowledge of stimulus condition did increase both the effects of element orientation and the effects of a meaningful sequence of stimulus elements. Whereas Harcum and Filion found none of

Fig. 6-20. Means of errors per exposure for one subject for letter patterns representing six conditions of stimulus directionality. See text for explanation. (Data from Harcum & Filion, 1963)

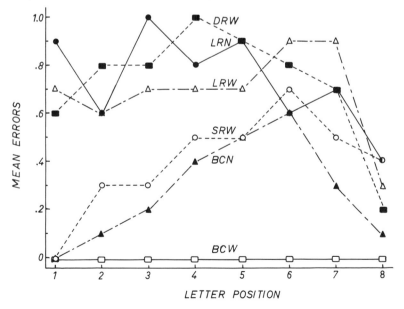

12 subjects who responded primarily to letter sequence, Harcum and Smith found 8 of 20 subjects who apparently did so. Most combinations of right-left hemifield differences and stimulus condition were represented by at least one subject. Nevertheless, the data curves were coherent for the individuals. Thus, preexposure set can augment the effects of stimulus directionality that depend upon postexposure perception of stimulus characteristics. This mechanism produces results similar to what would be expected on the basis of cerebral hemisphere dominance. As the group of subjects could be classified into several types on the basis of whether the inferred scanning was stable in direction or how it was affected by the experimental variables, there was again evidence for both task-specific and subject-specific processes.

Processing strategy

Response feedback. Camp and Harcum (1964) studied the effects of practice in the reproduction of binary patterns when the location of patterns was varied with respect to the fixation point. The results for three experiments using different response templates are shown in Figure 6-21, in which percentages of total errors are given for each element position for the chronologically first and second halves of the observations for those patterns bisected by the fixation point. The top panel shows the results when the patterns were reproduced on a 10-element responding template without any representation of the fixation marker, thus virtually forcing subjects to use the ends of the pattern for reference to localize the elements. Errors decreased with practice, but their relative distribution among elements did not change. Although this result is different from that of Harcum, Filion, and Dyer (1962), there were several differences in the conditions of the experiment. Probably the most important difference was that the subjects in the Camp and Harcum task did not have the initial difficulty in perceiving the end elements.

The second experiment of Camp and Harcum (1964), for which results are shown in the middle panel of Figure 6-21, employed a 20-element responding template with a reproduction of the fixation point between tenth and eleventh positions. The subjects were required to reproduce filled elements in relation to the fixation point, so they were virtually forced to orient the perception about the fixation point. The distributions of errors differed from those of the first experiment, particularly for the later observations. As Camp and Harcum concluded, the subjects changed their perceptual strategy to be consistent with the requirements of the task. In fact, the results were the same as those usually obtained with lateral exposure of letter arrays (Harcum, 1966b).

In a third experiment of Camp and Harcum (1964), the response template was identical to the stimulus configuration. The results

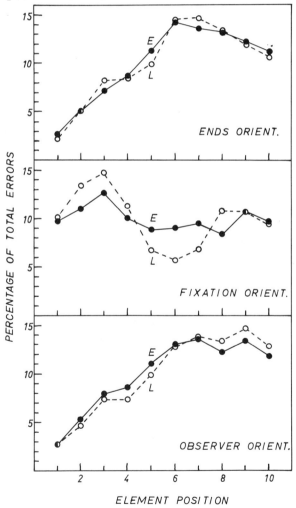

Fig. 6-21. Percentage of total errors at each element position for the early (E) and late (L) observations under three responding conditions. See text. (After Camp & Harcum, 1964)

(bottom panel of Fig. 6-21) essentially duplicated those of the first experiment, except that the curve became slightly more asymmetrical after practice. Thus, the subjects apparently preferred to use the ends of the pattern for orientation when they had a choice.

Other evidence. Sanders (1963) investigated reaction time in the enumeration of columns of dots presented 3° to the left and right of fixation under two different instructions. In one, the subject was instructed to attend and respond first to the stimulus on the left and then to the one on the right (i.e., successive treatment). In the other condition, he was instructed to attend to both stimuli before responding (i.e., grouping treatment). The grouping response was faster. However, the response to the right-hand stimulus was faster under successive instructions than would be predictable on the basis of two reaction times in sequence plus ocular excursion time. Therefore, even with the instructions for successive responding, some information about the right-hand stimulus was transmitted while the left stimulus was being processed. Further evidence that the grouping strategy was preferred by the subjects was the fact that it was difficult for them to use the successive strategy. Some subjects had to be discarded from the experiment because they were unable to follow these instructions. With increasing eccentricity of stimuli, in regions where eye and head movements become helpful to the subject, the tendency toward grouping and the advantage of grouping are reduced and finally disappear.

A group of aged veterans was given both serial-learning and pattern-perception tasks by McKenna and Harcum (1967). These subjects, 60 to 70 years old, participated in unpaced serial learning of word lists and in the reproduction of tachistoscopically presented binary patterns. An analysis of individual differences in frequency of errors of primacy, recency, and bowedness revealed some consistency across tasks. The major contributor to this consistency was relative primacy, measured by the tendency to err on the first (temporal or left spatial) item of each task. Primacy in tachistoscopic perception predicted total errors in serial learning. The correlation between total errors on the two tasks was not significant. The results indicated that a major difference between good and poor learners and perceivers was that the poorer subjects experienced difficulty in adopting a consistent strategy for organization of the material.

Discussion

Logically the mechanisms of bilateral exposure should involve all the mechanisms relevant to lateral exposure and at least one other. The additional mechanism could merely determine which lateral half of the projection system would achieve primacy, and the results would be an average of the individual lateral effects with greater weight to the effects from the so-called dominant side. Some results from the simpler tasks can apparently be explained in this way. For example, stimuli to the right or left of fixation may achieve primacy in perception because their projection to the cerebral cortex results in stronger excitation. Therefore, a possible phenomenal parallel of cerebral dominance would be greater apparent intensity of a stimulus when it is projected to one hemisphere than to the other. Dallenbach found, however, that sensory impact did not always correspond introspectively to cognitive clarity, although it usually did (e.g., Dallenbach, 1920; Glanville & Dallenbach, 1929).

Most of the results with bilateral exposure are not such simple averages or summations of results from lateral exposures. The configurations of results are frequently very dissimilar to the results of lateral exposure, indicating a response to the total configuration of the stimulus and its component parts. For example, the results of Crovitz et al. (1966) strongly support the argument that relative position, rather than retinal sensitivity, is the major determinant of distributions of errors among elements in such a task. Thus, all the information about visual information processing cannot be obtained from the lateral exposure condition.

Under ordinary circumstances the stimuli to the left of fixation are reproduced more accurately. Heron (1957) and Harcum (1957b) explained this by supposing that subjects' habits of reading from left to right take precedence over other factors, so that the elements on the left are favored by a primacy effect in a perceptual scanning mechanism. Some results of Braine (1968) support this conclusion. Braine (1968) used three nontachistoscopic tasks to investigate ordering perferences of Israeli subjects without including effects of visual field differences. Younger children tended to start naming arrays of forms and to arrange pictures making a story from right to left, whereas college students tended to start from the left. In number-cancellation and letter-cancellation tasks, however, the younger children were not as systematic. The college students tended to move back and forth across successive rows. Completion of either the number-cancellation or letter-cancellation task first tended to carry over in the direction of performing the second task, but much more often for the letter task, which, given first, showed a greater proportion of left-to-right responses. Therefore, there is evidence for subject-specific processess in bilateral exposure.

Seemingly trivial changes in the total configuration can change the way the stimulus components are used in the overall bilateral perceptual process. The workings of the process may also be very sensitive to situational variations that affect the subject's set (i.e., change the valence of the various stimulus components in the overall task). This can be illustrated by the findings of Fitzgerald and Marshall (1967), who investigated results of spatial grouping of stimulus elements and whether all elements or only those on one side of fixation were reported, as instructed after exposure. The usual superior perception to the left was found with full report, but with partial report right-field superiority was obtained, with the greater differential for arrays having differentially larger spacing about fixation. Since the stimuli on the left within each array on one side of fixation were always reproduced more accurately, the scanning interpretation would argue for a general left-to-right scanning of all patterns. Moreover, these curves for partial report were approximately parallel for left and right reproductions, contrary to the results of Harcum and Jones (1962) and Harcum and Finkel (1963) with lateral exposure. Also, the total report condition was much more accurate. These two facts indicate that the subjects were in fact processing the "irrelevant" information in partial report, as Mewhort (1967) has also concluded from a similar task. Thus, the conditions of the experiment apparently produced right-field superiority with the bilateral exposure of stimuli for a different reason than the scanning set (Harcum, 1967c); the mechanism seems more like the dominance set proposed by Harcum (1967c). This interpretation is supported by the fact that the effect of right-field superiority with partial report was greater when there was a gap in the spacing of the letters about fixation.

The laterality effects with bilateral exposure are usually the result of a temporospatial sequence of perceptual scanning, which favors the stimuli in one field by a primacy effect. As the directionality of this sequential scanning can vary with manipulation of both stimulus and response variables, the hemifield results are frequently task-specific.

CONTINUOUS PRESENTATION

Harcum (1964c, 1969b) tested the hypothesis that the usual hemifield differences in perceptual accuracy for bilateral tachistoscopic patterns are produced by a perceptual scanning of the pattern from end to end, causing one end to be favored by a primacy effect. "Continuous" patterns were designed to prevent this scanning by making the physical ends of the visual target extend so far peripherally that the stimulus configuration had no perceptible ends. As the only differential physical reference point for a sequential perceptual process is the fixation point, there should be no laterality effects if the scanning for such binary patterns has no consistent directional tendency. Harcum (1964c), attempting to use the above method, did not eliminate left-field superiority, or presumably the end-to-end scanning, with a 17-element pattern. Harcum (1969b), in a later experiment, used 28 blackened or open binary elements subtending a total visual angle about 13.4°. A fixation cross registered with a point between the two central elements. Before the experiment began, the 24 college student subjects were tested for eye dominance by a sighting test and for hand dominace (Crovitz & Zener, 1962).

The percentages of errors at each element position with left eye, right eye, or binocular viewing are shown in Figure 6-22. The subjects obviously did not discriminate the end elements at better than chance (50% errors). The distributions were symmetrical for each viewing condition, with minimums near fixation, similar to those usually found for lateral exposure of binary patterns (Camp & Harcum, 1964). This difference from the usual asymmetrical distribution of errors with bilateral exposure strongly supports the hypothesis that a primacy effect in a temporospatial perceptual process accounts for the usual hemifield difference with binary patterns. There obviously was no differential effect of hemisphere or eye dominance, as no subject exhibited a lateral difference in perceptual accuracy.

Although there was no overall laterality effect, the hypothesis also predicted no lateral difference within a single exposure. An additional analysis on single observations of individual subjects tested this prediction. There were, as predicted, significantly more cases of equal performance to the right and left than cases of a given lateral bias, and the frequencies of biases in either direction were not significantly different for any viewing condition.

White (1969a) achieved about the same result for reproduction of bilaterally exposed arrays of eight unrelated letters or digits with an exposure duration of 25 msec. Best performance appeared about fixation, with essentially symmetrically decreasing accuracy both left and right from fixation.

DISCUSSION

The causes of lateral asymmetry of function have been sought in the receptors because there are two eyes and in the central nervous system because there are two halves of the brain. The hemiretina has two possible bases for importance: its association with an eye and with a cerebral hemisphere. The potential sources of asymmetry interact in many ways with a multitude of other variables to foil simplistic solutions to the problems of lateral asymmetry in performance.

Types of dominance

Eye dominance
As mentioned previously, the various criterion measures of eye dominance fail to correlate consistently. Ogle's (1962) list of 10

Fig. 6-22. Percentage of errors for continuous binary patterns with left eye, right eye, and binocular viewing. (After Harcum, American Journal of Psychology, 1969b. Reprinted by permission of University of Illinois Press. Copyright 1969 by the Board of Trustees of the University of Illinois.)

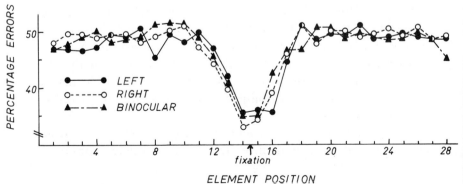

methods for defining ocular dominance include such diverse definitions of the dominant eye as that which has higher acuity, shows less heterophoric deviation under the cover test, has the more frequent image during binocular rivalry, aligns two points in space, and causes discomfort when occluded. Ogle pointed out that there is an important motor aspect of ocular dominance that maintains a stable visual space in spite of eye movements. Therefore, imbalance of the ocular muscles can produce asymmetries in hemifield performance, such as those discussed in this chapter (Ogle, personal communication). When the subject fixates on a given point binocularly, Ogle (1962) hypothesized, "it is the subjective egocentric spatial directions associated with the 'local signs' of the retina of the dominant eye and their connexions in the cortex that provide the spatial orientation of the observer. This is, therefore, a subjective directional dominance" (p. 410). He further assumed that the innervations to the ocular muscles that produce a movement also cause a sensory correction to maintain the stable spatial direction, as with conjugate eye movements. In binocular vision, the motor aspect and the sensory correction factor preferentially affect one eye, which is therefore the dominant eye. Evidence for this hypothesis was provided by the use of special apparatus, in which, by means of a half-silvered mirror, an apparently binocular image of a target was transmitted to one eye and reflected to the other. When the illumination was shifted to another target, the apparent distance to the object viewed through the mirror increased, but that object was now occluded from the other eye, which then had to make a conjugate movement for convergence on the new target. When the subject reported apparent movement of the target in the direction of the occluded eye, that eye was judged to be dominant. Results of this test correlated highly with results of a sighting test.

The Jasper and Raney (1937) phi test of lateral dominance also attempts to avoid the definition of dominance in terms of the subject's preference. It aims to assess dominance of the visual projection areas rather than the relative acuity of the peripheral receptors or of ocular motor control. In this test, the subject fixates on a spot of light, and then another light is flashed closer or farther away so that a double image of the new spot is produced, one to the right and one to the left. A dominant hemisphere is indicated by consistent perception of movement to one side, regardless of whether the second light is closer or farther away than the first light. Covariance of perceived movement with the eye stimulated indicates, of course, a peripheral mechanism of dominance by one eye.

Crovitz (1961) found a correlation between the preferred eye in a sighting test and the eye showing superior acuity. In a sample of 319 college students, acuity was greater in the left eye for subjects showing left-eye preference. Over twice as many subjects showed right-eye preference as showed left-eye preference. The numbers of subjects revealing no acuity differences, left-eye superiority, and right-eye superiority were, respectively, 87, 99, and 133.

Gabersek (1963) recorded the reaction time of eye movements for adults and children 6 to 15 years of age. Reaction time was the difference between the time the light appeared at the north, south, west, or east of a centrally fixated screen, and the beginning of the subject's eye movement toward it. The subjects were told in which direction they would have to look so that a decision time for choosing the proper direction would not be included. The average reaction time decreased significantly over ages 6 to 12 years. Children 12 to 15 years old responded faster than adults, but the difference was small. Horizontal movements were faster than vertical; movements toward the east and north were faster than those to the west and south, respectively. These differences increased with age from none at 6 years to most evident for adults. No relationship was found between these measures and hemispheric dominance.

Massa (1968) recorded eye movements when subjects reproduced horizontal patterns of blackened cells or numerals placed in different cells 1×6 matrixes, displayed for 30 or 100 msec. The horizontal extent of the matrix was 9°. With the numeral displays, the subjects were required either to indicate position only or to report both numeral and position. Both methods of report and both classes of stimuli produced fewer errors at the ends, with more errors to the east than west. The asymmetry was greater when numerals as well as positions had to be reported. Massa attributed the results to the effects of postexposure eye movements. Latency of eye movement was shorter for displays in which only position was to be reported, but the excursions were small (less than 0.5°). Numeral and position reports produced greater excursions of the eye than did position-only reports, "almost invariably" involving the sequence of fixations from center, east, center, west, and center – extending about 3° east and 2° west regardless of exposure duration.

The above results suggest that the importance of ocular dominance or monocular viewing may result from the effects of motor initation of eye movements. Crovitz and Daves (1962) did, in fact, find that postexposure eye movements tended to be directed toward the unoccluded eye.

Cerebral dominace

Specialization of function. The concept of cerebral hemisphere specialization (Semmes, 1968) implies that the hemispheres are differentially adapted to serve different functions, rather than that one hemisphere is universally prepotent. This conclusion is amply documented by evidence from studies of patients with brain lesions.

Temporal order. Rutschmann's (1966) results with paired flashes of light to the fovea and to the left or right periphery of the right eye produced greater asynchrony for four of the five subjects when the temporal hemiretina, rather than nasal hemiretina, was stimulated by the peripheral member. Therefore, transmission of stimulation to the higher centers appears to be faster for stimuli presented to the nasal hemiretina. Rutschmann also mentioned other data that indicated an effect of viewing eye.

Boynton and Corwin (1966) argued for the existence of a "simultaneity center" located in the left hemisphere. Because there were large individual differences in the ranges of temporal asynchronies that produced judgments of simultaneity, they speculated that the span of asynchrony that still produced the judgment of simultaneity could be influenced by the subjects' criteria for responding.

Sensory-tonic directionality. Takala (1951) concluded that the hemifield differences he obtained were not explicable in terms of asymmetries of the retina or differential eye movements. He inferred, rather, an asymmetry in the total behavior of the person caused by asymmetry of muscle tonus and related factors – a sensorimotor theory. Thus, he concluded, perception entails the location of stimuli in an asymmetrical coordinate system in which members of the body in addition to sensory pathways and brain centers participate. "Even when no eye movements are involved, e.g., when the stimulus has a short duration, the perceptual object is seen as if the eyes had moved during the observation, which is to say that the perceptual object acquires the quality of directedness" (p. 165).

van der Meer (1959) proposed that the asymmetry of performance can be at least partially explained in terms of the principles of innervation tonus and reafference. For example, the overestimation of objects in the right half of the visual field occurs because a right-directed tension causes that object to lie closer to the center of the dynamic field – presumably the center of attention. Examples of the sensory-tonic task are the location of the median plane in the dark (moves to the right), the location of the center of a visual field (shifts to the right and causes localization errors for objects in the unstructured field), location of the subjective vertical (somewhat inclined to the right), and judging the speed of horizontal movement (faster to the left, indicating a movement bias toward the right).

In order to discover the cause of perceptual asymmetry, van der Meer (1959) investigated whether children of elementary school age had a preference for left or right halves of space and if this preference was related to handedness. Kindergarten pupils, asked to draw someone going to a house, climbing up and going down a mountain, etc., tended to employ the side corresponding to the dominant hand – ordinarily the right.

According to van der Meer (1959), children begin at 9 years of age to prefer a perceptual orientation to the left side first, as a result of the habit in reading and writing to begin on the left. This is, however, an organization of the static field.

In another study, van der Meer (1959) visually presented single, short diagonal lines at various positions in the visual field to subjects from 9 to 18 years of age. The subjects were asked to report in which direction the line appeared to be moving. Children from 9 to 14 years of age had no preference for right or left movement. For subjects 15 years of age, significantly more reported rightward movement (53.9%) than leftward movement (20.4%). The rightward directional bias increased gradually to include over 70% of the 18-year-old subjects. Therefore, at about 16 years of age a new dynamic form of spatial structuring takes place. An educational component in this right-directedness in perception is indicated, as it did not appear among adults from a low educational level. Thus, dynamic organization in perception seems to be an effect of more than the reading habit alone because it seems to be critically dependent upon a sufficient level of mental development.

Sensory correction. The notion of sensory correction (Ogle, 1962) seems equivalent to the notion of reafference or corollary discharge (Teuber, 1960). Freedman (1968) proposed a similar theory of discordance information analysis that is more general than the proprioceptive change theory of Harris (1965) or the reafference theory (Held & Hein, 1958). This approach seems to hold the most promise theoretically.

The basic mechanisms

The basic mechanisms controlling lateral asymmetry of performance seem best described at the molar level as subject strategies or cognitive controls. Such strategies could have many origins resulting in some cases from stable characteristics of the subject (subject-specific mechanisms), from the experimental conditions (task-specific variables), and from semistable characteristics of the subjects (interaction of task-specific and subject-specific processes). This concept of a resultant strategy accounts for the great variability of relative hemifield performances in different studies.

Reading
The probable strategy of the subject in assimilating visual information is frequently inferred from his experience in reading. This relation of reading to tachistoscopic perception and to short-term memory is reflected in several current theories (e.g., Heron, 1957; Harcum, 1957a, 1967b). Some of the evidence for the relation between left-to-right reading and lateral asymmetries has been summarized above. LaGrone's (1942) results illustrate the difficulties in interpretation, however; he found no substantial relationship for any of his tachistoscopic tests, with bilateral as well as lateral exposure, to the number of regressive eye movements in reading. Thus, as Tinker (1946) concluded, regressive eye movements for fast readers do not necessarily indicate a poor strategy in reading. Instead, he concluded: "The particular eye-movement pattern employed is conditioned by the nature of the central processes" (p. 114). In place of the eye-movement concept, Tinker advocated the concept of "perceptual sequences." This argument is supported by the present discussion.

Schonell (1940) found more reversal errors in backward than in normal readers, but frequent occurrences of these errors by both types of readers before 9 years of age. He emphasized the magnitude of individual differences within each group, arguing that a structural basis would not produce both slow readers who showed no reversals and normal readers who did show reversals. As reversal errors are much less common in writing than in reading, the kinesthetic feedback from writing presumably provides an additional spatial clue to the proper directionality.

The orientation toward the left stimulus is developed early in young children. Marchbanks and Levin (1965) presented three- or five-letter nonwords to kindergarten and first-grade children and later asked them to select the most similar word from an array of words of equal length. Because the confusion stimuli were selected either for identical shape (i.e., comparable letters above and below the line) or for identical letters at one position within the word, the experimenters could determine which cue the children used in determining similarity. The subjects in both groups most frequently relied on the first letter, next on the last letter, and very little on shape. There were large individual differences. Kindergarten boys often responded on the basis of the last letter – presumably a recency effect – and first-grade girls often used the second letter after the first – presumably a greater primacy effect.

McFie (1952) examined 12 persons with reading disability and found frequent tendencies to reverse in form perception, and infrequent instances of clear cerebral dominance by Jasper and Raney's (1937) test. As these subjects showed few reversals in normal reading and writing, McFie concluded that their basic problem was in reproducing sequences because cerebral dominance was not established.

Orientation in space
The directionality of the perceptual sequence may be determined by the subject's orientation in space. Krise (1949) hypothesized that reversal errors in the perception of letters are due to "a problem of space perception, a confusion in the relationship between the figure and its ground" (p. 281). Kolers and Perkins (1969) concluded that interactions between a given transformation of letter orientation and a reading direction are very important. For

example, a congruence of the letter's "direction of facing" and the subject's direction of reading produces more accurate recognition.

Braine (1968) interpreted her findings of lateral asymmetries in Israeli children and college students in terms of differential degrees of attentiveness to stimuli, and parts of stimuli, at various lateral locations. She cited Takala's (1951) work to show that the left part of the visual field is attended to and recalled better than the right part. Although Takala's results were not influenced by eyedness, they were affected by handedness. Therefore, Braine concluded that experience in reading is not the determinant of hemifield differences. The effect is not simple lateral dominance, however, since different results have been obtained with lateral and bilateral exposure.

A theory of information translation

The theoretical implications of the foregoing description of results can be summarized first in terms of a machinelike model, giving primarily a molar description. The machine model follows suggestions by, for example, Broadbent (1958), Bryden (1967), Atkinson and Shiffrin (1968), Shiffrin and Atkinson (1969), Sperling (1967), and Masani (1967). A second level of analysis produces some speculations about what these mechanisms represent in neurological terms. These analyses follow closely a previous approach (Harcum, 1967c), although the earlier paper concentrated more on describing processes than on speculating about specific mechanisms.

The evidence indicates that the perception of tachistoscopic patterns of multiple elements includes a spatial and temporal analysis of persisting traces. Thus, a basic aspect of pattern perception is serial order of behavior. Because any basis of differentiation within the stimulus array could influence the order of analysis of elements, the important problem is to discover what determines which element will be processed first. This element then serves as the reference (anchor) point for the subsequent perceptual processes.

Lashley (1951) saw the importance of sequential processing of simultaneously presented spatial arrays, as well as spatial concepts, for the memory of items presented in temporal sequences. He proposed that sensory input is modified by body position, and stimulus traces can become associated with body referents as well as with other traces, and thus become more distinguishable. The spatial traces can be scanned by some other level of the nervous system and transformed into sequences. To illustrate this, Lashley pointed out that he could play a melody backward only by visualizing the music spatially and then reading the image backward. This idea of a scanning at a different level has also been proposed by Milner (1961).

Lashley (1951) concluded that this selective mechanism is relatively independent of motor units and the structure of thought. In speech, for example, the sequence of sounds must be programmed before the sentence begins because there is no time for sensory feedback to initiate the next response. The second event in the production of a serial movement, which Lashley called the syntax of the act, is a generalized pattern of integration habits, also having no intrinsic order, but which determines the order of individual components of the idea, as in language. Lashley cited translation from one language into another having a different sentence structure: "Somewhere between the reading and free translation, the German sentence is condensed, the word order reversed, and expanded again into the different temporal order of English" (p. 118).

Bryden's (1967) theory, following Lashley, included a model for sequential organization of behavior based primarily on data from tachistoscopic perception and dichotic listening. Bryden proposed that different stimuli arouse different traces, having both spatial and temporal components. If several traces are active simultaneously, the order of responses is controlled by a central system according to the identity of traces (e.g., verbal material) or the prevailing external set. This ordering mechanism inhibits responses to certain traces until their proper turn. A central decision (set) mechanism can determine whether an overt response is given or a covert response is maintained in rehearsal. This rehearsal consolidates traces through feedback to the auditory system, adding another temporal mechanism. It can disturb the pure temporal or spatial basis of organization and add other bases of organization (i.e., meaning).

Heron (1957), following Hebb, suggested that traces (cell assemblies) of the stimulation include central components of eye movements. Therefore, incipient central motoric responses prior to overt eye movements should facilitate the activation of the trace.

The stimulus trace is also facilitated by certain prior neural activity, which has been consistently activated previously through experience with reading. This suggestion was submitted to direct test by Harcum and Jones (1962) and Harcum and Finkel (1963), with positive results. Normally printed English words in lateral exposure in the right visual field were more accurately reported; the inferred primacy effect favoring the left half of these words was greater. Thus, the perceptual sequence was more consistent when the stimulus conditions led to an unambiguous reading sequence. Harcum and Finkel (1963) showed that the effect was not specific to left or right cerebral hemisphere. Mirror images of laterally presented words produced a mirror image of the results; a greater primacy effect for the right half of mirrored words, and greater overall accuracy, in the left field.

Harcum's (1967b) information-translation theory proposed that a sensory input is translated into the subject's conceptual or coding system, and thence into a language appropriate to the required response. The processes inferred in the reproduction of multiple elements include element discrimination, selective analysis of persisting traces, and the organization of information for mnemonic storage.

Input

Stimulus trace. The visual system generates what Harcum (1964d) called the s-factor in pattern reproduction. It produces, with binocular viewing, equal numbers of errors for visual patterns on opposite radiuses of the visual field – the vertical radiuses having least sensitivity and the horizontal the greatest. Ordinarily in tachistoscopic pattern recognition, the stimulus parameters (i.e., duration, area, contrast) would be adequate for detection of single elements. The end-segregation effect could result from the absence of spatial inhibition or masking effects of other elements from one direction, as performance for elements at the ends is, in fact, hindered by the addition of elements (Harcum, 1958b). Also, the unique location of the end elements permits more accurate localization within the overall pattern.

Perceptual accuracy is determined by element position within the pattern rather than by absolute retinal position (Harcum, 1964d). There is currently disagreement about whether the visual percept (Haber & Hershenson, 1965) can be distinguished independently of so-called postperceptual processes (Hershenson, 1969a,b). Haber (1966) believed that this may be possible, although Neisser (1967) denied that such a differentiation can be made introspectively. Harcum (1970b) refuted Hershenson's (1969a,b) claim of having demonstrated a method for measuring the so-called visual percept.

According to the information-translation hypothesis, while the stimulus trace is decaying, it is being "read" by a sequential processing mechanism – the r-factor (Harcum, 1957a). This direction of scanning depends upon the cognitive control of the subject. The traces of stimuli scanned first are favored by a primacy effect, producing differences in errors between opposite radiuses. For those meridians that are consistently scanned in one direction, the primacy effects consistently favor the stimuli in one hemifield. Mewhort et al. (1969), using a visual masking procedure, supported this internal scanning hypothesis. In their study, the perception of stimuli on the right of fixation was critically dependent upon prior accurate perception of the stimuli on the left.

Because the trace of an overt stimulus decays over time (Melton, 1963a; Neisser, 1967), the perceptual mechanisms encounter traces with different strengths, depending upon order in processing. The retinal trace may persist for only a few milliseconds (Sperling, 1967), whereas the complete trace (presumably including central components) may persist for much longer (Walker, 1958; Hebb, 1963; Melton, 1963b; Doerries & Harcum, 1967).

A basic assumption of this hypothesis is that sequential scanning of the stimulus traces employs the same order as that in which the eye would fixate the stimulus elements if the exposure were sufficiently long. Therefore, the perceptual process should be most efficient in a direction analogous to that of eye movements in reading (Heron, 1957). Thus, for bilateral patterns English readers would scan first toward the traces of the elements on the left and then toward the traces of elements on the right. Perceptual primacy would thus favor elements in the left hemifield. Arrays of forms lacking intrinsic directionality for informational scanning often do not, therefore, produce hemifield differences (e.g., Heron, 1957).

Any inhomogeneity within the stimulus configuration may attract the subject's attention, causing him to move the affected element up in the order for attention. Apparently the spatial dispersion of stimuli itself provides strong bases for organization. Miller (1963) pointed out how the isolation effect in serial learning could be accounted for by the macroprocess (strategy) postulates of Feigenbaum and Simon (1962). Harcum (1965a,b) applied the notion of strategy in perceiving to the isolation effect in pattern perception also; the perceptual scanning proceeds from a reference point after it has been selected. As Milner (1961) proposed, neural activity corresponding to a set to respond sequentially would also be necessary if the subject is to respond sequentially. As Bryden (1967) and Harcum (1967b) concluded, the spatial directionality of this scanning in tachistoscopic perception is determined by both input and central organization factors. The scanning process also might be different for subsequent scannings of a constant stimulus configuration because of feedback from the responses. This feedback would not be restricted to effects of overt responses, but to the perceptual response itself, if the trace persisted long enough.

In studies in which linear binary patterns were flashed at various orientations in the visual field (Harcum, 1964d), the smoothness of the function for the inclinations near vertical suggests that the scanning does not exhibit equivalent primacy effects from all starting points within a given cerebral hemisphere. Although the shape of this function is very similar for different subjects, the locations of maximums and minimums vary, usually with maximum of errors at some radius in the right hemifield. These results are more suggestive of a functional effect than of a stable difference between the two cerebral hemispheres.

Codification
The selective aspect of attention would be affected by what Atkinson and Shiffrin (1968) called "cognitive controls." These control mechanisms are primarily central in origin, rather than peripheral (Harcum, 1964a, 1967c). This directionality reflects a subject strategy because, as Ellis (1969) found, a primacy effect in a probe-type memory task is facilitated by the opportunity for the subject to rehearse. Primacy effects are not attributes exclusively of specific receptive systems; they have been found in various tasks. For example, Pollack (1953) found serial-position effects when subjects reproduced aurally presented message units. Harcum (1965a,b, 1966b) has argued that the perceptual task is a miniature

process of serial learning; Garner and Gottwald (1967) also supported this argument for overlapping mechanisms of perception and learning.

Glanzer and Clark (1963) showed that the reproducibility of a binary pattern becomes more difficult with increases in the number of words necessary to describe (encode) the pattern. Rommetviet et al. (1968) investigated rivalry of similar groups of letters presented tachistoscopically at corresponding locations in each eye. When a combination of rival letters produced a meaningful word of $n + 1$ letters, the subject frequently reported seeing that word, rather than a meaningless word with the two conflicting letters reported in the "incorrect" order. Thus, the ambiguity of letter location is resolved by an operation of the subject upon the stimulus input to generate a meaningful perception on the basis of "overlearned rules of word formation." This mechanism resembles figural synthesis (Neisser, 1967).

Harcum and Friedman (1963a) found an interaction between the subject's inferred strategy of perceiving and the temporospatial sequence in which multiple-element patterns are presented. This conclusion is supported by Gropper (1968). The size of individual differences has led, as described above, to the proposition of perceptual types (e.g., Harcum & Rabe, 1958a; Harcum & Dyer, 1962). The reality of such types in this connection is a most important phenomenon in this area of research, but its importance has not been generally recognized.

The dominance set, mentioned above, implies a general facilitation for all stimuli in one hemifield, usually the right. This mechanism is distinguished from scanning set in that it is more subject-specific and consequently less likely to be shifted by stimulus and situational characteristics. Depending upon conditions, it can overpower the scanning mechanism (Ayres & Harcum, 1962) or be overpowered by it (Harcum & Dyer, 1962), resulting in different directions of lateral difference. For example, if the mirror image of stimuli is presented in lateral exposure, a negative correlation between the results for normal and reversed stimuli would imply a scanning mechanism, as Harcum and Finkel (1963) obtained with multiple-letter stimuli. But a positive correlation between hemifield results with normal and reversed stimuli would imply a cerebral dominance mechanism, as Bryden (1966b) obtained with single-letter stimuli.

The dominance set is more likely to overpower the scanning mechanism for laterally exposed stimuli than for bilateral stimuli. For example, Orbach (1952, 1967) and Barton et al. (1965) often found superior recognition in the right visual field for Hebrew words in lateral exposure, contrary to the prediction from directional scanning of verbal material. Bilateral exposure of Hebrew words produced a clear superiority in the right field (Anderson, 1946), consistent with the scanning motion for words with sinistral directionality.

The above discussion implies that the stimulus impinges on continuously active central processes (Hebb, 1949), which are themselves continually being affected by sensory input (Hebb, 1963). Under ordinary conditions, the overt responding is delayed long enough to involve a mnemonic factor – the dynamic organization of serially presented elements. Therefore, stimuli that are physically salient or are more attense due to focal attention (Neisser, 1967) have an advantage over aggregated or undifferentiated elements.

The interpretation of the stimuli depends upon the existing patterns of neural activity. These neural patterns, corresponding to conceptualizations or codes, may be activated by feedback from the responses as well as by stimulation from the central state, by contextual cues, by instructions, and by the nominal stimulus. The total network of certain patterns of firings is thus a selective mechanism determining which concepts and codes are likely to be activated and consequently which responses are likely to be given. This process connotes another kind of scanning or searching process – something like scanning mnemonic storage (e.g., Yntema & Trask, 1963; Sternberg, 1967; Atkinson & Shiffrin, 1968). This is the heart of the translation process, having as its product a translated or interpreted input. Thus, the stimulus is coded or conceptualized on the basis of stimulus dimensions, distinctive features, or any other unifying principle. The translation is quick if the stimulus can activate a well-established pattern of firing.

The process of recognizing a familiar pattern in an unfamiliar (reversed) orientation would merely require a reversal of temporal characteristics of the neural patterns (Hebb, 1963). Hebb emphasized that the recognition of such stimuli is a part of the perceptual mechanism, requiring corollary neural activity, and is not a result of reorganization in the sensory areas of the system. Thus, the first scanning of the stimulus trace might have one temporospatial characteristic, but this scanning might yield a perception of stimulus attributes that would alter the temporospatial characteristics of subsequent scannings of these or succeeding traces. Results of Harcum and Filion (1963) and Harcum and Smith (1963) support this argument.

Greater perceptual accuracy has been obtained in the hemifield toward which the first postexposure fixation is directed (Bryden, 1961; Crovitz and Daves, 1962; Mandes, 1966). Thus, as Bryden (1961) proposed, the postexposure eye movement facilitates the perceptual process. This implies that the motor feedback adds excitation, which could facilitate the proper corollary neural activity for the veridical perception. The more congruent the stimulus directionality and the eye movement, the greater the facilitation, as in the case of verbal patterns. This is consistent with Bryden's (1960) notion of polarized trace systems for verbal stimuli in bilateral exposure, and with Dornbush and Winnick's (1965) finding of greater hemifield differences in lateral exposure for letters in greater approximation to English.

Output
Broadbent (1958) has related the effects of recall order and knowledge about recall order to two possible processes of memory. In a storage-first process, information is transmitted into permanent memory through a temporary storage system (S) before it enters a perceiving system (P). In an overload process, the information can go into either the S- or P-system first, with the S-system used only for overloads. The S-system can pass information simultaneously (parallel processing), but the P-system passes information successively only (sequential processing). Very brief exposures of familiar tasks would not require an extensive storage in the S-system, and the process would involve primarily P-mechanisms. Serial order effects may also be found with simultaneous stimulation in the storage-first process, even though instructions concerning order of report are delayed until after stimulation. In the overload process, if input order does not coincide with the order of the P-system, it may have to be transposed before the information can be processed. In such a case, element-position effects may be abolished when the instructions about order of recall are given only after stimulation. For example, when Kay and Poulton (1951) required

subjects to recall a serial list in the normal order, the serial-position curve was obtained if the subjects knew in advance that this recall condition would be required. When the last half of the list was to be recalled first, a dip in the serial-position curve was found at the first item of the second half of the list. When the subject did not know the order of recall in advance, the usual serial-position curve was still obtained when the items were recalled in order of presentation, but a symmetrical curve of errors was obtained with the displaced order of recall.

Lawrence and LaBerge (1956) also demonstrated the importance of the knowledge of recall order. They flashed two cards containing objects varying in number, shape, and color. Instructions to attend to one dimension, given prior to exposure, increased errors on the other dimensions. Instructions concerning the order in which dimensions were to be recalled produced more errors on the last dimensions reported. Therefore, Lawrence and LaBerge concluded that their results could be accounted for purely on the basis of recall order, without recourse to a concept of selective attention. However, Broadbent (1958) interpreted this evidence to show merely that the selective nature of perception does not occur at the moment of stimulation, but as the percept develops.

Postman (1964) also emphasized the importance of whether the subject is told to recall in a given sequence. Ordered recall and free recall may under some conditions produce the same results. The items reproduced first are favored in recall, but with free recall these are likely to be the items that were the last to be presented. If responding conditions are not specified, then the subjects reproduce their responses in a habitual way – a way of least effort. For example, the order of meaningful verbal responses would be constructed according to the syntax of the language (Deese, 1961; Schlosberg, 1965). With meaningless binary patterns, the subject consistently responds from left to right when given an option (Harcum, Hartman, & Smith, 1963).

The subject brings to the task learned tendencies to organize the material for recall. This strategy of organization for the specific task may be modified as he gains more experience in the situation. In tachistoscopic pattern perception, the Camp and Harcum (1964) study discovered a change in the distribution of errors among elements of the spatial pattern after practice, with a certain type of response recording. The subjects were apparently changing their strategy of perception in accordance with the requirements of the recording system. Thus, the response may affect even the original process of translation from the stimulus into the conceptual system of the organism. Harcum and Friedman (1963a) found that presenting elements of binary patterns in sequence interfered with reproduction of the pattern, presumably because the differential strengths of traces of elements interfered with the mnemonic organization.

A compatible stimulus-response relation (Fitts & Seeger, 1953) would be one in which the organization of the stimulus configuration required minimal coding or transformation to fit into the response requirement. If a concept or code for the stimulus did not exist in the repertoire of the subject, he would have difficulty in producing a quick and accurate response, particularly if he was not instructed about the responding condition until after the exposure. If a response configuration that was relatively incompatible with the stimulus configuration was required after stimulation, again the coding or translation process would be difficult.

SUMMARY

Evidence has been presented for the effect of task-specific factors on hemifield differences in perceptual accuracy for tachistoscopic patterns, suggesting that certain functional processes can overcome or mask the structural factors that underlie and shape these processes. This evidence suggests a functional concept of hemifield dominance, that is, a particular cerebral dominance for each perceptual task. As evidence has also been presented for subject-specific factors in lateral asymmetry, some restriction on this concept is obviously necessary. The central argument of this chapter is that the assimilation of items in a series is oriented and directed by organizational factors of memory. The general process is information translation, which incorporates mechanisms for taking in information and for organizing or coding it into the subject's mnemonic system for subsequent reproduction via a given method of responding. In the organizational process some items are marked by a cognitive operation as first, to be reproduced first. One basis for perceptual primacy is continually being pitted against other determinants of primacy, particularly with bilateral stimulation. Any inhomogeneity in the stimulus array that would be helpful in identifying and differentiating elements should influence their order of analysis in the information-processing mechanism. In an otherwise homogeneous array, the items at either end would be favored as anchors in the organization process, because of their intrinsic saliency (end segregation). The significant and difficult problem is to discover those factors that assign primacy to one end of the stimulus array rather than the other.

When the organization strategies based on stimulus attributes, subject sets, and response requirements facilitate one another, the resultant process is efficient, and the distribution of errors is highly articulated (Harcum & Jones, 1962). When the various bases for organizational strategies fail to supplement one another or conflict, perceptual accuracy is poor, and the distribution of errors is undifferentiated. Cerebral insult may cause deficit by disrupting any of several mechanisms, each of which is critical for successful completion of the information-translation process.

ACKNOWLEDGMENTS

The preparation of the original paper on which this chapter is based was supported by a PHS research grant (HD-00207-07) from the Institute of Child Welfare and Human Development, United States Public Health Service. The expanded version was prepared with the support of PHS Grant HD-00207-10. The author thanks Professor M. P. Bryden of the University of Waterloo for reading and making helpful comments on a draft of this manuscript. Gratitude is expressed also to colleagues and assistants at the College of William and Mary for their advice and assistance at all stages of this work and particularly to Randolph Spencer and Richard S. Mears for collection of data and other assistance with specific experiments.

NOTES

1 Because the writing of this chapter was completed in March 1970, the recent literature in the field is not included. Indeed, the amount of uncited recent work is substantial, because the interest in the problem of asymmetrical visual performance has rapidly increased. Nevertheless, for two reasons, the material of this chapter is not out of date: (1) the methodological problems that are discussed are still with us, and (2) the tenor of the results and the conclusions are still valid and generally supported by the more recent evidence. Nevertheless, all conclusions are based on the literature cited, and no substantive changes have been attempted in the final editing.

2 This conclusion differs from that of the earlier abstract (Miller & Harcum, 1963), which reported no effect of the scrambling. The earlier conclusion was based on a grouping of the data which, according to the present analysis, was inappropriate.

REFERENCES

Anderson, I. H. 1946. The effect of letter-position on range of apprehension scores, with special reference to reading disability. *Univ. Mich. School Educ. Bull. 18*:37–40.

Anderson, I., & Crosland, H. R. 1933. A method of measuring the effect of primacy of report in the range of attention experiment. *Am. J. Psychol. 45*:701–713.

Anderson, S. B., & Ross, S. 1955. Memory for items in a matrix. *Am. J. Psychol. 68*:595–604.

Atkinson, R. C., & Shiffrin, R. M. 1968. *Some Speculations on Storage and Retrieval Processes in Long-term Memory.* Institute for Mathematical Studies in the Social Sciences, Technical Report No. 127. Stanford, Cal.: Stanford University.

Aulhorn, O. 1948. Die Lesegeschwindigkeit als Funktion von Buchstaben und Zeilenläge. *Pflüger's Arch. 250*:12–25.

Ayres, J. J. B. 1966. Some artifactual causes of perceptual primacy. *J. Exp. Psychol. 71*:896–901.

Ayres, J. J., & Harcum, E. R. 1962. Directional response-bias in reproducing brief visual patterns. *Percept. Mot. Skills 14*:155–165.

Barton, M. I., Goodglass, H., & Shai, A. 1965. Differential recognition of tachistoscopically presented English and Hebrew words in right and left visual fields. *Percept. Mot. Skills 21*:431–437.

Berger, C. I. 1944. Stroke-width, form and horizontal spacing of numerals as determinants of the threshold of recognition. *J. Appl. Psychol. 28*:208–231.

Bisiach, E. 1965. Differenze nella trasmissione di informazioni visive verso ciascun emisfero cerebrale. *Riv. Patol. Nerv. Ment. 86*:802–813.
    1966. Differenze destra-sinistra nel riconoscimento di stimoli visivi complessi. *Riv. Patol. Nerv. Ment. 87*:393–396.

Bower, T. G. R. 1965. Visual selection: Scanning vs. filtering. *Psychonomic Sci. 3*:561–562.

Boynton, R. M., & Corwin, T. 1966. Support for the "simultaneity center" hypothesis concerning the temporal perception of point light flashes. Paper presented to the Eastern Psychological Association, New York.

Braine, L. G. 1968. Asymmetries of pattern perception observed in Israelis. *Neuropsychologia 6*:73–88.

Brandt, H. F. 1940. Ocular patterns and their psychological implications. *Am. J. Psychol. 53*:260–268.

Broadbent, D. E. 1958. *Perception and Communication.* New York: Pergamon Press.

Broerse, A. C., & Zwaan, E. J. 1966. The information value of initial letters in the identification of words. *J. Verbal Learning Verbal Behav.* 5:441-446.

Bryden, M. P. 1960. Tachistoscopic recognition of non-alphabetical material. *Can. J. Psychol.* 14:78-86.

——— 1961. The role of post-exposural eye movements in tachistoscopic perception. *Can. J. Psychol.* 15:220-225.

——— 1964. Tachistoscopic recognition and cerebral dominance. *Percept. Mot. Skills* 19:686.

——— 1965. Tachistoscopic recognition, handedness, and cerebral dominance. *Neuropsychologia* 3:1-8.

——— 1966a. Accuracy and order of report in tachistoscopic recognition. *Can. J. Psychol.* 20:262-272.

——— 1966b. Left-right differences in tachistoscopic recognition: Directional scanning or cerebral dominance? *Percept. Mot. Skills* 23:1127-1134.

——— 1967. A model for the sequential organization of behaviour. *Can. J. Psychol.* 21:37-56.

——— 1968. Symmetry of letters as a factor in tachistoscopic recognition. *Am. J. Psychol.* 81:513-524.

——— 1970. Left-right differences in tachistoscopic recognition as a function of familiarity and pattern orientation. *J. Exp. Psychol.* 84:120-122.

Bryden, M. P., Dick, A. O., & Mewhort, D. J. K. 1968. Tachistoscopic recognition of number sequences. *Can. J. Psychol.* 22:52-59.

Bryden, M. P., & Rainey, C. A. 1963. Left-right differences in tachistoscopic recognition. *J. Exp. Psychol.* 66:568-571.

Burke, R. S., & Dallenbach, K. M. 1924. Position vs. intensity as a determinant of attention of left-handed observers. *Am. J. Psychol.* 35:267-269.

Camp, D. S. 1960. Attensity gradients in the perception of binary patterns. Paper presented to the Virginia Academy of Science. Abstract: *Va. J. Sci.* 11:217-218.

Camp, D. S., & Harcum, E. R. 1964. Visual pattern perception with varied fixation locus and response recording. *Percept. Mot. Skills* 18:283-296.

Campbell, D. T., & Fiske, D. W. 1959. Convergent and discriminant validation by the multitrait-multimethod matrix. *Psychol. Bull.* 56:81-105.

Chaikin, J. D., Corbin, H. H., & Volkmann, J. 1962. Mapping a field of short-time visual search. *Science* 138:1327-1328.

Chen, L. K., & Carr, H. A. 1926. The ability of Chinese students to read in vertical and horizontal directions. *J. Exp. Psychol.* 9:110-117.

Clark, W. C. 1966. The *psyche* in psychophysics: A sensory-decision theory analysis of the effect of instructions on flicker sensitivity and response bias. *Psychol. Bull.* 65:358-366.

Corballis, M. C. 1964. Binocular interactions in letter recognition. *Aust J. Psychol.* 16:38-47.

Crosland, H. R. 1931. Letter-position effects, in the range of attention experiment, as affected by the number of letters in each exposure. *J. Exp. Psychol. 14:*477–507.

———— 1939. Superior elementary-school readers contrasted with inferior readers in letter-position, "range of attention," scores. *J. Educ. Res. 32:*410–427.

Crovitz, H. F. 1961. Differential acuity of the two eyes and the problem of ocular dominances. *Science 134:*614.

Crovitz, H. F., Daston, P. G., & Zener, K. E. 1959. Laterality and a phenomenon of localization. *Percept. Mot. Skills 9:*282.

Crovitz, H. F., & Daves, W. 1962. Tendencies to eye movement and perceptual accuracy. *J. Exp. Psychol. 63:*495–498.

Crovitz, H. F., & Friedman, L. A., 1967. Configurational letter spans. *J. Exp. Psychol. 73:*628–629.

Crovitz, H. F., & Lipscomb, D. B. 1963. Dominance of the temporal visual fields at a short duration of stimulation. *Am. J. Psychol. 76:*631–637.

Crovitz, H. F., & Schiffman, H. R. 1965. Visual field and the letter span. *J. Exp. Psychol. 70:*218–223.

Crovitz, H. F., Schiffman, H., Lipscomb, D. B., Posnik, G., Rees, J., Schaub, R., & Tripp, R. 1966. Identification and localization in the letter span. *Can. J. Psychol. 20:*455–461.

Crovitz, H. F., & Zener, K. 1962. A group-test for assessing hand- and eye-dominance. *Am. J. Psychol. 75:*271–276.

Dallenbach, K. M. 1920. Attributive vs. cognitive clearness. *J. Exp. Psychol., 3:*183–230.

———— 1923. Position vs. intensity as a determinant of clearness. *Am. J. Psychol. 34:*282–286.

Deese, J. 1961. From the isolated verbal unit to connected discourse. In C. N. Cofer (ed.). *Verbal Learning and Verbal Behavior*, pp. 11–31. New York: McGraw-Hill.

Derks, P. L., Cherry, R. L., & Larson, A. V. 1969. Effect of event proportion on the short-term perception of linear binary patterns. *J. Exp. Psychol. 79:*85–92.

Dick, A. O., & Mewhort, D. J. K. 1967. Order of report and processing in tachistoscopic recognition. *Percent. Psychopsysics 2:*573–576.

Doerries, L. E., & Harcum, E. R. 1967. Long-term traces of tachistoscopic word perception. *Percept. Mot. Skills 25:*25–35.

Dornbush, R. L., & Winnick, W. A. 1965. Right-left differences in tachistoscopic identification of paralogs as a function of order of approximation to English letter sequences. *Percept. Mot. Skills 20:*1222-1224.

Dyer, D. W., & Harcum, E. R. 1961. Visual perception of binary patterns by preschool children and by school children. *J. Educ. Psychol. 52:*161–165.

Efron, R. 1963. The effect of handedness on the perception of simultaneity and temporal order. *Brain 86:*261–284.

Egeth, H. 1967. Selective attention. *Psychol. Bull.* 67:41-57.
Ellis, N. R. 1969. Evidence for two storage processes in short-term memory. *J. Exp. Psychol.* 80:390-391.
Feigenbaum, E. A., & Simon, H. A. 1962. A theory of the serial position effect. *Br. J. Psychol.* 53:307-320.
Fitts, P. M., & Seeger, C. M. 1953. S-R compatibility: Spatial characteristics of stimulus and response codes. *J. Exp. Psychol.* 46:199-210.
Fitzgerald, R. E., & Marshall, A. J. 1967. Left-right field differences with partial report of letters. *Am. J. Psychol.* 80:370-376.
Forgays, D. G. 1953. The development of differential word recognition. *J. Exp. Psychol.* 45:165-168.
Freeburne, C. M., & Goldman, R. D. 1969. Left-right differences in tachistoscopic recognition as a function of order of report, expectancy, and training. *J. Exp. Psychol.* 79:570-572.
Freedman, S. J. 1968. On the mechanisms of perceptual compensation. In S. J. Freedman (ed.). *The Neuropsychology of Spatially Oriented Behavior,* pp. 231-240. Homewood, Ill.: Dorsey Press.
Friedman, H. 1968. Magnitude of experimental effect and a table for its rapid estimation. *Psychol. Bull.* 70:245-251.
Fudin, R. 1969a. Critique of Heron's directional-reading conflict theory of scanning. *Percept. Mot. Skills* 29:271-276.
  1969b. Recognition of alphabetical arrays presented in the right and left visual fields. *Percept. Mot. Skills* 29:15-22.
Gabersek, V. 1963. Le temps de réaction oculomoteur en fonction de l'âge. *J. Physiol.* 55:252.
Garner, W. R. 1962. *Uncertainty and Structure as Psychological Concepts.* New York: Wiley.
Garner, W. R., & Gottwald, R. L. 1967. Some perceptual factors in the learning of sequential patterns of binary events. *J. Verbal Learning Verbal Behav.* 6:582-589.
Garner, W. R., Hake, H. W., & Eriksen, C. W. 1956. Operationism and the concept of perception. *Psychol. Rev.* 63:149-159.
Gerjuoy, I. R., & Winters, J. J., Jr. 1966. Lateral preference for identical geometric forms 2. Retardates. *Percept. Psychophysics* 1:104-106.
Gilinsky, A. S. 1952. *A Review of Literature on the Relative Efficiency of the Dominant and the Non-dominant Eye. Wright Air Development Center* Technical Report 52-13.
Glanville, A. D., & Dallenbach, K. M. 1929. The range of attention. *Am. J. Psychol.* 41:207-236.
Glanzer, M. 1966. Encoding in the perceptual (visual) serial position effect. *J. Verbal Learning Verbal Behav.* 5:92-97.
Glanzer, M., & Clark, W. H. 1963. Accuracy of perceptual recall: An analysis of organization. *J. Verbal Learning Verbal Behav.* 1:289-299.
Goodglass, H., & Barton, M. 1963. Handedness and differential perception

of verbal stimuli in left and right visual fields. *Percept. Mot. Skills* 17:851–854.

Gottschalk, J., Bryden, M. P., & Rabinovitch, M. S. 1964. Spatial organization of children's responses to a pictoral display. *Child Dev.* 35:811–815.

Gropper, B. A. 1968. Effects of spatial-temporal ordering on short-term memory for arrays of digits. Doctoral dissertation, Johns Hopkins University.

Haber, R. N. 1966. Nature of the effect of set on perception. *Psychol. Rev.* 73:335–351.

Haber, R. N., & Hershenson, M. 1965. The effects of repeated brief exposures on the growth of a percept. *J. Exp. Psychol.* 69:40–46.

Harcum, E. R. 1957a. Three inferred factors in the visual recognition of binary targets. In J. W. Wulfeck & J. H. Taylor (eds.). *Form Discrimination as Related to Military Problems,* pp. 32–37. National Academy of Science–National Research Council Publication 561. Washington, D.C.: National Academy of Science.

1957b. Visual recognition along four meridians of the visual field: Preliminary experiments. University of Michigan, Project Michigan Report 2144-50-T.

1958a. Visual recognition along various meridians of the visual field. 2. Nine-element typewritten targets. University of Michigan Engineering Research Institute, Project Michigan Report 2144-293-T.

1958b. Visual recognition along various meridians of the visual field. 6. Eight-element and ten-element binary patterns. University of Michigan, Project Michigan Report 2144-303-T.

1958c. Visual recognition along various meridians of the visual field. 7. Effect of target length measured in angular units. University of Michigan, Project Michigan Report 2144-304-T.

1958d. Visual recognition along various meridians of the visual field. 9. Monocular and binocular recognition of patterns of squares and circles. University of Michigan, Project Michigan Report 2144-307-T.

1958e. Visual recognition along various meridians of the visual field. 10. Binary patterns of the letters "H" and "O." University of Michigan, Project Michigan Report 2144-308-T.

1958f. Visual recognition along various meridians of the visual field. 12. Acuity for open and blackened circles. University of Michigan, Project Michigan Report 2144-315-T.

1962. Recognition vs. recall of tachistoscopic patterns. *Percept. Mot. Skills* 15:238.

1964a. Effect of pre-recognition exposures on visual perception of words. *Percept. Mot. Skills* 16:99–104.

1964b. Effects of symmetry on the perception of tachistoscopic patterns. *Am. J. Psychol.* 77:600–606.

1964c. Interactive effects within visual patterns on the discriminability of individual elements. *J. Exp. Psychol. 68:*351–356.

1964d. *Reproduction of linear visual patterns tachistoscopically exposed in various orientations.* Williamsburg, Va., College of William and Mary.

1965a. An isolation effect in pattern perception similar to that in serial learning. *Percept. Mot. Skills 20:*1121–1130.

1965b. Pre-knowledge of isolation as a prerequisite for the isolation-effect. *Psychonomic Sci. 3:*443–444.

1966a. Mnemonic organization as a determinant of error-gradients in visual pattern perception. *Percept. Mot. Skills 22:*671–696. Monograph supplement 5-V22.

1966b. Visual hemifield differences as conflicts in direction of reading. *J. Exp. Psychol. 72:*479–480.

1967a. A note on "Some artifactual causes of perceptual primacy." *Psychonomic Sci. 8:*67–68.

1967b. Parallel functions of serial learning and tachistoscopic pattern perception. *Psychol. Rev. 74:*51–62.

1967c. Two possible mechanisms of differential set in tachistoscopic perception of multiple targets. *Percept. Mot. Skills 25:*289–304.

1968a. Hemifield differences in visual perception of redundant stimuli. *Can. J. Psychol. 22:*197–211.

1968b. A note on "Encoding in the perceptual (visual) serial position effect." *J. Verbal Learning Verbal Behav. 7:*275–277.

1968c. Perceptual serial-position curves with a frequently isolated element. *Am. J. Psychol. 81:*334–346.

1969a. Cognitive anchoring of different errors in continuous serial learning. *Psychol. Rep. 25:*79–82.

1969b. Continuous tachistoscopic patterns reproduced without laterality effects. *Am. J. Psychol. 82:*504–512.

1970a. Defining shape for perceptual element-position curves. *Psychol. Bull. 74:*362–372.

1970b. Perceptibility gradients for tachistoscopic patterns: Sensitivity or saliency? *Psychol. Rev. 77:*332–337.

Harcum, E. R., & Blackwell, H. R. 1958. Visual recognition along various meridians of the visual field. 11. Identification of the number of blackened circles presented. University of Michigan, Project Michigan Report 2144-314-T.

Harcum, E. R., & Dyer, D. W. 1962. Monocular and binocular reproduction of binary stimuli appearing right and left of fixation. *Am. J. Psychol. 75:*56–65.

Harcum, E. R., & Filion, R. D. L. 1963. Effects of stimulus reversals on lateral dominance in word recognition. *Percept. Mot. Skills 17:*779–794.

Harcum, E. R., Filion, R. D. L., & Dyer, D. W. 1962. Distribution of errors in tachistoscopic reproduction of binary patterns after practice. *Percept. Mot. Skills* 15:83–89.

Harcum, E. R., & Finkel, M. E. 1963. Explanation of Mishkin and Forgays' result as a directional-reading conflict. *Can. J. Psychol.* 17:224–234.

Harcum, E. R., & Friedman, S. M. 1963a. Reproduction of binary visual patterns having different element-presentation sequences. *J. Exp. Psychol.* 66:300–307.

1963b. Reversal reading by Israeli observers of visual patterns without intrinsic directionality. *Can. J. Psychol.* 17:361–369.

Harcum, E. R., Hartman, R. R., & Smith, N. F. 1963. Pre- vs. post-knowledge of required reproduction sequence for tachistoscopic patterns. *Can. J. Psychol.* 17:264–273.

Harcum, E. R., & Jones, M. L. 1962. Letter recognition within words flashed left and right of fixation. *Science* 138:444–445.

Harcum, E. R., & Rabe, A. 1958a. Visual recognition along various meridians of the visual field. 3. Patterns of blackened circles in an eight-circle template. University of Michigan, Project Michigan Report 2144-294-T.

1958b. Visual recognition along various meridians of the visual field. 4. Linear binary patterns at 36 orientations. University of Michigan, Project Michigan Report 2144-296-T.

Harcum, E. R., & Skrzypek, G. 1965. Configuration determinants in visual perception of binary patterns: Supplementary report. *Percept. Mot. Skills* 21:860–862.

Harcum, E. R., & Smith, N. F. 1963. Effect of pre-known stimulus-reversals on apparent cerebral dominance in word recognition. *Percep. Mot. Skills* 17:799–810.

1966. Stability of error distributions within tachistoscopic patterns. *Psychonomic Sci.* 6:287–288.

Harris, C. S. 1965. Perceptual adaptation to inverted, reversed, and displaced vision. *Psychol. Rev.* 72:419–444.

Harris, C. S., & Haber, R. N. 1963. Selective attention and coding in visual perception. *J. Exp. Psychol.* 65:328–333.

Haslerud, G. M., & Clark, R. E. 1957. On the redintegrative perception of words. *Am. J. Psychol.* 70:97–101.

Haslerud, G. M., & Motoyoshi, R. 1961. Direction of incidental learning in relation to Japanese and American reading habits. *Percept. Mot. Skills* 12:142.

Havens, L. L., & Foote, W. E. 1963. The effect of competition on visual duration threshold and its independence of stimulus frequency. *J. Exp. Psychol.* 65:6–11.

Hayashi, T., & Bryden, M. P. 1967. Ocular dominance and perceptual asymmetry. *Percept. Mot. Skills* 25:605–612.

Hays, W. L. 1965. *Statistics for Psychologists.* New York: Holt.

Hebb, D. O. 1949. *The Organization of Behavior.* New York: Wiley.
　1963. The semiautonomous process: Its nature and nurture. *Am. Psychol.* 18:16–27.
Held, R., & Hein, A. V. 1958. Adaptation of disarranged hand-eye coordination contingent upon re-afferent stimulation. *Percept. Mot. Skills* 8:87–90.
Heron, W. 1957. Perception as a function of retinal locus and attention. *Am. J. Psychol.* 70:38–48.
Hershenson, M. 1969a. Perception of letter arrays as a function of absolute retinal locus. *J. Exp. Psychol.* 80:201–202.
　1969b. Stimulus structure, cognitive structure, and the perception of letter arrays. *J. Exp. Psychol.* 79:327–335.
Hines, D., Satz, P., Schell, B., & Schmidlin, S. 1969. Differential recall of digits in the left and right visual half-fields under free and fixed order recall. *Neuropsychologia* 7:13–22.
Hirata, K., & Osaka, R. 1967. Tachistoscopic recognition of Japanese letter materials in left and right visual fields. *Psychologia* 10:7–18.
Horowitz, L. M., Chilian, P. C., & Dunningan, K. P. 1969. Word fragments and their redintegrative powers. *J. Exp. Psychol.* 80: 392–394.
Horowitz, L. M., White, M. A., & Atwood, D. W. 1968. Word fragments as aids to recall: The organization of a word. *J. Exp. Psychol.* 76: 219–226.
Jasper, H. H., & Raney, E. T. 1937. The phi test of lateral dominance. *Am. J. Psychol.* 49:450.
Jones, M. H., & Jones, F. N. 1947. The relationship of verbal reaction time to hemisphere of entry of a visual stimulus. *Am. Psychol.* 2:408.
Kaufman, E. L., Lord, M. W., Reese, T. W., & Volkmann, J. 1949. The discrimination of visual number. *Am. J. Psychol.* 62:498–525.
Kay, H., & Poulton, E. C. 1951. Anticipation in memorizing. *Br. J. Psychol.* 42:34–41.
Kephart, N. C., & Revesman, S. 1953. Measuring differences in speed performance. *Optom. Weekly* 44:1965–1967.
Kimura, D. 1959. The effect of letter position on recognition. *Can. J. Psychol.* 13:1–10.
　1966. Dual functional asymmetry of the brain in visual perception. *Neuropsychologia* 4:275–285.
Kirssin, J. E., & Harcum, E. R. 1967. Relative attensity of stimuli in left and right visual hemifields. *Percept. Mot. Skills* 24:807–822.
Klemmer, E. T. 1963. Perception of linear dot patterns. *J. Exp. Psychol.* 65: 468–473.
Knehr, C. A. 1941. The effects of monocular vision on measures of reading efficiency and perceptual span. *J. Exp. Psychol.* 29:133–154.
Koestler, A., & Jenkins, J. J. 1965. Inversion effects in the tachistoscopic perception of number sequences. *Psychonomic Sci.* 3:75–76.

Kolers, P. A., & Perkins, D. N. 1969. Orientation of letters and their speed of recognition. *Percept. Psychophysics 5:*275–280.

Krise, E. M. 1949. Reversals in reading: A problem in space perception. *Elementary School J. 49:*278–284.

L'Abate, L. 1960. Recognition of paired trigrams as a function of associative value and associative strength. *Science 131:*984–985.

LaGrone, C. W., Jr. 1942. An experimental study of the relationship of peripheral perception to factors in reading. *J. Exp. Educ. 11:*37–49.

LaGrone, C. W., & Holland, B. F. 1943. Accuracy of perception in peripheral vision in relation to dextrality, intelligence and reading ability. *Am. J. Psychol. 56:*592–598.

Lashley, K. S. 1951. The problem of serial order in behavior. In L. A. Jeffress (ed.). *Cerebral Mechanisms in Behavior.* pp. 112–136. New York: Wiley.

Lawrence, D. H., & Coles, G. R. 1954. Accuracy of recognition with alternatives before and after the stimulus. *J. Exp. Psychol. 47:*208–214.

Lawrence, D. H., & LaBerge, D. L. 1956. Relationship between recognition accuracy and order of reporting stimulus dimensions. *J. Exp. Psychol. 51:*12–18.

Mandes, E. J. 1966. A developmental study of the relation between visual field accuracy and eye movement directionality. Doctoral dissertation, George Washington University.

Marchbanks, G., & Levin, H. 1965. Cues by which children recognize words. *J. Educ. Psychol. 56:*57–61.

Markowitz, H., & Weitzman, D. O. 1969. Monocular recognition of letters and Landolt Cs in left and right visual hemifields. *J. Exp. Psychol. 79:* 187–189.

Masani, P. A. 1967. The attention mechanism and its relationship to postperceptual short-term storage. *Papers Psychol. 1*(2).

Massa, R. J. 1968. An investigation of human visual information transmission. In *Research Bulletin Number 18*, pp. 25–140. New York: American Association for the Blind.

Mathewson, J. W., Jr., Miller, J. C., Jr., and Crovitz, H. F. 1968. The letter span in space and time. *Psychonomic Sci. 11:*69–70.

McFie, J. 1952. Cerebral dominance in cases of reading disability. *J. Neurol. Psychiatry 15:*194–199.

McKenna, V. V., & Harcum, E. R. 1967. Strategies in serial learning and tachistoscopic perception of aged subjects. *Va. J. Sci. 18:*210 (abstract).

McKinney, J. P. 1967. Handedness, eyedness and perceptual stability of the left and right visual fields. *Neuropsychologia 5:*339–344.

Melton, A. W. 1963a. Comments on Professor Peterson's paper. In C. N. Cofer and B. S. Musgrave (eds.). *Verbal Behavior and Learning: Problems and Processes*, pp. 353–370. New York: McGraw-Hill.

1963b. Implications of short-term memory for a general theory of memory. *J. Verbal Learning Verbal Behav. 2:*1–21.

Melville, J. R. 1957. Word-length as a factor in differential recognition. *Am. J. Psychol. 70:*316–318.

Mewhort, D. J. K. 1966. Sequential redundancy and letter spacing as determinants of tachistoscopic recognition. *Can. J. Psychol. 20:*435–444.

1967. Familiarity of letter sequences, response uncertainty, and the tachistoscopic recognition experiment. *Can. J. Psychol. 21:*309–321.

Mewhort, D. J. K., Merikle, P. M., & Bryden, M. P. 1969. On the transfer from iconic to short-term memory. *J. Exp. Psychol. 81:*89–94.

Michon, J. A. 1964. Temporal structure of letter groups and span of perception. *Q. J. Exp. Psychol. 16:*232–240.

Miller, B., & Harcum, E. R. 1963. Left-right redundancy and the perception of visual patterns. *Va. J. Sci. 14:*272–273 (abstract).

Miller, G. A. 1963. Comments on Professor Postman's paper. In C. N. Cofer and B. S. Musgrave (eds.). *Verbal Behavior and Learning: Problems and Processes,* pp. 321–329. New York: McGraw-Hill.

Miller, G. A., Bruner, J. S., & Postman, L. 1954. Familiarity of letter sequences and tachistoscopic identification. *J. Gen. Psychol. 50:*129–139.

Milner, P. M. 1961. A neural mechanism for the immediate recall of sequences. *Kybernetik 1:*76–81.

Mishkin, M., & Forgays, D. G. 1952. Word recognition as a function of retinal locus. *J. Exp. Psychol. 43:*43–48.

Monty, R. A. 1968. Spatial encoding strategies in sequential short-term memory. *J. Exp. Psychol. 77:*506–508.

Müller, G. E., & Pilzecker, A. 1900. Experimentelle Beiträge zur Lehre vom Gedächtniss. *Z. Psychol. 1:*1–288.

Natsoulas, T. 1967. What are perceptual reports about? *Psychol. Bull. 67:* 249–272.

1968. Interpreting perceptual reports. *Psychol. Bull. 70:*575–591.

Neisser, U. 1967. *Cognitive Psychology.* New York: Appleton-Century-Crofts.

Ogle, K. N. 1962. Ocular dominance and binocular retinal rivalry. In H. Davson (ed.). *The Eye,* Vol. 4, pp. 409–417. New York: Academic Press.

Orbach, J. 1952. Retinal locus as a factor in the recognition of visually perceived words. *Am. J. Psychol. 65:*555–562.

1967. Differential recognition of Hebrew and English words in right and left visual fields as a function of cerebral dominance and reading habits. *Neuropsychologia 5:*127–134.

Overton, W., & Wiener, M. 1966. Visual field position and word-recognition threshold. *J. Exp. Psychol. 71:*249–253.

Penfield, W., & Roberts, L. 1959. *Speech and Brain-Mechanisms.* Princeton: Princeton University Press.

Pierce, J. 1963. Determinants of threshold for form. *Psychol. Bull. 60:*391–407.

Pillsbury, W. B. 1897. A study in apperception. *Am. J. Psychol. 8:*315–393.

Pollack, I. 1953. Assimilation of sequentially encoded information. *Am. J. Psychol.* 66:421-435.

Postman, L. 1964. Short-term memory and incidental learning. In A. W. Melton (ed.). *Categories of Human Learning*, pp. 145-201. New York: Academic Press.

Price, R. H. 1966. Signal-detection methods in personality and perception. *Psychol. Bull.* 66:55-62.

Regan, J. O., & Cropley, A. J. 1964. Directional set in serial, perceptual-motor tasks. *Percept. Mot. Skills* 19:579-586.

Rossi, G. F., & Rosadini, G. 1967. Experimental analysis of cerebral dominance in man. In F. L. Darley (ed.). *Brain Mechanisms Underlying Speech and Language*, pp. 167-184. New York: Grune & Stratton.

Rommetveit, R., Berkley, M., & Brøgger, J. 1968. Generation of words from stereoscopically presented non-word strings of letters. *Scand. J. Psychol.* 9:150-156.

Rutschmann, R. 1966. Perception of temporal order and relative visual latency. *Science* 152:1099-1101.

Sampson, H., & Spong, P. 1961. Handedness, eye-dominance, and immediate memory. *Q. J. Exp. Psychol.* 13:173-180.

Sanders, A. F. 1963. *The Selective Process in the Functional Visual Field.* Soesterberg, the Netherlands: Institute for Perception RVO-TNO.

Schlosberg, H. 1965. Time relations in serial visual perception. *Can. Psychol.* 6a:161-172.

Schonell, F. J. 1940. The relation of reading disability to handedness and certain ocular factors. *Br. J. Educ. Psychol.* 10:227-237.

Semmes, J. 1968. Hemispheric specialization: A possible clue to mechanism. *Neuropsychologia* 6:11-26.

Shiffrin, R. M., & Atkinson, R. C. 1969. Storage and retrieval processes in long-term memory. *Psychol. Rev.* 76:179-193.

Sperling, G. 1960. The information available in brief visual presentations. *Psychol. Monogr.* 74 (whole number 498).

1967. Successive approximations to a model for short-term memory. *Acta Psychol.* 27:285-292.

Sternberg, S. 1967. Two operations in character recognition: Some evidence from reaction-time measurements. *Percept. Psychophysics* 2:45-53.

Takala, M. 1951. Asymmetries of the Visual Space. *Ann. Acad. Sci. Fenn.* Series B, 72:1-175.

Tanner, W. P., Jr., & Swets, J. A. 1954. A decision-making theory of visual detection. *Psychol. Rev.* 61:401-409.

Terrace, H. S. 1959. The effects of retinal locus and attention on the perception of words. *J. Exp. Psychol.* 58:382-385.

Teuber, H.-L. 1960. Perception. In J. Field, H. W. Magoun, & V. E. Hall (eds.).

Handbook of Physiology, Vol. 3, Neurophysiology, pp. 1595-1668. Washington, D.C.: American Physiological Society.

Tinker, M. A. 1946. The study of eye movements in reading. Psychol. Bull. 43:93-120.

1955. Perceptual and oculomotor efficiency in reading materials in vertical and horizontal arrangements. Am. J. Psychol. 68:444-449.

van der Meer, H. C. 1959. Die links-rechts-Polarisation des phaenomenalen Raumes. Groningen: Wolters.

Wagner, J. 1918. Experimentelle Beiträge zür Psychologie des Lesens. Z. Psychol. 80:1-75.

Walker, E. L. 1958. Action decrement and its relation to storage. Psychol. Rev. 65:129-142.

Warrington, E. K., James, M., & Kinsbourne, M. 1966. Drawing disability in relation to laterality of cerebral lesion. Brain 89:53-82.

Warrington, E. K., Kinsbourne, M., & James, M. 1966. Uncertainty and transitional probability in the span of apprehension. Br. J. Psychol. 57:7-16.

Werner, H. 1937. Process and achievement: A basic problem of education and developmental psychology. Harvard Educ. Rev. 7:353-368.

Wertheim, T. 1894. Über die indirekte Sehschärfe. Z. Psychol. 7:172-187.

White, A. M., & Dallenbach, K. M. 1932. Position vs. intensity as a determinant of the attention of left-handed observers. Am. J. Psychol. 44:175-179.

White, M. J. 1969a. Identification and localization within digit and letter spans. Psychonomic Sci. 14:279-280.

1969b. Laterality differences in perception: A review. Psychol. Bull. 72:387-405.

Winnick, W. A., & Bruder, G. E. 1968. Signal detection approach to the study of retinal locus in tachistoscopic recognition. J. Exp. Psychol. 78:528-531.

Winnick, W. A., & Dornbush, R. L. 1965. Pre- and post-exposure processes in tachistoscopic identification. Percept. Mot. Skills 20:107-113.

Winnick, W. A., Luria, J., & Zukor, W. J. 1967. Two signal detection approaches to tachistoscopic recognition. Percept. Mot. Skills 24:795-803.

Winters, J. J., Jr., & Gerjuoy, I. R. 1966. Lateral preference for identical geometric forms. 1. Normals. Percept. Psychophysics 1:101-103.

Woodworth, R. S. 1938. Experimental Psychology. New York: Holt.

Wyatt, D. F., & Campbell, D. T. 1951. On the liability of stereotype or hypothesis. J. Abnorm. Soc. Psychol. 46:496-500.

Wyke, M., & Chorover, S. L. 1965. Comparison of spatial discrimination in the temporal and nasal sectors in the monocular visual field. Percept. Mot. Skills 20:1037-1045.

Wyke, M., & Ettlinger, G. 1961. Efficiency of recognition in left and right visual fields. Arch. Neurol. 5:659-665.

Yntema, D. B., & Trask, F. P. 1963. Recall as a search process. *J. Verbal Learning Verbal Behav.* 2:65–74.

Zangwill, O. L. 1960. *Cerebral Dominance and Its Relations to Psychological Function.* Edinburgh: Oliver & Boyd.

Zurif, E. B., & Bryden, M. P. 1969. Familial handedness and left-right differences in auditory and visual perception. *Neuropsychologia* 7:179–187.

7
Mapping cerebral functional space: competition and collaboration in human performance

MARCEL KINSBOURNE AND ROBERT E. HICKS

Human performance is a useful tool for revealing cerebral organization. The thesis of this chapter is that the *functional distance* between any two cerebral control centers decreases with the extent to which they collaborate on concordant tasks and with the extent to which they compete on discordant tasks. Thus, if effector A can be paired with either effector B (functionally close) or effector C (functionally distant), then the AB combination will more efficiently perform concordant movement sequences, whereas the AC combination will more effectively perform discordant movement sequences. First, some discussion of the neural centers underlying the performance variables presented.

CORTICAL INTERCONNECTIONS

The studies of performance presented here deal with vocalization and movements of the four limbs. There is evidence that neural connections are greatest between contralateral homologous limbs, somewhat more sparse between ipsilateral limbs, and indirect (i.e., requiring transsynaptic information flow) between diagonal limb pairings (i.e., right arm–left leg and left arm–right leg) (McCulloch, 1949; Krieg, 1963). Davis (1942a,b) measured the electromyographic (EMG) response amplitude of all four limbs when one limb was responding. Expressed as a percentage of the EMG of the responding limb, he found (1942a) 7.91% activity in the contralateral homologous limb, 3.6% in the ipsilateral limb, and 3.1% in the diagonally opposite limb. Cernacek (1961) had subjects flex and extend the index finger of one hand and then of the other. He measured the EMG of the flexor and extensor muscles of the fingers of both hands. All the subjects had contralateral (in homologous muscle groups) EMG activation during unilateral response, and this occurred on a mean of 77% of the 420 trials given each subject.

No direct connections exist between Broca's area (subserving speech) and any limb control centers in humans. However, McAdam and Whitaker (1971) found that during speech production tasks, greater electrical activity occurred over Broca's area than in corresponding areas of the right cerebral hemisphere, and they also found more pronounced activity in the left precentral gyrus (motor cortex) than in the right precentral gyrus. It is reasonable to assume that the control center for the right arm is functionally closer to the speech control center than any of the other limb control centers.

In summary, the functional distance is least between contralateral homologous limbs, intermediate between ipsilateral limbs, and greatest between diagonal limbs. The functional distance between vocal control and the various limbs is least for the right hand in right-handers. This is because within the motor strip, articulatory movement is lateral, leg movement medial, and arm movement intermedial in representation (Krieg, 1963).

INTERACTIONS BETWEEN LIMBS

Competition

The model predicts that the amount of concurrent interference in limb response should decrease directly with functional distance between the corresponding cerebral control centers. Welch (1898) had subjects perform rhythmical squeezing motions with one hand while the other hand attempted to maintain steady pull on a dynamometer. These tasks interacted, rendering the dynamometer pull variable. Similarly, Cohen (1970) found that a sequence of rhythmical alternating movements performed by one limb was modified when rapid movements or brief isometrical contractions were executed by the contralateral limb.

Briggs and Kinsbourne (1976) used all possible pairings of the four limbs in a concurrent dual-limb step-tracking task. In accordance with the functional distance model, they found that errors were least frequent for diagonal limb pairings, intermediate for ipsilateral limb pairings, and most frequent for contralateral homologous pairings. The same was true for speed of responding.

Collaboration

A response is faster if it is a repetition of the previous response than if it follows a different response (Bertelson, 1963, 1965). In addition, a response is faster if it follows another response made with the same hand than if it follows a response made with the other hand (Rabbitt, 1965).

The pattern of facilitation between responses of the limbs is consonant with what the functional space model predicts. Blyth (1962) examined all 12 of the possible cases in which responses one and two are made with different limbs. He found that the second response was faster when the two responses required successive movements with ipsilateral limbs or with contralateral homologous limbs than with diagonal limbs.

Practice on almost any motor task produces enhanced performance efficiency (i.e., learning occurs). Practice with one limb also produces enhanced performance efficiency in other limbs. The pattern of interlimb transfer of training is exactly predicted by the functional space model.

Cook (1933a) gave subjects 100 practice trials with the right hand in (visually guided) mirror tracing of a star-shaped maze. Then the left foot, right foot, and left hand were tested on the task. Another group was given original training with the left foot and had the other limbs tested for transfer. He found 89%, 81%, and 74% transfer for contralateral homologous, ipsilateral, and diagonal limbs, respectively. When only 10 original training trials were given (Cook, 1933b), the amounts of transfer obtained were 77%, 66%, and 55% for the three limb sequences. Cook (1934) used an irregular maze without visual guidance. The pattern of transfer was identical (90%, 80%, and 63%, given in the same order as above).

Ammons and Ammons (1970) provide a measure of commonality of response processes between the various limbs. They measured rotary pursuit performance for two successive 8-minute periods of continuous practice. There were 16 groups formed by the orthogonal pairings of limbs across the two sessions. Ammons and Ammons presented the correlation between sessions separately for each group. Based on these data, the mean correlations were: (1) for the four groups that used the same limb on both sessions, 0.85 (72% common variance); (2) for the four groups using contralateral homologous limbs on the two sessions, 0.76 (58% common variance); (3) for the four groups using ipsilateral limbs, 0.64 (40% common variance); and for the four groups using diagonal limbs, 0.56 (32% common variance).

The functional space model appears consonant with the results obtained in interference and in facilitation studies of the limbs.

With the exception of Briggs and Kinsbourne (1976), the preceding studies of human performance were primarily empirical investigations. No attempt was made to relate the results to known principles of cerebral organization. In contrast, the studies described below represent explicit attempts to test the functional distance model, first proposed by Kinsbourne and Cook (1971).

VOCAL-MANUAL INTERACTIONS

The functional distance model predicts that a concurrent verbal task will interfere with right-handed performance more than with left-handed performance (in right-handed subjects). This is because the right-hand control center is functionally closer than the left-hand control center to the speech area. This hypothesis has received support from a number of studies using varied tasks (Kinsbourne and Hicks, in press).

Kinsbourne and Cook (1971) had right-handed subjects balance a dowel rod on the index finger silently and while repeating sentences. They found that right-hand, but not left-hand, balancing was disrupted by the concurrent speech (i.e., balancing duration was decreased). Hicks (1975) replicated and extended this result. He found that increased phonetic difficulty of the verbalized material increased the interference with right-handed performance. Left-handed performance was unaffected by verbalization at both difficulty levels. The verbalization interfered with both hands of left-handers (who, as a group, have less clearly lateralized verbal functions).

Hicks (1975) also demonstrated that humming produced the same interference effects as speaking. Hicks et al. (1977) used finger-sequencing tasks and also found common (lateralized) interference effects for speaking and humming. The latter study also found that the amount of lateralized interference was inversely related to the response-response compatibility of the fingering sequence.

Hicks et al. (1975) found that concurrent verbal rehearsal produced more interference with finger movements of the right hand than of the left hand. This was true for silent as well as for vocalized (aloud) rehearsal.

Briggs (1975) found that the right hand made more errors than the left hand in a bimanual step-tracking task with a concurrent verbal task. The right hand was superior on control trials.

Kreuter et al. (1972) had a callosectomized subject tap with her index finger while performing verbal tasks. The disruption of tapping was greater for the right hand. Kinsbourne and McMurray (1975) and Kinsbourne and Hiscock (1977) found similar results in normal children.

In a recently completed experiment, Kinsbourne, La Casse, and Hicks (unpublished) used accomplished musicians. After much practice, the subjects were able simultaneously to play two orthogonal tunes on a piano; one with each hand. On some trials, they were also required to hum along with one of the tunes. Two types of errors (in playing) were observed: erroneous synchronization (substitution of a note that the other hand was simultaneously playing) and random errors (all errors other than synchronization errors). They found that random errors were invariant across conditions, but synchronization errors were reduced by half in the hand playing the tune that was being hummed compared with the other hand. However, accuracy was greatest when humming accompanied the right hand; it was somewhat less when humming accompanied the left hand. This outcome was predicted because the functional distance is less between the control centers for voice and the right hand. Thus, it is most advantageous for these two effectors to be concordantly active, rather than the more distant voice–left-hand pairing.

CONCLUSION

What is the mechanism of the functional distance effect? The following suggestion may have heuristic possibilities: When two tasks are performed concurrently, crosstalk is averted centrally by an inhibitory barrier (Powell & Mountcastle, 1959). Throwing up a barrier between control centers consumes neural capacity, leaving less available for expenditure on the two performances. If two control centers are richly and directly connected, this insulating process is more demanding than if they are sparsely and indirectly connected. Thus, at the neurophysiological level, it may at some time become possible to test the following prediction: A given cerebral control center is surrounded by more intense inhibition when an adjacent center is concurrently active than when a more distant center is concurrently active.

REFERENCES

Ammons, R. B., & Ammons, C. H. 1970. Decremental and related processes in skilled performance. In L. E. Smith (ed.). *Psychology of Motor Learning*. Proceedings of CIC Symposium on Psychology of Motor Learning, University of Iowa, October 10–12, 1969. Chicago: Athletic Institute.

Bertelson, P. 1963. S-R relationships and reaction-times to new versus repeated signals in a serial task. *J. Exp. Psychol.* 65:478–484.

1965. Serial choice reaction-time as a function of response versus signal-and-response repetition. *Nature* 206:217–218.

Blyth, K. W. 1962. Experiments on choice reactions with the hands and feet. Doctoral dissertation, University of Cambridge.

Briggs, G. G. 1975. A comparison of attentional and control shift models of the performance of concurrent tasks. *Acta Psychol.* 39:183–191.

Briggs, G. G., & Kinsbourne, M. 1976. Cerebral organization as revealed by multilimb tracking performance. Manuscript in preparation.

Cernacek, J. 1961. Contralateral motor irradiation-cerebral dominance: Its changes in hemiparesis. *Arch. Neurol.* 4:165–172.

Cohen, L. 1970. Interaction between limbs during bimanual voluntary activity. *Brain* 93:259–272.

Cook, T. W. 1933a. Studies in cross education. 1. Mirror tracing the star-shaped maze. *J. Exp. Psychol.* 16:144–160.

1933b. Studies in cross education. 2. Further experiments in mirror tracing the star-shaped maze. *J. Exp. Psychol.* 16:679–700.

1934. Studies in cross education. 3. Kinesthetic learning of an irregular pattern. *J. Exp. Psychol.* 17:749–762.

Davis, R. C. 1942a. The pattern of response in a tendon reflex. *J. Exp. Psychol.* 30:452–463.

1942b. The pattern of muscular action in simple voluntary movements. *J. Exp. Psychol.* 31:347–366.

Hicks, R. E. 1975. Intrahemispheric response competition between vocal and unimanual performance in normal adult human males. *J. Comp. Physiol. Psychol.* 89:50–60.

Hicks, R. E., Bradshaw, G. J., Kinsbourne, M., & Feigin, D. S. 1977. Vocal-manual trade-offs in hemispheric sharing of performance control in normal adult humans. *J. Mot. Behav.* In press.

Hicks, R. E., Provenzano, F. J., & Rybstein, E. D. 1975. Generalized and lateralized effects of concurrent verbal rehearsal upon performance of sequential movements of the fingers by the left and right hands. *Acta Psychol.* 39:119-130.

Kinsbourne, M., & Cook, J. 1971. Generalized and lateralized effects of concurrent verbalization on a unimanual skill. *Q. J. Exp. Psychol.* 23:341-345.

Kinsbourne, M., & Hicks, R. E., in press. Functional cerebral space: A model for overflow, transfer, and interference effects in human performance. In J. Requin (ed.). *Attention & Performance VII*. Hillsdale: L. Erlbaum Assoc.

Kinsbourne, M., and Hiscock, M. 1977. The development of cerebral dominance. In S. Segalowitz and F. A. Gruber, (eds.). *Language Development and Neurological Theory*. New York: Academic Press.

Kinsbourne, M., & McMurray, J. 1975. The effect of cerebral dominance on time sharing between speaking and tapping in preschool children. *Child Dev.* 46:240-242.

Kreuter, C., Kinsbourne, M., & Trevarthen, C. 1972. Are deconnected hemispheres independent channels? A preliminary study of the effect of unilateral loading on bilateral finger tapping. *Neuropsychologia* 10:453-461.

Krieg, W. J. S. 1963. *Connections of the Cerebral Cortex*. Evanston, Ill.: Brain Books.

McAdam, D. W., & Whitaker, H. A. 1971. Language production: Electroencephalographic localization in the normal human brain. *Science* 172:499-502.

McCulloch, W. S. 1949. Cortico-cortical connections. In P. C. Bucy (ed.). *The Precentral Motor Cortex*. Urbana: University of Illinois Press.

Powell, T. P. S., & Mountcastle, V. P. 1959. Some aspects of the functional organization of the cortex of the postcentral gyrus of the monkey, a correlation of findings obtained in a single unit analysis with cytoarchitecture. *Johns Hopkins Bull.* 105:133.

Rabbitt, P. M. A. 1965. Response facilitation on repetition of a limb movement. *Br. J. Psychol.* 56:303-304.

Welch, J. C. 1898. On the measurement of mental activity through muscular activity and the determination of a constant of attention. *Am. J. Physiol.* 1:283-306.

8
Lateral asymmetries revealed by simple reaction time

JAMES SWANSON, ALEXA LEDLOW,
AND MARCEL KINSBOURNE

This chapter presents a paradigm that has been a valuable source of data concerning cerebral asymmetry for over 65 years. The experimental task defining this paradigm was introduced by Poffenberger in 1912, and it relies on measuring reaction time (RT) to laterally presented stimuli. The initial version of the task minimized the cognitive components of human information processing in order to investigate the *structural* links between input and output operations. Accordingly, simple stimuli uncomplicated by meaning and a simple, prepared response are the basis for the experiments discussed here, and the simplest of cognitive processes (sensation or detection) are required to perform the task.

Why should the information gathered from such an uncompromising environment be reviewed? The originator of the subtraction method of reaction time analysis, Helmholtz (1850), realized that using reaction time to investigate *structures* in the nervous system produced such variable results that the technique was useless with respect to this purpose, and subsequent investigation confirmed his impression (Cattell & Dolley, 1895). Woodworth (1938) summarized this negative viewpoint:

> The failure of the RT experiment to furnish a measure of nerve conduction is instructive because it shows that the bulk of the time required for even so simple a response is consumed in the nerve centers and that this central process must be complex and variable. It is no more redirection of a nerve impulse from the sensory to the motor nerve.

Of course, the purpose of reviewing the data from the Poffenberger paradigm is to investigate the process that does consume the time measured – the central process described by Woodworth as complex and variable. The richness and variability of the data derived from the simple paradigm can be used to evaluate complex theories of lateral asymmetries.

A more complex paradigm is obtained when the simple nature of Poffenberger's paradigm is complicated by using stimuli that must be processed for their meaning before a response is required. Going from the detection to the discrimination requirement presents difficulties in the interpretation of RT data that were once thought to be insurmountable (Ach, 1905; Woodworth, 1938); and even though Sternberg (1966, 1969) revised the subtraction method for appropriate investigation of the "discrimination" reaction of Donders (1868), the application of the subtraction method in its discarded form is still common in the investigation of lateral asymmetries. A detailed evaluation of the technique and logic that have found widespread application in the area of laterality (e.g., Moscovitch, 1973) is needed to point out the inadequacies of the data base upon which some theories are constructed. Although information about cerebral organization – cortical localization of language output (Filbey & Gazzaniga, 1969) or language processing (Moscovitch & Catlin, 1970) – may be derived from the application of subtraction method and logic, its weakness (e.g., McKeever et al., 1975) should be emphasized.

HISTORY

Helmholtz (1850) invented the subtraction method of reaction time analysis over 100 years ago in an attempt to measure the speed of nerve conduction. The technique yielded a remarkably accurate estimate of nerve conduction time for the frog (26 meters/msec), but only owing to fortunate circumstances (see Boring, 1950, or Woodworth, 1938, for a discussion).

The subtraction method was extended into the cognitive domain by Donders (1868) in an attempt to measure the time taken by the mental events "stimulus detection," "stimulus discrimination," and "response selection." The subtraction of Donders's now classical a-, b-, and c-reactions was based on the assumption of "pure insertion." This method violated that assumption and was soon discarded. (See Woodworth, 1938, and Sternberg, 1969, for historical background on this point.) Sternberg (1966, 1967, 1969) made a critical revision of Donders's method and introduced the memory scanning paradigm now used to measure "memory comparison time" (or what Donders called "stimulus discrimination time"). This paper does not deal with experiments using this corrected application of the subtraction method.

The subtraction method was extended in its original direction by Poffenberger (1912), who refined Helmholtz's method of measuring nerve conduction time in an attempt to measure the time taken to traverse a synapse. His purpose was "to find a means of measuring the time lost in the transmission of an impulse through a synapse within the human nervous system, and to obtain an appropriate measure of this lost time." The synapse he chose to investigate was located in the corpus callosum. His method was to compare two types of reactions: (1) direct reactions involving reception of a visual stimulus by the visual cortex of one hemisphere and control of response by the motor cortex of the same hemisphere, and (2) indirect reactions involving reception of a visual stimulus by the motor cortex of the other hemisphere. Applying the logic of the subtraction method, Poffenberger reasoned that these two reaction types, along with careful counterbalancing, yielded average RTs reflecting nerve paths identical *except that the latter required crossing from one side of the brain to the other*. And so, by subtracting the two reactions, a measure of the time taken by the additional process (traversal of the additional synapse of the corpus callosum) could be obtained. Using this technique, Poffenberger estimated that interhemispheric transfer time (IHTT) was small – about 4 msec. A replication of Poffenberger's experiment by Berlucchi et al. (1971) confirmed this estimate, finding an IHTT of 3 msec. Poffenberger's experimental technique and data have assumed renewed importance recently as the role of the corpus callosum in interhemispheric integration of sensory and motor information has received its deserved attention. Use of the Poffenberger paradigm to investigate language lateralization is also discussed in this chapter.

MODELS

The hemisphere that controls movement of the fingers is presumed to be contralateral to the hand moved. Thus, when a stimulus is presented left of fixation and a left-hand response is required, the hemisphere that initially receives the stimulus also programs the response required; this is Poffenberger's *direct reaction*. But when a right-hand response to the same left-visual-field stimulus is required, control of the response is by the hemisphere not initially stimulated; this is an *indirect reaction*. It is well established that in most intact humans, the left hemisphere possesses exclusive control of the vocal apparatus required in making a simple verbal response that involves pronouncing a word. The right hand and verbal reaction, then, are programmed by the left hemisphere, and when these two response modes are activated in response to right-visual-field stimulation, they constitute direct reactions.

In the modern application of Poffenberger's IHTT paradigm, an important extension has been the requirement of a spoken word as a response rather than a manual key press. Because the control of speech is strongly lateralized in the left hemisphere, perhaps even more so than control of a finger movement of the right hand, Poffenberger's logic can be applied just as when a right-hand manual response is required. Filbey and Gazzaniga (1969) and Moscovitch and Catlin (1970) have provided data that show a right-visual-field advantage does exist when a verbal response is required.

The models proposed by Poffenberger (1912), Filbey and Gazzaniga (1969), and Moscovitch and Catlin (1970) are all based on known structrues (i.e., the corpus callosum) viewed as barriers that take time to cross and thus give rise to IHTT. Such models are here called the *structural models*, and they are contrasted to *attentional models*. An introduction to the class of attentional models is presented below, and the differential predictions with respect to response modality of the two classes of models are discussed later.

Kinsbourne (1970) introduced a model to account for laterality effects on the basis of response facilitation in relation to the direction of orientation. That model, initially related to accuracy of performance rather than speed of performance (Kinsbourne, 1973, 1974), predicts that any simple response controlled by the left hemisphere (e.g., a verbal response or a manual response to the right) should benefit from eccentric stimulation that elicits an orientation to the right side of space. This prediction and model are based on Sherrington's (1906) concept of reciprocal innervation applied to direction of lateral orientation. When the focus of attention is straight ahead, the hemispheric contralateral gaze centers are presumed to be in a state of mutually inhibitory balance. Eccentric stimulation elicits a corresponding orientation response

(OR) controlled by one hemisphere and inhibits the contralateral OR of the other. Responses that are compatible with the OR of the stimulated hemisphere should be favored, according to the model, in terms of speed and accuracy, whereas responses in the opposite direction should be at a disadvantage. A verbal response is compatible with a left hemisphere OR to the right side of space because it is also controlled by the left hemisphere, but it is incompatible with a right hemisphere OR to the left side of space.

Because the model presented by Kinsbourne is based on a process of orientation or attention, lateral differences in RT are interpreted as a reflection of additional processing time when the OR is incompatible with the lateralized response control. Thus, the explanation of the lateral differences in latency offered by this model is not that it represents the time taken to cross a structural link, but rather that it reflects the stimulus-directed compatibility effect within the Poffenberger paradigm. In the context of the orientational model, either information is exchanged between the hemispheres in a time so short that it cannot reasonably or reliably be detected by RT methods, or lateral processing takes place after interhemispheric integration of information and so a transfer time component is present in all response measures. In either case, the label IHTT applied to the lateral difference in RT obtained in paradigms similar to Poffenberger's is misleading.

CONTRASTING MODELS ON BASIS OF RESPONSE MODALITY

Filbey and Gazzaniga (1969) presented data that contradict Poffenberger's original finding. In their first experiment, they showed that a right-visual-field advantage does exist when a verbal response is required to indicate the presence or absence of a nonverbal stimulus presented unilaterally, but on the basis of a second experiment they claimed that when a manual response is required in the same paradigm, no lateral difference in RT is obtained. These data support the model proposed by Gazzaniga (1967), which asserted that manual responding is bilaterally controlled. Thus, while Filbey and Gazzaniga's first experiment is consistent with the extension of Poffenberger's model to incorporate the verbal response modality, their second experiment contradicts the original model based on manual responses. McKeever and Huling (1971) have pointed this out, but Gazzaniga (1971) has argued that the difference in response modality between the two instances of the paradigm leaves "very little common ground in the experiments." The results of Filbey and Gazzaniga's (1969) first experiment did, however, fit their preconceived anatomical notion, which differed from Poffenberger's in regarding both hemispheres as equal controllers of either hand (Gazzaniga, 1967). Thus, their view was that in the visual-manual paradigm, indirect reactions do not require commissure crossing and IHTT is not involved.

The combined results of the two experiments by Filbey and Gazzaniga (1969) are not supported by a study by Wallace (1971). He reported stimulus-response compatibility effects in RT (which may be interpreted as laterality effects) when manual responses were required in conditions similar to those of the first experiment of Filbey and Gazzaniga (1969). To demonstrate that the lateral differences in RT would disappear if the compatibility caused by responding toward the source of stimulation was removed, Wallace showed that no lateral difference in RT was obtained when a verbal response was required. Of course, these results are in direct opposition to those of Filbey and Gazzaniga (1969).

McKeever et al. (1975) reviewed data from several paradigms and concluded that vocal control asymmetry is not sufficient to produce reliable estimates of IHTT. They suggest that verbal processing of the stimulus must be required to obtain large and significant estimates of IHTT.

Because across-experiment comparisons have yielded such a confusing picture, the question of response modality and lateral differences in RT deserved an explicit test. Such a test was provided by Swanson and Ledlow (1974). They presented a comparison of lateral differences in RT obtained for verbal and manual response modalities from the same subjects. As their data were reported at a scientific meeting and are not readily available, the study is reported here in some detail.

A total of 12 subjects was tested in a simple RT task in which stimuli were projected in the right visual field on one-fourth of the trials and in the left visual field on one-fourth of the trials; blank slides (catch trials) were presented on the remaining half of the trials. A detection task was utilized. Subjects used one index finger (the right in one condition and the left in another) to press a response button in the manual response condition; a single word (*yes*) was required in the verbal response condition. Subjects were instructed to withhold the response if a blank slide was presented and to release the prepared response if the slide presented was not blank. The investigators found that the right-hand and verbal conditions yielded right-visual-field advantages in RT. But in a reversed pattern of left-hand response a left-visual-field advantage led to the significant interaction of type of response and stimulus location. Swanson and Ledlow (1974) obtained an average IHTT of 28.5 msec from the manual conditions and 27 msec from the verbal response condition. This direct comparison of IHTT values presents clear evidence for equivalence of lateralized control in terms of speed for a manual button press and a verbal response. Thus, Gazzaniga's (1967) contention that button pressing is not a lateralized function fails to receive support in these data.

The data from Swanson and Ledlow (1974) also suggest that the hand used to make manual responses should be treated as carefully as (and be expected to produce a lateral effect as large as) a verbal response. Right-visual-field performance superiority observed when a right-hand response is used cannot be unambiguously attributed to specialized *perceptual* processing within a hemisphere. For example, both Cohen (1972) and Geffen et al. (1972) reported lateral differences in RT from tasks in which right-hand responses were required or permitted to unilateral stimulation of both left visual field and right visual field. What can be inferred from such data? Both studies suggest that the left hemisphere is specialized for processing the stimuli (Posner name-identity pairs) used in the experiment. That may be true, but as response hand was not a counterbalanced factor, the hypothesis that the RT difference was due to a simple lateralized motor component attributable to right-hand responding cannot be discounted. The present data suggest that for the purposes of RT analysis a right-hand response should be treated as carefully as a verbal response.

To summarize, the data on response modality are consistent with Kinsbourne's (1970) model based on a stimulus-directed orientational bias as a mechanism for conferring the lateral difference in RT in this paradigm, as well as the structural model of Poffenberger (1912). They rule out the model proposed by Gazzaniga (1967; Filbey & Gazzaniga, 1969). The data of Swanson and Ledlow (1974) also conflict with the contention of McKeever et al. (1975) that verbal identification of stimuli is required to obtain reliable visual field differences. In their experiment, the requirement of a vocal or manual response to a nonverbal stimulus was sufficient to produce a large and significant difference.

Perhaps the estimates of IHTT obtained in the experiment by Swanson and Ledlow (1974) are suspect because they are larger than the classic results reported by Poffenberger (1912) and replicated by Berlucchi et al. (1971). Is there another underlying factor that determines the magnitude of IHTT?

VARIABILITY OF ESTIMATES OF IHTT

Substantial differences in the magnitude of the measured IHTT have been reported. For example, Poffenberger (1912) reported that direct reactions were only 4 msec faster than indirect reactions when a simple manual response (a key press) was required to a simple visual stimulus (light flash). Berlucchi et al. (1971) replicated Poffenberger's experiment using modern equipment and a larger number of subjects and confirmed the original results, finding a 3-msec advantage for direct reactions. However, Jeeves (1965) reported that direct reactions were 100 msec faster than indirect reactions for normal adult subjects, and thus found IHTT to be over 20 times the size initially reported. In a subsequent study, however, Jeeves and Dixon (1970) found no effect for IHTT. An intermediate magnitude was reported by Bradshaw and Perriment (1970), who found that direct reactions were 20 msec faster than indirect reactions in a well-controlled testing of 12 subjects. Still, this is 4 times the size of the effect reported by Poffenberger.

A similar discrepancy exists in the literature when experiments requiring a verbal response are considered. For example, Filbey and Gazzaniga's (1969) report of an average IHTT of 33 msec for 8 right-handed subjects is three times larger than the average IHTT reported by Moscovitch and Catlin (1970), who found an average IHTT of 10 msec for 7 right-handed subjects. Why should estimates of IHTT, which according to one class of theory represent invariant structural characteristics of neural pathways, differ so greatly?

Two suggestions have appeared in the literature. One suggestion, proposed by McKeever and Gill (1972), is based on the degree of lateralization of a visual stimulus as measured by the visual angle between the fixation point and the stimulus. This hypothesis had been tested by Poffenberger (1912) and Berlucchi et al. (1971) and rejected, and was even discounted by McKeever in later work (McKeever et al., 1975). The other suggestion to account for IHTT variability is based on subjects' uncertainty concerning where the stimulus is to be presented. Most evidence concerning this theory comes from experiments that did not use RT as a dependent variable (see White, 1969, for a review), but Carmon et al. (1972) have presented data on this topic in an RT paradigm slightly different from the Poffenberger paradigm. These data and theories are reviewed below.

When accuracy instead of speed of response is the dependent measure, a common design consideration concerns the subject's knowledge of *where* the stimulus is to be presented. Heron (1957) and Terrace (1959) presented the argument for controlling the subject's expectancy by informing the subject where the stimulus will be presented. This procedure is designed to rule out differential expectancy as an explanation for lateral differences and is necessary to determine if the observed differences are caused by poststimulus factors. The data from the RT experiments cited earlier suggest that the size of the observed IHTT may be affected by the subjects' knowledge of the location of the impending stimulus in both the visuomanual and visuoverbal paradigms. The experimenters who found small values of IHTT (Poffenberger, 1912; Moscovitch & Catlin, 1970; Berlucchi et al., 1971) used blocked presentation of stimuli. But the aforementioned experimenters who randomized the locus of stimulus presentation (Fibley & Gazzaniga, 1969; Bradshaw & Perriment, 1970) found relatively large estimates of IHTT.

The dichotomy of IHTT estimates derived when stimulus location is predictable or unpredictable is explained by the attentional model as follows. Only on a trial when the position of the stimulus source is unknown is an OR elicited. If the direction of this orientation is incompatible with the direction of the required overt motor response, a long delay is introduced. However, on a trial when the source of stimulation can be predicted with certainty, the OR can be suppressed by inhibition from the hemisphere that is preparing for response execution. The relatively large differences found between direct and indirect projection in experiments are caused by the effects of an OR inappropriate to the response direction. The differences reported when stimulus side is known may reflect actual IHTT, but the lack of precision of the RT method makes it likely that the appearance of such small differences is serendipitous.

Apparently only one study exists investigating the effect of uncertainty about stimulus location. Carmon et al. (1972) required verbal identification of visually presented Hebrew letters (by Hebrew readers) and found that IHTT was smaller when stimulus location was known than when stimulus location was unknown. However, their report can be criticized on three points: (1) the comparison of certain versus uncertain stimulus location was across experiments and was not a within-subject comparison, (2) the stimulus was presented off the fovea (at 8° visual angle), and (3) the requirement to read letters complicated interpretation in terms of IHTT.

An alternative to the explanation based on uncertainty about stimulus location is based on stimulus eccentricity: the degree of displacment of the stimulus from the fixation point. Poffenberger (1912) and Berlucchi et al. (1971) varied visual angle over wide ranges and found no effect of this variable on the magnitude of the observed estimate of IHTT when a manual response was required. The prediction, as stated by Berlucchi, is that a quantitative change in the underlying structure takes place as a function of the degree of lateral displacement because the callosal connections are richest for those areas representing the fovea and the richness or directness of the connections declines as the stimulus is projected at a greater eccentricity. A quantitative change in the barrier (callosal connections) should be reflected by a quantitative change in the time necessary to cross the barrier. However, this prediction was not supported by Berlucchi's or Poffenberger's data.

McKeever and Gill (1972) noted that across several experiments the magnitude of IHTT was inversely related to the degree of stimulus lateralization. This result is exactly the opposite of the prediction made (and discounted) by Poffenberger (1912) and Berlucchi et al. (1971). However, McKeever and Gill (1972) failed to note that in the experiments reviewed another factor (uncertainty about stimulus location) was also related to the magnitude of behaviorally measured IHTT. In an experiment to test their notions, McKeever and Gill (1972) presented support for their prediction that IHTT would vary inversely with the degree of lateralization. A large IHTT estimate (41 msec) was obtained for foveal presentation when letters were presented unilaterally for identification, but a smaller estimate (17 msec) was obtained when the same stimuli were presented 4° from central fixation. Stimulus location was uncertain.

The uncertainty factor and the displacement factor have considerable importance for the structural and attentional classes of models being considered. The orientational bias model proposed by Kinsbourne (1970) makes clear predictions in terms of the uncertainty factor. If attention is directed to a point in space even before a lateral stimulus is presented (by the subject's altering the inhibitory balance of the hemispheres by volition), then the advantage of a stimulus-directed orientation bias is lost. The prediction, then, is that the measure of IHTT obtained under the condition of location uncertainty will be large, and the measure of IHTT obtained under the condition of location certainty will be insignificant. The structural model, on the other hand, makes no differential prediction on the basis of the uncertainty factor; in both cases, the barrier to be traversed – the corpus callosum – is the same, and the trip should take the same time. The displacement factor, however, can assume importance only for a structural model.

As the available data in the literature do not present a clear picture, Swanson and Kinsbourne (1976) conducted an experiment incorporating degree of displacement and uncertainty about stimulus location in a within-subjects design.

Squares subtending 1° of visual angle were presented in four locations: (1) 2° left of fixation, (2) 2° right of fixation, (3) 6° left of fixation, or (4) 6° right of fixation. Half the squares presented at each location contained the letter a, and half were empty. Four subjects were instructed to respond with the right hand to indicate presence of the letter and with the left hand to indicate absence of the letter. Four other subjects responded with the opposite response hands. A response in each case involved pressing with the index finger a switch mounted in a tube held in the hand. Two conditions were run to vary uncertainty of stimulus location. In the uncertain condition, stimuli were equally likely to appear at any one of the four locations. A block of trials consisted of 32 stimuli presented at each location, half filled and half unfilled. In the certain condition, items at each location were blocked at presentation.

Responses were found to be about 30 msec faster in the certain than in the uncertain condition, and responses were faster to stimuli presented at 2° from fixation (365 msec) than for stimuli presented at 6° displacement (394 msec).

A significant interaction between hand and visual field indicated that direct reactions were faster than indirect reactions. The interaction of uncertainty condition and hand and visual field was also significant. The estimated value of IHTT from the uncertain condition was 20 msec, but when location was certain, an IHTT estimate of only 3 msec was obtained. Neither of the significant interactions was dependent upon the degree of lateral displacement. In the conditions of uncertainty, the estimates of IHTT obtained for 2° and 6° displacements were 21 and 18 msec, respectively. In the certainty conditions, the estimates were 5 msec for the 2° locations and 2 msec for the 6° locations.

Even though the overall effect of lateral displacement was large, no differential influence of this variable on direct and indirect reactions was obtained. This replicates the early finding of Poffenberger (1912) and the more recent results of Berlucchi et al. (1971) and fails to replicate the experiment reported by McKeever and Gill (1972). Failure to find an effect in accord with the known pattern of callosal connections between the visual cortexes makes it improbable that the differences measured are caused by transfer of visual input between hemispheres. Lateral displacement would, however, not be related to estimates of IHTT if the interhemispheric transfer processes being monitored were between the motor areas, as was assumed by Poffenberger.

The large effect of location uncertainty on direct-indirect differences replicates the effect reported by Carmon et al. (1972), which was also noted to exist in cross-experiment comparisons described earlier in this chapter. If stimulus location is foreknown to the subject, then the direct and indirect reactions described by Poffenberger (1912) differ only slightly, and IHTT estimates based on these times are small and often statistically insignificant. But, if the location of the stimulus is unknown to the subject before presentation, then the direct and indirect reaction times differ substantially. This finding casts doubt on the validity of the assumption that the RT disadvantage for indirect responses is caused by the time consumed by a callosal crossing between either the visual or the motor areas.

It may be recalled that one experiment with randomized presentations of left- and right-field stimuli requiring manual responses failed to produce large RT differences for direct and indirect responses. In that experiment (Filbey & Gazzaniga, 1969), the authors noted that large stimulus-response (S-R) compatibility effects occurred. Perhaps these effects overshadowed the expected laterality effects.

S-R COMPATIBILITY

Poffenberger (1912) recognized that "it is customary to react on the side from which the stimulus comes, whatever the nature of the stimulus may be." Perhaps the measure of IHTT reflects no more than a stimulus-directed orienting response that confers an advantage for responding toward the source of stimulation. Studies by Simon et al. (1970) and Wallace (1971) suggest that this is the case.

Neither the Simon et al. (1970) nor the Wallace (1971) study was designed to investigate IHTT, but each provides important information about the paradigm. The Wallace study used the visual modality for stimulus presentation; Simon et al. used auditory input. Because lateralization of visual input is unambiguous, the Wallace study is described in detail.

Wallace (1971) presented one of two figures (a circle or a square) in one of four locations: 3.5° left, right, above, or below the fixation point. A manual response was required, pressing a key with the left hand if a circle was presented and with the right hand if a square was presented. An additional response variable, originally introduced by Simon et al. (1970), required the subject to cross arms during half of the experiment so that the side of space to which a response was directed changed but the side of the body making the response remained the same. With reference to the Poffenberger paradigm, the arms-uncrossed condition yielded a measure of IHTT of over 50 msec by subtracting the direct reactions from the indirect reactions. If interhemispheric transfer was responsible for this time difference, it should have remained in the other condition (arms crossed) when the same combinations of visual field and hand were compared. However, in the arms-crossed condition an estimate of IHTT of -52 msec was obtained: Poffenberger's direct reactions took longer than indirect reactions.

Of course, Wallace (1971) offered an accurate account of this finding in terms of S-R compatibility (Deininger & Fitts, 1955). The direction of the response, not the hand making the response, was the important variable. This explanation is consistent with the attentional model proposed by Kinsbourne (1970), which predicts that an orientation to the right side of space will facilitate a response to that side of space and inhibit or interfere with a response to the left side of space, regardless of the hand used to make the response. The results of Simon et al. (1970) are also consistent with the prediction of the attentional model.

In a second experiment, Wallace (1971) used the same stimulus conditions but required a vocal response (*tip* for a circle and *tap* for a square). Unfortunately for consistency with respect to measurement of IHTT, a small (6 msec) estimate of IHTT can be obtained from his data, and this difference in direct (right stimuli) versus indirect (left stimuli) reactions is not statistically significant.

The two experiments reported by Wallace (1971) involved a between-subject comparison, and the outcome of the second experiment may have been influenced by an unidentified source of noise. Swanson and Kinsbourne (1976) conducted a second experiment incorporating a within-subject design to compare verbal and manual response in Wallace's paradigm.

The stimuli used in this experiment were an empty square and a square filled with an x. The stimuli were presented in random order 2° left or right of fixation; a response by one hand was required to indicate presence of the letter inside the square and a response by the other to indicate absence. The hand assignment was switched for each subject after 2 days of responding and the experiment continued for an additional 2 days. On each day a vocal response condition was also given, a *yes* indicating presence and a *no* indicating absence for 2 days, and the opposite during the other 2 days. On each day, each subject responded in an arms-crossed and an arms-uncrossed condition.

No difference was found between IHTT obtained from the responses of the right hand in the arms-uncrossed condition and the vocal responses. These responses and the left-hand responses in the arms-uncrossed condition yielded estimates of IHTT of over 20 msec. These results are consistent with the data reported earlier for randomized presentation of stimulus location and with the data on vocal and manual response modalities. The data obtained from the arms-uncrossed condition replicate the findings of Wallace (1971) and in the context of the Poffenberger paradigm yield an estimate of −40 msec for IHTT.

One must conclude that in both the Wallace (1971) experiment and the present data, any attempt to apply the simple switchboard model to measure IHTT is unreasonable. Basically, an S-R compatibility effect is large enough completely to overshadow the IHTT effect.

CONCLUSIONS

The conclusion about measurement of IHTT by RT methods must be similar to the conclusion drawn by Helmholtz (1850) and Woodworth (1938) about measuring nerve conduction time: The method is too variable to provide a good estimate of these physiological reaction times.

How can one measure IHTT? Woodworth (1938) predicted the future use of "'brain waves' as indicators of the beginning or end of a mental process." Bremer (1958) measured IHTT using single cell recordings in the cat and obtained estimates of IHTT of less than 10 msec. Thus, by measuring responses to laterally presented stimuli at the cortical level, major sources of variability of RT methods are bypassed, and physiological reaction time accurately measured. A secondary interhemispheric transfer process, not to be confused with the results obtained by Bremer (1958), involves a *delayed* response. This delayed response has been described in detail by Rutledge and Kennedy (1961), who also obtained data from single cell recordings in cats, and its magnitude is around 40 to 50 msec after the initial IHTT process (which they also found to be less than 10 msec).

More recent work with humans has been done utilizing averaged evoked potentials obtained from surface electrodes. Ledlow (1976) showed that IHTT can be measured by presenting a visual stimulus in one lateral field and recording simultaneously from each hemisphere. By averaging the evoked responses from several trials, the on-going or background electrical activity of the electroencephalogram, which was not synchronized with stimulus presentation (i.e., noise), was canceled out and the time-locked components (i.e., signal) were enhanced. Her results show that even in a complex RT task requiring cognitive processing in which differences in RT for direct and indirect responses are not obtained, a clear measurement of IHTT from averaged evoked potentials can be obtained. Such cortical RTs disclose an advantage of direct versus indirect projection of approximately 15 msec. In this case, direct projections are with respect to visual field–hemisphere combination, as the RT responses were measured directly over the visual association areas rather than from manual or verbal responses.

In Ledlow's experiment, uncertainty about stimulus location was another factor. Location uncertainty increased the latency of the averaged evoked response only when indirect reactions were considered. The indirect-direct latency difference was thereby reduced by foreknowledge of stimulus location. Thus, even at this level of measurement, cognitive differences dependent upon attention as well as structural differences apparently contribute to the magnitude of IHTT. This research indicates that evoked potentials hold some promise of revealing IHTT, but further investigations must be made before a conclusion can be reached on what the time differences found by this procedure really reflect.

What do the variable data from the Poffenberger paradigm tell us about laterality? The most important point of this review is the emphasis on a complex, central mechanism as the source of most lateral differences in RT. The simple view that crossing a structural link (such as the corpus callosum) produces significant difference in RT can be held only when all other sources of variation (uncertainty of location, S-R compatibility, specialized cognitive processing, etc.) are held constant. Whether such factors can in fact be controlled is questionable. Thus, crossing the structural link does take time, but the time is short and is overshadowed by other factors that involve how the subject distributes attention before stimulus presentation and how the stimulus directs attention after presentation.

REFERENCES

Ach, N. 1905. Uber die Willenstatigkeit und das Denken. Cited by R. S. Woodworth. *Experimental Psychology*. New York: Holt, 1938.

Berlucchi, G., Heron, W., Hyman, R., Rizzolati, G., & Umiltà, C. 1971. Simple reaction times of ipsilateral and contralateral hand to lateralized visual stimuli. *Brain* 94:419–430.

Boring, E. G. 1950. *A History of Experimental Psychology*. New York: Appleton-Century-Crofts.

Bradshaw, J. L., & Perriment, A. D. 1970. Laterality effects and choice reaction time in a unimanual two-finger task. *Percept. Psychophysics* 7:185–188.

Bremer, F. 1958. Physiology of the corpus callosum. *Res. Publ. Assoc. Res. Nerv. and Ment. Dis.* 36:424-428.

Carmon, R., Nachshon, I., Iseroff, A., & Kleiner, M. 1972. Visual field differences in reaction times to Hebrew letters. *Psychonomic Sci.* 28:222-224.

Cattell, J. McK., & Dolley, C. S. 1895. On reaction-times and the velocity of the nervous impulse. *Natl. Acad. Sci. Mem.* 7:391-415.

Cohen, G. 1972. Hemispheric differences in a letter classification task. *Percept. Psychophysics* 11:139-142.

Deininger, R. L., & Fitts, P. M. 1955. Stimulus-response compatibility, information theory, and perceptual-motor performance. In H. Quastler (ed.). *Information Theory in Psychology*. New York: Free Press.

Donders, F. C. 1868. Over de snelheid van panische processen. Translated by W. G. Koster. *Acta Psychol.* 30:412-431, 1969.

Filbey, R. A., & Gazzaniga, M. S. 1969. Splitting the normal brain with reaction time. *Psychonomic Sci.* 17:335-336.

Gazzaniga, M. S. 1967. The split brain in man. *Sci. Am.* 217:24-29.

———. 1971. Reply to McKeever and Huling. *Psychonomic Sci.* 22:222-223.

Geffen, G., Bradshaw, J. L., & Nettleton, N. C. 1972. Hemispheric asymmetry: Verbal and spatial encoding of visual stimuli. *J. Exp. Psychol.* 95:25-31.

Helmholtz, H. v. 1850. Uber die Methoden, kleinste Zeitheile zu messen, unde ihre Andwerdung Fur physiologische Zwecke. *Philos. Magazine J. Sci.*, series 4, 6:313-325, 1853.

Heron, W. 1957. Perception as a function of retinal locus and attention. *Am. J. Psychol.* 70:38-48.

Jeeves, M. A. 1965. Psychological studies of three cases of congenital agenesis of the corpus callosum. In E. G. Ettlinger (ed.). *Functions of the Corpus Callosum*. London: Churchill.

Jeeves, M. A., & Dixon, N. F. 1970. Hemispheric differences in response rates to visual stimuli. *Psychonomic Sci.* 20:249-251.

Kinsbourne, M. 1970. The cerebral basis of lateral asymmetries in attention. *Acta Psychol.* 33:193-201.

———. 1973. The control of attention by interaction between the cerebral hemispheres. In S. Kornblum (ed.). *Attention and Performance IV*. New York: Academic Press.

———. 1974. Mechanisms of hemispheric interaction in man. In M. Kinsbourne & W. L. Smith (eds.). *Hemispheric Disconnection and Cerebral Function*. Springfield, Ill.: Thomas.

Ledlow, A. 1976. A reaction-time and evoked-potential investigation of lateral asymmetries in a stimulus classification task. Unpublished doctoral dissertation, University of Texas, Austin.

McKeever, W. F., & Gill, K. M. 1972. Interhemispheric transfer time for visual stimulus information varies as a function of the retinal locus of stimulation. *Psychonomic Sci.* 26:308–310.

McKeever, W. F., Gill, K. M., & VanDeventer, A. D. 1975. Letter versus dot stimuli as tools for "splitting the normal brain with reaction time." *Q. J. Exp. Psychol.* 27:363–374.

McKeever, W. F., & Huling, M. D. 1971. A note on Filbey and Gazzaniga's "splitting the brain with reaction time." *Psychonomic Sci.* 22:222.

Moscovitch, M. 1973. Language and the cerebral hemispheres: Reaction-time studies and their implications for models of cerebral dominance. In P. Pliner, T. Alloway, & L. Krames (eds.), *Communication & Affect: Language & Thought.* New York: Academic Press.

Moscovitch, M., & Catlin, J. 1970. Interhemispheric transmission of information: Measurement in normal man. *Psychonomic Sci.* 18:211–213.

Poffenberger, A. T. 1912. Reaction time to retinal stimulation with special reference to the time lost in conduction through nerve centers. *Arch. Psychol.* 23:1–73.

Rutledge, L. T., & Kennedy, T. T. 1961. Brain-stem and cortical interaction in the interhemispheric delayed response. *Exp. Neurol.* 4:470–483.

Sherrington, C. 1906. *Integrative Action of the Nervous System.* New Haven: Yale University Press.

Simon, J. R., Hinrichs, J. V., & Croft, J. L. 1970. Auditory S-R compatibility: Reaction time as a function of ear-hand correspondence and ear-response-location correspondence. *J. Exp. Psychol.* 86:97–102.

Sternberg, S. 1966. High-speed scanning in human memory. *Science* 153:652–654.

1967. Two operations in character recognition: Some evidence from reaction-time measurements. *Percept. Psychophysics* 2:45–53.

1969. The discovery of processing stages: Extension of Donders' method. *Acta Psychol.* 30:276–315.

Swanson, J. M., & Kinsbourne, M. 1976. S-R compatibility and interhemispheric transfer time. Paper presented at American Psychological Association meeting, Washington.

Swanson, J. M., & Ledlow, A. 1974. Unilateral input, attention, and performance in RT experiments. Paper presented at American Psychological Association meeting, New Orleans.

Terrace, H. 1959. The effects of retinal locus and attention on the perception of words. *J. Exp. Psychol.* 58:382–385.

Wallace, R. J. 1971. S-R compatibility and the idea of a response code. *J. Exp. Psychol.* 88:354–360.

White, M. J. 1969. Laterality differences in perception: A review. *Psychol. Bull.* 72:387–405.

Woodworth, R. S. 1938. *Experimental Psychology.* New York: Holt.

9
Asymmetry of electrophysiological phenomena and its relation to behavior in humans

GAIL R. MARSH

There is now a substantial body of literature in neurology and neuropsychology showing that the two cerebral hemispheres in man are different in function. The question naturally arises, therefore, whether electrophysiological indicators of brain function, such as the electroencephalogram (EEG), as recorded from the scalp, reflect this difference in function. Many earlier studies were done in the context of exploring the newfound "Berger rhythm" looking for a correlate of cerebral dominance (the concept that the hemisphere opposite the dominant hand is specialized for language). However, this conceptualization had little to contribute to an understanding of the function of the "minor" hemisphere. The more broadly based concept of lateralization of function, which affirms that the minor hemisphere also has specialized functions, is now the general context within which most work is done. This has allowed the search for EEG correlates of psychological function to proceed in both hemispheres.

ELECTROPHYSIOLOGICAL METHODS

Since its adoption in 1958, the International 10-20 system (Jasper, 1958) for noting electrode recording sites has seen increasing use. This notation system is used in the following discussion when possible. Figure 9-1 illustrates how the left hemisphere lies with regard to the electrode sites. The left hemisphere has odd-numeral designations; right has even. Midline positions are designated by the subscript z.

Recording bioelectrical signals from the scalp requires recording between two points. Ideally, recording should be between a site of

Asymmetry of electrophysiological phenomena

interest and some "silent" reference area. Unfortunately, no silent area exists. A reference site off the head is satisfactory, but is more difficult to apply because a circuit to cancel the potent signal from the heart must be arranged. Reference points on the head can contribute their own signals, but this can be used to advantage in some cases. If one wishes to eliminate some background activity that might be common to two areas, while looking at what might be possible differences, then recording between them accomplishes that purpose because only the *difference* in potential between the recording sites is seen as a signal. Another approach to canceling signals of no interest is to tie those two or more areas together (such as the two earlobes) and use them as a joint reference for recording a third area of interest. In comparisons of electrophysiological phenomena in the left and right hemispheres, the choice of a reference site that does not bias the case is essential.

ALPHA BAND ACTIVITY

Activity in the alpha frequency band (8 to 13 Hz) has often been a dependent variable in EEG research because it has the distinctive quality of occurring in well-defined bursts and stands out from other activity in high-amplitude, rhythmical waves (Fig. 9-2).

Fig. 9-1. Human brain viewed from the left, showing the main landmarks in the 10–20 system for noting electrode sites. The subscript z denotes a midline position. A_1 marks the left earlobe, and Cb_1 the left mastoid. The large-stippled section is Broca's area; the small-stippled area, Wernicke's area. The *dotted line* underlying C_3 denotes the central sulcus.

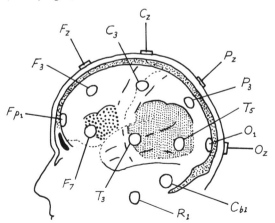

Measurements of alpha activity are usually restricted to the parietal, occipital, and temporal portions of the head, where the activity is most commonly found. Alpha activity seems to correlate well with activities associated with relaxation, such as sitting quietly with eyes closed. For this reason it has often been interpreted as a sign of inhibition of activity in the areas where it is found. Beta activity (13 to 30 Hz) is equally distinctive, being of low amplitude and occurring during aroused, alert conditions; it is frequently interpreted as a sign of activated cortex. It has often been studied as the opposite of alpha activity after being termed the alpha-blocking response by Adrian and Matthews (1934). (For further exploration and interpretation of alpha activity see Andersen & Andersson, 1968.)

Early reports (Raney, 1939; Strauss et al., 1943) mentioned a greater amount of alpha activity occurring in the right hemisphere of most subjects. One report ascertained that 58% of right-handed subjects had EEG activity of longer duration and larger amplitude in the nondominant (right) hemisphere (Cornil & Gastaut, 1947). The clinical literature lends support, for there is a general expectation of finding more, but not strikingly more, alpha activity over the right hemisphere (Hill & Parr, 1963, p. 238). It should be noted, however, that striking asymmetry may be a sign of disease (Strauss et al., 1943).

A study conducted in our laboratory by Steven Schaaf examining the phenomenon of habituation of the alpha-blocking response (Wilson, 1962; Wells, 1963) has shown a greater return of alpha activity in the right parietal area (P_4 as referenced to linked ears) compared with the left (P_3) after exposure to a bright light. An examination of the eye dominance and "purity" of handedness of

Fig. 9-2. Electroencephalograms from the parietal area showing the readily identifiable alpha-wave bursts, the usual left-right asymmetry of abundance, and an alpha-blocking response to a strong light turned on at the *arrow* are illustrated above.

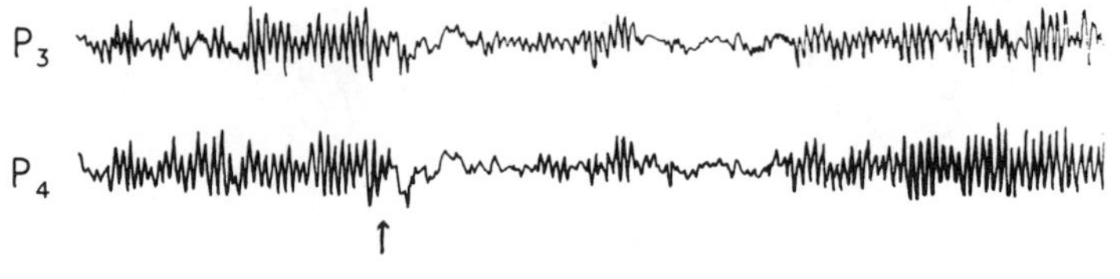

the subjects brought other tendencies to light. All the subjects who showed right-eye dominance and had no left-hand-dominant close relatives showed greater sustained alpha blocking in the left hemisphere. All left-eye-dominant subjects showed the reverse. Subjects affirming some left-hand-dominant members in the immediate family gave mixed results.

Attempts to demonstrate that asymmetry in alpha activity is tied to handedness (used as the most obvious sign of lateral dominance) have not been entirely successful. There have been some negative reports, suggesting that asymmetry may not always be found, or if found, may not always correlate with the handedness of the subject. Glanville and Antonitis (1955) used a large number of subjects (50) and ascertained handedness by a well-standardized test (Jastak, 1939). Occipital placements (O_1 and O_2 referenced to the right ear) showed no difference in proportion of the EEG record displaying alpha or amplitude of alpha activity in the overall group. The same result was also found for selected groups of strongly right-handed and strongly left-handed subjects. The choice of only the right ear as reference may have been unfortunate, as it would suppress any alpha activity common to the right leads (O_2 and the ear reference). Provins and Cunliffe (1972) could find no consistent difference between left and right hemisphere recordings from the parietal area (against a parietal midline reference) using such dependent measures as abundance of alpha (8 to 12 Hz), beta (13 to 20 Hz), or total EEG activity, in either a right-handed or a left-handed group. But a comparison of the two groups did find that the right-handed subjects had the greater abundance of alpha activity. Further, if only right hemisphere activity between the groups was compared, the right-handed group had greater alpha activity. As these investigators obtained excellent (0.90 or better) test-retest correlations of amount of within-hemisphere activity and tested the extent of right- or left-handedness in each subject, the negative results must be seriously considered.

Recent experiments have begun to explore whether hemisphere asymmetries are associated with specific cognitive processes. Because the left hemisphere seems predominant in the performance of verbal, analytical, and mathematical operations, one might argue that the left hemisphere would be more activated during these kinds of tasks. Conversely, because the right hemisphere appears predominant in processing music, spatial information, and visual imagery, one might expect greater activation in the right hemisphere when the subject is engaged in such psychological operations. Morgan et al. (1971), using only right-handed subjects, reported finding more alpha activity in the occipital area of the right hemisphere (referenced against C_z) and noted that this condition was maintained during all tasks for most subjects. However, this asymmetry was significantly increased during verbal-analytic tasks and decreased during visuospatial tasks. A similar experiment by Galin and Ornstein (1972) also showed greater power (i.e., amplitude × duration) in the EEG (from 1 to 35 Hz) over the right hemisphere at both temporal (T_3 and T_4) and parietal sites (P_3 and P_4, all referenced against C_z) during verbal tasks as opposed to spatial ones. The results were essentially the same whether the subject had to perform motor acts as a part of the response or only mentally process the information. The reliability of these findings seems good, as Galin and Ornstein found the same result reliably over days in the same subjects, both men and women. A follow-up study (Doyle et al., 1974) found alpha activity to be that aspect of the EEG most changed by cognitive task changes. Further, the temporal sites demonstrated more asymmetries than did the parietal. Interestingly, when analysis was restricted to the alpha band, motor tasks showed greater asymmetries than did nonmotor tasks.

McKee et al. (1973) extended Galin and Ornstein's work by testing with three increasingly difficult language tasks and finding that the corresponding left-right asymmetry in alpha abundance in both the temporal and parietal areas (referenced to C_z) increased significantly. As expected, the alpha activity was smallest in the left hemisphere. For comparison they also included a musical task that decreased the disparity between left-sided and right-sided alpha abundance more than any verbal task. Unfortunately, the reporting of only left-right ratios leaves open the question of whether decreases in alpha abundance are caused by decreases in alpha abundance in the right hemisphere or increases in the left hemisphere, or possibly both. Total alpha abundance was reported to be constant throughout all tasks, however, indicating no overall change in arousal level.

Schwartz et al. (1974) have reported a study combining both left and right hemisphere tasks with several levels of difficulty. Their subjects whistled a song, spoke the words, or sang the melody with the words while occipital alpha activity was recorded (referenced to C_z). The whistling significantly decreased alpha activity in the right hemisphere; speaking had the opposite effect. Singing (combining both words and melody) provided an effect midway between whistling and speaking. All effects were enhanced by doing all tasks at double speed and/or by selecting a group of pure right-handed subjects by excluding those subjects with family members who were left-handed.

A test of differences between the hemispheres in all frequency bands during (1) noise, (2) music, (3) verbal input, (4) arithmetic computation, (5) patterned vision, and (6) diffuse visual stimulation conditions found few differences between the hemispheres (Giannitrapani, 1971). Alpha activity during the visual tasks was more suppressed in the right hemisphere (O_2) than in the left (O_1). In the temporal sites (T_3 and T_4) only patterned vision had the same effect. In these same temporal sites noise and computation increased beta activity in the right more than in the left hemisphere. Interestingly, the left prefrontal area showed more increase in beta activity than the homologous right area. These changes were viewed as the operation of a scanning mechanism that searches for structure in the input to the various regions.

We have also attempted to relate asymmetry of alpha activity to stage of learning and learning ability in a small number of subjects (12) (Chartock et al., 1975). The ratio of the amount of time during which alpha activity was present in the right parietal region (referenced to linked mastoids) to the combined time of alpha activity production for both parietal areas was examined so as to compare the early and late portions of a task involving learning (1) a list of verbal associates and (2) a series of dot patterns. There was a significant correlation between the change in learning rate and the change in alpha rhythm asymmetry as the verbal task progressed into overtraining. There were no significant correlations during the spatial task, however. It is important to note that significant correlations to behavior were obtained only after the initial asymmetry had been removed from the data by looking only at the change from early learning to final learning conditions. The normal tonic asymmetry did not seem to be tied to the learning task.

There has also been some interest in the question of whether the two hemispheres produce synchronous alpha waves or whether one hemisphere normally leads or lags behind the other. Lindsley

(1940) reported that persons with a greater degree of laterality had more synchronous alpha activity. This was especially true of subjects who were both right-handed and right-eyed.

Cohn (1948) reported that the nondominant hemisphere usually leads in the initiation of a burst of alpha activity. However, such relationships may be true only for portions of the scalp, for Giannitrapani et al. (1966) found significant differences between hemispheres in prefrontal and frontal sites (left lobe leading) during sleep and in parietal and occipital areas (right lobe leading) during wakefulness (recording against a linked ear reference). However, these relationships were found only in left-handed subjects and were linked to the strength of handedness. Children were less differentiated than adults. Lindsley (1938) had also reported several independent focuses of alpha activity in each hemisphere at temporal and occipital sites, which could lead to differences between experiments depending upon what sites on the scalp were tested.

Several studies have shown that, when compared over periods of time 1 or 2 minutes in duration, there was no enduring significant lead or lag by either hemisphere (Brazier & Casby, 1952; Barlow et al., 1964; Walter et al., 1966). However, Liske et al. (1967) demonstrated a significant (albeit small) lead time for the right hemisphere of 0.83 msec (alpha activity ranges from 125 to 83 msec per cycle). Moreover, this effect seems to occur through a skewed distribution of lead times, for there is a strong tendency for the right hemisphere to lead by a greater extent when it does lead. As one might expect, the corpus callosum has been shown to aid in the synchronization of the hemispheres (Garoutte & Aird, 1958; Berlucchi, 1966; Green & Russell, 1966). One subject with agenesis of the corpus callosum showed a marked lag of the left hemisphere (Aird & Garoutte, 1958).

Quantification of the EEG has always been a problem. Comparing EEG output from each hemisphere using the usual power measures (taking both amplitude and wave duration into account) inevitably leads to reliance upon presence or absence of alpha activity, as this rhythm produces such large-amplitude activity with relatively long wave duration. In an alert adult, frequencies below 7 Hz are seldom encountered, often leaving beta activity as the alternative to alpha (Berkhout, 1965). Of course, beta activity is the opposite of alpha in terms of power measures. But using more overt measures of alpha activity, such as alpha abundance or percentage of time or duration, still imposes a limitation in interpretation because one can speak only of relative degrees of activation in the areas being measured. There exists in these measures no bioelectrical index of the processing of information from any particular stimulus. Nor can one speak, except in the broadest of terms, about how information is being processed during any particular task.

PHOTIC DRIVING

One alternative to recording the raw EEG is to record the EEG during special stimulation conditions, such as repetitive light flashing, where the posterior EEG is known to follow the stimulus frequency with rhythmical waves (*photic driving*), or to use the more finely controlled sine wave modulated light, which produces much the same result. Two early reports on asymmetry in this response are in conflict: One found that flickering light in one visual half-field produces photic driving only in the opposite hemisphere (Adrian, 1943); the other reported that it produces driving in both hemispheres (Toman, 1941). Two recent experiments conclude that photic driving in both hemispheres can be achieved by binocular, monocular, or hemiretinal stimulation (Freedman, 1963; Lansing & Thomas, 1964). Further, the photic driving showed an asymmetry in most subjects, and this greater reactivity of one hemisphere was preserved no matter whether the left or right visual half-field was stimulated. To what extent such asymmetry is based on carefully determined hand or eye dominance remains to be tested. Freedman (1963), using only right-handed subjects, found the greatest amount of photic driving in the right hemisphere recording from parietal to occipital sites. Lansing and Thomas (1964), recording in the same manner, reported the largest amount of photic driving in the dominant (left) hemisphere, but unfortunately, they also found most of their subjects had a predominance of alpha activity in a resting state in this same hemisphere, which runs contrary to most reports. In a second experiment they found that subjects selected for either left or right hemisphere asymmetry under full field stimulation usually maintained the same asymmetry with either left or right half-field stimulation. Intrinsic factors, not which hemisphere was receiving direct stimulation, thus seem to govern this response.

One report (Crowell et al., 1973) has shown a strong tendency toward unilateral photic driving in some human neonates. A significant portion of the newborns did not respond to the photic stimulus, suggesting that this may be a first stage of development, with unilateral response coming later before maturing to a bilateral response. How this response changes during maturation remains to be examined. Kooi et al. (1957) have reported that 5% of the normal population has an asymmetry of 50% or greater. Knowing whether this is an extension of an early maturational difference could enhance our understanding of this phenomenon.

The one report in the literature (Pfefferbaum & Buchsbaum, 1971) that has centered on EEG recordings of hemisphere asymmetry in response to differing strengths of sine wave modulated light found no asymmetry in right-handed subjects. However, a group of left-handed subjects showed an increased asymmetry (stronger left occipital response) with increasing strength of modulation. Unfortunately, no detailed investigation was made of the handedness of the subjects other than their stated hand preference.

AVERAGE EVOKED POTENTIALS

Another electrophysiological phenomenon that can demonstrate hemisphere asymmetry is the electrical activity elicited by presen-

Fig. 9-3. Average evoked potentials recorded from the temporal area, elicited by a dim red light at the *arrow* are illustrated above. The usual components mentioned in hemisphere asymmetries can be seen: (1) negative (80 to 120 msec), (2) positive (140 to 240 msec), and (3) positive (270 to 400 msec).

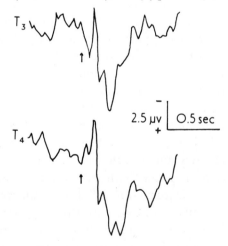

tation of stimuli. This activity is generally of smaller magnitude than the on-going EEG activity and must be extracted from the scalp-recorded EEG by summing together several short portions (e.g., 500 msec) of the EEG record following several stimulus presentations. Random portions of a normal, unstimulated EEG record tend to sum to zero because the varying electrical activity is not tied to any one specific external or internal event. However, when just the portions of the EEG time-locked to the repeated presentation of a stimulus are summed, the electrophysiological responses elicited by the stimulus are accumulated, and the electrical events not tied to the stimulus tend to cancel toward zero. A typical average evoked potential (AEP) to a light flash can be seen in Figure 9-3.

Visual average evoked potentials

AEPs elicited by flashes of light have been shown to have an asymmetry in bright, normal children (Rhodes et al., 1969). The AEP recorded over the right central region (C_4 with ear reference) was larger than the AEP from the left central region (C_3) in the components from 100 to 200 msec. This relationship was found in children with high IQs, but a group of retarded children did not show such asymmetry. The occipital region showed no asymmetry in either group. The same results have been reported (Bigum et al., 1970) in a comparison of normal and mongoloid children. Handedness was not considered as a variable in either study. In a group of adults selected for a marked asymmetry between C_3 and C_4, this asymmetry was eliminated by ingestion of 3 oz of alcohol (Lewis et al., 1970). Both AEPs were reduced in amplitude, but the reduction was more severe on the right. No differences were found between O_1 and O_2. Interpretation of changes in these wave components was in terms of changes in attention or arousal. However, none of the mechanisms described would give rise to asymmetrical responses.

A second study (Richlin et al., 1971) examining the AEP to light flashes at O_1 and O_2 (referenced to linked ears) in normal and retarded children reported a larger amplitude AEP late component in the right hemisphere for the normal group and the reverse asymmetry for the retarded group. These investigators further reported that the latency of a positive-going component

(P_2) was greater in the left hemisphere than in the right for normal children, but not different in the retarded. Whether these differences are due to the difference between normal and retarded children or to the fact that a large percentage of the retarded group was of mixed hand preference remains to be determined. Gott and Boyarsky (1972) have also reported slower AEPs in the dominant hemisphere (either O_1 or O_2) in both left- and right-handed subjects. But this was found only with unilateral stimulation: bilateral stimulation elicited no differences.

Schenkenberg (Schenkenberg, 1970; Schenkenberg & Dustman, 1970) found the right hemisphere AEP to light stimulation was larger than that of the left (C_3 and C_4 referenced to linked ears) in a population of normal subjects ranging from children to old people. A similar asymmetry was found at O_1 and O_2 in young subjects (through adolescence). The difference occurred in a late component between 100 and 200 msec after the light flash. However, AEPs to clicks and brief shocks to the finger did not elicit such asymmetries at O_1 and O_2, C_3 and C_4, or F_3 and F_4, where the light-elicited AEPs had been recorded. There was no check for handedness.

Handedness apparently influences AEP asymmetry. Eason et al. (1967) have reported that left-handed men have a greater response in the right occipital area (O_2 referenced to the right ear) than in the left (O_1 to left ear) when each hemisphere is stimulated separately by stimulation of the left or right visual field, respectively. The stimulus was a simple light flash, to which right-handed men showed no asymmetry. However, in a follow-up study examining handedness and eye dominance in a more quantitative fashion (Culver et al., 1970), the difference between left- and right-handed groups failed to reach significant levels. The difference between right and left hemispheres under input from left and right visual fields, as found previously, was now shown in all subjects. This effect was even stronger in left-eyed subjects and stronger yet if these were also left-handed. This second study used only women subjects, whereas the first study used only men; this raises the question of sex differences. Also the second study used monocular stimulation as opposed to the binocular stimulation used in the first study. It may be recalled that Pfefferbaum and Buchsbaum (1971) reported results similar to the above left-handed effect using modulated light as a stimulus. Friedlander (1971) has previously pointed out the association between handedness and eye dominance. However, mechanisms underlying the presently reported asymmetries have yet to be proposed. Furthermore, Martin (1970) and others have shown that binocular input may be affected more by one eye than by the other and that this effect is based on something other than a stable eye dominance.

When essentially meaningless stimuli, such as diffuse light flashes, are used, the question can be raised why these stimuli should be received differentially in the two hemispheres unless they are conceived as being "spatial" stimuli. Thus, such simple stimuli fall heir to some of the same interpretation problems as the earlier indexes of EEG alpha activity, and experimenters must rely on comparisons between groups (such as IQ grouping) to gain meaningful results.

Another approach, allowing within-subject comparisons, is to use meaningful stimuli that are thought to be selectively processed in one hemisphere or the other. An experiment using this strategy has demonstrated that verbal stimuli elicit stronger AEPs in the left hemisphere, and spatial stimuli produce stronger AEPs in the right. Buchsbaum and Fedio (1969) delivered three different types of stimuli – words, patterns, and random dots – to the subject's visual field centered around a fixation point. All subjects claimed right hand and eye dominance. Amplitude differences for various components of the AEP were not reported, as the authors were using a measuring index that summarized all portions of the AEP in one measure. Recordings from the left (O_1) and right (O_2) hemispheres (referenced to ipsilateral ears) demonstrated that word and design stimuli produced AEPs that were more different from each other at O_1 than at O_2. Also the latency to a prominent positive peak (190 to 280 msec latency) was consistently earlier on both sides to both word and design stimuli. However, within each hemisphere taken alone the AEPs to words were more similar and had their positive peaks earlier than the AEPs to designs. Interestingly, the discrimination index used by these investigators to differentiate between AEPs did not reach a peak until about 400 msec after the delivery of the stimuli. This is even longer than decision reaction times to these same stimuli obtained from a different group of subjects.

In a second study (Buchsbaum & Fedio, 1970), words or random dots were projected into either the left or the right visual half-field.

Again, both left and right hemispheres produced AEPs (at O_1 and O_2, respectively) that were significantly different between words and nonsense patterns, and this was especially marked for the left (dominant) hemisphere. Also the difference in AEP to the two classes of stimuli was enhanced when the AEP was recorded from the hemisphere with the most direct neuronal connection to the stimulated hemiretinas. A third report (Fedio & Buchsbaum, 1971) concerning patients who had undergone left and right temporal lobe resections indicated that loss of the left or right temporal lobe led to a loss of AEP differentiation to words or dots, respectively. The injured hemisphere tended to have the most undifferentiated response to all stimuli.

Vella et al. (1972) have reported an effect that was selective for the right hemisphere. Using only right-handed subjects, they presented a checkerboard pattern briefly to only one hemiretina at a time. The stimulus always produced the largest response in the right posterior temporal cortex (recorded T_6 to O_2 as compared with T_5 to O_1), while producing no asymmetry in the primary cortical area (O_2 to O_z as compared with O_1 to O_z). The asymmetry was obtained under all conditions, but was strongest when stimulation was by the more indirect route. The asymmetrical portion of the AEP was a large positive wave falling at approximately 200 msec. Interestingly, the asymmetry could not be obtained with the less complex stimulus of a simple light flash. Failure to find asymmetry in the case of a less complex visual stimulus could be due to the smaller late components of the AEP elicited by such stimuli (Spehlmann, 1965; Nash & Fleming, 1970).

Not all studies have reported hemisphere asymmetries. Shelburne (1972) presented strings of three letters, with each letter presented for 1 sec. Each string spelled either an English word or a nonsense syllable. The subject had to indicate as soon as possible to the experimenter whether the string of letters formed a nonsense syllable or a word. The third letter was always the crucial letter and it gave the largest amplitude positive components (falling at 450 to 550 msec), most prominently seen at C_z (referenced to linked ears). Recordings from P_3 and P_4 in right-handed subjects showed no significant differences. Switching to a new set of words where the first letter was the crucial letter also switched the large positive component of the AEP to that letter. The positive component, perhaps owing to eye movements, then arose at the much shorter latency of 165 to 320 msec. Except for this, the results were the same as in the first experiment. These experiments were repeated in a group of 10- and 11-year-old children with similar results (Shelburne, 1973).

Another visual stimulus experiment that failed to demonstrate a difference between hemispheres was conducted by Seales (1973) using 10 right-handed subjects, 4 of whom were left-eye-dominant. The experiment used pairs of letters as stimuli (Ab, EE, etc.) with three different instructional sets given to the subject: (1) press a key as soon as any stimuli are presented, (2) press one of two keys to indicate if the two stimuli are of the same or different size, and (3) press one of two keys to indicate if the two stimuli are of the same or different name. AEPs were recorded from O_1 and O_2 and two placements approximately halfway between T_3 or T_4 and P_z. Reference was linked earlobes. A positive-going wave with a peak at approximately 270 msec was not seen to be different under any of the three conditions between placements over Wernicke's area on the left and the symmetrically opposite position on the right. However, this peak was decreased in amplitude and increased in duration with increased task complexity. Another component, a negative-going wave with a latency of 500 to 750 msec, did tend to show a greater amplitude over Wernicke's area on the left. The decision reaction times in the two more difficult processing tasks fell between 800 and 1100 msec.

We have carried out a similar experiment (Poon et al., 1976) with the same letter stimuli as Seales under two conditions: (1) performing a simple detection task of pressing a key as soon as the stimuli were seen to appear, and (2) processing the stimuli to respond same or different (by pressing a key) in a categorization match (both vowels or both consonants or nonmatching). The AEPs recorded from T_3 and T_4 against a linked mastoid reference in right-handed, right-eyed subjects showed asymmetries at two points on the wave form. A negative component at 80 to 100 msec was significantly larger in the left hemisphere, and a positive component at 160 to 200 msec was significantly larger in the right hemisphere during the demanding verbal task. Only the component at 160 to 200 msec was significantly different in the simple task. The difference between these results and Seales's (1973) may derive from the complexity of the task, the differing electrode placements, or the differing eye dominance of the subjects.

Another experiment in the visual mode designed to test different levels of verbal and nonverbal information load on the two hemispheres was carried out in our laboratory (Bell, 1973). A verbal task with low verbal demands required the subject to determine whether two words, presented tachistoscopically 1.4 sec apart, were the same or different. An analogous low-level nonverbal task (facial recognition) was presented in the same manner, using two faces in place of the words. In a second task, involving higher-level information processing, the verbal component required the subject to decide if two words presented 1.4 sec apart were synonymous. As a high-level facial recognition task, the subject was presented with faces from different perspectives and asked to decide if the faces were of the same or different persons. Amplitudes of the AEPs from left and right temporal areas were measured between a negative peak at 90 to 120 msec and a positive peak at 240 to 400 msec. No difference between left and right hemispheres was reliably produced. However, there was a significant reduction in amplitude of the AEP in the higher-level processing tasks.

Auditory average evoked potentials

AEPs elicited by spoken words are unequal in size when the left is compared with the right. Morrell and Salamy (1971) recorded from a left frontal area over the site of speech production (Broca's area), a left central area over the motor area for jaw movement, and a left temporoparietal area predominant in speech interpretation (Wernicke's area). Electrodes also were placed symmetrically over the right hemisphere. AEPs elicited by five different spoken nonsense words, randomly delivered 50 times each, were recorded with the subjects repeating the word after each presentation. The left hemisphere responses were 40% larger in an early negative component falling at a latency of about 90 msec. This asymmetry was especially marked over the temporoparietal area. In the right hemisphere the largest response was seen in the central lead and was interpreted as an antecedent to jaw movements, and not linked to linguistic phenomena. The authors suggested that the early negative component may represent the spread of information from deeper cortical layers to more superficial ones. Because the cortex is thicker on the left than on the right in the area of the temporoparietal site (Geschwind & Levitsky, 1968), the larger left hemisphere response may be a reflection of this more potent source. However, a control using some other nonverbal stimulus would have allowed more substantial conclusions; this study does not permit the conclusion that *only* verbal auditory stimuli elicit larger AEPs on the left side. It should be noted that if response bias has any effect in experiments such as this, then its impact could be great, because the subject is heavily biased in favor of expecting a continuing flow of speech input and output. That such bias may operate is seen in the work of Matsumiya et al. (1972), reviewed below.

An experiment by Wood et al. (1971) compared AEP asymmetries elicited by spoken phonemes in which the subjects were required to detect phonemic or tonal differences. The choice of stimuli was especially ingenious in that the phoneme discrimination (used to engage the left hemisphere) was between /ba/ and /da/, and the tone discrimination was between two differing fundamental frequencies of the spoken phoneme /ba/, thus tapping a right hemisphere function. Wood et al. recorded evoked potentials from both central (C_3 and C_4) and temporal (T_3 and T_4) sites to a linked ear reference. The results showed that the AEPs generated in the left hemisphere differed between the speech and tonal conditions. However, there was no difference between the AEPs elicited in the two conditions in the right hemisphere. Both central and temporal sites showed the same effect, with the largest difference occurring in a negative-positive component falling approximately over the interval from 100 to 200 msec. The authors took this as evidence that added linguistic information processing was carried out in the left hemisphere but not in the right. That both central and temporal sites showed very similar AEPs and hemisphere asymmetry effects suggested that large areas of cortex are involved in speech analysis.

Matsumiya et al. (1972) had subjects perform several different perceptual tasks. There was no difference between AEPs recorded from the scalp overlying Wernicke's area on the left and a similar placement on the right (referenced to P_3 and P_4, respectively) when a subject was tallying occurrence of words or sound-effect noises randomly interspersed. But there were larger evoked potentials in the left hemisphere when the subject was listening to words that formed a set of instructions and when he listened to sound effects and had to categorize them into four different types. Again the portion of the AEP that was most different between left and right hemispheres occurred at a latency of approximately 100 msec, but in this case was a positive-going component. In most subjects this latency preceded the end of the word stimuli, which averaged a total duration of about 485 msec. The authors felt that this could be an indication of a *response set* prior to the complete delivery of the stimulus. They interpreted the results to indicate that the significance or meaningfulness of the stimuli for the subject, as imposed by the task at hand, contributed to the asymmetry, not merely the fact that the stimulus was either verbal or nonverbal. The findings may also indicate that the left hemisphere can act as a discriminator or director in processing auditory stimuli. One flaw in this experiment makes interpretation difficult. The subjects reported they attached verbal labels to the nonverbal stimuli, thus possibly reducing the effect to one of various degrees of verbal loading. Also the rapidity with which words can be decoded or identified is an important unknown affecting the interpretation of these findings.

Another experiment in the auditory mode including both verbal and nonverbal stimuli, but not testing effects of continuing speech, was conducted by Molfese (1972). His subjects showed differences between left and right scalp responses (over Wernicke's area referenced to the unlinked earlobes) to musical sounds, noise, syllables, and words. The left hemisphere gave larger responses to the verbal stimuli, and the right showed larger responses to the nonverbal stimuli. The asymmetrical AEP components were an early negative wave (at a latency of 100 msec) and a positive peak (at a latency of about 160 msec). These results were shown to hold for children and infants as well as adults, indicating early differentiation of the brain with regard to different types of stimuli. However, the latencies to the measured components tended to be longer in the children and even longer in infants; for example, the early negative in the adult fell at about 100 msec, and in the infant at about 450 msec. That such differential hemispheric activity is present long before language develops suggests that lateralized mechanisms may be functional very early in life.

Another study that looked at the lateralization of auditory AEPs to verbal and nonverbal sounds used a different approach to the effect of meaningfulness. Greenberg and Graham (1970) studied the AEPs to low-pass filtered plosive consonants and to electronically shaped piano notes where the subject had to learn to identify these sounds. They recorded from T_3 and T_4 referenced only to the right earlobe. The AEPs were compared in terms of their spectral components and found to display little asymmetry except that the left hemisphere normally gave rise to responses of greater amplitude. There was an exception in that the asymmetry was severely diminished in response to well-learned stimuli. To what extent the asymmetry reflects arousal, attentional, or other such changes possibly related to learning a task remains a question.

CONTINGENT NEGATIVE VARIATION

Several of the above-described studies have demonstrated that a response set adopted by the subject can influence his AEP. In a perceptual experiment Kinsbourne (1970) has shown that subjects can adopt sets that affect the usual difference between ability to detect verbal or nonverbal material in the left and right visual fields. Specifically, it was shown that while holding a list of six words in memory the ability to detect a gap in a square was increased in the right visual half-field and reduced in the left visual half-field.

To investigate the possibility that such perceptual sets could produce an electrophysiological correlate prior to the time of presentation of the stimuli to be processed, we conducted a series of studies in our laboratory measuring a subject's perceptual set by means of the contingent negative variation (CNV) (Marsh & Thompson, 1973). The CNV, a slow negative shift in brain potential occurring in the interval between two paired stimuli, had previously been shown to reflect a subject's attentional set, arousal, or expectancy (Tecce, 1972). It was originally anticipated that during verbal presentations a larger CNV would appear over the left hemisphere than over the right. However, when subjects were tested using a procedure that first illuminated a small lightbulb as a ready signal at the fixation point, followed in 1.4 sec by a word flashed on the screen, there was no greater CNV elicited over the left hemisphere than over the right. A similar series of presentations of nonverbal stimuli (slanted lines), where the subject had to judge the slant of the line, also produced no asymmetry.

On the hypothesis that perhaps subjects adopted and held a response set throughout a block of trials, a second experiment was conducted in which the verbal and spatial stimuli were randomly intermixed so as to break up the blocks of trials. In this procedure a differential warning signal had to be used. A green warning light preceded a verbal presentation, and a red warning light preceded a spatial presentation. Use of such a randomized presentation procedure produced asymmetry at both temporal and parietal sites (T_3, T_4, P_3, and P_4, recorded against a linked mastoid reference). Interestingly, the asymmetry for the verbal material was seen largely in the temporal regions, with T_4 larger; that for the nonverbal was observed in the parietal regions, with P_3 larger. However, the difference in perceptual error score between the half-fields was significant only for the verbal material.

In a second experiment using a more difficult nonverbal task (dot location), the perceptual asymmetries were strengthened, but the left-right CNV asymmetries could not be replicated in either the temporal or the parietal locations, although the trends in the data were in the same direction as in the prior experiment. In this case, the CNV amplitudes were very small, which may reflect the difficulty of the task (Delse et al., 1972). The load the task places on the subject may be very important, especially in situations where the underlying brain processes are only vaguely understood at present.

Butler and Glass (1971) measured the CNV generated by giving the subject a warning signal followed by a tachistoscopically presented number that the subject added in an on-going string of eight such dot-number presentations before reporting the sum. They recorded from F_3, F_4, C_3, and C_4 (referenced to A_1 or A_2, respectively), and reported greater CNV production in the left hemisphere (location unspecified). Such CNV production appears to be evidence of a preparation on the part of the left hemisphere for the reception and processing of the arithmetic stimulus that is to follow the warning signal. There seems no rational basis as yet on which to resolve the differences between the above experiments.

An interesting study failing to find asymmetry in CNV production in humans with corpus callosal section has been reported by Gazzaniga (1972). He presented tachistoscopically in the left visual half-field either a *1* or a *0*. The subject was to respond by pressing a key when he heard a tone after a 1-sec pause if he had been presented with a *1*, but to make no response if it had been a

0. Both hemispheres were reported to show about equal CNV responses at approximately C_3 and C_4 recording locations. Because the information was being presented to the right hemisphere, the left hand could make correct responses even though the left (speaking) hemisphere could report no knowledge of what had been presented. A lack of asymmetry in this case raises the question whether the corpus callosum participates in the generation of some of the previously reported asymmetrical phenomena.

The studies of AEP and CNV seem to show the left hemisphere acting in a directive role and showing more distinctiveness in its response. That this should be true even for nonverbal stimuli refutes any claim that anatomy is the dominant factor in producing the AEP asymmetries.

SPEECH PRODUCTION

Larger asymmetries may be shown by using tasks in which the subject is performing some output function rather than a processing or an input function. Bruce (1973) demonstrated greater asymmetry of function when subjects had to sing or speak while performing a task rather than merely perceiving tachistoscopically presented stimuli. Thus on these grounds alone, we might expect speech production to elicit asymmetry of bioelectrical phenomena.

Speech production has long been known to be highly lateralized. However, electrophysiological evidence of such asymmetrical function is only of recent origin (McAdam & Whitaker, 1971). The recordings were made against a linked ear reference. Areas overlying both Broca's area and the left central motor strip showed a greater negative shift ("readiness" potential) preceding a spoken letter than the corresponding sites on the right. The greatest asymmetry was found between the bases of the frontal gyruses (Broca's area on the left) as expected. During a similar period preceding coughing and spitting movements, no asymmetries were found. Within the left hemisphere Broca's area tended to give greater response than the central site for speech stimuli, but on the right the largest response tended to come from the central motor strip preceding the nonverbal jaw and lip movements. This reflects linguistic and motor contributions of left and right hemisphere, respectively, in language production.

A recent report (Low et al., 1976) has also claimed that one can predict which hemisphere is dominant for speech in epileptic patients by recording CNV responses over Broca's area (reference was opposite earlobe) while the patient performs the task of speaking words on cue with a warning stimulus given 1.4 sec before the word. Hypothesizing that larger CNV production would occur in the dominant hemisphere prior to speech led to correct prediction of laterality of speech in 10 of 11 subjects as checked by carotid injection of sodium amytal (Wada & Rasmussen, 1960) prior to necessary surgery. However, testing the assumption that the larger CNV would occur in the speech-dominating hemisphere in two groups of 11 normal pure right-handed and 11 normal pure left-handed subjects (Arnett, 1967) showed mixed results. Only 8 of the right-handed group showed a greater left hemisphere CNV and only 6 of the left-handed group showed right hemisphere dominance. No check was made on eye dominance, nor was the more conclusive amytal test performed on this group to further check the purity of the lateralization of function in the normal volunteers. If the findings in the patient group can be substantiated, this technique could prove quite useful in the clinic.

CONCLUSIONS

It appears that the issue of how alpha abundance is distributed is partially, but not completely, settled. The nondominant hemisphere seems to produce more abundant alpha activity. Perhaps a more searching examination of hand and eye dominance in future studies will enhance understanding of how specific brain rhythms are distributed with regard to cerebral lateralization. White (1969), reviewing perceptual asymmetries, has also noted the need for better examination of both hand and eye dominance in laterality experiments.

This same issue of hand and eye dominance can be raised with regard to the data on AEP to simple stimuli. But at least for the present those experiments exploring the effects of verbal and nonverbal tasks seem conclusive in demonstrating some asymmetrical bioelectrical responses during the processing of the information. The most responsive AEP components seem to be at 100 to 200 msec in latency. However, the results have not yet resolved what aspect of the stimulus is being processed or what mechanisms may be involved. The data support the parietal and temporal regions of the head as being the active sites, and as yet allow little further anatomical refinement.

Anticipation of making various responses or being required to perform certain discriminations may also give rise to asymmetrical function. A great deal of further work is necessary to clarify the governing mechanisms that give rise to these interesting electrophysiological phenomena.

REFERENCES

Adrian, E. D. 1943. Doyne memorial lecture: Dominance of vision. *Trans. Ophthal. Soc. U.K. 63*:194-207.

Adrian, E. D., & Matthews, B. H. C. 1934. The Berger rhythm: Potential changes from the occipital lobes in man. *Brain 57*:355-385.

Aird, R. B., & Garoutte, B. 1958. Studies on the "cerebral pace-maker." *Neurology (Minneap.) 8*:581-589.

Andersen, P., & Andersson, A. 1968. *Physiological Basis of the Alpha Rhythm*. New York: Appleton-Century-Crofts.

Arnett, M. 1967. The binomial distribution of right, mixed and left handedness. *Q. J. Exp. Psychol. 19*:327-333.

Barlow, J. S., Rovit, R. L., & Gloor, P. 1964. Correlation analysis of EEG changes induced by unilateral intracarotid injection of Amobarbital. *Electroencephalogr. Clin. Neurophysiol. 16*:213-220.

Bell, S. K. 1973. The effect of task and stimulus on the cortical evoked potential. Master's thesis, Duke University.

Berkhout, J. 1965. Comparative frequency distributions of large, small amplitude rhythms of the human electroencephalogram. *Electroencephalogr. Clin. Neurophysiol. 19*:598-600.

Berlucchi, G. 1966. Electroencephalographic studies in "split brain" cats. *Electroencephalogr. Clin. Neurophysiol. 20*:348-356.

Bigum, H. B., Dustman, R. E., & Beck, E. C. 1970. Visual and somatosensory evoked responses from mongoloid and normal children. *Electroencephalogr. Clin. Neurophysiol. 28*:576-585.

Brazier, M. A. B., & Casby, J. U. 1952. Crosscorrelation and autocorrelation studies of electroencephalographic potentials. *Electroencephalogr. Clin. Neurophysiol. 4*:201-211.

Bruce, R. 1973. The role of attention in perceptual asymmetries. Doctoral dissertation, Duke University.

Buchsbaum, M., & Fedio, P. 1969. Visual information and evoked responses from the left and right hemispheres. *Electroencephalogr. Clin. Neurophysiol. 26*:266-272.

1970. Hemispheric differences in evoked potentials to verbal and nonverbal stimuli in the left and right visual fields. *Physiol. Behav.* 5:207-210.

Butler, S. R., & Glass, A. 1971. Interhemispheric asymmetry of contingent negative variation during numeric operations. *Electroencephalogr. Clin. Neurophysiol.* 30:366 (abstract).

Chartock, H. E., Glassman, P. R., Poon, L. W., & Marsh, G. R. 1975. Changes in alpha rhythm asymmetry during learning of verbal and visuospatial tasks. *Physiol. Behav.* 15:237-239.

Cohn, R. 1948. The occipital alpha rhythm: A study of phase variations. *J. Neurophysiol.* 11:31-37.

Cornil, L., & Gastaut, H. 1947. Etude électroencephalographique de la dominance sensorielle d'un hemisphere cérébral. *Presse Med.* 37:421-422.

Crowell, D. H., Jones, R. H., Kapuniai, L. E., & Nakagawa, J. K. 1973. Unilateral cortical activity in newborn humans: An early index of cerebral dominance? *Science* 180:205-208.

Culver, C. M., Tanley, J. C., & Eason, R. G. 1970. Evoked cortical potentials: Relation to hand dominance and eye dominance. *Percept. Mot. Skills* 30:407-414.

Delse, F. C., Marsh, G. R., & Thompson, L. W. 1972. CNV correlates of task difficulty and accuracy of pitch discrimination. *Psychophysiology* 9:53-62.

Doyle, J. C., Ornstein, R., & Galin, D. 1974. Lateral specialization of cognitive mode. 2. EEG frequency analysis. *Psychophysiology* 11:567-578.

Eason, R. G., Groves, P., White, C. T., & Oden, D. 1967. Evoked cortical potentials: Relation to visual field and handedness. *Science* 156:1643-1646.

Fedio, P., & Buchsbaum, M. 1971. Unilateral temporal lobectomy and changes in evoked responses during recognition of verbal and nonverbal material in the left and right visual fields. *Neuropsychologia* 9:261-271.

Freedman, N. L. 1963. Bilateral differences in the human occipital electroencephalogram with unilateral photic driving. *Science* 142:598-599.

Friedlander, W. J. 1971. Some aspects of eyedness. *Cortex* 7:357-371.

Galin, D., & Ornstein, R. 1972. Lateral specialization of cognitive mode: An EEG study. *Psychophysiology* 9:412-418.

Garoutte, B., & Aird, R. B. 1958. Studies on the cortical pacemaker: Synchrony and asynchrony of bilaterally recorded alpha and beta activity. *Electroencephalogr. Clin. Neurophysiol.* 10:259-268.

Gazzaniga, M. S. 1972. One brain – Two minds? *Am. Sci.* 60:311-317.

Geschwind, N., & Levitsky, W. 1968. Human brain: Left-right asymmetries in temporal speech region. *Science* 161:186-187.

Giannitrapani, D. 1971. Scanning mechanisms and the EEG. *Electroencephalogr. Clin. Neurophysiol.* 30:139-146.

Giannitrapani, D., Sorkin, A. I., & Enenstein, J. 1966. Laterality preference of children and adults as related to interhemispheric EEG phase activity. *J. Neurol. Sci. 3*:139-150.

Glanville, A. D., & Antonitis, J. J. 1955. The relationship between occipital alpha activity and laterality. *J. Exp. Psychol. 49*:294-299.

Gott, P. S., & Boyarsky, L. L. 1972. The relation of cerebral dominance and handedness to visual evoked potentials. *J. Neurobiol. 3*:65-77.

Green, J. B., & Russell, D. J. 1966. Electroencephalographic asymmetry with midline cyst and deficient corpus callosum. *Neurology (Minneap.) 16*: 541-545.

Greenberg, H. J., & Graham, J. T. 1970. Electroencephalographic changes during learning of speech and nonspeech stimuli. *J. Verbal Learning Verbal Behav. 9*:274-281.

Hill, D., & Parr, G. 1963. *Electroencephalography.* New York: Macmillan.

Jasper, H. H. 1958. The ten twenty electrode system of the international federation. *Electroencephalogr. Clin. Neurophysiol. 10*:371-375.

Jastak, J. 1939. The ambigraph laterality test. *J. Appl. Psychol. 23*:473-487.

Kinsbourne, M. 1970. The cerebral basis of lateral asymmetries in attention. *Acta Psychol. 33*:193-201.

Kooi, K. A., Eckman, H. G., & Thomas, M. H. 1957. Observations on the response to photic stimulation in organic cerebral dysfunction. *Electroencephalogr. Clin. Neurophysiol. 9*:239-250.

Lansing, R. W., & Thomas, H. 1964. The laterality of photic driving in normal adults. *Electroencephalogr. Clin. Neurophysiol. 16*:290-294.

Lewis, E. G., Dustman, R. E., & Beck, E. C. 1970. The effects of alcohol on visual and somato-sensory evoked responses. *Electroencephalogr. Clin. Neurophysiol. 28*:202-205.

Lindsley, D. B. 1938. Foci of activity of the alpha rhythm in the human electro-encephalogram. *J. Exp. Psychol. 23*:159-171.

1940. Bilateral differences in brain potentials from the two cerebral hemispheres in relation to laterality and stuttering. *J. Exp. Psychol. 26*:211-225.

Liske, E., Hughes, H. M., & Stowe, D. E. 1967. Cross-correlation of human alpha activity: Normative data. *Electroencephalogr. Clin. Neurophysiol. 22*:429-436.

Low, M. D., Wada, J. A., & Fox, M. 1976. Electroencephalographic localization of conative aspects of language production in the human brain. In W. C. McCallum and J. R. Knott (eds.). *The Responsive Brain*, pp. 165-168. Bristol: John Wright.

Marsh, G. R., & Thompson, L. W. 1973. Effect of verbal and non-verbal psychological set on hemispheric asymmetries in the CNV. In W. C. McCallum and J. R. Knott (eds.). *Event Related Slow Potentials of the Brain: Their Relation to Behavior*, pp. 195-200. Amsterdam: Elsevier.

Martin, J. I. 1970. Effects of binocular fusion and binocular rivalry on cortically evoked potentials. *Electroencephalogr. Clin. Neurophysiol. 28:*190–201.

Matsumiya, Y., Tagliasco, B., Lombroso, C. T., & Goodglass, H. 1972. Auditory evoked response: Meaningfulness of stimuli and interhemispheric asymmetry. *Science 175:*790–792.

McAdam, D. W., & Whitaker, H. A. 1971. Language production: Electroencephalographic localization in the normal human brain. *Science 172:* 499–502.

McKee, G., Humphrey, B., & McAdam, D. W. 1973. Scaled lateralization of alpha activity during linguistic and musical tasks. *Psychophysiology 10:*441–443.

Molfese, D. L. 1972. Cerebral asymmetry in infants, children and adults: Auditory evoked responses to speech and noise stimuli. Doctoral dissertation, Pennsylvania State University.

Morgan, A. H., McDonald, P. J., & MacDonald, H. 1971. Differences in bilateral alpha activity as a function of experimental task, with a note on lateral eye movements and hypnotizability. *Neuropsychologia 9:*459–469.

Morrell, J. K., & Salamy, J. G. 1971. Hemispheric asymmetry of electrocortical responses to speech stimuli. *Science 174:*164–166.

Nash, M. D., & Fleming, D. E. 1970. The dimensions of stimulus complexity and the visual evoked potential. *Neuropsychologia 8:*171–177.

Pfefferbaum, A., & Buchsbaum, M. 1971. Handedness and cortical hemisphere effects in sine wave stimulated evoked responses. *Neuropsychologia 9:* 237–240.

Poon, L. W., Thompson, L. W., & Marsh, G. R. 1976. Average evoked potential changes as a function of processing complexity. *Psychophysiology 13:*43–49.

Provins, K. A., & Cunliffe, P. 1972. The relationship between EEG activity and handedness. *Cortex 8:*136–146.

Raney, E. T. 1939. Brain potentials and lateral dominance in identical twins. *J. Exp. Psychol. 24:*21–39.

Rhodes, L. E., Dustman, R. E., & Beck, E. C. 1969. The visual evoked response: A comparison of bright and dull children. *Electroencephalogr. Clin. Neurophysiol. 27:*364–372.

Richlin, M., Weisinger, M., Weinstein, S., Giannini, M., & Morganstern, M. 1971. Interhemispheric asymmetries of evoked cortical responses in retarded and normal children. *Cortex 7:*98–105.

Schenkenberg, T. 1970. Visual, auditory, and somatosensory evoked responses of normal subjects from childhood to senescence. Doctoral dissertation, University of Utah.

Schenkenberg, T., & Dustman, R. E. 1970. Visual, auditory, and somatosensory evoked response changes related to age, hemisphere, and sex.

In *Proceedings, 78th Annual Convention, American Psychological Association*, pp. 183-184.

Schwartz, G. E., Davidson, R. J., Maer, F., & Bromfield, E. 1974. Patterns of hemispheric dominance in musical, verbal and spatial tasks. *Psychophysiology 11*:227 (abstract).

Seales, D. M. 1973. A study of information processing and cerebral lateralization of function in man by means of averaged visually evoked potentials. Doctoral Dissertation, University of California, Los Angeles.

Shelburne, S. A., Jr. 1972. Visual evoked responses to word and nonsense syllable stimuli. *Electroencephalogr. Clin. Neurophysiol. 32*:17-25.

1973. Visual evoked responses to language stimuli in normal children. *Electroencephalogr. Clin. Neurophysiol. 34*:135-143.

Spehlmann, R. 1965. The averaged electrical responses to diffuse and to patterned light in the human. *Electroencephalogr. Clin. Neurophysiol. 19*:560-569.

Strauss, H., Liberson, W. T., & Meltzer, T. 1943. Electroencephalographic studies: Bilateral differences in alpha activity in cases with and without cerebral pathology. *J. Mt. Sinai Hosp. 9*:957-962.

Tecce, J. J. 1972. Contingent negative variation (CNV) and psychological processes in man. *Psychol. Bull. 77*:73-108.

Toman, J. 1941. Flicker potentials and the alpha rhythm in man. *J. Neurophysiol. 4*:51-61.

Vella, E. J., Butler, S. R., & Glass, A. 1972. Electrical correlate of right hemisphere function. *Nature (New Biol.) 236*:125-126.

Wada, J., & Rasmussen, T. 1960. Intra-carotid injection of sodium amytal for the lateralization of cerebral speech dominance. *J. Neurosurg. 17*:266-282.

Walter, D. O., Rhodes, J. M., Brown, D., & Adey, W. R. 1966. Comprehensive spectral analysis of human EEG generators in posterior cerebral regions. *Electroencephalogr. Clin. Neurophysiol. 20*:224-237.

Wells, C. E. 1963. Alpha wave responsiveness to light in man. In G. H. Glaser (ed.). *EEG and Behavior*, pp. 27-59. New York: Basic Books.

White, M. 1969. Laterality differences in perception: A review. *Psychol. Bull. 72*:387-405.

Wilson, S. 1962. Electrocortical reactivity in young and aged adults. Doctoral dissertation, George Peabody College for Teachers.

Wood, C. C., Goff, W. R., & Day, R. 1971. Auditory evoked potentials during speech perception. *Science 173*:1248-1251.

10
Weber on sensory asymmetry

J. D. MOLLON

Ernst Heinrich Weber's Latin monograph of 1834 (Fig. 10-1), widely celebrated but seldom read, contains many diverse and interesting observations besides the statement of Weber's law. The passages translated here correspond to pages 84–85, 92–94, and 119–122 of *De pulsu, resorptione, auditu et tactu*.

On the difference between the right and left sides of the body in the ability to judge weight by touch

We should not omit to mention that there is often a difference between the right and left sides in the sense of touch. Just as the two sides are unequal in muscular strength, so too, according to my experiments, they differ in cutaneous sensitivity. Whereas, however, in most men the right side exceeds the left in muscular strength, we find the contrary if we measure tactual sensitivity [*subtilitatem tactus*] on the right and left sides of the body by placing weights on various parts and keeping these parts passive and motionless. For the same weight seems heavier to most subjects when it is placed on the left and lighter when it is placed on the right; and different weights often appear equal if the heavier is placed on the right and the lighter on the left, whereas they appear unequal if the heavier is placed on the left and the lighter on the right. This phenomenon seems to arise from the fact that the sensitivity of the left side is, in most subjects, finer [*subtilior*] that that of the right.[1]

I was myself the subject when this observation was first made by my colleague Seyffarth,[2] Professor at Leipzig. Seyffarth once before gave me helpful advice when I was investigating wave theory, and now in this enquiry he likewise drew my attention to the

Fig. 10-1. Title page of Weber's monograph.

DE PULSU, RESORPTIONE, AUDITU ET TACTU.

ANNOTATIONES ANATOMICAE ET PHYSIOLOGICAE

AUCTORE

ERNESTO HENRICO WEBER

ANATOMIAE PROFESSORE IN UNIVERSIT. LITERARUM LIPSIENSI.

LIPSIAE
PROSTAT APUD C. F. KOEHLER.
1834.

discrepancy between left and right. So considerable is this difference in my own body that it shows itself not only on my hands but also on the soles of my feet and other places; and it significantly hinders the accurate comparison of weights that are placed one on each side of the body.

I am certainly not claiming that in all subjects the left side enjoys a keener weight perception [*ad percipienda pondera subtilior sensu praeditum esse*]. I have placed weights on the hands of many subjects and examined their sense of touch. I found only that in the majority of subjects the left hand has the more delicate sense of touch; in some the right is more sensitive, and in a few no difference is observed between right and left hands.

Out of 14 subjects, varying in sex and age and engaged in various studies and occupations, 11 perceived the same weight as heavier when it lay on the left hand rather than on the right; in 2 subjects the contrary held; and in just 1 there was no clear difference between right and left.

On the cause of the difference between the right and left sides of the human body in the perception of weight

Why, in the case of most subjects, does the same weight appear to press less hard on the right hand than on the left? Many, perhaps, would seek the answer to this question in the hardness and thickness of the skin, which have been increased by work and repeated pressure.[3] For most people use their right hand more often than their left and we may suspect that the sense of touch is thereby blunted in the right hand.

Others perhaps will think that the cause is to be sought in muscular differences between the right and left. They believe that a weight lifted by hand seems heavier on that side on which the muscles, being weaker, require a greater effort of will for the contraction necessary to raise the imposed weight.

However, both explanations are wrong. The first is wrong because thickness of the cuticle does not impair the perception of weight. On the heel and on the sole of the foot the skin is thicker than anywhere else and all thermal sensation and two-point discrimination [*perceptio distantiae crurum circini has corporis partes tangentium*] are remarkably impaired; yet the comparison of two weights is there performed readily and precisely. The other explanation, which is based on the muscular weakness of the left side, is not able to explain what we wish to explain. For there would then be a difference in weight perception only if weights were lifted and not if they were placed on the passive hand.

So, since we lack any other explanation, it is likely that the difference lies in the structure of the sensory nerves. Just as the muscles of the right side are thicker than those of the left and thus more powerful, so it is not impossible that the sensory nerves on the left should be more sensitive than those on the right.

In this way one can readily explain why the perception of weight by touch is more delicate on the left than on the right, not only on the hand but also on the foot and shoulder. For these latter parts are not practiced in the perception and judgment of weights and yet they show the same difference as is found in the hands.

Some time ago I explored the question of whether the right-left difference that has been recorded in muscular strength is also found in the nervous system and in the perceptual faculty [*sentiendi facultatem*]. At the time I thought the eye to be an organ singularly suited for experiments on this problem and I commended such a research project to Holke.[4] For I had at some time examined the journals of Tauber,[5] a spectacle-maker who was a Master of Liberal Arts and had formerly been secretary to Hindenburg, Professor of Physics at Leipzig. I discovered that over a number of years Tauber had examined the eyes of many people who were buying spectacles from him: He had measured their visual acuity and optimal reading distance and he had made quantitative records of the difference between the eyes. I hoped it would be worth the work involved if a thorough analysis of these observations were to show whether both eyes usually enjoy the same acuity at a given distance or whether one has better acuity than the other.

Holke collated and analyzed Tauber's records, but they show that there is no clear difference between right and left eyes in this respect.

Having in vain sought this difference in the visual organ, I was the more surprised when I unexpectedly observed it in the tactual modality on the left side.[6]

I did hope that I might observe the same difference between right and left hands by another method, if, that is, I were to place two different weights simultaneously on the flexor surface of the fingers and were to record the least difference in the weights that could be confidently detected. I suspected that these differences would be greater on the right hand and smaller on the left.

These experiments were therefore instigated as follows. I placed a weight of 32 drams on the caput of the metacarpal bone of the right index finger and a second weight, slightly lighter, on the caput of the subject's little finger. I asked the subject to compare the weights as accurately as possible, first holding his hand still, then lifting it together with the weight. The same experiment was then repeated on the left hand of the same subject. My hypothesis, however, was wrong: When this second method was used there was no evidence that the left hand was more sensitive.

On thermal sensitivity

... I come now to a quite remarkable observation concerning another error to which we are subject when we judge temperature: *Fluids that are equally hot or equally cold do not affect the left hand in the same way as the right: in the case of most subjects a greater sense of cold or of heat is aroused in the left hand.* This difference between right and left is the more worthy of our attention because I have myself shown in other experiments that equal weights placed simultaneously on the right and left hands arouse, in most subjects, a greater sense of pressure in the left than in the right.

First I used two accurate thermometers to examine the intrinsic temperature of the hands in a number of subjects. The subjects' hands had been identically covered [*iisdem tegumentis tectae*] for a long time, or in general had been exposed to the same conditions. The subjects then took hold of the two thermometers, totally enclosing them. In this way it was found that immediately after the beginning of the experiment the mercury rose 0.5° or 1° higher on the left than on the right. This difference in temperature declined as the experiment was continued, so that at the end of an extended experiment the temperature in the two hands was found to be equal (i.e., differing by 0.33° or 0.25°). The hands were found to have a temperature of 28.5°, 29°, 29.5° or, at the most, 29.66° on Réaumur's scale;[7] a slightly higher temperature was observed transiently, now on the left, now on the right. Next I filled two large wooden vessels with warm water and checked their temperature with two thermometers that were permanently immersed in the water. Then I asked a colleague, who did not know the temperature, to immerse both index fingers or both hands in the water, simultaneously and to the same depth; and then to make a judgment of the temperature of the water.

In this way, all the while monitoring the thermometers, I demonstrated the result mentioned above, viz., *that most subjects felt the water observed with the left hand to be hotter than that judged with the right hand*, even though the temperature was exactly the same. If, without my colleague's knowledge, the positions of the vessels were interchanged, then it was always the water in which the left hand was immersed that seemed the warmer.

Moreover, the same result was observed if the water to be judged with the right hand was 0.5° or 1° warmer than the water to be judged with the left hand.

In order to illustrate these claims with examples, I now present some experimental results [Table 10-1].

If the water has a temperature of 19° or less, it imparts to the hand a sensation of coldness rather than of warmth. If right and left hands are simultaneously immersed in such water, a fiercer sensation of cold is imparted to the left than to the right. And so we see that not only the addition of heat but also its deficiency excites the left hand more forcibly [Table 10-2].

Of course, we should not expect it to be true of all subjects that the left hand is more sensitive to heat than is the right; but my experiments show that it is so in most cases. If now we look for the reasons for these findings... some may emphasize the fact men-

Table 10-1.

Part of body immersed in water	Temperature of this water (°R)	Part of body immersed in water	Temperature of this water	Parts in which sensation indicated greater temperature
Left hand	33.5	Right hand	34	Left hand
Left index finger	33	Right index finger	34	Left finger
Little finger of left hand	31.5	Little finger of right hand	32	Left finger, but uncertainly
Left hand	31	Right hand	32.33	Neither hand
Left index finger	29	Right index finger	31	Neither finger
Left hand	29.5	Right hand	29.5	Left hand

tioned above, that in the case of most people the two hands are not of the same temperature, the left being commonly found to be warmer.

However, the falsity of this explanation is shown by the fact that the same result is observed in both cold and warm water. If the left hand were more sensitive to cold water because of its own greater warmth, then it should necessarily be less sensitive to warm water than the right; for the intrinsic temperature of the water would be less different from the temperature of the left hand than it would be from that of the colder right hand. But this is not so, since the left hand is affected more keenly than the right, whether one immerses one's hand in warmer water or colder. So we must have recourse to another explanation. The palm of the right hand is covered with thicker skin than that of the left. Since the thicker skin is less easily penetrated by heat or cold, it presumably has two effects. First, if the bulb of a thermometer is enclosed in the right hand, a given quantity of heat does not flow so quickly into the thermometer as it does from the left hand. Second, if the hands are immersed, a given quantity of heat or cold does not pass so quickly from warm or cold water into the right hand as it does into the left. This explanation is supported by another phenomenon that I have observed on the palm and the dorsum of the hands. A moderately large wooden vessel is filled with water at a temperature of $+9.5°$ and the two hands are spread out and immersed. The ulnar surface is turned downward and the radial surface upward and the two hands are held a little way from each other. At the same time I attend to whether the dorsum or the palm is more strongly stimulated by the cold. I always find that at first the water

Table 10-2.

Part of body immersed in water	Temperature of this water (°R)	Part of the body immersed in water	Temperature of this water (°R)	Parts in which sensation indicated a greater coldness
Thumb and left index finger	10.5	Thumb and right index finger	10.5	Left fingers
Same fingers	11.75	Same fingers	10.66	Right fingers
Same fingers	12	Same fingers	12	Left fingers
Same fingers	12	Same fingers	11.8	Left fingers

touching the dorsum of my hand seems colder than the water between my two hands; but when 10, or 15, or 24 seconds[8] have passed, the sensation of cold gradually lessens in the dorsum of the hand and increases in the palm, so that finally the water between the two hands seems to be colder than the water touching the backs of the hands. I observe the same if the water is less cold. If indeed I use warm water, I find that at first the water touching the dorsum of my hand feels hotter but after some seconds have passed the palm seems more strongly stimulated by the heat. I have observed this in the case of a colleague as well as in my own case when the water had a temperature of 38.5°.

ACKNOWLEDGMENTS

The translator is grateful to Helen Ross and Kirsti Simonsuuri for discussion of several textual points. A full translation of *De tactu* is being prepared by Dr. Ross and will be published for the Experimental Psychology Society by Academic Press.

NOTES

1 Cf. J. Semmes, S. Weinstein, L. Ghent, and H.-L. Teuber, *Somatosensory Changes After Penetrating Brain Wounds in Man* (Cambridge: Harvard University Press, 1960), and A. Carmon, D. E. Bilstrom, and A. L. Benton, *Cortex* 5:27–35, 1969.
2 Probably Gustav Seyffarth (1796–1885), the Egyptologist.
3 Cf. S. Weinstein and E. A. Sersen, *J. Comp. Physiol. Psychol.* 54:665–669, 1961, and S. Weinstein, *Am. J. Psychol.* 76:475–497, 1963.
4 See Ferd. Aug. Holke. *Disquisitio de acie oculi dextri et sinistri in mille ducentis hominibus sexu, aetate et vitae ratione examinata.* Diss. inaug. Lipsiae 1830 apud. Leop. Voss. 4. (Weber's note)
5 Gottfried Tauber (1766–1825).
6 The experiment on weight discrimination that Weber is about to discuss would have offered a rather closer analogy. In these passages he does not clearly distinguish sensitivity and resolution: The left hand is simply "*subtilior.*"
7 1° Réaumur equals 1.25° Centigrade. The zero point is the same.
8 *Sexagesimis*: See *Corrigenda* at end of 1834 edition.

PART IV

COMPARATIVE STUDIES

11
Manipulative strategies of baboons and origins of cerebral asymmetry[1]

COLWYN TREVARTHEN

The properties of the mind prove the functional unity of the brain. I perceive what I attend to where it is, I understand what it means, I will to act on it, with a single awareness. The same unity of consciousness may be inferred in others. If a man appears to act with intention contrary to his perceptions of the world, we conclude his brain is damaged and its functions disordered. Even the insane may be mentally coherent, though their concepts are unreal or socially abnormal. Each of us normally acts and perceives in one world in which each object has one identity, each act has one immediate goal. Events not accounted for in experience must be assimilated into the single system of intelligence if they are to be perceived. In the course of life each individual maintains one line of conscious experience, one memory, one set of skills, and his changing emotions involve him as a whole.

THE FUNCTION OF BRAIN SYMMETRY

Although mind is one set of functions, the brain is formed in two equal-sized, mirror-symmetrical halves. Descartes thought mind and brain to be separate, but acknowledged this one-mind – two-brains dilemma and attempted to find a rational solution to it.[2] He reasoned as follows: The brain, and not the heart, is the material organ of the soul (or mind). The immaterial soul must, therefore, communicate with and integrate the twin organs of the body and brain through the action of some median, unpaired structure of the brain. He then proceeded to locate the organic agent of the soul in the tiny pineal gland, chosen because it is situated at the geometrical center of the brain.

Modern students of the brain accept no less than Descartes that its functions are fully integrated; but, with few exceptions, they leave aside the anatomical problem of mental unity of function. Taking account of knowledge that brain functions often require physiological interactions between nerve cells that are widely separated in the nervous system, they tend to disregard brain symmetry. They appear to believe that there are interconnections in sufficient abundance to link the parts together whatever shape the whole system has.

It is probable that the precise and near-universal brain bisymmetry is a manifestation of the central organizing principle, the common brain code, which ensures and maintains functional unity in relation to the external world, even in the most complex cerebral processes of man (see Young, 1962, and accompanying discussion by Kuypers and Nauta). This point of view has the immediate advantage of giving the most obvious morphological feature of the brain a functional significance. It also indicates an approach to the problem of cerebral asymmetry.

Cerebral asymmetry on a large scale, although it probably has ancient roots, is a recent evolutionary adaptation of forebrain function – a characteristic of certain "higher" cerebral processes of adult man. It is most evident in the neural mechanisms of language, which are strongly lateralized in the left hemisphere in most individuals over the age of 10 years (Penfield & Roberts, 1959; Zangwill, 1960a,b; Basser, 1962; Ajuriaguerra & Hécaen, 1964; Hécaen & Ajuriaguerra, 1964). Lateralization of language was first revealed by comparison of the effects on speech, writing, and comprehension of language of unilateral brain lesions in each of the two hemispheres, but it is also seen in subtle motor or perceptual asymmetries in normal people (Teuber, 1962; Milner, 1962; Kimura, 1967; White, 1969) and has been brought out by electrical stimulation of the cortex of conscious human subjects (Penfield & Rasmussen, 1950). The phenomenon of lateralization of language, associated with manual dominance, has encouraged the belief that mental functions are unified by being concentrated in one-half of the brain and richly interconnected or "associated" there. But even the recently discovered differences in anatomy between the half of the brain supposed to hold all the higher intelligence (and consciousness) and the other half in which there are held to be only unconscious, automatic, or "physiological" functions are clearly local enlargements in a basically bisymmetrical plan (Geschwind & Levitsky, 1968).*[1]

Tests revealing psychological functions specifically sensitive to right hemisphere damage (Patterson & Zangwill, 1944; McFie et al., 1950; Hécaen et al., 1956; Zangwill, 1960a; Milner, 1962; Arrigoni & De Renzi, 1964; Hécaen, 1969) have led to an alternative view: that there are two complementary realms of mentation that are segregated, one in each half of the brain. This supposes that a partnership obtains. However, once again, the structural basis for this differentiation must be at a level of histological refinement or circuitry of which we have inadequate knowledge. The nerve cells of the two hemispheres appear indistinguishable, and they are grouped on the two sides in finely mirror-symmetrical, and therefore basically very similar, configurations over most of the cortex. Moreover, a reciprocal hemispheric specialization merely brings us back where we were – attempting to find ways of integrating separate and unlike cerebral functions. The differences must be complementary within the functioning whole.

The corpus callosum supplies the principal direct link between the cerebral hemispheres. In man it is by far the largest fiber tract in the central nervous system. Research by Sperry's group has shown that sectioning of the corpus callosum and the other forebrain commissures produces dissociation of two organic systems of learning, awareness, and volition (Sperry, 1967, 1968; Sperry et al., 1969). The functions of the cerebral hemispheres appear to be closely equal on the two sides in animals, but markedly unequal in man, and the thesis of complementary systems of cognition is supported. However, though the separation may permanently affect intelligence in ways too subtle to have been revealed as yet, surgical disconnection of the hemispheres by itself does not disrupt skillful voluntary behavior; nor does it shatter consciousness. Elaborate test procedures are required to demonstrate clearly the remarkable effects of separation. We are forced to conclude that the brain stem and beyond, including the body itself, constitute an integrative machine capable of keeping the split brains one, even for the highest functions.*

Studies on split-brained subjects have been taken to support Penfield's theory (Penfield, 1954) of an integrative center for motor control in the brain stem (see Schrier & Sperry, 1959; Myers et al., 1962; Gazzaniga, 1966). But this thesis is inadequate to account for the integration of higher mental functions; it is contradicted by experiments showing that cerebral bisection dissociates higher functions, including skillful movements, if orientation is appropriately controlled. For a full explanation of cerebral integration it is not sufficient to retreat to a notion of a centrencephalic or brain stem locus for mind. The available evidence indicates that the cerebral hemispheres in which the higher functions are located must be organized as functionally compatible components of a larger system whose actions can be regulated and integrated by way of relatively meager circuits of the lower brain stem. They must be organized with respect to the way in which a bisymmetrical body can act on the world and at the same time be appropriately receptive to the feedback reflected from the environment on bisymmetrical receptors as a consequence of every act.

We conclude that the highest brain functions do not have intrinsic and autonomous unity at the level of the cerebrum or in the brain stem. The hemispheres are components of a hierarchically organized system built on a single fundamental principle of design, and they are integrated together in part as a result of their application through acts of the body on the environment. In the integration of cerebral functions the bisymmetrical central nervous system is in resonant relationship with the bisymmetrical body and its field of action.

This chapter presents information concerning the structure of skillful manipulatory behavior in baboons and the results of a preliminary experiment showing the effects of brain bisection on this behavior. This leads to discussion of the theory that brain bisymmetry is the basis of a fundamental orienting function from which even the higher cognitive functions are derived. According to this view, cerebral asymmetry of function depends on the underlying operation of such bisymmetrically organized orienting functions.

The concept of a *behavioral field* is employed. This is defined as the totality of regulated movement an animal can make, together with the sensory consequences predicted by the brain of the animal for these acts. It is essentially the *Umwelt* of von Uexküll (1937). If the body of the animal is bisymmetrical in form, as are the bodies of almost all freely locomotor forms, then the behavioral field will also be bisymmetrical. A central portion of the behavioral field can be distinguished as that portion in which specific and consummatory behavioral commitments are generally made after orientation is completed. Acts in the central behavioral field require closer specification and more detailed discrimination of goals than do acts of bodily displacement or orientation.

The essential, primitive interdependence between the central nervous system, the body, and the field of behavioral action in the environment – all three built on the same plan of bisymmetry – can be seen most clearly in the behavior of a lower vertebrate. Therefore, the visuomotor mechanisms of fish and amphibia is described first to supply a prototype of hemispheric organization in primates. I hope by this means to establish the general significance of neural symmetry and to elucidate the circumstances in response to which functional asymmetry evolved.

Brain morphology and the behavioral field

The brain of a fish or an amphibian is a bisymmetrical chain of laterally paired lobes, each pair primarily concerned with one exteroceptive or proprioceptive field, all interconnected via the brain stem, and all in more or less direct communication with the motor system of the brain stem and spinal cord. Bisymmetry is almost universal. The brain of the adult flounder, which has abandoned normal progression for an asymmetrical posture adapted to lying on the bottom, is as asymmetrical as the body itself.

The visual brain at this level of phylogeny is mainly located in the superior colliculi and adjacent midbrain tegmentum, with some collateral projections to the hypothalamus and to primitive and rudimentary diencephalic visual areas.

Experiments with fish, amphibia, and cats show that each half of the midbrain is a motor-integrating mechanism governing total patterns of movement that employ all the musculature of the body. Electrical stimulation of one-half of the brain evokes complex motor synergies; stimulation of the other half tends to produce mirror effects, and these patterns of movement may be indistinguishable from spontaneous whole-body orientations to natural objects (Apter, 1946; Hess, 1949; Akert, 1949; Ewert, 1967a). Ewert (1967a) described these evoked movements as constituting a *Richtfeld* or "aiming field" centered on the body. The same motor mechanisms are accessible to influences from many sensory fields, and the nervous system is integrated so that the various modalities define together a unique spatial context for behavior (Fig. 11-1*A*). This context is coincident with the orientational part of the behavioral field as it is defined above.*

Neural mechanisms for three kinds of acts have been located by electrical stimulation in the midbrain and diencephalon and confirmed by surgical studies. Bisymmetrical or near-midline stimulations produce symmetrical *defensive reactions*: arrest of movement, cowering, or undirected flight. Lateral points of the midbrain govern *orientations or taxic movements* to or from the opposite half of the external behavioral space. Stimulation in other anterior

Fig. 11-1. Relation between brain morphology and the behavioral field. *A*. Selected frames from a movie. A frog takes a mealworm dropped at 112° left of the midline (*m*), outside the binocular field (*b–b*). A two-step rotation brings the prey into the binocular central field (*c–c*). After a pause, the frog strikes. *B*. Mapping bisymmetrical behavioral field on the brain. *Left*. The trajectory of the worm relative to the frog shows successive orientation and approach. *Right*. The optic tectum of a frog, showing the topographical projection of the visual field with monocular, binocular, and binocular central areas. The path of the image of the worm near the horizontal meridian is indicated by *arrows*. It is in both halves of the brain when the worm is in the binocular central field. *com.*, commissural connections that contribute to the binocular mapping in this field (*A*, from a film made by D. Ingle. *B*, based on Gaze & Jacobson, 1962a,b; Jacobson, 1962; Schneider, 1954).

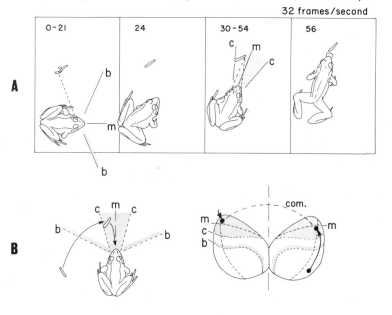

and midline areas, particularly after elicitation of an orienting movement, trigger a stereotyped and symmetrical fixation, then a *consummatory response*: biting, snapping, or tongue flicking, as if to an imagined prey located at a near point not far from the midline. Functional combinations of these effects can also be elicited by focal stimulation or disrupted by focal lesions (Hess, 1949; Sprague et al., 1961; Ewert, 1967a,b).

Representatives of all major vertebrate groups have the same precise topographical neuronal mapping from retina to the roof of the midbrain. Because the eye is normally maintained in a standard position relative to the body, this neural mapping from retina to brain is also, at least when the body is in its usual posture, a mapping from points in the visual field outside the body into the brain. Calibrated compensatory adjustments would permit this map to function independently of active changes of body posture. It is highly significant that the orientation of the map on the brain and the relationship between the visual projection and the spatially organized midbrain motor system are constant. Throughout the vertebrates, in spite of varied positions of eyes in the head (varied frontality), the bisymmetrical, body-centered visual field is always mapped in register with the bisymmetrical neural field of the midbrain and thence onto the efferent system and motor apparatus (Trevarthen, 1968). The area of orienting movements corresponds with the lateral monocular fields. The region of consummatory responses corresponds with the central binocular field.

It is, therefore, possible to draw a diagram of the behavioral field, representing sensory and motor fields outside the body, on the surface of the optic tectum (Fig. 1*B*). This appears to be the prototype of a universal system mapping the bisymmetrical behavioral field onto the brain.

Total-body orientation movement remains the province of the midbrain motor system throughout the mammals. In guinea pigs, hamsters, rats, cats, and monkeys, ablation of the superior colliculi of the midbrain clearly interferes with large-scale orienting behavior as well as with vision of space (Schneider, 1967; Trevarthen, 1968). The forebrain visual mechanism has evolved in conjunction with refined oculomotor and manipulatory movements made under visual guidance within the central part of the behavioral field. Yet, removal of cortical visual projection areas, though increasingly incapacitating up the phylogenic series as the relative size of the visual cortex increases, does not completely abolish visual guidance of the gross movements of reaching or locomotion, even in the monkey (Denny-Brown & Chambers, 1955; Humphrey & Weiskrantz, 1967).

The evolution of the forebrain visual mechanism has added a specific capacity for identifying objects and resolving patterns without replacing the primitive process of locating and orienting to visual stimuli. The basic layout of the midbrain visuomotor mechanism seen in the frog is retained in elaborated form in the forebrain of mammals, and the same bisymmetrical spatial context is employed. In monkeys and in man each cerebral hemisphere has most complete excitatory and organizing control over acts performed by the contralateral forelimbs and is most receptive to sensory information from the contralateral half of behavioral space. There is, however, a high degree of bilateral representation, so that a monkey or a man with one hemisphere removed can obtain cerebral control over the actions of all his body and can orient voluntarily to either half of space. The lateral spatial sensory fields for orientations by the ipsilateral limbs are represented in topographically or somatotopically organized parts of the cortical mechanism. Other parts are concerned with the central behavioral space where both forelimbs are employed (Trevarthen, 1970a).

Relation between schemata for complex behavior and the spatial context

Although even the most intricate mating rituals or territorial adaptations of lower forms are bound to a particular succession of orientations to present stimuli, signs defined within relatively narrow limits, perception and volition in the higher forms give predictive control that is largely freed from the body-centered orientation frame. The activities are guided by, and adjusted to match, the spatial and temporal occurrence of events in relation to the body, but their goal is defined in terms free of these particular temporary relationships.[3]

Manipulation is a skilled motor function that depends on the existence of psychological structures or schemata defining the form and mechanical functions of objects manipulated. The forelimb of a toad is one of an equal pair of props employed in locomotion and in postural regulations in reciprocal mirror patterns. In contrast, the hand of a monkey may cease to be a locomotor structure and become a tool employed intelligently to manipulate an object in a unique way according to structure inherent in the object. He manipulates an object once he is oriented to it, with posture adjusted so the object lies within his central behavioral field. There he can readily apprehend the appearance, sound, weight, texture, taste, and odor of an object and so define a rich identity for it. Either hand can be used simply to grasp an object. In manipulation the two hands rarely perform mirror-symmetrical roles. Usually there is polarization favoring one hand, and the other is relegated to a supportive, context-establishing role. The hand that dominates is the temporary agent for the whole organism.*

The class of manipulatory functions, much larger in man than in other primates, can be compared to language (DeLaguna, 1963). When a man is speaking to another, the whole of him speaks, and he talks of things that belong in a common field or context as independent as possible of his unique self-centered view of the world. Communication between animals depends upon a reduction in the behavioral isolation of the individuals within private body-centered frames of reference. It is the special power of language in man that carries this freeing from personal context to its ultimate.

The nature of the complex functional relationships between the strictly bisymmetrical body-centered orientational field and the central field within which unique, decentered, and asymmetrical perceptual and motor schemata are created is shown in the way a monkey behaves when he learns a fine manipulative skill. The next section describes the important steps in the acquisition of such a skill and shows that the actively differentiated bimanual schema, which is essentially asymmetrical, is also lateralized more in one-half of the cerebrum.

EXPERIMENTS WITH BABOONS

Acquisition of manipulative skill by normal baboons

Method

Five young baboons – Sophie, Marguerite, Jojo, Daphne, and Coco – were trained to open a problem box to get food. Each subject worked in a cage that confined its body to a standard position in relation to the task while the subject was responding (see Fig.

11-8). The manipulations required to open the box were made symmetrical with respect to the body axis of the subject by intertrial reversal of the left-right orientation of the problem box (Position A and Position B) on an unpredictable, balanced schedule. The response movements were made mainly with the lower arm, wrist, and fingers, with minimum displacement of the shoulders and trunk.

To see the box, the monkey placed his face against a fitting mask and then reached his arms either side of a small breastplate. This arrangement permitted free arm movements over a wide space, but the experimenter was able to restrict the responses to one arm and hand by closing a door on one side. The face mask with two eyeholes like a pair of spectacles gave similar control for vision.

A record was kept of the hand used for each step in opening the box. The time taken (work time) was measured by an electronic timer started by the initial contact of the monkey's hand with the hook (H) and stopped by the opening of a microswitch (M) at the

Fig. 11-2. Manipulations (a to h) used in opening the problem box with the left hand. See text for explanation.

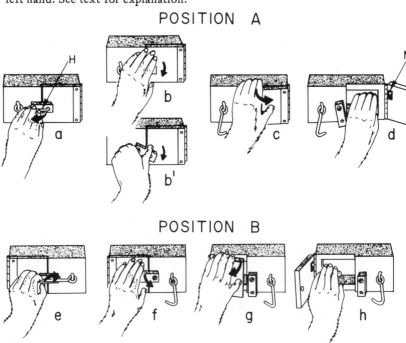

Manipulative strategies of baboons

start of the movement of the lid (Fig. 11-2). Further details of the performance were recorded with a movie camera, operating at 25 frames per second, aimed from vertically above the box. A flash electronically synchronized with the camera stopped the image during exposure of the film, the flash frequency being double the shutter frequency to eliminate disturbance to the subject by flickering of the light falling on the box.

The monkeys learned to open the box with both eyes open and free choice of hands. Each quickly learned to open the lid, then to turn the cleat to free the lid; at this point the hook was inserted at the start of each trial and timing commenced. Removal of the hook required the subject to make a precisely controlled movement – a slight pull or push applied horizontally caused it to block. It was necessary at first to encourage the subject by partially freeing the hook. In general, the monkeys were attentive to demonstrations by the experimenter of how the hook could be lifted to free the cleat. They clearly imitated. After the first few successful tries on the first day with the complete task, no further assistance was necessary.

Details of behavior during learning

The design of the box favored use of both hands. Though it could be opened in either orientation by one hand, some movements were awkward, particularly those requiring translation of a hand to the midline (Fig. 11-2). Lifting the lid in Position A, lifting the hook in Position B, and turning the cleat in Position A were difficult for the left hand. All these movements were easily performed by the right hand. The task could be performed more easily and more quickly by use of the two hands in partnership, with strategies such as those illustrated in Fig. 11-3. (The drawings of hand

Fig. 11-3. Two-handed strategies for opening the problem box.

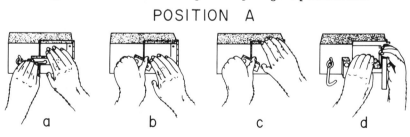

movements were obtained by tracing from movie frames of skilled subjects.)

Although all subjects initially made unskilled attempts to open the box with both hands, when more practiced they performed primarily with one hand. The second hand either served to assist in certain steps or dropped out entirely. Final habits, when learning had become stabilized and when the movements were swift and regular, are shown in Table 11-1. Fig. 11-4 shows the course of learning and the gradual equilibration of a skillful bimanual habit for one subject, Daphne.

The initial participation of the subject was clearly based upon his knowledge of the approximate locus in space where the food was hidden (i.e., that the food was in the box). In learning how to open the box quickly, the impulse to move directly to the lid concealing the food was subordinated to a plan to move the cleat, and then pulling at the cleat was postponed if the hook remained in place. The task had to be structured by differentiating and correctly ordering the successive acts required to reach the food.

For a naive subject, success in opening the box was the result of many rapidly repeated, stereotyped, and poorly adapted movements. Often force was employed in attempts to pull and push the parts free. Pushing movements with the palm and strong pulls with hooked fingers and fingernails were employed. It was observed that the action was broken into brief periods of contact with the task and periods in which the hand was withdrawn. Each approach

Table 11-1. *Final habits and speed of performance for alternative orientations of problem box*

Subject	Position A			Position B		
	Dominant hand	Bimanual strategy[a]	Median time for 32 trials (msec)	Dominant hand	Bimanual strategy[a]	Median time for 32 trials (msec)
Sophie	Left	None	600	Left	None	750
Marguerite	Right	None	570	Right	R L L R	460
Jojo	Left	L R R L	670	Left	None	600
Daphne	Right	L R R R	740	Right	R R L R	480
Coco	Right	L R R R	390	Right	R R L R	580

[a]The symbols used to describe the bimanual strategies show hand used (left, L; right, R) for each of the four steps in obtaining the reward: lifting the hook, turning the cleat, lifting the lid, and taking the peanut. See Figures 11-3 and 11-4.

or withdrawal appeared as a regulated gesture occupying about 0.25 sec, and there was frequently a break of the same duration after failure to move a part or when an erroneous gesture was made. Sometimes it could be observed that in this space of time the hand was held aside to allow visual inspection, but at other times the pause appeared to be occupied with "taking in" touch information. After such a pause, the next move to the box frequently showed evidence of a change of plan incorporating new information. The observed changes show that learning consisted in the formulation of strategies for forays of action, with increasingly accurate predictions of the sensory consequences of each group of movements or period of contact with the task.

When the experimenter inserted the hook for the first time, this immediately caused the subject to be more cautious or hesitant and exploratory, even though a few futile attempts were still made to turn the cleat by force without removing the hook. Thus, the subject became more attentive and visually curious about many aspects of the task before he knew how to act effectively. Fruitless

Fig. 11-4. Learning curves for the two orientations of the box. Subject Daphne. Median times and interquartile range for groups of 16 trials. For Position A only one strategy was used. For Position B two rival strategies alternated, then equilibrium was attained favoring the bimanual strategy. L, left hand; R, right hand. See text and Table 11-1.

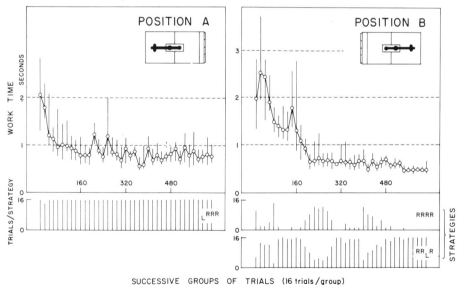

repetition of an ineffectual strategy was not characteristic of normal learning. Though errors and omissions continued to be made over many days of training, their correction became increasingly rapid and automatized. When the task was well learned, corrections were made without a pause for observation showing that expert proficiency depends upon advanced formulation of "rules" permitting detection and correction of common errors along with the formulations essential to the control of the correct gestures.

The smooth flow of skilled movements in the performance of well-learned manipulations for opening the box gave evidence of continuous sensory guidance for finger and wrist movements as well as for the length and direction of reaching with the whole arm and for the transitions from one gesture to the next. The practiced manipulatory steps took 150 to 200 msec and passed without pause into one another so that the first three steps – seizing and lifting the hook, turning the cleat, and lifting the lid – occupied as little as 400 msec. Little force was used, and the critical movements were made with fine motions of the fingers and rotations of the hand and wrist. These observations are best explained by the hypothesis that initially independent movements become assimilated into a continuous pattern or strategy comprehending the task, defining the action and predicting its reafferent consequences.

Bimanual patterns
At the start of learning, two hands were often used together in forceful attempts to open the box by coupled action in a whole-body movement involving rotation of the shoulders. Later, the movements of the hands sometimes appeared to compete, and occasionally there was interference with one hand by the other as if the two were independently controlled for an instant. However, such independent action was rare and of short duration, during periods when many small exploratory gestures were made in rapid succession and neither hand was well controlled. As has been said, familiarity with the task soon led to more restrained patterns of exploratory movement localized in the fingers and wrist, with one hand favored over the other. Learning was accomplished primarily by one hand. Afterward, the second hand was recruited to give assistance at points where, apparently, some insight or organizing mechanism led the subject to shift spontaneously, in several cases, to the easier strategy (Table 11-1). The cooperation of the two

hands became, thenceforth, more and more rapid and well synchronized. What were initially separate gestures detached in time became closely combined parts of a single pattern of movements by a process of assimilation like that described for improvement of the skill as a whole (Fig. 11-4). The speed of practiced two-handed patterns of activity shows that the central coordination for the two hands occurs in a system that can become as unified as the one concerned with the coordination of muscle contractions to move one limb. Practiced bimanual activity is not merely a chain of alternated one-handed steps. Two hands can open the box in less time than it can be done with one hand simply because the two hands play their parts overlapping in time (Fig. 11-5).

Role of subordinate hand
It is important to note that the subordinate hand performed both less frequently and with less skill, being relegated to the relatively undifferentiated steps in the task as learning proceeded. With stabilization of the skill, fine finger and wrist movements of the dominant hand were applied to all the more difficult steps, except by subjects Daphne and Coco, who were more nearly ambidexterous than the other three subjects and employed the subordinate left hand to remove the hook in Position A. It will be recalled that it is awkward to do this step with the right hand because to do so involves a center-to-opposite-side displacement of the whole arm.

Fig. 11-5. Speed of performance for spontaneous one-handed and two-handed strategies. Subject Marguerite. See text and Table 11-1.

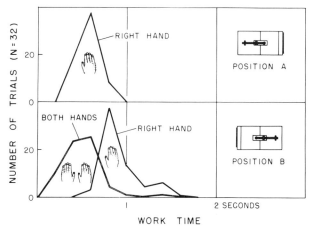

Nevertheless, the remaining three of the five subjects did make this more awkward gesture and so preserved use of the dominant hand. On all other occasions when the subordinate hand was employed, the manipulation made was a simple one requiring little finger dexterity; it took the form of a swift "pattern aside" movement with all fingers held together, accomplished with a twist and an arm displacement (see Fig. 11-2: Manipulation f was done by Jojo with the right hand; g was performed by Marguerite, Daphne, and Coco with the subordinate left hand, and the mirror pattern of g was done by Jojo with the right hand).

Table 11-2 shows the distribution of the eight main manipulative gestures shown in Figure 11-2 (combined with the mirror configurations of each) in the practiced performances of all five subjects. From this it is possible to determine which of the alternative gestures is the more easily performed, and in every case it is this gesture which is made by a recruited subordinate hand to replace a more awkward step by the dominant hand. There is no such difference for removal of the reward; here only the dominant hand was used.

Intermanual equivalence

The extent to which learning had occurred in neuronal mechanisms associated with each of the hands was tested by observing responses restricted to one hand at a time. As Figure 11-6 shows, the performance with each hand depended on how much this hand had been used in the spontaneous habits acquired for opening the box. The nonpreferred hand was least skillful in the case of the monkey Sophie, who had showed a 100% left-handedness in learning the

Table 11-2. *Distribution of eight main manipulative gestures (a to h) shown in Figure 11-2 (combined with the mirror configurations of each) in the practiced performances of all five subjects*

	Hook			Cleat			Lid			Reward			Total
	a	e	Total	$b+b'$	f	Total	c	g	Total	d	h	Total	
Dominant hand	5	3	8	3	5	8	1	5	6	5	5	10	32
Subordinate hand	2	—	2	—	2	2	—	4	4	—	—	—	8
Total	7	3	10	3	7	10	1	9	10	5	5	10	40

task. For Coco, the subject who had learned to perform the task with the two hands employed spontaneously to approximately equal extents, each of the two hands was lacking in skill. The other subjects fell between, except that there appears to be an additional factor influencing the distribution of skill between the hands in the case of two subjects given markedly different degrees of overtraining. Marguerite was allowed to practice for over 1000 trials after attaining a consistently high level of proficiency; Jojo learned the task in approximately 400 trials and was tested for efficiency with each hand separately without overtraining. Apparently overtraining favors adaptation to the task with benefit to both hands, and so the difference between preferred and nonpreferred hand is reduced. Put another way, overtraining favored the establishment of a transferable or generalized schema for the task. With this schema, the two hands showed a greater degree of motor equivalence, but none of the subjects showed immediate and complete transfer of the skill between the hands.

Fig. 11-6. Comparison of work times for five subjects when free to use both hands (*F*), forced to use the preferred hand only (*P*), or forced to use the nonpreferred hand only (*N*). Each column shows the interquartile range and medians (*black bars*) for 32 trials. The percentage preference and direction of preference for each subject are indicated at bottom.

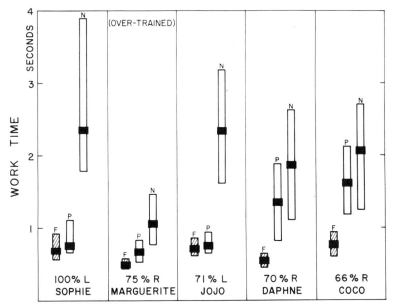

Training of the nonpreferred hand was not continued beyond the few trials necessary to measure the initial level of proficiency, because it was desired to avoid delateralizing the spontaneously acquired habit. However, in the few trials given, there was a marked improvement in proficiency. In each case, the subordinate hand, though initially unable to perform some of the steps of the task quickly and with the most appropriate and economic gestures, was able to acquire skill by accelerated learning. Some aspects of the experience gained primarily with the dominant hand were transferred to the subordinate hand. The errors made were of a kind suggesting difficulty with the correct shaping of the pattern of muscular contractions to move the wrist and hand properly rather than difficulty in visualizing the task correctly or attending to its components. In every case, the first efforts were correctly directed to the hook, in each of the two orientations of the box, even when the hand in use had not previously been employed to lift the hook. Similarly, other steps were attempted in correct sequence even though the unfamiliar gestures were made awkwardly. There was transfer to the inexperienced hand of conceptual elements and strategies derived from past experience, but refinements of motor control required active practice of the muscles directly concerned, especially for the distal segments of the limb, and did not transfer as readily between the limbs.

Presumably, most of the intermanual transfer of skill depended on retention of *visual recognition* of the task and *visual orientation* to the parts of the task. The responses were aimed visually, even though they were not precisely guided in a skilled choice of movements. Tests made with vision restricted to one eye at a time led to little change when two hands were in use or when activity was restricted to either of the hands alone. The two eyes appear perfectly equivalent as avenues for sensory information for this task.

It can be concluded from the above observations that subordination during the acquistion of a lateralized manipulative skill involves use of one hand for less skilled kinds of movements at points in the overall program of the task where, because of spatial constraints, the dominant hand fails. The subordinate hand is, however, still capable of rapid improvement if the dominant hand is prevented from acting. The two hands become integrated in a bimanual strategy by a process of competitive interaction and progressive spatial polarization, and this limits the practice of the subordinate hand in precise movement under visual and touch guidance.

Effects of surgical separation of hemisphere on acquired skill

Two of the trained baboons (Marguerite, who was right-handed, and Jojo, who was left-handed) have been observed following division of the optic chiasm and section of the anterior commissure and corpus callosum.[4,5] These animals showed disruption of the acquired bimanual control following surgery. When tested within a few days of surgery, both subjects first performed the task spontaneously with the dominant hand alone; the subordinate hand remained inert even for steps in the manipulation that had been performed by it before surgery. The dominant hand was initially incapable of filling these gaps in the performance of the tasks.

Some of the first reaching movements to the hook were misjudged as to depth, a defect simply accounted for by loss of binocular parallax owing to chiasm section. These errors were corrected promptly and they rapidly became rare. On other trials, mistaken reaching to the cleat with neglect of the hook in one orientation of the box, and other comparable errors, could be attributed to the loss of one-half of the visual field. This loss may have been the result of a spontaneous restriction of the integrative activity underlying the voluntary response to the hemisphere opposite to the moving hand. Certain redundantly repeated movements, particularly to the cleat, appeared to be caused by insufficiency of afferent information when the gesture involved movements of the hand across the midline in the direction of the neglected half-field. Similar losses resulting from disturbance of feedback integration are discussed in detail later.

In addition to the above, there were more complex efforts of surgery on central integrative processes. Marguerite and Jojo were efficient and nearly as fast as before surgery when using the more experienced and still spontaneously preferred hand. However, in some trials a response that had been most frequently performed by the subordinate hand before surgery was completely omitted. This suggests that the automatic strategy of movements for one hand tended to allow spaces for the responses made habitually by the other hand. While the subordinate hand was inert, these spaces led to omissions and required corrections that were occasionally performed very late and very awkwardly by the dominant hand (Fig. 11-7, manipulations *e* and *f*).

When forced to use the subordinate hand alone, both subjects worked inefficiently. Again, there were errors of depth estimation and of hemispatial neglect, but in addition the responses were frequently hesitant and inaccurately directed, and sometimes the hand felt over the task briefly as if visual guidance were absent. At the start of forced performance with the unpreferred subordinate hand, the hitherto dominant hand pushed at the panel barring the opening in front of it and made unsuccessful efforts to get to the box. Following such attempt, the movements of the subordinate hand were poorly integrated and slow. Some steps were omitted. The most efficiently performed steps included those performed well by this hand before surgery, but occasionally these responses were poorly executed as well. There was a steady improvement in the performance of the second hand with practice. For Marguerite and Jojo the deficiencies of skill were observed in the absence of any indications of weakness or malcoordination of the same limb in performing other delicate tasks requiring eye-hand coordination for following moving objects and for precision reaching and grasping;

Fig. 11-7. Abnormal performance on problem box task following split-brain surgery.

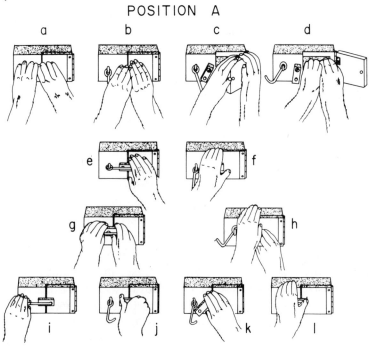

that is, the defective motor coordination was specific to the manipulation learned under controlled conditions just before surgery and associated with a spontaneous set to use the dominant hand.

After the first postsurgical period in which the responses were made entirely with the dominant hand, Marguerite and Jojo made a few spontaneous gestures with the subordinate hand, and thereafter many trials included responses made by the two hands. The first moves of the subordinate hand appeared after failure of the preferred hand to execute a response immediately. Thereafter, many redundant bimanual responses occurred with imperfect timing and shaping of gestures of the two hands. It was striking how the coordination of the dominant hand always deteriorated when the subordinate hand became active. This may have been caused, in part, by disturbance of the processes of control in the hemisphere contralateral to the dominant hand by unanticipated visual or tactile feedback from activity of the subordinate hand. In many cases, however, the effect was more like a momentary inhibition of blocking of activity in one side of the brain by regulatory influences of central origin passing through the undivided brain stem (see Discussion).

Bimanual responses of both subjects included many double attempts at one or all steps in the task (Fig. 11-7, manipulations *a* to *d*). Often the two hands were directed in distinctly abnormal, symmetrical, and synchronous reaching and grasping movements toward the end of the hook in the center of the task (Fig. 11-8). Occasionally, the two hands were then withdrawn in mirror arcs, only one hand carrying the hook. On other occasions, both hands turned the cleat or the lid together in an arc to one side (Fig. 11-7, manipulations *b* and *c*). In many trials, both hands pounced on the food to struggle in the opening of the box which was too small for both hands to reach inside at once (Fig. 11-7, manipulation *d*). Double responses with prolonged conflict occurred sporadically; more commonly one hand would quickly withdraw to one side when it touched the other hand, or even before contact, presumably under the influence of the visual image of the other hand entering the field.

In postsurgery bimanual performance, some correctly chosen complementary gestures of the two hands (see Fig. 11-3) were mistimed, or one hand performed so roughly that the combined responses failed. Frequently, for example, the rapid sweeping movement downward in the midline by the fingers of one hand to turn the cleat was made before the other hand had removed the hook, and this latter hand was roughly knocked aside (cf. Fig. 11-3, manipulation *a*). Sometimes when one hand became correctly placed to open the lid, the other hand, en route to the cleat, abandoned this course to pull at the lid also.

Although poorly coordinated bimanual responses were made by both surgically treated monkeys, and performance was initially very confused, both subjects overcame this. At first, control of the two hands appeared to be by acquisition of habits alternating activity rhythmically from side to side between them. The swaying rhythm generated may have aided interhemispheric communication via peripheral channels and resembled a locomotor pattern. Eventually the two hands became so well synchronized in their performance that some trials were executed as quickly and as smoothly as before surgery. Such high levels of proficiency were not completely stable, and lapses to discoordinated patterns occurred.

Generally, the postsurgery bimanual strategies were built upon the bimanual habits developed before surgery, but some new alternations of the hands were acquired. It seems most likely that this acquisition of bimanual control after the corpus callosum had been cut depended in part upon each hemisphere's learning to

Fig. 11-8. Three successive frames from a movie showing symmetrical reaching to the hook at start of trial. Subject Marguerite, after split-brain surgery.

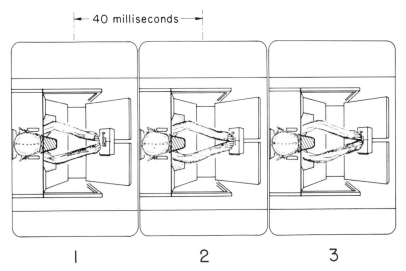

anticipate the actions of the other, much as drilled soldiers or piano duettists acquire closely collaborative movement patterns by attending to and eventually anticipating correctly the visual, auditory, and other cues transmitted between them. In the case of a split-brained individual, even if there are no means for immediate cross-integration of visual information between the two hands, the separated cerebral hemispheres are still within one body whose parts remain in communication by mechanical contact at least, and postural movements would produce correlated reafferent information, some of which would cross-communicate between the two halves (see Discussion).

It should be noted that though precise and rapid bimanual coordination for opening the problem box was reacquired by the split-brained baboons with difficulty after a period of bimanual anarchy, these same animals performed many familiar bimanual coordinations surprisingly well the day after surgery when they were free in the home cage. For example, they both held a jar suspended from a string in one hand while removing nuts from it with the preferred hand.[5]

Distribution of skill in the hemispheres

Closing off one eye of a split-brained monkey (optic chiasm and interhemispheric commissures sectioned midsagittally) limits direct visual input to one hemisphere. Tests with the trained subjects after surgery showed that when vision is restricted to the left eye, both hands retain the ability to respond to the manipulative task, but the ipsilateral (left) hand becomes less active and, when it does respond, makes errors of reaching or gestures that are inappropriate or misdirected. This accords with results seen in discrimination tests of split-brained monkeys where the monkey expresses his choice between visual patterns by simply pushing a lever or displacing a wooden block over a food well (Downer, 1959; Trevarthen, 1962a; Gazzaniga, 1964). Despite these imperfections of ipsilateral eye-hand coordination, which include errors not attributable to visual field losses, bimanual performance may be little affected by restriction of vision to one or the other eye. Again, the defects of visual control are probably compensated for by improved use of somesthetic or proprioceptive cues.

Marguerite, the baboon who learned the problem box with 1000 trials of overtraining, was tested with each of the four possible eye-hand pairs in isolation. These tests revealed a distinctly asymmetrical distribution of the skill in the forebrain (Fig. 11-9). In comparison with the preoperative levels of performance, the times taken to open the box immediately after surgery were a little longer when no restraints were imposed, but, as we have seen, there was immediate improvement in this, mainly owing to increase of control of responses made by the nonpreferred left hand. Of the one-eye-one-hand combinations, the contralateral left-eye–right-hand pair was least affected by surgery. Performance by this pair was about the same as preoperatively, though a little slower. The ipsilateral right-eye–right hand pair was much slower, and there were a few highly discoordinated responses. A gradual improvement occurred with practice. When forced to use the left hand alone with either eye, Marguerite performed poorly. At first the left hand was

Fig. 11-9. Presurgery (*white bars*) and postsurgery (*hatched bars*) performance on problem box task in all eye-hand combinations. Subject Marguerite. Medians and interquartile ranges for groups of 32 trials. Probability levels from X^2 test. *N.S.*, not significant; *p*, preferred hand; *c*, contralateral eye; *i*, ipsilateral eye.

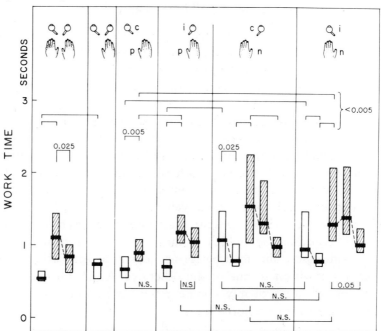

not significantly slower in opening the box when guided by the left eye than when guided by the right eye, but the latter combination showed a greater immediate improvement in performance, and this gain was continued steadily and rapidly. With the ipsilateral combination, left eye–left hand, coordination was more erratic and improvement less significant. Thus, one can say that though the right-eye–left-hand combination appeared to retain the skill least after surgical separation of visual and interhemispheric pathways, this combination was able to *learn* quickly in isolation. In 96 test trials with each combination, the contralateral pair showed steady gain in proficiency, and the ipsilateral pair, favoring bihemispheric control, performed erratically and was susceptible to sudden decline in proficiency with distraction or fatigue.

These results for this subject are summarized in diagrammatic form in Figure 11-10. From this we can form a picture of the retention of strategies for the complex manipulative skill among cell populations in the hemispheres. In the left part of Figure 11-10, the time scores per groups of 32 trials for all four eye-hand combinations are represented as circles whose radiuses equal the difference between the median times to open the box and an arbitrarily chosen limit time of 2 sec, which approximates the score of an almost naive subject. In other words, the radiuses of the circles are

Fig. 11-10. Diagram of retention of manipulative skill. Subject Marguerite. *Left.* Postsurgery retention and immediate effects of further practice. *Right.* Presurgery distribution of learning changes. See text for explanation.

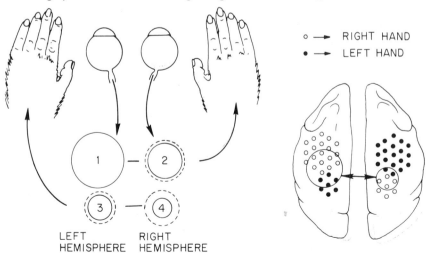

proportional to the gain in time achieved by learning, before surgery, for each eye-hand system. The effect of postsurgery learning is indicated by the broken-line circles, which show the way the divided engram developed with practice. For the right-eye–hand combination (2), the effect in 32 trials of more practice is shown; for the left hand, the two broken-line circles show the overall gain in 64 trials with each eye (3, 4). The dominant left-eye–right-hand combination (1) did not show improvement; it retained the practiced skill with undiminished strength. On the right of Figure 11-10 is a simple schematic representation of these results in terms of cortical units recruited by learning to the control of the manipulative movements from among asymmetrically arranged groups of units pertaining to the two hands. The assumption is made that, related to its commitment to guidance for movements in one-half of the space surrounding the body, the left hemisphere contains more "right-hand" units (white dots) than "left-hand" units (black dots). Conversely, the right hemisphere has more units representing the left hand. In the case of Marguerite, the skill was learned mainly with the right hand with some assistance from the left hand. The learning, therefore, had changed the adapting memory units whose cortical elements were more numerous for the right hand than for the left.

The postsurgery retention scores for all four eye-hand pairs suggest that the right hemisphere had an incomplete representation of the recently acquired skill. The scores fit the assumption that transcallosal communication during learning enabled memory units for left and right limbs to be consolidated *in the same proportions* in both hemispheres, but more abundantly in the left hemisphere, which thus had attained a functional dominance for this task. The latent population of left-hand units in the right hemisphere, which were not recruited while the corpus callosum was intact, could now effect a rapid adaptation, and so there was relatively impressive learning by the right-eye–left-hand combination. Both ipsilateral combinations, unbalanced with respect to the visual versus nonvisual cues available to the two halves of the divided population, and lacking a reserve of units in hemisphere receiving visual afference, showed less extensive and less constant improvement.*

Cerebral asymmetry for control of manual responses was also shown by Marguerite in a reaction time test. Trained before surgery, the baboon learned to release a start pedal to touch one of the two small neon contact switches quickly when it was illuminated 0.5 to 2.0 sec after the start pedal had been depressed. Reaction times from the appearance of the stimulus to the release of the start pedal were equal for left and right eye before surgery. The preferred right hand attained very steady performance, with a median reaction time of close to 250 msec. Reaction times with the left hand rapidly became equal to this or even slightly shorter.

After surgery, reaction times of the left hand remained unchanged when the stimulus was made visible to either the left or the right eye alone. There was a slight improvement over presurgery times as a result of practice. With the right hand, however, there was a significant difference between contralateral and ipsilateral eye-hand combinations. The reaction times for the left-eye–right-hand combination were 20 to 30 msec shorter than those for the right-eye–right-hand combination (Fig. 11-11).

This result recalls the findings with human subjects that cortical lateralization of somesthetic control may be more evident for the dominant right hand than for the left. Though the right hand is consistently more affected by left hemisphere lesions, touch discrimination by the left hand is frequently affected by lesions of the ipsilateral sensorimotor region, and bilateral lesions may have a greater effect on the left hand than on the right (Semmes et al., 1960; Weinstein, 1962).

Results of the reaction time test can be interpreted in terms of lateralization in the left hemisphere of nonvisual components

Fig. 11-11. Postsurgery reaction times. Subject Marguerite. *Dotted areas* show frequency distribution curves for ipsilateral eye-hand combinations. Each distribution curve is based on 192 trials.

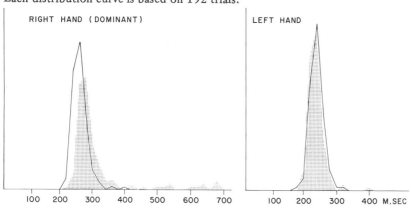

"readying" the right hand. Restriction of visual input to the right side of the brain resulted in inhibition or deprivation of this element of right-hand mobilization and weakened response.*

DISCUSSION

Hierarchy of functions in use of hands

Experimental studies of the brain mechanisms controlling hand movements in primates have produced what appear to be discordant results because different forms of activity, dependent on reafferent guidance from different receptor fields, have not been distinguished (Schiller, 1957). The simplest functions of the hands of a higher primate differ little from the use of forepaws by the most primitive land vertebrate, but within the primates there is extensive evolution of more complex and more delicately regulated uses of the hands. The prosimians grab moving objects rapidly and accurately, but they are incapable of holding a small object precisely with opposed thumb and forefinger or of moving elements of a mechanical system separately with fine extensions and flexions of individual fingers (Bishop, 1964). Manipulation with dextrous combination of individual digits is peculiar to the higher primates (Schiller, 1957; Napier, 1961). In man, manipulation is far richer than in any primate, and it is facilitated by bipedal standing and walking. In the development of the human infant, coordinated displacement of the hands for locomotor placing, grabbing, grasping, and fine manipulation appear at different times as distinct though related functions. This development undoubtedly reflects successive maturation of different brain mechanisms having convergent control over hand movements.

Both in evolution and in development, the more complex manipulative functions appear integrated within a single orientational framework or spatial context that governs placement of the hand in the space around the body. The skills of manipulation attain freedom from the spinal mechanisms governing quadrupedal patterns and the brain stem mechanisms controlling postural and spatial orientation, but remain in intimate association with them.

Hierarchical organization is also seen in the attention governing the selective application of the hands to objects in the space surrounding the individual. For example, with respect to vision, there is evidence for two main levels of perception process in the primate brain: one, largely unconscious, regulates immediate orientation in ambient space surrounding the body; the other performs the focal analysis and resolution of detail essential to perceptual recognition of objects remote from the body and to specifying complex programs of refined and discriminatory action appropriate to them. Simple manual placing or reaching can be regulated in the ambient field, but manipulation requires both focal visual attention to detail and scanning or displacement of this attention over the task in an intelligent strategy (Trevarthen, 1968). The latter cognitive aspect distinguishes both skillful manipulation and appropriate visual perception of manipulable objects.

Symmetry in field of action in hand and in brain

The acts in which the hands of a primate participate constitute a hierarchy of functions in a bisymmetrical field of action.

In locomotion, the total of all potential orienting adjustments forms a bisymmetrical ensemble. Similarly, activities combining reaching and grasping with one or two hands and postural and locomotor movements can be as frequently oriented to left or to right or can be symmetrically aligned in the midplane of behavioral space. There is no requirement for asymmetry of functional control for these acts.

For reaching with the hands, the field of action is, again, bisymmetrical about the body. Either hand can reach across in the space in front of the body for a certain distance beyond the midline to grasp an object. Thus, there is a bimanual reaching field as well as two more lateral fields where only the ipsilateral limb can reach.

In fine manipulation, in contrast to all these simpler acts, a small object is picked up and oriented to favor use of one of the two hands at the center of attention, within numerous complementary sensory fields. The use of one hand more frequently than the other introduces a special functional asymmetry into the motor field. Any asymmetry of motor strategy depends on the assumption of an orientational or postural control over the relationship between body and object. Manual preference for skills may be considered an evolutionary outcome of the introduction of such consistently asymmetrical motor functions into spontaneous patterns of behavior (Fig. 11-12).

Cutting across the vertical hierarchy of the brain mechanisms governing hand movements is an inherent relationship between bisymmetry at any level of the brain and orientations in the space centered on the body. It is suggested in the introduction to this chapter that this bisymmetry should be regarded as the common plan, determined in morphogenesis, that ties together the various levels in the hierarchy of sensorimotor control. A converse relationship obtains between bilateral duplication, with representation of functions for both halves of the behavioral field in each half of the brain, or asymmetry of function in the cerebral hemispheres, and the psychological processes by which the structure or meaning of identities or goals is analyzed independently of the body-centered orientation space.[6]

Control of manual responses by brain

The effects of cerebral ablations on the behavior of primates show that control of voluntary movements of either hand is neither localized in any one cortical area nor confined to one-half of the brain. Each hemisphere regulates the movements of both limbs. However, corticospinal connections mediate the fine adjustments

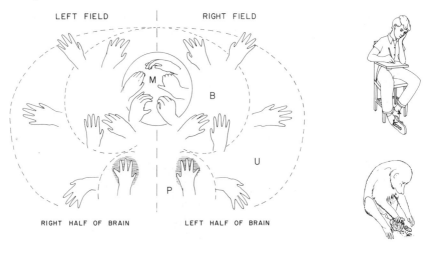

Fig. 11-12. *Left.* Field of hand use for a higher primate. *M*, manipulation; *B*, bimanual reaching and grasping; *U*, unimanual reaching and grasping; *P*, locomotor placing. *Right.* Manipulation and praxis while the body and lower limbs are passive and the posture is subservient to focal attention to the goal.

of the contralateral hand without exercising a comparable control over the ipsilateral hand. Two mechanisms of cortical motor control can be distinguished. These appear to differ in their relationship to the proximodistal hierarchy of limb segments. One governs coordinated orientations of limbs in relation to the whole body; the other controls immediate local adjustments of the distal segments of the hand in fine grasp or manipulation. Any manual response is likely to bring both these mechanisms and large areas of the cortex into action.

Recent neuroanatomical studies have disclosed that the motoneurons and their associated interneuronal and sensory fields in the spinal cord are grouped according to a consistent somatotopic map, and the same map is reiterated at higher levels of the brain. There are thus formed two parallel efferent systems (the precision motor system and the core motor system) throughout the length of the central nervous system, differentiated according to their terminations in the distal and in the proximal muscles of the body (Brodal, 1963; Kuypers, 1964; Eldred & Buchwald, 1967).

The *core motor system* of the body axis and proximal limb segments controls the axial and proximal muscles from many cortical areas through the medial reticular formation of the brain stem, the vestibular nuclei, and interneuron groups in the intermediate zone of the spinal cord. The tectobulbar and tectospinal pathways also join this medial system, which appears to subserve whole-body patterns of motor activity in the regulation of posture and locomotion. Each hemisphere projects to both sides of the body through multisynaptic routes.

In the *precision motor system* corticospinal and corticorubrospinal projections lead from each hemisphere by crossed tracts to the contralateral interneurons and motoneurons of the distal musculature of the limbs. In the brain stem, fibers from the motor cortex terminate in the lateral reticular formation in relation to the motor nuclei of the muscles of the face and mouth. This motor system of discrete and refined movements is a lateral column of neurons in the motor region of brain stem and spinal cord and the cortical cells bordering the central sulcus. It has long been known that the cortical motor cells are grouped according to a somatotopic plan that can be plotted by local stimulation. Microelectrode studies demonstrate exceedingly fine somatotopic architecture in this cortex and an equally precise projection to the appropriate motoneuron pools in the cord (Asanuma & Sakata, 1967).

The fact that there are two cortical projections to the motor system of discrete movements explains why cutting the pyramids or ablating the precentral motor area does not abolish these movements. If the pyramidal and corticorubrospinal projections of a monkey are both removed, discrete and skillful movements of the face and extremities are permanently lost but postural and locomotor coordination and large-scale voluntary behavior remain (Lawrence & Kuypers, 1965).

A correlation in evolution and ontogeny between skillful use of the distal limb segments and the presence of crossed corticomotoneuron connections that are direct or nearly direct (Kuypers, 1964) does not mean that neural regulation of voluntary manipulation of objects and other kinds of skilled movement is independent of the less direct cortical projections. Sensory receptor zones, the intermediary zones, and the motor zones of brain stem and cord all receive descending projections. Each central efferent command is accompanied by a corollary regulation of sensory intake by special motor systems of orientation and attention: the oculomotor system, the head-orienting system, the γ-efferent system that controls the sensitivity of stretch receptors in the skeletal muscles, and others. Projections from the sensorimotor cortex are also distributed to the cuneate and gracile nuclei, thalamus, midbrain tectum, vestibular system, and cerebellum as well as to the spinal sensory relays. These various projections undoubtedly preset the spinal mechanisms of reflexive control for discrete movements and adjust the integration of these movements to those of postural coordination and receptor orientation (Bernstein, 1967).

Close observation of voluntary behavior, whether in the form of automatic locomotor patterns or specialized technical skills, shows that the control of such movements cannot be by reflex coordination alone (Lashley, 1951; Berstein, 1967). The smooth, accurate, and rapidly accommodating movements of spontaneous activity require *prepatterning* of whole motor sequences and concurrent adjustment of receptor processes at many levels simultaneously. Further evidence for schematic patterning of *motor preparatory sets* comes from ablation studies. Learned manipulative tasks are performed intelligently, though clumsily, after the motor cortex and adjacent areas have been ablated (Lashley, 1924).

The motor capacities that remain, though they are imperfectly differentiated, can be guided to make the correct combination of acts by cortical and subcortical mechanisms of orientation. The process may resemble that which makes it possible for a monkey with deafferented arms to reach and grasp a peanut swiftly and accurately after he has become oriented to it visually (Bossom & Ommaya, 1968). Under normal conditions the skillful patterning and proper serial ordering of the steps of a learned function require correct anticipatory perception of the location and space structure of the task and adjustment of discrete distal movements, well individuated and precisely executed, that actually perform the task with speed and economy in immediate relation to its mechanics. Ablation of sensorimotor cortex or bilateral pyramid section may destroy the latter part of the executive skill completely, but leave intact the capacity to solve the task with the right plan while employing only crudely differentiated and unresponsive distal movements (Lashley, 1924; Lawrence & Kuypers, 1965).

More extensive impairment of acts by the forelimbs, including disruption of reaching to objects located in the visual field, follows bilateral removal of areas of posterior parietal and frontal cortex. Unilateral lesions, if large, may cause inappropriate acts or neglect of goals by both limbs within the contralateral half of behavioral space. Only inconsistent, weak, and diffuse, but spatially organized, patterns of limb movement can be obtained by direct electrical stimulation of these cortical areas (Denny-Brown, 1966). It is said that such parietal and frontal cortex lesions deform manipulation in large part by interfering with the orienting of the hand in conjunction with the arm in reaching to or withdrawing from a visual or tactile stimulus. These disturbances in the spatial coordination of acts – instances of a phenomenon called "amorphosynthesis" by Denny-Brown et al. 1952, (see Denny-Brown & Chambers, 1958) – remove an essential foundation from fine praxis as well as the processes of perception. Lesions of the inferior temporal lobe or prefrontal cortex appear to impair performance even more indirectly by affecting the selective focalization of attention in relation to more complex schemata that bear no direct relationship to overt orientation in the body-centered orientational field.*

It appears established that cortical regulation of voluntary hand movements in primates includes very widespread control – in brain stem, cerebellum, and spinal cord – over the processing of sensory reafference and the prepatterning of posture and locomotion that underlie use of the hands. Within this frame is established the more direct and more refined or discriminatory control over the small movements of the distal segments that is essential to the performance of a delicate skill. The latter function needs cortical patterning of motor activity via the more direct efferent projection to the distal limb segments (precision motor system). Refined manipulation is severely depleted if this system is destroyed. The bilaterally organized core motor system, with multisynaptic transmission of cortical influences through widespread reticular circuits in brain stem and cord, appears to sustain the schemata by which sensory orientation and deployment of refined focal praxis are regulated in close association.

Such a design in the motor system is compatible with a distinction between specific patterning of focalized motor and perceptual activity and the formulation of orientational acts that bring a manipulative task into standard relationship with the specialized organs of praxis. It suggests that each cerebral hemisphere governs whole patterns of motility through orienting schemata and, furthermore, that elaborations of such schemata are lateralized when the control of a skill is distributed asymmetrically in the cerebrum.*

Voluntary movements after cerebral bisection

When the cerebral cortical commissures of a monkey are cut, each of the separated hemispheres recognizes sensory patterns and learns independently of the other to guide voluntary responses with respect to these patterns (Myers, 1956; Sperry, 1961). At the same time, an animal with the forebrain divided remains well coordinated in locomotor and orienting responses and gives few signs in his free behavior of the reduced connectivity of his brain. Coordination breaks down only when attention and motor patterning are concentrated about the hands or when general motility has been stopped in concentration on a local task.

The two hands of a split-brained subject may show discoordination, but they can also be well coordinated in the performance of simpler, large-scale learned responses (Ettlinger & Morton, 1963). An object located by one hand, and oriented to without vision, can be reached for accurately by the other hand even when the brain has been split through the forebrain and the midbrain roof (Mark & Sperry, 1968.) A monkey whose brain has been sectioned through the chiasm and corpus callosum can reach to an object located visually by one eye with the ipsilateral hand, even when the primary visual and manual centers are confined to opposite cortexes by reciprocal cortical ablations of their mirror complements (Myers et al., 1962; Bossom & Hamilton, 1963; Black & Myers, 1965; Hamilton & Hamilton, 1966; Gazzaniga, 1966). These findings show that visual and manual locating, with the associated oculomotor and manual orienting functions, remains bilaterally integrated and interconnected after cerebral commissurotomy, although perceptual recognition of visual or haptic identities takes place in two separate cerebral mechanisms.*

The two hemispheres of a monkey can be shown to differ in their potential for direction of response orientation. Monkeys with the optic chiasm cut and the forebrain bisected, and with vision restricted to one eye, show a spontaneous preference for the contralateral eye-hand pair while they are choosing between learned visual symbols to direct simple pushing or grasping responses of one arm and hand at a time (Downer, 1959; Trevarthen, 1965). This may occur when everything in the visual field but the symbols important for choice of the manual response is visible to both eyes (Trevarthen, 1965). These monkeys may subsequently show variable but sometimes pronounced dyspraxia when they are forced to use ipsilateral eye-hand pairs on the same tasks (Downer, 1959; Trevarthen, 1962, 1965; Gazzaniga, 1964, 1966).

Comparable effects occur in oculomotor exploration, but these have been little studied. The monkey whose chiasm and corpus callosum have been sectioned has two separate mirror oculomotor mechanisms for changes of visual fixation. Each eye of such an animal is, at least for a time, defective for reorientational glances (the flick-back in optokinetic nystagmus) in the direction of the blind homolateral field; but each can follow in all directions once locked onto a moving target (Pasik & Pasik, 1964). Each of the separated hemispheres can, in these circumstances, make visual recentering only to the opposite half of space outside the body; that is, into the opposite half of the visual field. This produces unilateral neglect in the visual field, but a split-brained animal learns to look in both directions when only one hemisphere is receiving visual input (Trevarthen, 1965).

Studies on split-brained subjects show that the monkey has two equipotential cerebral systems for perceiving and learning the identity of stimulus configurations and for directing whole-body coordinations of activity. At the same time each hemisphere is most attentive to the opposite half of space and better able to guide praxic responses and perceptual acquisition of information into this same space by exercising a directing or orienting influence over one-half of the mirror-symmetrical distal motor mechanism of the body.

Mirror-orienting functions and duplication of voluntary control are reflected in the post-surgery behavior of split-brained subjects outside experimental situations (Kennard & Watts, 1934). The baboons used in the study reported here were also observed free in a large cage immediately after surgery. Defects in spontaneous motor coordination of the hands were recorded on movie film.

Dissociated reaching of the two hands in duplicate and synchronous aim to the same or different objects was observed to depend on the symmetry of posture and on reduced axial motility. Double volition was quickly resolved once reafference permitted recoordination of the two cerebral motor systems that had been disconnected by commissurotomy. Hence, episodes of this kind were brief. Periodic neglect of one hand, with arrest of its activity while attention was focused on the other hand, gave clear evidence of temporary lateralization of cerebral control to one side at the expense of the other. Sometimes this one-sidedness of voluntary control was followed by conflict of action between the two hands; for example, on several occasions for both subjects, a brief tug-of-war was observed between the two hands for possession of a single piece of food. Marguerite and Jojo were very similar in the post-surgery behaviors. (See Trevarthen, 1965, discussion pp. 103–106, for observations made with Marguerite shortly after surgery.)

These effects of cerebral commissurotomy leave no doubt that normally each hemisphere of a monkey is concerned with the initiation of use of the contralateral hand more than the ipsilateral member and that bimanual coordination is partly effected by the corpus callosum under normal circumstances. They also show that hand movements and visual orientation may be influenced by adjustments in brain stem mechanisms capable of regulating functions in the two hemispheres after the latter have been deconnected surgically. Voluntary activity is unified for whole-body activity and egocentric orientation, but local and refined acts are dissociated into two mirror systems that can only be associated through reorientations.*

Psychological functions after cerebral commissurotomy in man

Studies of patients in whom commissurotomy has been performed to control epileptic seizures show that the findings described above for sub-human primates apply to man as well (Sperry et al., 1969). After surgical section of all cortical commissures, overall motor coordination of a human subject may be little affected, but each half of the body away from the face and trunk, and each half of the visual field, appears to have access to its own cognitive system, and the active perceptual processes of the two systems are not in communication (Geschwind & Kaplan, 1962; Gazzaniga et al., 1963, 1965; Sperry, 1968).*

These same studies also demonstrate the unique importance of asymmetry or cerebral lateralization of the psychological functions peculiar to man. The higher functions are consistently developed in markedly different ways in the two hemispheres. Thus, commissurotomized subjects are usually able to speak or write down the name of an object seen only if it is presented in the right half of the visual field. In other tests, the same object is seen only in the left visual field, and then the patient can pair it with a similar object felt with the left hand. Therefore, the object may be perceived visually in the left field, but the percept cannot be brought out in writing or speech (Gazzaniga et al., 1962, 1965; Geschwind & Kaplan, 1962; Geschwind, 1965; Gazzaniga & Sperry, 1967. In contrast, the right visual half-field and the right hand, when compared with their left counterparts, have deficits in capacity to integrate behavior to conform with non-verbal, configurational, or constructional patterns in either visual or tactile perceptions (Gazzaniga et al., 1962, 1963; Bogen & Gazzaniga, 1965; Levy-Agresti & Sperry, 1968; Bogen, 1969).

It is concluded from these and similar tests that, in general, the dominant left hemisphere is specialized to formulate language functions and calculation with numbers and simple mathematical symbols, and the right is superior in integrating the perception and use of spatial gestalts, seen or felt (Sperry et al., 1969). Hemispheric lateralization of function is seen as an evolutionary outcome of competition between rival developmental adjustments in higher intelligence (Levy-Agresti & Sperry, 1968; Levy, 1969). As indicated in the introduction to this chapter, the lateralization of functions demonstrated in commissurotomized subjects is in agreement with a large body of knowledge of the effects of lesions in one or the other of the two hemispheres of man.

Although spontaneous coordination of the whole body remains generally normal after cerebral commissurotomy, discoordinated involuntary movements of the limbs, comparable to those described for split-brained monkeys, are observed in different degrees with different patients. Episodes of dissociated voluntary activity of the two hands or transitory silence or inertia of one limb while the other is active are not infrequently seen; but bimanual coordination is retained to a high degree, especially in well-learned automatic performance (Smith & Akelaitis, 1942; Akelaitis, 1944; Gazzaniga et al., 1962, 1967; Trevarthen & Sperry, 1973). Most commissurotomized subjects show a tendency to neglect the left leg and arm from time to time, and apraxia of the left side may be observed, particularly after a period in which the subject has been generally inactive while concentrating on speech or manipulation with the right hand. There may be a variable degree of idiokinetic apraxia for the left hand, which Liepman held would follow from disconnecting the right hemishpere from the "language and thought" mechanisms of the left hemisphere (Liepman, 1905; Geschwind, 1965; Gazzaniga et al., 1967). Left-sided apraxia is particularly evident in the period immediately after surgery (Bogen, 1969). Temporary neglect of the right hand or even right-sided apraxia may also be observed.

The observations indicate that, although separate and differently organized mirror hemispheric mechanisms exist for the initiation of voluntary motor control of the limbs, and although these may be dissociated after commissurotomy, alternative mechanisms are available for bilateral coordination that may be increasingly effective in overcoming post-surgical defects in patterning of spontaneous behavior. The alternative mechanisms have been described as ipsilateral control projections originating in each hemisphere, and it is held that such systems are weaker, more variable and plastic, and more susceptible to modification as a result of extra-callosal cerebral damage (Gazzaniga et al., 1967; Sperry et al., 1969). It is probable, however, that the cerebral mechanisms in each hemisphere that have control over the ipsilateral limb are part of a bilateralized system of different function that may be in full command when appropriately employed and that, moreover, is in reciprocal communication with neural mechanisms of similar general organization in the brain stem. For this kind of motor function complex interactions take place between the hemispheres, evidently by way of subhemispheric regions of the brain (Trevarthen, 1970c).*

Observations of manual capacities in split-brained human subjects have brought out the distinction between cross-lateralized cerebral control of well-differentiated movements governing palm and fingers and bilateral control in each hemisphere of movements involving proximal segments of the limb (Gazzaniga et al., 1967). Similarly, input and perceptual analysis of touch, temperature, and articulation-receptor functions of the hands are more completely separated than are equivalent sensory functions of the proximal limb segments, trunk, and face (Gazzaniga et al., 1963; Sperry et al., 1969).

Variable performance and differences between individual subjects indicate that after commissurotomy the mechanisms unifying orientation and attention are highly labile. Both the degree of control of each hemisphere over the ipsilateral limb and the degree of independence of function of the two limbs vary widely. This intrinsic variation appears to reflect competitive orientational adjustments within a central neural mechanism that includes subhemispheric components (Trevarthen & Sperry, 1973). Competitive perceptual interaction is found between stimuli applied synchronously to the two hands of commissurotomized subjects (Gazzaniga et al., 1963); these are comparable to interactions obtained with binaural presentation of conflicting auditory stimuli (Milner et al., 1968).

Tests on commissurotomized subjects show a hierarchical organization of visuomotor functions and functional interaction between mechanisms spontaneously distributing attention to left and right in the visual field. If attention is appropriately balanced and distributed toward the periphery (i.e., not sharply focalized at a center of attention), primitive motion-dependent perception of visual space may become unified (Trevarthen &

Sperry, 1973). Thus, perceptions in left and right halves of the visual field are not wholly independent or separate, but cross-unified in some measure, presumably through mediation of visual projections to the midbrain (Trevarthen, 1970b; Trevarthen & Sperry, 1973). The ambient visual field that provides a context for localization of attention and for orientation of gaze appears to be bilaterally represented in each hemisphere as well as in the brain stem.

The hierarchical organization of visual functions brought out in the above-described test leads to an interpretation of the effects of cerebral commissurotomy in terms of an attention process that alternates activity in the brain between systems governing orientation in an extended ambient space and those by which goals for perception and movement are focalized, narrowed down, and accurately formulated.

Spontaneous, "ballistic" patterning of whole-body activity and adjustment to proprioceptive sensory reafference caused by this activity remain well coordinated in commissurotomized subjects because the neural control system governing axial and proximal motor elements is cross-integrated at subhemispheric levels. Anatomical studies and studies of brain-damaged subjects have shown that each hemisphere has representation for these functions and that the cerebral component of the core motor system is in close reciprocal communication with the motor integrating system in brain stem and spinal cord.

On the other hand, the precision motor functions of speech and of fine haptic exploration and manipulation are controlled by neural systems that are essentially cortical. These functions require focalized attention of receptors for touch, vision, and audition and depend on sensory discrimination of high acuity. Each hemisphere projects more completely to the contralateral hand in control of its refined movements, and each is capable of attending fully to and perceiving clearly the objects explored with the same hand. Objects grasped in the other hand are attended to with less ease and are poorly perceived by this hemisphere. There is a comparable lateralization of refined oculomotor function. Each hemisphere perceives clearly only that detailed visual material which requires oculomotor orientation toward the contralateral half of behavioral space for its full visualization with foveal fixation.

In man, the midline and unitary functions of speech articulation, the movements of which are rapid and extremely precise, are also lateralized – but only on one side of the cerebrum. This shows that mirror spatial complementarity does not determine which motor functions are lateralized. The essential features appear to be precision and speed of action and complexity of serial patterning of performance with fast afferent guidance. Where the organs of refined praxis are paired, cerebral lateralization follows the mirror segregation of orienting functions. The fact that the more highly developed manipulatory skills of right-handed man are lateralized in the same hemisphere as the dominant control for speech and language means that the most common genetic determination of cerebral lateralization in man favors location of patterning for the most rapid and complex chains of movement in the left cerebral hemisphere. It is not excluded that underlying all the left hemisphere functions, including language, whether or not it is developed through reading, may also be a basic precision orienting function that distributes focal attention most effectively from left to right of complex objects of comprehension.*

Differentiation of a strategy for manipulative skill

The study of manipulation in baboons gave information about the sequence of sensorimotor functions involved in learning to open the problem box quickly. The spatial relationship between the box and the subject was controlled by placing the box in front on the midline and by balanced reversal of its left-right orientation. In spite of this, it was found that increase in skill and speed correlated with an increase in the asymmetry of performance. Thus the "skill" acquired by the baboons is an example of an asymmetrical motor function adapted to a small, complex object. It was a specific pattern of rapid hand movements brought to precisely measured contact with parts of the task in proper serial order under visual and tactile control.

In the early steps of learning, tactile and proprioceptive exploration by the hands was dependent on visual orientation. Vision served to locate the task, and what the eyes saw of the shape of the task the hands put into practice. The subsequent attainment of more precise movements was the result of more careful visual orientation and of direct experience of the feel of the box and its movable parts. Presumably, once the dimensions and mechanics of the task were learned through direct contact, dependence upon critical visual attention to its parts became less. Any visual guidance required would need to be fast and direct to control the finger movements.

Two distinct trends were seen during learning. The early movements were less responsive and less well oriented. Subsequently, there was a progressive refinement of aiming of the hands to significant mechanical elements of the task and a parallel change in the type of movement. Large and forceful acts were eliminated, and generalized pulling, pushing, and grabbing movements, which required adjustment after contact, were replaced by more localized, more precise and delicate movements of wrist and fingers performed in a smooth rhythm. Speed and precision were attained by close matching of continuous hand displacements and wrist and finger flexions to the mechanics of the box in each orientation. These practiced movements showed prepatterning on the basis of visual perception, being partly formed before contact, while the hand was still being projected to the box. The narrow dispersion of performance times with practiced subjects shows that the routines of movement, schematically determined by the brain in advance of the act, were highly regular in cadence and rate, as well as form (c.f. Bernstein, 1967).

Both hands could do the initial unpracticed movements about equally well. Later, in spite of the symmetry of the task over a run of trials, all five baboons showed a polarization of action favoring one hand as the skill became elaborated. Every effective pattern of finger movement of one hand could have been matched by a mirror pattern for the opposite hand in another trial, but the responses actually made were asymmetrical. Careful attention to the structure of the task was correlated with increased attention to the actions of one hand.

The highest speed of performance was obtained by a smooth figure of movements coupling the two hands as intimately as the components of one limb were integrated for its actions. Thus, a high degree of intermanual coordination was achieved within a refined asymmetrical strategy in which the most precise movements were made by the most active preferred hand. Tests for the capacity to transfer the learned strategy between the hands to mirror motor mechanisms showed that, although orientations to the significant features of the task did transfer, specific refined movement patterns were not immediately available to the unpracticed subordinate hand. In other words, the subordinate hand was coupled in a bimanual program that nevertheless left the two hands to make independent adjustments with respect to specific steps in the task.

Bisection of the cerebrum resulted in dissociation of manual responses without disturbing locomotion and without visible deterioration in the coordination of fine grasping or manipulation by either hand separately. The defects of coordination appeared to be of two kinds. Periodic neglect of one-half of the task while responses were being made spontaneously by only one hand gave evidence of a one-sided perceptual or attentional defect or spatial agnosia, probably mainly visual and probably indicating that one cerebral hemisphere was partly unconscious as the other controlled the responding hand. Confused or confabulatory responses were indicative of a more complex deficiency of control, possibly the result of simultaneous unrelated motor programming activity of the hemispheres, both of which were evidently capable of some influence on the movements of each hand.

These findings indicate that each hand was primarily controlled by a different hemisphere and that the bimanual coordination of the subject before surgery depended on commissural connections. However, bimanual coordination was rapidly regained, and the results reported here of tests of all eye-hand combinations of one subject after surgery show that the schema governing the serial and spatial ordering of fine movement of both hands in response to sight of the box in one of the two orientations was laid down at the end of learning the skill in one side of the brain more fully than in the other. Apparently, this dominant hemisphere carried the general temporospatial program for guidance of *both* hands, even though it could not at first cause the subordinate hand to act properly in its turn. In consequence, cutting the corpus callosum not only interfered with bimanual coordination and permitted dissociated organization of approaches of the hands to the task, it also deprived the hemisphere opposite the subordinate hand of an orienting and timing command from the dominant hemisphere. Before surgery, this command, transmitted through the corpus callosum or facilitated by way of the callosum, had activated and organized the responses of both the subordinate hand and the dominant hand in relation to the task.

The above-described data on the progress of skill acquisition and the subsequent effects of surgery favor the view that the skill governing manipulation of the problem box had two components. A global schema comprising a set of rules for appropriate orientations of the two hands together to elements in a functional order was probably founded mainly on visual orientations and on a visual image of the form of the task. Precise execution of the appropriate movements of each hand, with accurate timing of successive steps, depended upon a more localized, hand-specific sequence of rules favoring rapid adjustment to the successive reafferences – visual, auditory, and tactile – consequent on contact and manipulation of the task by the fingers. The control of this fine manipulation was more cross-lateralized from the start, each cerebral hemisphere controlling the contralateral limb. The former function, general visual aiming of attention, was bilateral at first, but became more concentrated in one hemisphere as the asymmetry of the strategy of manipulation became consolidated in the service of more refined and effective movements.

If this is the correct interpretation, then visual recognition and comprehension of the task as well as accurate motor orientation to its parts were better performed or localized in one hemisphere of the brain after practice. In addition, voluntary use of each hand required visual activation of somesthetic mechanisms located in the hemisphere contralateral to each hand. This would explain why, when visual adjustments were experimentally confined to the dominant hemisphere after cerebral commissurotomy and optic chiasm section, the ipsilateral hand was initially inactive or extremely clumsy, as if ill prepared to act or to make corrections.

Cerebral dominance in relation to evolution of skill

Skillful, intelligent behavior and cerebral lateralization of function show analogous evolutionary and ontogenetic development. Evidently the cerebral functions of higher intelligence work on a principle that favors asymmetrical localization of their neural mechanism in the brain. Moreover, experience and learning play a significant role in the maturation of cerebral lateralization, whereever it occurs.

Limb preferences are found in many mammals, but one-sided skillful responses are not conspicuous below primates. Individual cats or monkeys show consistent preferences in spontaneous use of the forelimbs to perform simple manipulations. These preferences may be reversed by ablations of contralateral cortex. The early manifestation and stability of the preferences and their resistance to reversal by training indicate that they may have an innate foundation (Warren et al., 1967). There is, however, no clear-cut evidence of a consistent genetically determined side of lateralization in any subhuman species (Brookshire & Warren, 1962; Ettlinger & Moffett, 1964). With monkeys, the lateralization is stronger for the execution of more difficult tasks (Warren, 1958).*

The two hemispheres of a cat or a monkey are remarkably similar in their potential for visual or somesthetic learning when they are isolated from each other by surgery (Sperry et al., 1956; Ebner & Myers, 1962). Only one experimental study of cerebral lateralization of visual processes has been reported (Gazzaniga, 1963). Monkeys trained with horizontally paired visual stimuli before midline surgery and tested after division of the optic chiasm and forebrain retained the learned discriminations better through one eye than through the other eye, even when the two forelimbs had been given equal practice during learning. This indicates an intrinsic asymmetry for visual retention in the two hemispheres of the normal brain, or a primitive cerebral dominance, unless the unequal performances were a consequence of a chance *postsurgical* readjustment of oculomotor posture, to compensate for one of the two half-field defects produced by dividing the chiasm (see Trevarthen, 1965).

There is evidence that transcallosal communication of learning is neither immediate nor complete in cats or monkeys. More difficult discriminations (as determined by the speed with which they are learned in the first place) transfer interocularly in chiasm-sectioned subjects less well than easy ones (Myers, 1955; Ebner & Myers, 1962), and intrahemispheric transfer of learning between the eyes is more efficient than interhemispheric (transcallosal) transfer. Such imperfect equilibration between the two cortexes could conceivably lead to automatic concentration of learning for more complex skills in one or the other side of the cerebrum. Any slight initial bias of handedness during acquisition would favor one side over the other, and as long as transfer of the neuronal changes permitting retention of learning were incomplete, the effect would be cumulative. Later practice would start with the lateralized engram as a foundation (see Mishkin, 1962). It is probable, however, that incomplete transcallosal communication may reflect cerebral adjustments lateralizing brain activity, rather than cause them (see Pasik & Pasik, 1964; Butler, 1968).

Human neonates show behavioral signs of cerebral asymmetry from birth (Gesell & Ames, 1947; Turkewitz et al., 1965), but consistent lateralization of function becomes more apparent much later in life. There is clearly a complex interaction of intrinsic and environmental determination for cerebral dominance, with the foundation undoubtedly genetic. A child obtains language progressively, and this development is found to be a product of both a complex, intrinsically determined maturational process that is completed years after birth and concurrent adaptation by learning (Lenneberg, 1967). There is a rapid development beginning about the age of 2 years. This is after the infant has exhibited a lateralization for grasping and manipulating with one hand; in the great majority of individuals the right hand is used more often and more skillfully than the left by the end of the first year (Lippman, 1927; Gesell & Ames, 1947; Flament, 1945). By that time, the infant is well able to adapt grasping movements of the fingers precisely to the shape of objects inspected visually and is able to handle objects in a number of exploratory or testing maneuvers (Halverson, 1931; Twitchell, 1965).* The effects of early cerebral trauma lead to the conclusion that clear lateralization for language begins at about 3 years and is completed within a few years after 10 years of age (Basser, 1962). Lateralization of processes of perception for speech sounds follows a comparable course (Kimura, 1967).

That a general function may regulate both manipulative actions and language is indicated by the work of genetic psychology on the development of thought and language and their correspondences with the growth of motor intelligence (Piaget, 1954; Vygotsky, 1962; DeLaguna, 1963; Bruner et al., 1966). Lateralization of manipulative practice and lateralization of language are consequences of an inborn asymmetry in cerebral organization of the mechanisms that effect orientation to complex tasks, but it remains uncertain if there is a close functional bond between them (Zangwill, 1960a; Hécaen & Ajuriaguerra, 1963).

Evidence has been found for extensive plasticity in basic cerebral organization during development.* In addition to the lateralization of higher functions differently in the two hemispheres, even the dissociation of mirror-symmetrical cerebral orienting functions in the occipital visual areas may require exposure to environmental stimuli for its normal maturation. Evidently the visual projection areas attain the classic mirror topography, with left visual field projected to the right hemisphere, and vice versa, at a postnatal stage in the maturation of the brain, each hemisphere becoming increasingly patterned after birth to control orientation to the contralateral visual field. Initially, each hemisphere appears to have a much greater potential for sensorimotor integration throughout the field. It has been found that a human being lacking a corpus callosum through genetic anomaly (agenesis) subsequently shows none of the typical effects of surgical commissurotomy on visual functions (Sperry, 1968). It appears that the patterning of precision sensory analyzing mechanisms may be determined by postnatal exercise and that the corpus callosum normally plays a role in the process of adaptation involved. Experiments with neonatal kittens have also shown that the neural organization of visual mechanisms of the cortex is highly plastic (Hubel & Wiesel, 1965).*

In the light of the evidence for plasticity of cortical organization for visual functions, hand use, and speech, it seems likely that when a manipulative skill is acquired by practice, changes occur in a brain mechanism related to all these functions – one that provides for the development of precise fixation on visual detail; or, over a longer time, the acquisition of speech and language. We have found that, in the baboon, the neural determination of the correct combination of gestures and responses for a manipulative task takes place at first in a bilaterally duplicated cerebral system that serves to orient and bring together individual sensorimotor focalizing functions. These, in turn, direct the precise adjustments of eyes and hands together in the center of the behavioral field while the subject pays close attention to the task. The changes observed can be compared to changes in the sensorimotor coordination that appear in the course of the growth of human intelligence in infancy.

CONCLUSIONS

Comparative studies lead to the conclusion that the brain is bisymmetrical in relation to its function for orienting acts in the space for behavior defined by the symmetrically mobile body. All the more intelligent kinds of actions involve standardization, and generally stabilization, of the reorienting of the body (Fig. 11-12). Attention is shown by an arrest of large body movement and by refined orienting activity in one place. The process of immediate reorientation of the body, on which primitive sensorimotor intelligence is founded, becomes regulated and subordinated to the processes of intellectual schematization, innate and acquired, by which local features and identities are defined in abstraction from egocentric space and within an established context of surroundings. The cognitive schemata are not organized, as primary spatial orientation is, by counterbalancing leftward- against rightward-turning tendencies. Therefore, the neural mechanism is structured in increasing independence of neuroanatomical bisymmetry.

Despite the consistent asymmetry in the anatomical substrate for complex, highly practiced cerebral functions, the fact that the brain remains morphologically bisymmetrical to the degree it does, even in the cerebrum of man, leads to the further conclusion that the functions which are asymmetrical, or nonsymmetrical, when mature are built up by selection in the symmetrical mechanisms first laid down by brain morphogenesis. The brain may grow progressively more asymmetrical in functional organization by accommodation to specific complex regularities or invariants in the environment. However, the basis for cerebral asymmetry is laid down before birth in the fetus. The selection process begins there.

Apparently the growth of lateralized cerebral mechanisms takes place within that sector of the cortex of a hemisphere which develops from the start with projective affinity to both halves of the body and with sensory projections pertaining to both halves of egocentric space. Anatomically, the bilateral organization of cortical areas in one hemisphere is associated with commissural connections over the corpus callosum to the other hemisphere. Callosal communication is strongest between bilaterally organized association cortexes of the two hemispheres, and the connections effect transfer of complementary spatial components (Sperry, 1962; Trevarthen, 1970a). It is possible that cerebral asymmetry of function is always acquired by a process of competitive selection of neural elements from a bisymmetrical ensemble, with the aid of an inherent bias channeling the process of exposure to selective environmental influences. Levy (1969) has suggested that differentiation of lateralized hemispheric functions is competitive in man, processes associated with the maturation of speech and language in one hemisphere causing displacement of less elaborate psychological processes, such as synthetic appreciation of gestalts, into the other hemisphere.*

The steps by which a baboon attains a specific manipulative skill, with lateralization of hand movement to a dominant hand and with concentration of schematic control in the contralateral hemisphere, appear to be representative of a general process of differentiation by which cerebral asymmetry of function is acquired in the service of refined behavioral adjustment.

POSTSCRIPT

This chapter was written in 1968. It was revised and a review of the cerebral motor system added in 1970. Now, there is much new information that, in general, supports the position taken. Most important are the confirmation, by Brinkman and Kuypers, of the distinction between proximal and distal motor control for arms and hands in the rhesus monkey and the demonstration by Brinkman that the split-brain operation does indeed divide only the projective, visually elicited motor control over fine finger movements. She has also shown how important touch reafference may be in overcoming and concealing this defect (Brinkman & Kuypers, 1972, 1973; Brinkman, 1974).

The training box and methods of surgery used in the experiments with baboons have been described in detail since this chapter was written (Trevarthen, 1972).

Recent experiments with human commissurotomized patients have proved that a unified cerebral mechanism, capable of orienting the whole body and of directing the left or right arm and hand in either half of space, survives hemisphere deconnection (Trevarthen & Sperry, 1973; Trevarthen, 1974a,b). Moreover, vigilance experiments and many incidental observations on attention processes in these subjects have shown that a single cerebral system still regulates the functions of perception and of preparation for movement that achieve elaborate form separately in the two hemispheres. It follows that lateralization of a psychological function in one-half of the intact brain may involve asymmetrical or lateralized directives from the brain stem, as well as interhemispheric influences carried by the corpus callosum.

The theory of a map of the behavioral field in the midbrain is confirmed and elaborated by studies of the visuomotor mechanisms of lower vertebrates and mammals reviewed by Ingle and Sprague (1975).

Studies with human commissurotomized patients indicate that perceptual and cognitive specializations of the hemispheres revolve about contrasting strategies for information pickup as well as different programs for control of skilled movement. Hemispheric functions are regulated in association with alternative modes of operation within the brain stem that are capable of exercising a metacontrol over the higher processes of consciousness (Sperry, 1970, 1974; Levy, 1974; Trevarthen, 1974a,b, 1975; Levy & Trevarthen, 1976).

Current explanations of manual preference in man emphasize a distinction between ballistic and guided movements of the hands (Flowers, 1975). This is also used in studies of visuomanual coordination of split-brained baboons by Paillard and Beaubaton (1975) and to explain developments of reaching and bimanual coordination in human infants (McDonnell, 1975; Bresson et al., 1976). Preilowsky (1972) has examined deficiencies of bimanual coordination and visuomanual guidance in epileptics with anterior partial transection of the corpus callosum. He suggests that the frontal lobes and callosal links between them are important in guiding bimanual movements.

There is evidence that the dominant hand of man is inherently more precisely guided by vision (as in writing) or by audition (as in bowing a violin) (see Flowers, 1975; Levy, 1976). The left hemisphere is specialized in most subjects for phonetic guidance of speech (Zaidel, 1976; Levy & Trevarthen, 1977). These findings suggest that the hemispheres become different by segregating the modalities of reafference that guide various precise distal motor operations. Alternatively, they may separate afferent regulation of rapid sequences of movement, as in speech, reading, writing, or gesturing, from coordinated synchronous arrays of action and detection, as in feeling braille or recognizing faces (Kimura & Vanderwolf, 1970; Hermelin & O'Connor, 1971; Levy et al., 1972; Kimura, 1973a,b). Both the above interpretations of hemispheric lateralization of function infer specialization of the associative or integrative bilateral mechanisms of the cortex in each hemisphere, cortex that is rich in commissural interconnections across the midline of the brain.

New evidence that anatomical and physiological differences between the human cerebral hemispheres develop before birth has been reviewed by Corballis and Beale (1976), Hécaen (1976), and Levy (1976). These differences correlate with lateralized psychological functions that attain maturity long after birth. It is important to note that there are reciprocal enlargements, of unknown function, in the hemisphere that lacks enlargements related to language. Corballis and Beale (1976) and Levy (1976) discuss genetic theories to explain inheritance of handedness and anatomical and psychological correlates of both left-handedness and right hemisphere lateralization for speech and language.

Research with monkeys still fails to establish a genetic regulation for manual preference in subhuman primates (Hamilton et al., 1974).

Brain structures of visual perception are now known to undergo elaborate self-differentiation prior to birth and shortly after birth, but correct segregation of competing elements depends on exposure to environmental stimulation at critical periods (Barlow, 1975; Trevarthen, 1977). Segregation of ocular dominance territories in the lateral geniculate and striate cortex of monkeys, pre- and postnatally, is essential to the development of binocular stereopsis (Hubel et al., 1977). This may offer a model for the development of complementary manual functions that segregate between the hemispheres with practice of skills. Observations on the development of lateralized speech and language functions and effects of brain lesions at different ages would appear to fit the same model (Hécaen, 1976; Zaidel, 1976).

ACKNOWLEDGMENTS

The experiments with baboons were carried out at the Institut de Neurophysiologie et Psychophysiologie of the C.N.R.S. at Marseille, France, with the aid of a U.S.P.H.S. Post-Doctoral Fellowship (MH-19, 674-03), and under the direction of Professor Jacques Paillard, whom I wish to thank for excellent facilities and much encouragement and help.

The experiments were performed in collaboration with Mrs. Dorothy Paul and Mme. Sylvie Requin.

I thank Dr. David Ingle of Boston City Hospital for permission to use his films of frogs feeding.

Preparation of the manuscript was supported in part by Grant No. 1 P01 MH-12623 from the National Institute of Mental Health, and in part by Grant No. 1 R01 HD-03049 from the National Institute of Child Health and

Human Development, both to Harvard University, Center for Cognitive Studies.

NOTES

1 This chapter was written in 1968 and revised in 1970. Today, much new information supports the position taken; the most important is discussed briefly in the Postscript at the end of the chapter. Throughout the text, an asterisk indicates areas where material added in the Postscript is relevant.

2 *How we know that this gland is the main seat of the soul.* The reason which persuades me that the soul cannot have any other seat in all the body than this gland wherein to exercise its functions immediately, is that the other parts of our brain are all of them double; and inasmuch as we have but one solitary and simple thought of one particular thing at one and the same moment, it must necessarily be the case that there must somewhere be a place where the two images which come to us by the two eyes, where the two other impressions which proceed from a single object by means of the double organs of the other senses, can unite before arriving at the soul, in order that they may not represent to it two objects instead of one. [René Descartes, *The Passions of the Soul*, Part I, Article XXXII, published in French in 1649; English translation by Elizabeth S. Haldane and G. R. T. Ross, in *The Philosophical Works of Descartes*, London, Cambridge University Press, 1967.]

3 Theories of intelligence ascribe the freedom from body-centered space of motor acts of orientation ("operational space," von Uexküll, 1937) to compensating processes. Schemata extending behavioral control outward into the world require this compensation. The theories of active perception, such as those of von Holst (1954) and Gibson (1966), hold that the attainment of an accurate percept depends on the formulation by the brain of a compensating adjustment of receptor functions. Each act is accompanied by a neural prediction of transformations in the sensory field appropriate to it. A stable object is perceived when the transformations of the patterns of stimulation it creates at the receptors are accounted for by the brain, either by anticipation of the effect of movements made by the perceiver or by anticipation of effects inherent in the object itself. Von Holst replaced the "unconscious inference" of Helmholtz by a mechanism in which a central command generates an "efference copy" with each formulation of an act and so compensates for the effects of active shifts of receptors. Gibson emphasized that the perceptual constructions exist independently of the sense impressions, which are incidental accompaniments of perceiving. Perceptions are based upon

invariant information selected from the environment (in which it is inherent) by an active process.

According to Piaget's theory of perception, space with objects in it becomes centered on one object of attention, and this leads to distortions that are corrected by systematic coordination of different "centerings," in other words, by decentralization (Piaget, 1937, 1950). Decentralization permits the establishment of what Piaget called perceptual schemata or "groupings" through "motor operations." Thus, again, perception is held to be originally dependent upon the same process by which acts of decentering or refocusing are formulated. The phenomenon of grouping is extended in Piaget's schema beyond overt behavior to include logical operations by which conceptual *groupement* is attained.

The logical operations of Piaget recall the "creative principle" upon which Descartes based his explanation of the unique achievements of language in man (Chomsky, 1966). In essence, they are freed from the "mechanical," "sensorimotor" constraints of successive overt behavioral orientations.

Of historical importance to the problem of schematic brain functions and activity produced by the brain are the theories of Head (1926) and Bartlett (1932). According to Head, the brain holds schemata that describe a continuous record of bodily transformations (changes in body image) produced by spontaneous actions. The brain attains vigilance by attuning the highest levels of discriminating functions in the cortex in relation to the body schema. Cerebral injuries cause disruption of perception or language or skillful activity by interfering with this underlying schematization. Bartlett extended Head's concepts to explain the activity of remembering and its relation to visible forms of activity during the acquisition of experience. Further extensions of this approach are to be found in the work of Craik (1943).

4 Sterile surgery was performed under halothane (Fluothane) or pentobarbital (Nembutal) anesthesia. After removal of a large dorsal cap of bone, the dura was incised over one hemisphere and turned back to allow slight separation of the hemispheres and exposure, segment by segment, of the corpus callosum and other midline connections. Cutting was performed with fine suction cannulas under a binocular Zeiss operating microscope. Commissures and chiasm were sectioned in one operation. In every case, access for surgery between the hemispheres was obtained by opening the dura over the hemisphere contralateral to the preferred hand in an attempt to minimize injury to the cortex opposite the least-preferred hand. For further information on the surgical technique, see Trevarthen (1972).

5 Subjects Marguerite and Jojo recovered rapidly from the effects of anesthesia, and there were no behavioral signs of injury to cortex or thalamus. They were tested 3 days after surgery. All observed abnormalities follow-

ing surgery took the form of changes of manual coordination or visual orientation that are described fully in the text.

Marguerite was sacrificed at 6 years and Jojo at 2.5 years after surgery. Their brains were immediately perfused with physiological saline followed by 10% Formalin through a cardiac cannula in the left ventricle. For further fixation, the whole brains were cut in 1-cm blocks along frontal stereotaxic planes and stored in Formalin. They were subsequently sectioned and stained in the histology laboratory of the Institut du Neurophysiologie et Psychophysiologie at Marseille. Alternate sections at 250μ separation were stained for cells (cresyl violet) and myelin throughout the extent of the corpus callosum.

Marguerite's brain showed complete midline division of the corpus callosum, anterior commissure, and optic chiasm. The incision through the septum and chiasm deviated 0.5 to 1.0 mm left of the midline. The left fornix was demyelinated and may have been nonfunctional. The left cingulate gyrus was reduced 50% over its anterior three-quarters. Other cortex was normal in both hemispheres. Both lateral geniculate bodies showed absence of cells in layers corresponding to the respective contralateral eyes. The other layers were normal.

Jojo had closely comparable histology indicating complete midline commissurotomy and chiasm section without damage to cortex or thalamus, with the following exceptions. A threadlike bridge containing dorsal and ventral strands of myelinated axons remained in the extreme tip of the genu. The line of surgery entered the left lateral ventricle at the level of the septum. The left fornix was reduced by 50%. A lesion 2 mm long penetrated the ventral half of the right cingulate gyrus above the genu. The chiasm and anterior commissure were cleanly sectioned in the midline. The lateral geniculate nuclei showed absence of cells in layers corresponding to the crossing fibers of the chiasm.

6 The fundamental relationship between brain bisymmetry and the symmetry of the motor apparatus has been clearly expressed by Ramón y Cajal (1955). Ernst Mach, who established experimentally that perception may be influenced by preparation for action, has also given a clear description of the connection between bisymmetry and the function of orientation (Mach, 1959). This is the central idea in Loeb's theory of tropisms (Loeb, 1918). Mach also discussed the secondary independence that identity-determining perception processes and intelligence attain with respect to this symmetry. Young (1962) has restated this distinction between simpler behavior strictly related to a bisymmetrical brain map and more complex, more intelligent behavior freed from this map. He has employed concepts of computer engineering to distinguish the properties of analogical mapping systems in the brain from those employing a process comparable to digital computation. He has not openly discussed

the role of specific motor structures in linking the organism to the world outside the body and in determining bisymmetry of behavior and, hence, functional bisymmetry in the central nervous system.

REFERENCES

Ajuriaguerra, J. de, & Hécaen, H. 1964. *Le cortex cérébral: Etude neuropsycho-pathologique.* Paris: Masson.

Akelaitis, A. J. 1944. A study of gnosis, praxis and language following section of the corpus callosum and anterior commissure. *J. Neurosurg.* 1:94–102.

Akert, K. 1949. Der visuelle Greifreflex. *Helv. Physiol. Pharmacol. Acta 7:* 112–134.

Apter, J. T. 1946. Eye movements following strychninization of the superior colliculus of cats. *J. Neurophysiol.* 9:73–86.

Arrigoni, G., & De Renzi, E. 1964. Constructional apraxias and hemispheric locus of lesion. *Cortex* 1:170–197.

Asanuma, H., & Sakata, H. 1967. Functional organization of a cortical efferent system examined with focal depth stimulation in cats. *J. Neurophysiol.* 30:35–54.

Barlow, H. B. 1975. Visual experience and cortical development. *Nature 258:* 199–204.

Bartlett, F. C. 1932. *Remembering.* London: Cambridge University Press.

Basser, L. S. 1962. Hemiplegia of early onset and the faculty of speech with special reference to the effects of hemispherectomy. *Brain* 85:427–460.

Bernstein, N. 1967. *The Co-ordination and Regulation of Movements.* Oxford: Pergamon Press.

Bishop, A. 1964. Use of the hand in lower primates. In J. Buettner-Janusch (ed.). *Evolutionary and Genetic Biology of Primates,* vol. 2, pp. 133–225. New York: Academic Press.

Black, P., & Myers, R. E. 1965. A neurological investigation of eye-hand control in the chimpanzee. In G. Ettlinger (ed.). *Functions of the Corpus Callosum,* pp. 47–59. Ciba Foundation Study Group No. 26. London: Churchill.

Bogen, J. 1969. The other side of the brain. I. Dysgraphia and dyscopia following cerebral commissurotomy. *Bull. Los Angeles Neurol. Soc.* 34:73–105.

Bogen, J. E., & Gazzaniga, N. S. 1965. Cerebral commissurotomy in man: Minor hemisphere dominance for certain visuospatial functions. *J. Neurosurg.* 23:394–399.

Bossom, J., & Hamilton, C. R. 1963. Interocular transfer of prism-altered co-ordinations in split-brain monkeys. *J. Comp. Physiol. Psychol.* 56:769–774.

Bossom, J., & Ommaya, A. K. 1968. Visuo-motor adaptation (to prismatic transformation of the retinal image) in monkeys with bilateral dorsal rhizotomy. *Brain 91:*161–172.

Bresson, F., Maury, L., Pieraut-le Bonniec, G., & de Schonen, S. 1977. Organization and lateralization of reaching in infants: An instance of asymmetric function in hands collaboration. *Neuropsychologia 14:*311–320.

Brinkman, J. 1974. Split-brain monkeys: Cerebral control of contralateral and ipsilateral arm, hand and finger movements. Doctoral dissertation, Erasmus University, Rotterdam.

Brinkman, J., & Kuypers, H.G.J.M. 1972. Split-brain monkeys: Cerebral control of ipsilateral and contralateral arm, hand and finger movements. *Science 176:*536–539.

1973. Cerebral control of contralateral and ipsilateral arm, hand and finger movements in the split-brain rhesus monkey. *Brain 96:*653–674.

Brodal, A. 1963. Some data and perspectives on the anatomy of the so-called "extrapyramidal system." *Acta Neurol Scand. 39* (Suppl. 4):17–38.

Brookshire, K. H., & Warren, J. M. 1962. The generality and consistency of handedness in monkeys. *Anim. Behav. 10:*222–227.

Bruner, J. S., Olver, R. R., Greenfield, P. M., et al. 1966. *Studies in Cognitive Growth.* New York: Wiley.

Butler, C. R. 1968. A memory record for visual discrimination habits produced in both cerebral hemispheres of a monkey when only one hemisphere has received direct visual information. *Brain Res. 10:*152–167.

Chomsky, N. 1966. *Cartesian Linguistics.* New York: Harper & Row.

Corballis, M. C., & Beale, I. L. 1976. *The Psychology of Left and Right.* Hillsdale, N.J.: Erlbaum.

Craik, K. J. W. 1943. *The Nature of Explanation.* London: Cambridge University Press.

DeLaguna, G. A. 1963. *Speech: Its Function and Development.* Bloomington: Indiana University Press.

Denny-Brown, D. 1966. *The Cerebral Control of Movement.* The Sherrington Lectures, VIII. Liverpool: Liverpool University Press.

Denny-Brown, D., & Chambers, R. A. 1955. Visuo-motor function in the cerebral cortex. *J. Nerv. Ment. Dis. 121:*288–299.

1958. The parietal lobe and behavior. *Res. Publ. Assoc. Nerv. Ment. Dis. 36:*35–118.

Denny-Brown, D., Meyer, J. S., & Horenstein, S. 1952. The significance of perceptual rivalry resulting from parietal lesions. *Brain 75:*433–471.

Downer, J. L. de C. 1959. Changes in visually guided behavior following midsagittal division of optic chiasm and corpus callosum in monkey (*Macaca mulatta*). *Brain 82:*251–259.

Ebner, F. F., & Meyers, R. E. 1962. Corpus callosum and the interhemispheric transmission of tactual learning. *J. Neurophysiol. 25:*380–391.

Eldred, E., & Buchwald, J. 1967. Central nervous system: Motor mechanisms. *Annu. Rev. Physiol. 29:*573-606.

Ettlinger, G., & Moffett, A. 1964. Lateral preferences in the monkey. *Nature 204* (4958):606.

Ettlinger, G., & Morton, H. B. 1963. Callosal section: Its effects on performance of a bimanual skill. *Science 139:*485-486.

Ewert, J.-P. 1967a. Elektrische Reizung des retinalen Projektionsfeldes im mittelhirn der Erdkröte (*Bufo bufo* L.). *Pflügers Arch. 295:*90-98.

1967b. Untersuchungen über die Anteile zentral nervöser Aktionen an der taxisspezifischen Ermüdung beim Beutefang der Erdkröte (*Bufo bufo* L.). *Z. Verg. Physiol. 57:*263-298.

Flament, F. 1975. *Coordination et prevalence manuelles chez les nourrissons.* Paris: Editions C.N.R.S.

Flowers, K. 1975. Handedness and controlled movement. *Brt. J. Psychol. 66:* 39-52.

Gaze, R. M., & Jacobson, M. 1962a. The projection of the binocular visual field on the optic tecta of the frog. *Q. J. Exp. Physiol. 47:*273-280.

1962b. The path from the retina to the ipsilateral optic tectum of the frog. *J. Physiol. 165:*73-74.

Gazzaniga, M. S. 1963. Effects of commissurotomy on a preoperatively learned visual discrimination. *Exp. Neurol. 8:*14-19.

1964. Cerebral mechanisms involved in ipsilateral eye-hand use in split brain monkeys. *Exp. Neurol. 10:*148-155.

1966. Visuo-motor integration in split-brain monkeys with other cerebral lesions. *Exp. Neurol. 16:*289-298.

Gazzaniga, M. S., Bogen, J. E., & Sperry, R. W. 1962. Some functional effects of sectioning the cerebral commissures in man. *Proc. Natl. Acad. Sci. U.S.A. 48:*1765-1769.

1963. Laterality effects in somesthesis following cerebral commissurotomy in man. *Neuropsychologia 1:*209-215.

1965. Observations on visual perception after disconnection of the cerebral hemispheres in man. *Brain 88:*221-236.

1967. Dyspraxia following division of the cerebral commissures. *Arch. Neurol. 16:*606-612.

Gazzaniga, M. S., & Sperry, R. W. 1967. Language after section of the cerebral commissures. *Brain 90:*131-148.

Geschwind, N. 1965. Disconnection syndromes in animals and man. 1 & 2. *Brain 88:*237-294, 585-644.

Geschwind, N., & Kaplan, E. 1962. A human cerebral deconnection syndrome. *Neurology* (Minneap.) *12:*675-685.

Geschwind, N., & Levitsky, W. 1968. Human brain: Left-right asymmetries in temporal speech region. *Science 161:*186-187.

Gesell, A., & Ames, L. B. 1947. The development of handedness. *J. Genet. Psychol. 70:*155-175.

Gibson, J. J. 1966. *The Senses Considered as Perceptual Systems*. Boston: Houghton-Mifflin.

Halverson, H. H. 1931. An experimental study of prehension in infants by means of systematic cinema records. *Genet. Psychol. Monogr. 10*:107-286.

Hamilton, C. R., & Hamilton, C. L. 1966. Reaching abilities of split-brain monkeys when using different eye-hand combinations. Unpublished manuscript, California Institute of Technology.

Hamilton, C. R., Tieman, S. B., & Fariell, W. S. 1974. Cerebral dominance in monkeys? *Neuropsychologia 12*:193-197.

Head, H. 1926. *Aphasia and Kindred Disorders of Speech*. London: Cambridge University Press.

Hécaen, H. 1976. Acquired aphasia in children and the ontogenesis of hemispheric functional specialization. *Brain Lang. 3*:114-134.

Hécaen, H., & Ajuriaguerra, J. de 1964. *Left-handedness: Manual Superiority and Cerebral Dominance*. New York: Grune & Stratton.

Hécaen, H., Penfield, W., Bertrand, C., & Malmo, R. 1956. The syndrome of apractognosia due to lesions of the minor cerebral hemisphere. *Arch. Neurol. Psychiatry 75*:400-434.

Hermelin, B., & O'Conner, N. 1971. Functional asymmetry in the reading of Braille. *Neuropsychologia 9*:431-435.

Hess, W. R. 1949. *Das Zwischenhirn*. Basel: Schwabe.

Holst, E. von 1954. Relations between the central nervous system and the peripheral organs. *Br. J. Anim. Behav. 2*:89-94,

Hubel, D. H., & Wiesel, T. N. 1965. Binocular interaction in the striate cortex of kittens reared with artificial squint. *J. Neurophysiol. 28*:1041-1059.

Hubel, D. H., Wiesel, T. N., & Le Vay, S. 1977. Plasticity of ocular dominance columns in monkey striate cortex. *Proc. R. Soc. (Lond.) Ser. B.* (in press).

Humphrey, N. K., & Weiskrantz, L. 1967. Vision in monkeys after removal of the striate cortex. *Nature* (Lond.) *215*:595-597.

Ingle, D., & Sprague, J. M. (eds.). 1975. Sensorimotor function of the midbrain tectum. *Neurosci. Res. Progr. Bull. 13* (2): 169-288.

Jacobson, M. 1962. The representation of the retina on the optic tectum of the frog. Correlations between retino-tectal magnification factor and retinal ganglion cell count. *Q. J. Exp. Physiol. 47*:170-178.

Kennard, M., & Watts, J. W. 1934. The effect of section of the corpus callosum on the motor performance of monkeys. *J. Nerv. Ment. Dis. 79*: 159-169.

Kimura, D. 1967. Functional asymmetry of the brain in dichotic listening. *Cortex 3*:163-178.

1973a. Manual activity during speaking. I. Right handers. *Neuropsychologia 11*:45-50.

1973b. Manual activity during speaking. II. Left handers. *Neuropsychologia* 11:51–55.
Kimura, D., & Vanderwolf, C. H. 1970. The relation between hand preference and the performance of individual finger movements by left and right hands. *Brain* 93:769–774.
Kuypers, H.G.J.M. 1964. The descending pathways to the spinal cord, their anatomy and function. In J. C. Eccles & J. P. Schadé (eds.). *Progress in Brain Research*, Vol. II, *Organization of the Spinal Cord*. Amsterdam: Elsevier.
Lashley, K. S. 1924. Studies of cerebral function in learning. 5. The retention of motor habits after destruction of the so-called motor areas of primates. *Arch. Neurol. Psychiatry (Chicago)* 12:249–276.
  1951. The problem of serial order in behavior. In L. A. Jeffress (ed.). *Cerebral Mechanisms in Behavior: The Hixon Symposium*. New York: Wiley.
Lawrence, D. G., & Kuypers, H.G.J.M. 1965. Pyramidal and non-pyramidal pathways in monkeys: Anatomical and functional correlation. *Science* 148:973–975.
Lenneberg, E. H. 1967. *Biological Foundations of Language*. New York: Wiley.
Levy, J. 1969. Possible basis for the evolution of lateral specialization of the human brain. *Nature* 224:614–615.
  1974. Psychobiological implications of bilateral asymmetry. In S. Dimond & J. G. Beaumont (eds.). *Hemisphere Function in the Human Brain*. London: Paul Elek.
  1976. A review of evidence for a genetic component in the determination of handedness. *Behav. Genet.* 6:429–453.
Levy, J., & Trevarthen, C. 1976. Metacontrol of hemispheric function in human split-brain patients. *J. Exp. Psychol. (Hum. Percept. Perf.)* 2:299–312.
  1977. Perceptual, semantic and phonetic aspects of elementary language processes in split-brain patients. *Brain* 100:105–118.
Levy, J., Trevarthen, C., & Sperry, R. W. 1972. Perception of bilateral chimeric figures following hemispheric deconnexion. *Brain* 95:61–78.
Levy-Agresti, J., & Sperry, R. W. 1968. Different perceptual capacities in major and minor hemispheres. *Proc. Natl. Acad. Sci. U.S.A.* 61:1151.
Liepman, H. 1905. Die Linke Hemisphäre und das Handeln. *Munch. Med. Wochenschr.* 52:2322–2326, 2375–2378.
Lippman, H. S. 1927. Certain behavior responses in early infancy. *J. Genet. Psychol.* 34:424–440.
Loeb, J. 1918. *Forced Movements, Tropisms, and Animal Conduct*, Chapter II. Philadelphia: Lippincott.
Mach, E. 1959. *The Analysis of Sensations, and the Relation of the Physical to the Psychological*. New York: Dover. Republication of 1914 edition, translated by C. M. Williams and edited and revised by S. Waterlow.

Mark, R. F., & Sperry R. W. 1968. Bimanual coordination in monkeys. *Exp. Neurol. 21:*92–104.

McDonnell, P. M. 1975. The development of visually guided reaching. *Percept. Psychophys. 18:*181–185.

McFie, J., Piercy, M. F., & Zangwill, O. L. 1950. Visual spatial agnosia associated with lesions of the right cerebral hemisphere. *Brain 73:*167–190.

Milner, B. 1962. Laterality effects in audition. In V. B. Mountcastle (ed.). *Interhemispheric Relations and Cerebral Dominance*, pp. 177–196. Baltimore: Johns Hopkins Press.

Milner, B., Taylor, L., & Sperry, R. W. 1968. Lateralized suppression of dichotically presented digits after commissural section in man. *Science 161:* 184–185.

Mishkin, M. 1962. Discussion. In V. B. Mountcastle (ed.). *Interhemispheric Relations and Cerebral Dominance*, pp. 101–107. Baltimore: Johns Hopkins Press.

Myers, R. E. 1955. Interocular transfer of pattern discrimination in cats following section of crossed optic fibers. *J. Comp. Physiol. Psychol. 48:* 470–473.

1956. Function of corpus callosum in interocular transfer. *Brain 79:*358–363.

Myers, R. E., Sperry R. W., & McCurdy, N. M. 1962. Neural mechanisms in visual guidance of limb movement. *Arch. Neurol. 7:*195–202.

Napier, J. R. 1961. Prehensility and opposability in the hands of primates. In J. E. Harris (ed.). *Vertebrate Locomotion: Symposium Number 5*, pp. 115–132. London: Zoological Society.

Paillard, J., & Beaubaton, D. 1975. Triggered and guided components of visual reaching. Their dissociation in split-brain monkeys. In M. Shahami (ed.). *The Motor System: Neuropsychology and Muscle Mechanisms.* The Hague: Elsevier.

Pasik, T., & Pasik, P. 1964. Optokinetic nystagmus: An unlearned response altered by section of chiasma and corpus callosum in monkeys. *Nature 203:*609–611.

Patterson, A., & Zangwill, O. L. 1944. Disorders of visual space perception associated with lesions of the right cerebral hemisphere. *Brain 67:*331–358.

Penfield, W. 1954. Mechanisms of voluntary movement. *Brain 77:*1–17.

Penfield, W., & Rasmussen, T. 1950. *The Cerebral Cortex of Man.* New York: Macmillan.

Penfield, W., & Roberts, L. 1959. *Speech and Brain Mechanisms.* Princeton: Princeton University Press.

Piaget, J. 1937. *La construction du réal chex l'enfant.* Neuchatel: Delachaux & Niestlé.

1950. *The Psychology of Intelligence.* Translated by M. Piercy & D. E. Berlyne. London: Routledge & Kegan Paul.

1953. *The Origins of Intelligence in Children*. London: Routledge & Kegan Paul.

Preilowski, B. F. B. 1972. Possible contribution of the anterior forebrain commissures to bilateral motor coordination. *Neuropsychologia* 10:267–277.

Ramón y Cajal, S. 1955. *Histologie du système nerveux de l'homme et des vertébrés*. Madrid: Instituto Ramón y Cajal.

Schiller, P. H. 1957. Innate motor action as a basis of learning: Manipulative patterns in the chimpanzee (1949). Translated by C. H. Schiller. In C. H. Schiller (ed.). *Instinctive Behavior: The Development of a Modern Concept*. New York: International Universities Press.

Schneider, D. 1954. Das Gesichtsfeld und der Fixiervorgang bei Einheimischen Anuren. *Z. Vergl. Physiol.* 36:147–164.

Schneider, G. E. 1967. Contrasting visuomotor functions of tectum and cortex in the golden hamster. *Psychol. Forsch.* 31:52–62.

Schrier, A. M., & Sperry, R. W. 1959. Visuo-motor integration in split-brain cats. *Science* 129:1275–1276.

Semmes, J., Weinstein, S., Ghent, L., & Teuber, H. L. 1960. *Somatosensory Changes after Penetrating Brain Wounds in Man*. Cambridge, Mass: Harvard University Press.

Smith, K. U., & Akelaitis, A. J. 1942. Studies on the corpus callosum. 1. Lateral dominance in behavior and bilateral motor coordination in man before and after partial and complete section of the corpus callosum. *Arch. Neurol. Psychiatry* 47:519–543.

Sperry, R. W. 1961. Cerebral organization and behavior. *Science* 133:1749–1757.

  1967. Mental unity following surgical disconnection of the cerebral hemispheres. *Harvey Lectures*, Series 62. New York: Academic Press.

  1968. Hemisphere deconnection and unity in conscious awareness. *Am. Psychol.* 23:723–733.

  1970. Perception in absence of the neocortical commissures. *Percept. Dis. Res. Publ. Assoc. Res. Nerv. Ment. Dis.* 48:123–138.

  1974. Lateral specialization in the surgically separated hemispheres. In F. O. Schmitt & R. G. Worden (eds.). *The Neurosciences: Third Study Program*. Cambridge, Mass: M.I.T. Press.

Sperry, R. W., Gazzaniga, M. S., & Bogen, J. E. 1969. Interhemispheric relationships: The neocortical commissures: Syndromes of hemisphere disconnection. In P. J. Vinken & G. W. Bruyn (eds.). *Handbook of Clinical Neurology*, Chapter 14. Amsterdam: North-Holland.

Sperry, R. W., Stamm, J. S., & Miner, N. 1956. Relearning tests for interocular transfer following division of optic chiasma and corpus callosum in cats. *J. Comp. Physiol. Psychol.* 49:529–533.

Sprague, J. M., Chambers, W. W., & Stellar, E. 1961. Attentive, affective and adaptive behavior in the cat. *Science* 133:165–173.

Teuber, H.-L. 1962. Effects of brain wounds implicating right or left hemisphere in man: Hemisphere differences and hemisphere interaction in vision, audition, and somesthesis. In V. B. Mountcastle (ed.). *Interhemispheric Relations and Cerebral Dominance.* Baltimore: John Hopkins Press.

Trevarthen, C. B. 1962. Studies on visual learning in split-brain monkeys. Doctoral dissertation, California Institute of Technology.

　1965. Functional interactions between the cerebral hemispheres of the split-brain monkey. In E. G. Ettlinger (ed.). *Functions of the Corpus Callosum,* pp. 24–40. Ciba Foundation Study Group No. 20. London: Churchill.

　1968. Two mechanisms of vision in primates. *Psychol. Forsch. 31*:299–337.

　1970a. Brain bisymmetry and the role of the corpus callosum in behavior and conscious experience. In J. Cernacek & F. Podivinsky (eds.). *Proceedings of the Colloquim on Interhemispheric Relations.* Bratislava: Slovak Academy of Sciences.

　1970b. Experimental evidence for a brain-stem contribution to visual perception in man. *Brain Behav. Evol. 3*:338–352.

　1970c. Cerebral-midbrain relations reflected in split-brain studies of higher integrative functions. Paper presented at the 19th International Congress of Psychology, London.

　1972. Specialized lesions: The split-brain technique. In R. D. Meyers (ed.). *Methods in Psychobiology: Laboratory Techniques in Neuropsychology.* New York: Academic Press.

　1974a. Analysis of cerebral activities that generate and regulate consciousness in commissurotomy patients. S. J. Dimond & J. C. Beaumont (eds.). *Hemisphere Function in the Human Brain.* London: Paul Elek.

　1974b. Functional relations of disconnected hemispheres with the brain stem and with each other: Monkey and man. In M. Kinsbourne and W. L. Smith (eds.). *Hemispheric Disconnection and Cerebral Function.* Springfield, Ill.: Thomas.

　1975. Psychological activities after forebrain commissurotomy in man: Concepts and methodological hurdles in testing. In F. Michel & B. Schott (eds.). *Les Syndromes de disconnexion calleuse chez l'homme.* Lyon: Hôpital Neurologique.

Trevarthen, C., & Sperry, R. W. 1973. Perceptual unity of the ambient visual field in human commissurotomy patients. *Brain 96*:547–570.

Turkewitz, G., Gordon, E. W., & Birch, H. 1965. Head turning in the human neonate: Spontaneous patterns. *J. Genet. Psychol. 107*:143–158.

Twitchell, T. E. 1965. The automatic grasping responses of infants. *Neuropsychologia 3*:247–259.

Uexküll, J. von 1937. A stroll through the worlds of animals and men. Translated by C. H. Schiller. In C. H. Schiller (ed.). *Instinctive Behavior: The*

Development of a Modern Concept. New York: International Universities Press, 1957.

Warren, J. M. 1958. The development of paw preference in cats and monkeys. *J. Genet. Psychol. 93:*229-236.

Warren, J. M., Abplanalp, J. M., & Warren, H. B. 1967. The development of handedness in cats and rhesus monkeys. In H. W. Stevenson (ed.). *Early Behavior: Comparative and Developmental Approaches.* New York: Wiley.

Weinstein, S. 1962. Differences in effects of brain wounds implicating right or left hemispheres: Differential effects on certain intellectual and complex perceptual functions. In V. B. Mountcastle (ed.). *Interhemispheric Relations and Cerebral Dominance*, pp. 159-176. Baltimore: Johns Hopkins Press.

White, M. J. 1969. Laterality differences in perception: A review. *Psychol. Bull. 72:*387-405.

Young, J. Z. 1962. Why do we have two brains? In V. B. Mountcastle (ed.). *Interhemispheric Relations and Cerebral Dominance*, pp. 7-25. Baltimore: Johns Hopkins Press.

Zaidel, E. 1976. Auditory language comprehension in the right hemisphere following cerebral commissurotomy and hemispherectomy: A comparison with child language and aphasia. In E. Zurif & A. Caramazza (eds.). *The Acquisition and Breakdown of Language: Parallels and Divergences.* Baltimore: Johns Hopkins Press.

Zangwill, O. L. 1960a. *Cerebral Dominance and Its Relation to Psychological Function.* Edinburgh: Oliver & Boyd.

1960b. Speech. In J. Field (ed.). *Handbook of Physiology*, Sect. 1, *Neurophysiology*, Vol. III. Washington, D.C.: American Physiological Society.

12
Dichotic listening and the development of linguistic processes

M. P. BRYDEN AND F. ALLARD

The dichotic listening procedure holds great promise as a method for isolating the perceptual and cognitive functions of the two hemispheres and for providing information about the course of development of such functions. This chapter examines critically the studies using the dichotic listening procedure with children and suggests some new directions for developmental research on lateralization.

STUDIES ON DICHOTIC LISTENING

The first developmental study employing dichotic listening was that of Kimura (1963). She presented children with dichotic lists consisting of from one to three pairs of numbers and instructed them to report as many numbers as they could, in any order they chose – a free-recall procedure. She found a clear right-ear superiority as early as age 4, and the effect failed to reach statistical significance only in the 7- and 9-year-old girls. These results suggest a very early lateralization of language functions.

Subsequent developmental studies have employed the same technique: presentation of two or three pairs of numbers or words for free recall. By and large, the results of these studies corroborate Kimura's initial findings. For example, Kimura (1967) replicated her original findings on children of a somewhat lower socioeconomic level, reporting a right-ear superiority at all ages from 5 to 8 years, except for the 5-year-old boys. Knox and Kimura (1970) also found right-ear superiority at all ages from 5 to 8, and Bever (1971) observed a significant right-ear superiority in a sample of 195 children ranging in age from 2.5 to 5.5 years.

Geffner and Hochberg (1971) found a right-ear superiority at all ages from 4 to 7 years in children from middle socioeconomic levels, but no significant right-ear advantage until age 7 in children from lower socioeconomic levels. Sommers and Taylor (1972) reported a large right-ear superiority in normal 5- and 6-year-olds, but not in children whose speech onset had been delayed.

With somewhat older children, Satz et al. (1971) found a very large right-ear superiority at both age 8 and 11, as did Ling (1971) in a sample of children ranging in age from 6 to 15. Other studies of children of a similar age have produced somewhat equivocal results. For example, Zurif and Carson (1970) found only a trend toward a right-ear superiority in normal fourth-grade children, and Witelson's (1962) results failed even to approach statistical significance. Bryden (1970) found a general right-ear superiority in children ranging in age from 7 to 12, but reported that the difference between left- and right-handed subjects observed in adults (Bryden, 1965; Zurif & Bryden, 1969) did not emerge until fourth grade (age 10) in girls and sixth grade (age 12) in boys.

The most striking exception to the general rule of a right-ear superiority for dichotic lists is found in the study of Inglis and Sykes (1967). Like Kimura, they used material varying in length from one to three pairs. Their subjects ranged from 5 to 10 years of age. They found a right-ear superiority only at ages 6 and 9, and then only for the three-pair material.

Despite occasional failures, the majority of dichotic listening studies on children have shown a measurable right-ear superiority, even in the youngest age groups tested. It is unfortunate, however, that the vast majority of experiments on children have involved the presentation of lists of words for free recall. As a procedure for determining the source of the laterality effect, free recall leaves much to be desired (Bryden, 1967; Inglis & Sykes, 1967). When the successive pairs of items are presented fairly rapidly, most subjects report the items presented to one ear followed by the items presented to the other ear. This has been termed the *ear order of report* (Bryden, 1962, 1967). At slower rates, it becomes more common to find the subject reporting first one pair of items and then the next, in a *temporal order of report*. There is some evidence that the relative frequency of these report strategies changes with age (Witelson & Rabinovitch, 1971; Bryden, 1972).

The lack of control over report order implied in the free-recall procedure makes it possible for many different factors to influence accuracy. For instance, a subject may elect, simply by preference, to report the items from the right ear before giving those from the left ear. By so doing, he relegates the left-ear items to late in the report sequence, when he is less likely to be able to remember them, and thus produces a large right-ear superiority that is a measure of nothing more than the fact that he chose to report the right-ear items first. In fact, a right-ear starting preference has been observed in a number of studies with adults, but when it is controlled by requiring an equal number of left-ear and right-ear starts, the right-ear superiority remains (Bryden, 1967).

Even without a starting-ear bias, the order of report can influence laterality scores profoundly. A subject using the ear order of report maximizes time differences between reporting items from one ear and reporting items from the other, and thus tends to show a large difference between ears on each trial. A subject using the temporal order of report, on the other hand, minimizes the time differences between reporting items from the two ears, and thus shows a small laterality effect on any one trial. Although this may average out over many trials, it remains a potential source of uncontrolled variability when the free-recall procedure is used.

The use of lists introduces the question of memory load as well. It is possible, for example, that items presented to the left ear are not harder to perceive, but are more difficult to remember, than are items presented to the right ear. Supporting the argument that memory effects may be important is the fact that the only signs of a right-ear superiority in the Inglis and Sykes (1967) study came with the three-pair material. However, in a simple free-recall experiment, difficulty level and memory load are hopelessly confounded. Existing data do not permit any firm statements about the role of memory factors. Although Bryden (1967) argued against their importance in adult studies, it is worth noting that short-term memory abilities change dramatically during childhood (Inglis & Sykes, 1967).

To illustrate some of the problems noted above, consider some data from an unpublished free-recall study done in our laboratory. In this study, 72 right-handed children were tested, 12 boys and 12 girls, at each of three grade levels. Analysis of variance indicated

Dichotic listening and linguistic processes

a significant right-ear superiority and no interaction of ear and age. However, a somewhat different picture emerged when the subjects were considered individually. Figure 12-1 shows, in the left panel, the percentage of subjects at each grade level exhibiting a right-ear superiority in terms of a raw accuracy measure. This figure indicates that more subjects were superior on the right-ear items than on the left-ear items at all grade levels, although a binomial test reached significance only on the sixth-graders. In the center panel, corresponding figures for initiating report with an item presented to the right ear are shown. More than two-thirds of the second-grade subjects started with a right-ear item more often than with a left-ear item, but in the later grades this effect was small. Finally, in the right panel, a corrected

Fig. 12-1. Performance of children at three grade levels on a dichotic free-recall task. Each point summarizes data from 24 subjects. *Left panel*: proportion of subjects with right-ear superiority when uncorrected data are used. *Center panel*: proportion of subjects initiating report more frequently with a right-ear item. *Right panel*: proportion of children with right-ear superiority when data have been corrected to take starting bias into account.

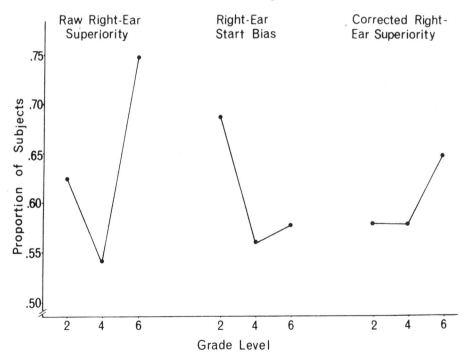

laterality measure is shown. This measure was derived in the manner used by Bryden (1963): Average performance was determined for all trials on which the subject started with a right-ear item, and separately for those trials on which the subject first reported a left-ear item. The two separate measures were then summed to give an indication of performance had the subject initiated report equally often with an item presented to either ear. With this measure, a greater proportion of sixth-graders showed right-ear superiority.

One interpretation of these data is that second-graders showed a right-ear superiority because of a starting-ear bias, and sixth-graders showed this superiority because of a more basic perceptual difference between inputs to the two ears. This argument suggests that the corrected measure is the most appropriate, and that laterality develops between the ages of 8 and 12. Such an analysis points out some of the potential pitfalls of employing a multiple-item free-recall task to assess laterality effects in children.

A second more general question of data analysis is also raised by Figure 12-1. Based on the raw score analysis, only 64% of the subjects showed a right-ear superiority. Yet evidence suggests that in some 95% to 99% of right-handed adults the left hemisphere is dominant for speech (Milner et al., 1964; Rossi & Rosadini, 1967; Roberts, 1969). Granted that the dichotic listening procedure is not a perfect measure of speech lateralization – in adults some 80% to 85% show a right-ear superiority, not the much higher figure that would be expected from a direct assessment of speech lateralization – the lower figure obtained in children is more consistent with notions of a developing laterality than with the concept of a fully developed lateralization.

Unfortunately, it has been more commonplace to report laterality effects in terms of the mean difference between left and right ears rather than in terms of the proportion of subjects showing a right-ear superiority. In fact, both sets of data are required for a clear understanding of the processes involved. Those studies in which individual subject data are reported are in general agreement with the data shown in Figure 12-1. Bever (1971) found that 67% of a sample of children ranging in age from 2.5 to 5.5 years showed a right-ear superiority, and Ling (1971) gave a figure of 73% for a group ranging in age from 6 to 15. In a striking exception, Sommers and Taylor (1972) found that all 10 of their normal 5- and 6-year-olds were more accurate on the right ear. Although these data suggest that a smaller proportion of children than adults have a dominant right-ear, they are still too sparse to permit any firm conclusions. It is more to the point simply to argue that individual data as well as group means should be made available in future studies.

In the preceding pages, we have argued that procedures involving the free recall of multiple-item lists introduce too many sources of variability and potential contamination to provide more than a gross assessment of the development of cerebral lateralization. An alternative is to reduce the dichotic listening situation to its simplest, by pairing two items contrasting in some specified way. This approach has been used to great advantage in adults by Studdert-Kennedy and Shankweiler (1970), who paired consonant-vowel-consonant syllables differing only in the medial vowel or in the initial or terminal stop consonant. They found a laterality effect for consonants, but not for vowels, and have developed the beginnings of a theory of lateralization based on their data.

Although many investigators have included single pairs of numbers in their free-recall tasks, accuracy is usually so high in this condition that little information about laterality can be gained (Inglis & Sykes, 1967). We have been able to find only three studies on children in which more difficult single-pair tasks have been employed. Knox and Kimura (1970) used a task in which two words, differing by a single phoneme, were paired dichotically. In a group of children ranging in age from 5 to 8, they found a significant right-ear superiority and no interactions with either age or sex. They did not, however, report scores for different age levels or give any indication of individual performance. Sommers and Taylor (1972) used a similar task: All 10 of their normal 5- and 6-year-olds and 8 of their 10 subjects whose speech onset was delayed showed a right-ear superiority.

Bryden et al. (1973) presented natural speech consonant-vowel syllables differing only in the initial stop consonant to 12 boys and 12 girls at each of five grade levels from kindergarten (age 6) to eighth grade (age 14). Although there were sex differences, in that the girls showed a clear laterality effect by the fourth grade and the boys lagged considerably behind, Figure 12-2 shows data for both sexes combined. Only in grades 4, 6, and 8 do significantly more subjects show right-ear superiority. These data, in contrast to those of Knox and Kimura (1970) and Sommers and Taylor (1972), suggest a gradual development of cerebral lateralization, approximating the adult state by the eighth grade. One should remember that the distinction between elements of a dichotic pair in this study, which employed only the six stop consonants /b/, /p/, /d/, /t/, /k/, and /g/ as initial phonemes paired with the vowel /a/, was much more carefully controlled with respect to both nature and location than the distinctions in the Knox and Kimura (1970) and Sommers and Taylor (1972) studies.

In summary, dichotic listening studies on children indicate that some linguistic functions are lateralized at a very early age. Because of the difficulties in interpreting free-recall data, it is not entirely clear to what extent these are perceptual differences and to what extent they depend on memory factors or even response or attentional biases irrelevant to cerebral speech lateralization. The success

Fig. 12-2. Development of lateral asymmetry in perception of stop consonants. *Left*: proportion of subjects showing right-ear superiority. *Right*: mean right-ear superiority.

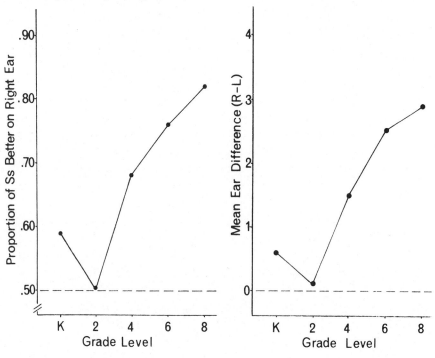

of studies employing word pairs that differ in only a single phoneme suggests that there are true perceptual differences between ears in some children as early as age 5. However, at the present time the data are not sufficiently precise for us to determine whether these differences are as universal as those found in adults. The results of one study, confining the contrast between ears to stop consonants, suggests that processing of stop consonants by the left hemisphere develops slowly, reaching adult level only at or near the time of puberty.

THE DEVELOPMENT OF LATERALITY

What do the data from dichotic listening studies indicate about the development of language laterality? Certain speculations seem appropriate at the present time.

Laterality effects in adults reflect hemispheric specialization for perceptual processing. The perceptual specialization of the left hemisphere extends to all types of language material: lists of words (Bryden, 1967) or numbers (Kimura, 1961), sound shadowing connected prose (Treisman & Geffen, 1968), portions of backward speech (Kimura & Folb, 1968), single phonemes (Studdert-Kennedy & Shankweiler, 1970), and even Morse code for experienced operators (Papcun et al., 1974) – in fact, virtually any type of acoustic material with a linguistic referent. In children, as we have seen, laterality effects are observed when words are presented dichotically, especially when more than one pair of words or digits is involved. However, laterality effects are not found in young children when single pairs of consonant-vowel syllables contrasting only in initial stop consonants are presented dichotically (Bryden et al., 1973). This is a very curious finding. One would expect that phonemes, universally considered the building blocks of speech, would be one of the first language skills to become lateralized. At least three reasons for this finding seem possible: (1) the Bryden et al. results may not be replicable, (2) the task demands of multiple-list experiments may be heavily loaded to demonstrate and even magnify laterality effects, and (3) different aspects of speech perception may actually lateralize at different times and because they involve so many factors, multiple-list experiments may pick up the earliest elements to become lateralized.

The problems with multiple lists and free-recall procedures are discussed earlier in this chapter. As we have seen, the real danger in multiple-list experiments that do not control or consider strategies of report is that a spurious laterality effect may be found. This point has been made time and time again by Inglis (e.g., Inglis & Sykes, 1967). If a subject shows no perceptual laterality, but has a strong tendency to initiate report from the right ear, an "artificial" laterality effect will appear because left-ear material is held longer in memory. Thus, an effect of short-term memory may easily be misinterpreted as a laterality effect. This problem is easily avoided by using controlled report order (e.g., precued partial report) or by scoring trials as a function of starting ear and correcting for the starting bias.

As well as list length, the meaningfulness of the elements composing the dichotic pairs may be important. It is well known that meaningful material can be perceived correctly in noise more readily than single syllables. The left hemisphere may be better able to take advantage of the sequential constraints in language than the right hemisphere, resulting in a greater laterality effect with meaningful words. This could easily be determined by comparing the magnitude of laterality effects for two types of dichotic sounds with the same phonetic differences; for example, a group of nonsense syllables with no meaning, such as *ba, da, ga,* and a group of words like *bore, door, gore.*

Yet another possible influence in differing estimates of onset of lateralization is selective attention. A child must be able to disentangle the simultaneously occurring sounds in order to process them. Lists or words may act to lock attention more effectively onto one sound, then the other. With syllables differing only in the initial 100 msec, children may have trouble determining that two sounds are there. That something of this type is a factor in laterality effects is suggested by the almost random guessing that occurred for the second responses in the Bryden et al. (1973) study. Children in this study seemed to be hearing only one of the sounds presented. The ability of children to attend selectively to dichotic inputs has not been fully investigated.

Kinsbourne (1970) has suggested that "verbal" thinking may be an important element in laterality. He argued that any kind of verbal set induces greater left-hemisphere activity and hence a greater sensitivity to right-ear stimuli. Lists of digits or words would produce a greater verbal set than meaningless syllables in both children and adults and should result in increased laterality effects. Zurif and Sait (1970) have demonstrated with adults that the more language-like the elements of a dichotic list, the greater the laterality. They presented dichotic lists of syllables read as a list or read as a sentence, and found greater laterality with sentence-like material.

Different components of language abilities may well prove to be lateralized at different times. The Geffner and Hochberg (1971) finding that lateralization is delayed in lower socioeconomic groups may indicate that the greater the experience with language, the earlier this lateralization takes place. Developmental studies controlling order of report, selective attention, list length, and the meaningfulness of stimulus elements would distribute the importance of the various factors, as well as provide some estimate of the age or the stage in language development at which lateralization is observed.

If it is true that different linguistic processes become lateralized at different times, it is interesting to speculate on why phonemes are the last linguistic component to demonstrate laterality. It would be expected that phonemes, being the elements of speech, would be the first things to be lateralized. Obviously, children are able to understand and use phonemic distinctions well before phonemes show any evidence of lateralization. Eimas et al. (1971) have shown that 1- and 4-month-old infants are able to distinguish between stop consonants differing in voicing. The acoustic stimulus in the distinctive feature "voicing" is a very subtle temporal cue, involving the presence or absence of a low-frequency "voicing bar" during the initial moments of the consonant burst. After habituation to a /b/ sound, the babies showed increased activity when the sound was changed to /p/. Adult listeners are unable to differentiate degrees of "/b/ness": Varying the onset of the voicing bar relative to the beginning of the consonant does not change the quality of the /b/ heard. Subjects report /b/ until the voicing bar lags by 25 msec; then they report hearing /p/. This same "categorical perception" was found for infants. Recovery from habituation was not found if the second sound fell within the same category as the first, as judged by adult values. When the second sound fell into another category, the babies showed recovery from habituation.

The Eimas et al. (1971) finding may be contrasted with the idea of language development proposed by Lenneberg (1967). According to Lenneberg, speech is a biological function and subject to the maturation of the brain. Brain damage to the speech areas before puberty can be compensated for by the developing brain. Damage to the speech areas after puberty results in loss of speech; the brain has lost its ability to compensate. Our data on lateralization of phonemes agree with the time course proposed by Lenneberg, not only in age of occurrence, but also in showing a sex difference that could possibly occur because of the earlier onset of puberty in girls than boys.

Eimas's data indicate that phonemic perception is present in very young children; Lenneberg's maturational concept and our data suggest a much later age of lateralization. Perhaps phoneme perception, then, is bilaterally represented until the final stages of lateralization. This would allow the rebuilding of language skills in the right hemisphere, as suggested by Lenneberg. Hill (1972) has pointed out that it is impossible for an adult to acquire a new language without a foreign accent, although children are quite able to do so. Eguchi and Hirsch (1969) have shown that variability in the pronouncing of vowel sounds is much greater in children than in adults. The variability in children's performance decreases to adult values at about age 11, around the same time as lateralization of phonemes and the end of brain flexibility in the reacquisition of speech.

It may well be that the developing brain possesses a great potential range for the detection of phonemes, much the same as the visual cortex of the brain of the cat possesses a great potential range for the detection of orientation (Hubel & Wiesel, 1963). As the brain matures, phonemes present in the environment are sharpened much the same way as orientations present in the environment are sharpened in the cat's cortex (Blakemore & Cooper, 1970; Hirsch & Spinelli, 1970). These suggestions, of course, remain highly speculative until methodological considerations – report order, meaningfulness of sounds, etc. – confirm or deny the existing estimates of age of onset of lateralization.

ACKNOWLEDGMENT

The preparation of this chapter was aided by Grant No. A-95 from the National Research Council of Canada to M. P. Bryden.

REFERENCES

Bever, T. G. 1971. The nature of cerebral dominance in speech behavior of the child and adult. In R. Huxley & E. Ingram (eds.). *Language Acquisition: Models and Methods.* New York: Academic Press.

Blakemore, C., & Cooper, C. F. 1970. Development of the brain depends on the visual environment. *Nature 228:*477-478.

Bryden, M. P. 1962. Order of report in dichotic listening. *Can. J. Psychol. 16:*291-299.

1963. Ear preference in auditory perception. *J. Exp. Psychol. 65:*103-105.

1965. Tachistoscopic recognition, handedness, and cerebral dominance. *Neuropsychologia 3:*1-8.

1967. An evaluation of some models of laterality effects in dichotic listening. *Acta Oto-laryngol. 63:*595-604.

1970. Laterality effects in dichotic listening: Relations with handedness and reading ability in children. *Neuropsychologia 8:*443-450.

1972. Perceptual strategies, attention, and memory in dichotic listening. Research Report 43, University of Waterloo.

Bryden, M. P., Allard, F., & Scarpino, F. 1973. The development of language lateralization and speech perception. Unpublished manuscript, University of Waterloo.

Eguchi, S., & Hirsh, I. J. 1969. Development of speech sounds in children. *Acta Oto-laryngol.* Suppl. 257.

Eimas, P. D., Sigueland, E. R., Jusczyk, P., & Vigorito, J. 1971. Speech perception in infants. *Science 171:*304-306.

Geffner, D. S., & Hochberg, I. 1971. Ear laterality performance of children from low and middle socioeconomic levels on a verbal dichotic listening task. *Cortex 7:*193-203.

Hill, J. H. 1972. On the evolutionary foundations of language. *Am. Anthropol. 74:*308-317.

Hirsch, H. V. B., & Spinelli, D. N. 1970. Visual experience modifies distribution of horizontally and vertically oriented receptive fields of cats. *Science 869:*168-171.

Hubel, D. H., & Wiesel, T. N. 1963. Receptive fields of cells in striate cortex of very young, visually inexperienced kittens. *J. Neurophysiol. 26:*994-1002.

Inglis, J., & Sykes, D. H. 1967. Some sources of variation in dichotic listening performance in children. *J. Exp. Child Psychol. 55:*480-488.

Kimura, D. 1961. Cerebral dominance and the perception of verbal stimuli. *Can. J. Psychol. 15:*166-171.

1963. Speech lateralization in young children as determined by an auditory test. *J. Comp. Physiol. Psychol. 56:*899-902.

1967. Functional asymmetry of the brain in dichotic listening. *Cortex 3:* 163-178.

Kimura, D., & Folb, S. 1968. Neural processing of backwards-speech sounds. *Science 161*:395-396.

Kinsbourne, M. 1970. Cerebral basis of asymmetries in attention. *Acta Psychol. 33*:193-201.

Knox, C., & Kimura, D. 1970. Cerebral processing of nonverbal sounds in boys and girls. *Neuropsychologia 8*:227-237.

Lenneberg, E. 1967. *Biological Foundations of Language.* New York: Wiley.

Ling, A. H. 1971. Dichotic listening in hearing-impaired children. *J. Speech Hear. Res. 14*:793-803.

Milner, B., Branch, C., & Rasmussen, T. 1964. Observations on cerebral dominance. In A. V. S. de Reuck & M. O'Conner (eds.). *Ciba Symposium on Disorders of Language.* London: Churchill.

Papcun, G., Krashen, S., Terbeek, D., Remington, R., & Harshman, R. 1974. Is the left hemisphere specialized for speech, language, and/or something else? *J. Acoust. Soc. Am. 55*:328-333.

Roberts, L. 1969. Aphasia, apraxia and agnosia in abnormal states of cerebral dominance. In P. J. Vinken & G. W. Bruyn (eds.). *Handbook of Clinical Neurology*, Vol. 4. Amsterdam: North-Holland.

Rossi, G. F., & Rosadini, G. 1967. Experimental analysis of cerebral dominance in man. In F. L. Darley (ed.). *Brain Mechanisms Underlying Speech and Language.* New York: Grune & Stratton.

Satz, P., Rardin, D., & Ross, J. 1971. An evaluation of a theory of specific developmental dyslexia. *Child Dev. 42*:2009-2021.

Sommers, R. K., & Taylor, M. L. 1972. Cerebral speech dominance in language-disordered and normal children. *Cortex 8*:224-232.

Studdert-Kennedy, M., & Shankweiler, D. 1970. Hemispheric specialization for speech perception. *J. Acoust. Soc. Am. 48*:579-594.

Treisman, A., & Geffen, G. 1968. Selective attention in perceiving and responding to speech messages. *Q. J. Exp. Psychol. 20*:139-149.

Witelson, S. 1962. Perception of auditory stimuli in children with learning problems. Unpublished manuscript, McGill University.

Witelson, S., & Rabinovitch, M. S. 1971. Children's recall strategies in dichotic listening. *J. Exp. Child Psychol. 12*:106-113.

Zurif, E. B., & Bryden, M. P. 1969. Familial handedness and left-right differences in auditory and visual perception. *Neuropsychologia 7*:179-187.

Zurif, E. B., & Carson, C. 1970. Dyslexia in relation to cerebral dominance and temporal analysis. *Neuropsychologia 8*:351-361.

Zurif, E. B., & Sait, P. E. 1970. The role of syntax in dichotic listening. *Neuropsychologia 8*:239-244.

13
Sex differences in spatial ability: possible environmental, genetic, and neurological factors

LAUREN JULIUS HARRIS

The scientific study of spatial ability is surely as old as Galton's investigations of imagery, following which hundreds – perhaps thousands – of tests of the "spatial sense" have been devised and used. *Spatial ability* has been variously defined: "to move, turn, twist, or rotate an object or objects and to recognize a new appearance or position after the prescribed manipulation has been performed" (Guilford, 1947); "to make discriminations as to the direction of motion such as up and down, left and right, and in and out" (Guilford, 1947); "to recognize the identity of an object when it is seen from different angles" (Thurstone, 1950); "to think about those spatial relations in which the body orientation of the observer is an essential part of the problem" (Thurstone, 1950); "to perceive spatial patterns accurately and to compare them with each other" (French, 1951). Each characterization implies mental imagery, but of a distinctly kinetic rather than static kind. This kinetic quality is evidenced in the demonstration that the time required to determine whether two differently oriented objects have identical shapes is a linearly increasing function of the angular difference in the objects' portrayed orientations and is no longer for a rotation in depth than for a rotation in the picture plane (Shepard & Metzler, 1971).

The spatial sense has been of particular interest to differential psychologists because the most persistent of individual differences on multifactor tests of psychological functioning is a sex difference in spatial ability. Males have decidedly better spatial skill than females. Indeed, on a number of tests, only 20% to 25% of females exceed the average performance of males.

The fact of the male's superior spatial ability is not in dispute; but the explanation is. In light of evidence implicating a critical role for the right cerebral hemisphere – particularly the temporal, parietal, and occipital areas – in spatial perception, the most pertinent question to raise, in the context of the present book, is whether sex differences in cerebral organization and functioning underlie the male's greater spatial ability. This chapter traces lines of evidence bearing on this possibility. Of course, we are past the stage – or should be – of arguing whether human variation is a product of nature *or* nurture. Instead we accept the view, expressed by Bock and Vandenberg (1968, p. 233), that "the variation of observable characteristics, whether physiological or behavioral, is the outcome of a lengthy sequence of interactions between the genetic material and the environment." Any neurological account, therefore, must be interwoven with an examination of evidence for the roles of environmental and genetic factors, our task being not to set these kinds of explanations *against* each other as though we simply had to choose one among them, but to try to understand how they complement one another.

ILLUSTRATIVE STUDIES

To convey some sense of the scope and extent of sex differences in spatial ability, let us begin by reviewing some of the kinds of tasks in which the sex difference has appeared.

Recall and detection of shapes

Among the most familiar cognitive tasks on which sex differences are known to exist is the embedded figures test. The test makes use of modified Gottschaldt figures in which the subject first looks at and tries to remember a simple geometrical form, and then, with the form no longer in view, tries to find it in a complex geometrical figure. In 5- to 10-year-old children, sex differences in speed and accuracy are typically absent (Witkin et al., 1954; Goodenough & Eagle, 1963; Karp & Konstadt, 1963; Corah, 1965; Bigelow, 1971; Graves & Koziol, 1971; Keogh & Ryan, 1971), though where differences appear, they tend to favor boys (Chateau, 1959; Witkin et al., 1967). In older children, 12 to 18 years, the male's superior skill begins to emerge more reliably (Fiebert, 1967;

Schwartz & Karp, 1967; Witkin et al., 1967; Okonji, 1969), until by age 18 through middle age, it appears routinely (Newbigging, 1954; Andrieux, 1955; Bennett, 1956; Bieri et al., 1958; Goodnow, 1962; Corah, 1965; Goldstein & Chance, 1965; Schwartz & Karp, 1967).

The critical skill on the embedded figures test typically is assumed to be the detection or disembedding of the simple form from the embedding field. This assumption has led to the test's characterization (Witkin et al., 1954) as a measure of cognitive differentiation, analytic style, or field independence. But males seem to be better in the simple recollection and visual coding as well as the disembedding of shapes. A recent experiment makes the point dramatically: College students were instructed to proceed mentally through the alphabet from A to Z, counting the number of letters containing a curve in their upper-case form. As no information about the sound constituting the name of a letter could assist in deciding whether its printed form contains a curve, the investigators thought it reasonable to consider this task purely visual. Significantly more men than women gave the correct answer (Coltheart et al., 1975).

Mental rotation and identification

The aforementioned tests appear to require primarily static visual imagery. As we saw in the definitions quoted earlier, spatial ability more commonly presupposes an ability in kinetic mental manipulation. Perhaps the most familiar example is the spatial subtest of the Differential Aptitude Test (Bennett et al., 1959). On this test, the subject must visually construct a three-dimensional figure from a two-dimensional pattern, remember the three-dimensional image and match it to perspective drawings of alternative objects, and, after locating a correct object, visualize the rotation of the object in three-dimensional space and match it with other objects. Again, males excel, at least in the age range 11 years through college age (Flanagan et al., 1961; Vandenberg et al., 1968; Hartlage, 1970).

Another more kinetic test is block counting; for example, from a two-dimensional representation of a three-dimensional stack of blocks, to count the number of block surfaces visible from perspectives different from one's own. Boys again excel (e.g., Stafford, 1961). The differences are maintained at least through young adulthood. Book (1932) tested 950 college-age men and women and found that only 23.9% of the women scored better than the average man.

Geometrical and mathematical skill

It has been suggested that the male's greater spatial visualizing ability also underlies his superior academic achievement in geometry. Saad and Storer (1960) reported that fifth-grade boys had markedly better understanding than girls of geometrical concepts and principles. And Gästrin (1940; cited in Smith, 1964) reported superior performance by high school boys in solid geometry. Smith (1964, p. 123) has proposed that the difference "may be another manifestation of sex difference in spatial ability, reflecting a greater capacity on the part of boys to perceive, recognize, and assimilate patterns within the conceptual structure of mathematics." This interpretation is supported by studies demonstrating a relationship between geometrical ability and performance on standard spatial visualization tests (e.g., Siegvald, 1944; cited in Smith, 1964).

The male's greater skill appears less reliably in arithmetic and algebra (Saad & Storer, 1960). And in the more mechanical parts of mathematics, such as numerical addition, the sex differences may be negligible. Performance on standard spatial tests is known to correlate higher with geometry tests than with marks in arithmetic and algebra (Smith, 1960) and, on factor-analytic studies, to load higher on geometry than on algebra and higher on algebra than on mechanical arithmetic (Barakat, 1951). Thus, the male's superior skill seems to be clearest in those disciplines for which spatial ability is most critical – disciplines that are genuinely mathematical as distinct from those involving more mechanical computational processes.

Chess

In a discussion of sex differences in standard tests of spatial visualization and mental transformation, it is tempting to consider the game of chess. Throughout the world it has been and remains a game so much dominated by males that the best women players compete in separate tournaments (Byrne, 1975). As Nash (1970) has observed, not even in the Soviet Union, where chess is practically the national game, is there a woman chess player of grand master status. One wonders whether chess involves a strong spatial element, therefore making it more difficult for females. The grand master Samuel Reshevsky has implied that his chess skill is a manifestation of high spatial ability. Reshevsky had been a tremendous prodigy and when he was 8 years of age – before formal schooling – he did excellently on psychological tests involving visualization of form and memory for digits and outstandingly on tests requiring fitting and dissecting shapes (Reshevsky, 1948).

Such anecdotal evidence is strengthened by experimental analysis. In the 1940s, de Groot (1965) studied some of the finest players in the world in an attempt to discover how they differed from weaker players. He found one major difference: Masters could reconstruct a chess position almost without error after seeing it for only about 5 seconds; players below the Master level could not. But when the same pieces were placed randomly on the board, recall was equally poor for Masters and weaker players. Masters, therefore, have no better short-term memory than weaker players. Chase and Simon (1973) have confirmed de Groot's findings and propose that what the Master perceives during the brief exposure of a *coherent* position is

> familiar or meaningful constellations of pieces that are already structured for him in memory, so that all he has to do is store the label or internal name of each such structure in short-term memory. At recall, then, the Master simply uses the label to retrieve the structure from long-term memory. With a normal memory span of about 5 to 7 chunks (Miller, 1956), the Master must be perceiving about 4 or 5 pieces per chunk in order to recall about 25 pieces. [Chase & Simon, 1973, p. 217].

It is this ability to perceive familiar patterns of pieces quickly that Chase and Simon posited to be the basic ability underlying chess skill. Because some of the recognizable chess patterns will be relevant, others not, to the player's analysis, the player must construct "a more concrete internal representation of the relevant patterns in the mind's eye, and then modify these patterns to reflect the consequences of making the evoked move" (p. 269). Chase and Simon likened the information-processing operations required for such cognitive acts to mental rotation processes (e.g., Shepard & Metzler, 1971) or those processes believed to be involved in solving cube-painting and cube-cutting puzzles (Baylor, 1971).

Sense of direction

Perhaps the commonest expression of spatial ability is in orientation in space – the sense of direction. We all must learn to find our way about, first in our own surrounds, then in new places. We consult maps, ask directions, and occasionally get lost. To orient oneself in space requires memory of and then movement or transformation of a spatial layout, but now a critical part of the layout is one's own body. There is much anecdotal evidence that males have a better sense of direction than females, and there is a fair amount of corroborating evidence from psychological tests.

Visual mazes

The instrument most commonly used to test direction sense is the visual maze. In an early study with young adults (Book, 1932), the subjects had to trace the correct path through a variety of mazes with a pencil. Only 17.4% of the women's scores were above the average score of the men. As for children, the earliest reported sex difference may be Porteus's (1918) first normative study of his Porteus Maze Test, in which he compared the performance of 453 normal 7- to 14-year-old children on the maze test and the Binet Intelligence Test. Inasmuch as Porteus designed his test as an alternative for the heavily verbal Binet test, we can imagine his consternation to find that the girls scored significantly below the boys. The difference persisted in many subsequent investigations, and in 1965 (p. 115), Porteus cited repetitions of his first experiment on 210 separate groups, providing 105 separate comparisons of males and females numbering in excess of 10,000 subjects. In 99 comparisons, the males held the advantage.

The sex difference is not peculiar to a pencil tracing task. Among school-age children, boys also excel in slot mazes (McNemar, 1942) and, among 12-year-olds, foot mazes (Batalla, 1943). And again, as with other spatial tests, sex differences tend to be less reliable in younger children. For example, Mattson (1933) found that 5- and 6-year-old boys and girls were equally able to roll a ball through a maze with different numbers of blinds. McGinnis (1929) failed to find sex differences in a similar task with 3- to 5-year-olds.

It is noteworthy that sex differences in maze performance have been reported in another species. Research consistently shows that male rats are superior to females in maze-learning tasks (Hubbert, 1915; Sadownikova-Koltzova, 1926; Tryon, 1931; McNemar &

Stone, 1932; Tomlin and Stone, 1933; Cowley and Griesel, 1963; Barnes et al., 1966; Barrett & Ray, 1970; Dawson, 1972).

Tactual mazes
All the aforementioned tests are visual, so the subjects, whether rats or human beings, could see where they were going. Males also excel on tactual mazes where vision is prevented. Langhorne (1948) tested 68 college-age men and 34 college-age women on a maze built on a concentric principle, with progress to be made from the inside out. All the subjects learned the maze to criterion and then were tested on the maze in the same position or after it had been rotated 90°, 180°, or 270°. The men averaged fewer trials to criterion, less time, and fewer errors on all positions than the women. Furthermore, though both the men's and women's performance was hampered by the change in maze position, the women's performance was more severely disrupted.

Pattern walking
In the case of visual mazes, one clue to the sex difference, at least in human beings, is the possibility that males make better use of visual cues than females do. This possibility is suggested by a study of pattern walking (Keogh, 1971). The subjects – 75 boys and 60 girls, all 8 to 9 years of age – were compared on two tasks: making pencil copies of simple designs (e.g., circle, triangle) and designs consisting of combinations of the simple designs (e.g., two circles touching, a triangle within a circle), and later, making the same patterns by walking. The walking test was carried out under three conditions designed to represent increasing levels of available reference points. In one test, the children walked out the pattern on the unmarked floor of a large room; in another, on a 9-×9-ft plain mat; and in still another, in a 9-×9-ft sandbox that left visible footprints. The boys and girls were equally accurate in their drawings and their walked patterns on the unmarked floor. But only the boys improved in pattern walking across the two remaining conditions as more visual cues became available, so that the boys were significantly more accurate than the girls in the latter two conditions.

Map reading
Map reading is yet another likely measure of direction sense. One standardized instrument is the Road Map Test of Direction Sense

(Money et al., 1965), which consists of a schematic outline map of several city blocks with a standard route through the streets. The subject must imagine himself following the route and, without turning the map, must tell whether each turn on the route would be to his left or right. When this test was administered to over 1000 normal children between the ages of 7 and 18 years, both boys and girls showed roughly parallel improvements across age, but the boys were slightly better than the girls at ages 7 to 10 and 15 to 18, and substantially better at ages 11 to 14. The most difficult parts of the road map test are the "come-back-and-turn" discriminations, which require the subject to imagine himself turned around so that his left and right are reversed – in other words, mirror-image trials. Performance was poorest on these trials, especially for the girls.

Left-right discrimination
In a test of direction sense like the road map test, a critical factor may be the skill with which the subject has mastered the basic left-right discrimination. In this regard, the sex differences found are surprising, inasmuch as they have not appeared in other developmental studies (e.g., Harris, 1972, Long & Looft, 1972). These studies, however, required the naming of left and right body parts or following simple spatial directions like drawing a cross at the bottom left-hand corner of the paper. They did not place a premium on speed of discrimination as the road map test seems to; nor did they require so many sudden shifts in perspective. On left-right discrimination tasks that emphasize speed and require perspective shift, sex differences appear. Baken and Putnam (1974) showed university students slide projections of body parts in different orientations (Laterality Discrimination Test; Culver, 1969) and asked them to identify each as left or right. Each slide was shown for 3 seconds with 3 seconds for response. Men averaged fewer than five errors for 32 pictures; women averaged nearly eight.

Aiming and tracking
In basic motor skills, such as static balance (maintaining one's balance on an unstable surface) and dynamic balance (walking forward and backward along the length of a narrow beam), children improve with age, but generally there are no differences between boys and girls (Bayley, 1935; Goodenough & Smart, 1935; McCaskill & Wellman, 1938; Seils, 1951; Drowatzky & Zuccato, 1967).

Indeed, where sex differences are reported, girls have been found to excel (Hanson, 1965; DeOreo & Wade, 1971; cited by DeOreo & Wade, 1971, p. 334), perhaps because of maturational differences. But on motor tasks requiring the coordination of one's own movements to a target, boys do better.

A common example of this coordination is aiming and throwing an object. Among 5- to 7-year-olds, boys can throw farther and more accurately than girls (Jenkins, 1930). The difference appears in children as young as 3.5 years (Gesell et al., 1940). The boy's superior ability probably stems from his more efficient throwing style. Gesell et al. (1940) found that up to about 4 years of age, both boys and girls generally threw with an overhand movement in which the hand describes a vertical arc. Later the ball was directed from the side of the shoulder on a more or less vertical plane. This change in the course of the throwing hand appeared to be correlated with the act of shifting the weight from the right to the left foot and typically was evident in boys by 5 years of age, but in girls by 6 years of age, and then only in some girls.

There also are sex differences on the rotary pursuit task – a different kind of aiming task requiring coordination of the hand to a continuously moving target. On this task, it is unlikely that males and females differ in previous practice or that physical strength and posture play significant roles. Yet again males do better. Ammons et al. (1955) tested 350 9- to 18-year-olds and found that in every age group, the boys stayed on target longer, on the average, than the girls. Furthermore, the boys improved across the age range tested, but the girls improved only to age 15 and then got worse.

These sex differences have been confirmed in young adults (Buxton & Grant, 1939; Archer, 1958; Noble et al., 1958; Noble, 1969; Noble & Noble, 1972). Generally, in subjects without previous experience in this task, males initially are better than females, develop the skill faster, and reach a higher maximum performance.

Another device used to test the coordination of body movement to rapid changes in direction of a target is the Toronto Complex Coordinator. The subject faces a display panel placed at eye level in a vertical position at right angles to his line of sight. A red ring is made to appear on the display panel in any of 49 different positions (17 rows and 17 columns). Directly below the subject and vertical to the floor is an airplane-type control stick ("joy stick") that can be moved in left-right and forward-backward directions, as well as simultaneously in both axes. Moving it causes a green disk to move on the screen. The subject must match the disk with the ring (i.e., move the disk into the ring and keep it there for a moment). As soon as a match is made, a new ring appears on the screen so rapidly that the disk is located in the position of the last ring when the new ring appears. The subject then must move the disk to the new ring, and so on through several minutes of practice. The horizontal and vertical movements of the disk on the display board can be associated with various directions of movement of the control stick. On a standard task, left and right movements of the control stick move the disk to the left and right, and movements toward and away from the subject move the disk down and up. For a reverse task, the subject's movements move the disk in the reverse direction. Finally, on an out-of-phase task, horizontal movements of the stick move the disk vertically; vertical stick movements move it horizontally.

With this procedure, Shephard et al. (1962) tested 420 subjects (10 males and 10 females at 5, 10, 20, 30, 40, 50, and 70 years of age). As expected, the subjects made the most matches in the standard task and the least in the out-of-phase task. On other measures of performance, including number of vertical and horizontal movement errors, persistence of errors, and latency, in addition to number of matches, there was improvement to about age 40 and then a drop in performance; but at every age the males' mean scores were higher than the females'. The margin of difference was usually substantial, particularly on number of matches, where it nearly always was 25% or more. Finally, just as Langhorne's (1948) women subjects' performance on a tactual maze suffered more than men's by change in position of the maze pattern, the females on the Toronto Complex Coordinator did relatively worse than the males in the reverse task than in the standard task. At ages 5, 10, 20, 40, and 50 years, the difference between the males' and females' mean number of matches was greater in the reverse than in the standard task.

Rod-and-frame test

Another test that could be characterized – though unconventionally – as a measure of direction sense is the rod-and-frame test. The task is to adjust a luminescent rod to the vertical when the rod is inside a luminescent square frame, which itself is tilted, and when the subject may be tilted as well.

Compared with the embedded figures test, with which the rod-and-frame test is usually associated and administered, sex differences on the rod-and-frame test appear earlier and more reliably. In children between 5 and 10 years of age, superior performance by boys has been reported in several investigations (Witkin et al., 1967; Canavan, 1969; Graves & Koziol, 1971; Keogh & Ryan, 1971); beyond that age, male superiority is routine (Gross, 1959; Witkin et al., 1962, 1967; Kato, 1965; Fiebert, 1967; Schwartz & Karp, 1967; Okonji, 1969; Bogo et al., 1970; Morf et al., 1971; Saarni, 1973; Silverman et al., 1973).

Geographical knowledge
If females have poorer directional sense than males, one might expect them to do less well on more general tests of directional sense and geographical ability. At least two studies bear this out. Lord (1941) reported tests on 173 boys and 144 girls in fifth through eighth grade in rural Michigan. The children were to name the four cardinal and four intermediate orientations of an arrow placed on a desk before them, to indicate the direction in which known cities lay, to locate familiar places in town by telling on which side of the street each is found, to say which way streets ran, and to keep track of directions during an actual automobile trip. The boys were consistently better than the girls. For instance, 55% of the boys but only 38% of the girls correctly designated all eight directions of the arrow.

A more recent survey of nearly 2000 children, this time from both urban and rural parts of Michigan (Bettis, 1974), gave similar results. Each child answered a 49-item multiple-choice test including questions on map reading (e.g., interpretation of distances, traffic and population movement, direction of river flow); knowledge of place names on maps; knowledge of geographical facts (e.g., origin of swamps, nature of the land surface); and graph reading. Boys had higher average scores than girls on 42 questions, equal on 3, lower on 4.

It is hard to identify precisely the skills underlying good direction sense. In maze tracing, at least where rats are concerned, it has been suggested that the "maze-bright" rat develops spatial or directional hypotheses about the solution, and the "maze-dull" rat relies more on visual cues (Krechevsky, 1933). To put this another way (Rosenzweig et al., 1960), good performance requires ignoring dominant visual cues, for example, an entrance immediately adjacent to one's present position that ends in a cul-de-sac, in favor of those less obvious cues specifying location space.

Perhaps the ability to ignore an immediately adjacent "entrance" partly underlay the boys' greater accuracy of performance in Keogh's (1971) pattern-walking tests described earlier. Girls often tried to walk complex patterns in a single continuous line rather than in separate units. They also walked hesitantly, made rounded corners and imprecise angles, did not coordinate starting and stopping points, and left patterns incomplete. The boys, by contrast, made precise angles and corners, were more accurate in starting and stopping points, clearly indicated when a pattern was completed, and more often reduced complex patterns to their subunits, completing one part of the design and then pausing before starting the next part.

Keogh herself suggests that the pattern-walking test, insofar as it requires some abstraction and organization of parts from an embedding field, is another correlate of cognitive differentiation or field independence. Thus the boys' superiority reflects their greater field independence. However, only 2 of the 10 figures used were embedded. That the boys were superior only in walking and not the paper-and-pencil task, and realized more advantage from visual cues, suggests that the task contains a visuospatial component and requires the kind of direction sense measured on the road map test (Money et al., 1965).

However sex differences in sense of direction are to be explained, we must not overgeneralize our conclusions. At least in animal studies, male superiority in maze learning seems to be tied to the specific spatial components of the task and does not reflect a sex difference in learning tasks in general. This principle is nicely illustrated in Barrett and Ray's (1970) finding that the same male rats that were superior to females in maze learning were not superior in avoidance learning in a shuttlebox.

Piagetian tests

Finally, let us consider certain Piagetian tests. Many are like conventional spatial visualization tasks, so that sex differences might be expected. One such set was given to children in the first through fourth grades (Tuddenham, 1971). The child had to choose from several photographs the one showing how a model of a small farm would look from different vantage points (perspectives), choose from several pictures of flat patterns those that could be folded to produce simple three-dimensional forms (geometrical forms), and place a toy car painted a different color on each side at various places on a spiral track (tracks). The fourth-graders also had to construct block buildings from plans and front elevations (house-plans). Except for the tracks task, the boys' mean score was higher then the girls' for every test.

Such tests are not universally deemed spatial visualization tests, or at least the tests are not used to assess spatial ability per se. The perspectives task, for example, more typically is used to determine when a child's perceptions come to be decentered, that is, when he can understand that his own spatial perspective is not necessarily shared by others. Clearly, the sex differences on this task do not imply sex differences in decentration in this sense, but only in the skill by which the decentration is measured.

That spatial visualization is involved in these tasks is further suggested by the finding that in patients with neurological damage, performance is differentially affected by lesion location. Butters and Barton (1970) gave the perspectives task to patients with left or right parietal damage and patients with left frontal or temporal lesions. The patients with right parietal damage did significantly worse than the others.[1]

Logical conservation
Other Piagetian tests that seem to have spatial elements, though as spatial tasks they seem less demanding, are the various tests of conservation. In the test of conservation of number, chips of one color are lined up parallel to an equal number of chips of a different color. After the child confirms that there are an equal number of each color, the chips of one set are bunched together, and the others are separated. The child then must compare the number of each color. To test conservation of distance, the child is shown two car tracks, one forming a straight line, the other segmented at right angles. The segmented track, though much longer (if straightened out), represents the same distance from one point to another when laid on a board. The child must move a toy car the same distance on the straight track as the experimenter moves a second car on the segmented track. To test conservation of area, the child is given two identical green boards ("fields"), each with a toy cow

placed in the center. After the child confirms that both cows have the same amount of grass to eat, barns are added successively to each field, being bunched closely in one field and spread out over the entire board in the other. After each addition of barns, the child is asked whether the cows in the two fields have the same amount of grass to eat.

Such tasks are presumed to measure the development of logical, not spatial, skills, and so sex differences have rarely been assessed. Indeed, for the child who *has* the conservation principle of number, distance, or area, the physical appearance of the chips, tracks, or fields is irrelevant. Since the child knows that nothing has been added or subtracted unequally, he knows that the numbers, distances, and areas have not changed. But in the child for whom the conservation principle is still imperfect, the spatial aspects of the tasks become critical so that individual differences in spatial ability become important.

In most cases, where scores for boys and girls have been reported separately on such tasks, the difference is nonsignificant (Pratoomraj & Johnson, 1966; Brainerd, 1971; Harris & Allen, 1971; Gelman & Weinberg, 1972; Gruen & Vore, 1972); but where sex differences are found, the male is ahead. There are three such reports for the 6- to 9-year age range (Goldschmid, 1967; Hooper, 1969; Tuddenham, 1970), and one for adults for the more difficult test of conservation of volume (Graves, 1972). Goldschmid's (1967) results are specially impressive, since all the children – 38 boys and 41 girls, all in the first and second grades – were matched almost perfectly for age, IQ, and vocabulary. The boys scored higher than the girls on every one of 10 tasks, significantly for 2 individual tasks and total score.

Water level test

The Piagetian task on which sex differences appear most reliably is the water level test. In the classic demonstration (Piaget & Inhelder, 1956), the child is shown a bottle partly filled with water, asked to notice the position of the water in the bottle, and then to predict, either by pointing or by making a drawing, where the water will be when the bottle is tipped. According to Piaget and Inhelder (1956), the principle that the water level will remain horizontal is mastered by about 12 years of age. This appears to be so, but for boys more than for girls (Thomas, 1971; Thomas & Hummel, 1972; Liben, 1973). Among adults, women also lag (Rebelsky, 1964; Morris, 1971); indeed, it now has been estimated that about 50% of college women do not know the principle (Thomas, 1971; Thomas & Hummel, 1972; cited in Thomas et al., 1973).

In all earlier studies the subjects had to draw or otherwise construct the predicted waterline. The sex difference persists even with a presumably simpler multiple-choice recognition procedure. Harris et al. (1975) gave sixth-graders and college students a variety of questions each requiring the subjects to pick that one of four drawings which correctly represented the water level of a tilted container set on a table – in other words, to pick the *one* drawing that showed the water to be level with the table surface. The males in both age groups did substantially better than the females. The same test has been repeated with first- through sixth-graders (A. Anderson, unpublished data, Michigan State University, 1975). The boys were ahead at every age, though significantly so only by the fourth grade. The boys' scores improved over this period from 25% to 52% correct; the change for girls was from 22% to only 28%.

Many cognitive developmental psychologists are puzzled by these findings because they conceive of the principle that horizontality is invariant as strictly a milestone in logical, analytical thinking like the principles assessed in tests of abstract reasoning. On these other tasks, adults, for the most part, perform at their expected level of competence, and men are no better than women. Piaget and Inhelder themselves, however, were well aware of the spatial element, as their discussion of children's performance shows:

> Now although it is doubtful whether failure to predict horizontality . . . is by itself proof of inability to conceive of a coordinate system – since it could be due to lack of interest, inattention, and so on – the repeated difficulty in appreciating the material facts themselves carries an entirely different implication. It undoubtedly indicates an inability to evaluate the perceptual data in terms of the orientation of lines and planes, and thereby suggests a failure on the part of coordination as such. What indeed is a system of

coordinates but a series of comparisons between objects in different positions and orientations? [Piaget & Inhelder, 1956, p. 390].

Consistent with this view is the finding that for high school seniors, performance on a test of horizontality is correlated significantly with performance on the spatial subtest of the Differential Aptitude Test (L. S. Liben, unpublished data).

Auditory perception

If, underlying males' superior spatial ability, there are sex differences in hemispheric functioning, it is pertinent to ask whether sex differences in spatial ability are confined to visual and tactual perception or whether they also appear in other perceptual systems for which hemispheric asymmetries are known to exist. One such system is audition, so the question may be raised whether males also excel in certain auditory skills known to be subserved by the right hemisphere. Evidence bearing on this question has been reported by Knox and Kimura (1970). In one experiment, 80 right-handed children between 5 and 8 years of age were presented dichotic pairs of verbal stimuli (digits) and pairs of environmental sounds (e.g., dog barking–dishwashing; phone dialing–clock ticking; children playing–car starting). For the digits, there was no sex difference; boys and girls at every age showed roughly the same right-ear advantage, and accuracy (combined-ear) scores increased significantly with age. For the environmental sounds, both boys and girls at every age showed a left-ear advantage, with the ear differences, across all subjects, significant. However, the boys' combined left-ear plus right-ear scores also exceeded the girls' at every age, with the overall difference significant.[2]

In a second experiment dichotic pairs of these same environmental sounds and also pairs of animal sounds were presented to 120 other right-handed 5- to 8-year-olds. For the environmental sounds, the left-ear advantage was found for all groups except the 5-year-old girls, though this time, the boys' and girls' overall scores did not significantly differ. For the animal sounds, the left-ear advantage was found for all groups except 8-year-old boys and 7-year-old girls. But the boys' combined-ear scores exceeded the girls' at every age, and across age were significantly greater.

The animal sounds also were presented to 27 2.5- to 5-year-olds.

This time the sounds, each 4 sec in duration, were played monaurally on a tape recorder, and after each one, the child was asked to name or describe the animal he had heard. The boys identified significantly more sounds than the girls.

Inasmuch as these demonstrations give evidence of ear asymmetry effects in young children for both verbal and nonverbal sounds (though not consistently), they constitute a contribution to the developmental neuropsychology of audition. The sex difference in overall performance for the nonverbal sounds, however, raises the more interesting question of the basis of sex differences in other auditory-cognitive skills. What follows is frankly shameless speculation, but it nonetheless will put the question of audition, spatial skill, and sex differences into an interesting theoretical perspective.

Music composition
One of the most remarkable things about music composition is that it is a field almost totally dominated by men, both historically and in contemporary times, and in all compositional modes from classical to jazz to the pop, film, and theater music of the day. By contrast, in music *performance* women have left a mark of accomplishment, from the middle half of the eighteenth century to the current day, nearly equal to men's. (The exception today is jazz, and it may be critical that this is a form of music combining technical-executionary and creative modes in equal measure.) To make the point stronger, it can be noted that until the twentieth century many outstanding male composers also were outstanding performing artists, many with independent careers as performers; examples are Bach, Mozart, Beethoven, Mendelssohn, Chopin, Liszt, and Bizet. A comparable list of women performers would contain no names of successful or especially talented composers.

Why the discrepancy, in the case of women, between composition and performance? It is possible that music composition involves cognitive skills subserved predominantly by the right cerebral hemisphere and, therefore, like visuospatial skills, stronger in males than females.

There is no direct evidence of right hemisphere specialization for *compositional* skill, though there is evidence of right hemisphere specialization for certain elements of music perception probably critical for composition. One source of evidence is the clinical neuropsychological literature. For example, performance on the Timbre and Tonal Memory subtests of the Seashore test is depressed by right but not by left temporal lobectomy (Milner, 1962). The same asymmetry exists for orchestrated melodies with a dichotic listening procedure. Patients with right temporal lobectomies are significantly worse than those with left lobectomies in recognizing melodies presented dichotically, but are better in recognition of verbal material (Shankweiler, 1966). The same asymmetries appear in normal persons. Kimura (1964) presented dichotic pairs of spoken digits and melodies to right-handed young women. The melodies were solo passages from concertos in the baroque style. For melodies, 16 of 20 subjects had higher left-ear recognition scores; for digits, 15 of the same 20 subjects had higher right-ear scores.

Similar hemisphere differences are evidenced after barbiturization of the cerebral hemispheres through the injection of sodium amytal into the carotid artery.[3] In one demonstration (Bogen & Gordon, 1971), six right-handed patients who had not had brain surgery were asked to sing familiar songs, such as "Happy Birthday" and "London Bridge," before, during, and after the injection, but to substitute "la la la" for the words. With right hemisphere barbiturization, singing was grossly disturbed in all cases, the most typical effect being monotonicity with a few unnatural pitch changes. Speech, however, remained generally unaffected in intelligibility and rhythmicity, save for some slurring of words and the presence of monotonicity.

The same asymmetries are revealed when brain activity is directly monitored. Schwartz et al. (1973) recorded alpha activity from the left and right occipital regions in normal right-handed young adults while they whistled songs, recited song lyrics, and sang – the three tasks being performed successively in 1-minute segments in counterbalanced order. Whistling produced relative right hemisphere dominance, talking produced relative left hemisphere dominance, and singing (the task presumably combining both linguistic and musical modes) produced comparable alpha blocking in both hemispheres.

Of course, singing in tune, and timbre and tonal memory, are not the same as composing. The only neurological evidence with respect to composition is indirect: two case reports suggesting that musical ideas, like spatial abilities, are *not* lost after brain injury that is confined to the left side. The composer in the first case was Maurice Ravel. In 1932, in his fifty-seventh year and at the summit of his creative powers, Ravel struck his head in an automobile accident, became aphasic, and died 5 years later, 10 days after an operation from which he never regained consciousness. According to Alajounine (1948), who examined Ravel, the aphasia was of the Wernicke type, "of moderate intensity, without any trace of paralysis, without hemianopia, but with an ideomotor apractic component" (p. 232). It apparently left Ravel's memory, judgment, aesthetic taste, and musical thinking relatively unimpaired, for Ravel immediately recognized tunes he had known before and perfectly recognized his own work. The recognition was not vague.

> He immediately notices the lightest mistake in the playing: several parts of the "Tombeau de Couperin" were first correctly played, and then with minor errors (either as to notes or rhythm). He immediately protested and demanded a perfect accuracy. . . . my piano [Alajounine's] had become somewhat out of tune. The patient noticed it and demonstrated the dissonance by playing two notes one octave apart, thus showing again the preservation of sound recognition and valuation. [Alajounine, 1948, pp. 232-233]

Note reading and writing, however, were extremely faulty, and therein lay the full tragedy. Ravel could no longer write out his musical ideas.[4]

The more recent case is of the Russian composer, V. G. Shebalin (Luria et al., 1965). In 1953, at age 51, Shebalin sustained a vascular lesion of the left hemisphere, including the temporal lobe. He suffered from a severe and predominantly sensory asphasia, but – luckier than Ravel – "preserved his musical abilities . . . and executed a number of outstanding compositions" over the next several years (Luria et al., 1965, p. 288). Luria presents this remarkable case as evidence that "phonematic and musical (prosodic) organization of acoustic perception and memory are included in different systems and have as their basis different cortical structures" (p. 292).

Spatial elements in music

If music perception and composition, like visuospatial perception, is predominantly a right hemisphere skill, how is music "spatial"? How are the respective specializations of the left and right cerebral hemispheres for linguistic and visuospatial skills matched, in audition, for linguistic and *auditory*-spatial skills? A close parallel to such clearly spatial *visual* tasks as dot localization or depth discrimination might be a demonstration of superior auditory localization (i.e., location of a sound in space) by the right hemisphere. Such a demonstration does not seem to have been made, but the question remains open. The demonstrated parallels, instead, are only approximate. For instance, there is the sense in which certain kinds of sounds naturally seem to belong in certain parts of space relative to other sounds. Why are high notes and low notes *called* high notes and low notes? Why does the flute seem to float above the bassoon, and the violin above the 'cello? Pitch and resonance seem to impart "spatial" qualities to sound in more than a merely metaphorical sense. The possibilities are well expressed in the following passage:

> Pitches often do not appear on a single plane in our mind's image of the piece, as it grows and as we comprehend it. On the contrary, one sound or group of sounds tends to stand out, to demand more of our attention. Other sounds recede, become obscured by still other sounds. The concept "masking" ... is good here ... "Depth" is almost as useful. Indeed it is convenient to describe certain passages of music as separating into a foreground, background, and places in between. [McDermott, 1972, p. 492] [5]

Consider, too, the formal structure of music. Music has been likened to woven fabric, with the "warp" the simultaneous sounds forming chords or sonorities like the threads in cloth that run lengthwise (vertically), and the "woof" the threads that run crosswise (horizontally) like the successive sounds forming melodies. This vertical-horizontal spatial structure that is musical "texture" figures critically in the musical form called *canon*, or counterpoint (from *punctus contra punctum*, or "note against note," or, by extension, "melody against melody"). A canon, then, is a contrapuntal device in which an extended melody, stated in one part, is imitated usually shortly afterward in one or more other parts, so that the normal contrapuntal texture of horizontal (melodic) and vertical (harmonic) relationships is strengthened by diagonal "threads" that connect the places of imitation. The canon can be inverted, in which case the principle of mirror reflection is applied, either with the mirror placed underneath the music or, less commonly, at the end (retrograde form).

Sex differences in spatial ability

There are many other examples of spatial structure in music, including cross-relations (appearance in different voices of two tones of mutually contradictory character that normally are placed as a melodic progression in one voice); cross-rhythms (simultaneous use of conflicting rhythmic patterns); bitonality, and polytonality.

Perhaps the ability to recognize and to execute and, above all, to create, a melodic pattern is a spatial ability not unlike the visual detection of an embedded figure or the mental rotation of a geometrical form so as to anticipate how it will look from a different spatial perspective. The recognition of counterpoint or of variations on a theme may depend on similar abilities to disembed a figure from a complex background, to remember it, and to follow it through a variety of transformations.

Here are many problems ripe for study by experimental and developmental psychologists. Tests to detect the various possible spatial components in musical structure could be devised and administered to normal children and adults. If these auditory elements are "spatial," performance should correlate significantly on analogous visuospatial tests, and the usual sex differences would be expected to appear. Long-term correlates with musical skill, particularly compositional skill, could be determined as well.[6]

ROLE OF EXPERIENCE

Socialization theory

Why do males have better spatial skill than females? An influential view about psychological sex differences in general is that they are based on different life experiences for males and females that are culturally prescribed; in particular, that the culture expects male activity and female passivity and selectively reinforces for conformity to sex-role stereotypes (Kagan, 1964). In the case of spatial ability, it therefore has been supposed that society encourages boys, more than girls, to engage in activities that sharpen spatial skills. Presumably, these activities include the exploration and manipulation of objects. According to Piaget, the mental image of a spatial form is originally the interiorization of the movements of exploration with respect to the form. Consequently, children with greater experience in manipulation and visual-tactile exploration would tend to have, and Inhelder (Tanner & Inhelder, 1958) reported do have, better spatial representation skill.

The socialization argument has taken a number of different forms and has appealed to several different kinds of evidence.

Age of first appearance of sex differences in spatial skill
We have reviewed studies in which, among children as young as 4 to 6 years, boys have shown superior spatial ability, but generally, sex differences on *standard* spatial tests are not clearly expressed until the middle and sometimes even the later childhood years. In the case of the embedded figures test, it is likely that the standard task is too difficult for young children, creating a floor on performance. However, in simplified versions of the test (e.g., Children's Embedded Figures Test, Karp & Konstadt, 1963), which use meaningful complex figures and which ensure that the complex figure is initially perceived as a whole, sex differences still have failed to appear reliably in children under 10 years of age. Several such studies were cited earlier (e.g., Goodenough & Eagle, 1963; Karp & Konstadt, 1963; Corah, 1965). The inference to be drawn from such reports, according to the environmentalists, is that several years are required before the *culturally* prescribed different learning experiences for boys and girls become expressed in actual skill differences.

Effects of training
If sex differences in spatial skill reflect different experiences prescribed by society, then one also would expect male superiority to be more pronounced and consistent on certain spatial tests such as mechanical comprehension (e.g., naming the direction of movement of interlocking gears) than others such as the Space Relations subtest of the Differential Aptitude Test. There is such a tendency (Anastasi & Foley, 1953). Presumably, the test of mechanical comprehension is sex stereotyped and so tests what is a more practiced skill for males, and the more abstract Spatial Relations Subtest is more nearly equally unfamiliar to males and females. The implication is that spatial skill is trainable and that girls have lacked this training but could be brought up to the male level if allowed to share in the appropriate experiences.

There is some evidence that spatial skill *can* be improved with training. For example, Blade and Watson (1955) demonstrated a significant improvement in performance on spatial tasks in students after 1 year of engineering studies – and no increase in other students or in students refused admission to engineering school and not doing any other advanced studies during this time.

Presumably, certain engineering courses are more facilitatory than others. Churchill et al. (1942) demonstrated significantly greater improvement on the Surface Development Test by engineering students after 9 weeks' training in a drafting class than by engineering students taking a course in water purification. The Surface Development Test, developed by Thurstone (1938), involves matching similar parts for drawings shown in two dimensions and in three-dimensional perspective. Similar skills are involved in drafting and blueprint reading. The water purification course taught principles of electricity and mechanics as well as water purification and, presumably, was little related to spatial visualization.

The engineering students in these studies were young men, perhaps already self-selected for some minimal level of spatial skill. The question is whether children, and girls in particular, could also benefit from such training. Apparently they can. Brinkmann (1966) gave eighth-grade children (14 boys and 13 girls) programmed training designed to teach the visualization of spatial relations. The program actually was a short course in elementary geometry and included such topics as basic elements of point, set, line, angle, simple plane figures, and simple solids. The program stressed problem solving, employing behaviors such as pattern folding and manipulation of solid objects. Training took place over a 3-week period during class periods normally devoted to mathematics. A control group of 27 children matched in age and sex received their regular mathematics exercises during these periods. All the children took the Space Relations Subtest of the Differential Aptitude Test before and after the training period. On the pretest, the groups did not differ. But on the posttest, the experimental subjects improved substantially; the control subjects, only slightly and insignificantly. As for sex differences, in the experimental group, boys and girls did not differ on the posttest, leading Brinkman to conclude that "girls can at least hold their own when provided with the opportunity to learn something about a particular area in which they are often assumed to possess less ability" (p. 184). Unfortunately, separate pretest scores for boys and girls are not reported, no information is given about sex differences in the control subjects, and follow-up data are lacking. Finally, Rovet

(1975) has reported significant improvement by third-grade children on an object-rotation task like Shepard and Metzler's (1971) after the children either watched an animated film showing one figure being rotated into congruence with another or after they were trained to make the same rotations with real blocks. Sex differences, however, were not mentioned.

Cross-cultural studies

It has been argued that effective training of spatial skill inheres in the day-to-day experiences provided by the physical environment. A study comparing Eastern Canadian Eskimos from Baffin Island with people of the Temne tribe of Sierra Leone (Africa) is frequently cited in support of this view (Berry, 1966). The environments and experiences of these two groups differ markedly. The Temne land offers a far greater variety of visual stimulation than the Eskimo environment. But the Temne are farmers, who rarely need to leave their villages; the Eskimos are hunters, who must travel widely on sea, land, and along the coasts. Unlike the Temne, the Eskimo, merely to survive, must be able to isolate slight variations in visual stimulation from a relatively featureless array and to organize these details into a spatial awareness of his present location in relation to objects around him. Finally, Eskimos are skilled in arts and crafts and map making, skills quite lacking in Temne culture.

Berry (1966) has asserted that these cultural and environmental differences are reflected in performance on various perceptual and spatial tasks (Kohs Blocks, Embedded Figures Test, Morrisby Shapes) as well as on a test of the ability to detect small gaps in tachistoscopically projected simple geometrical forms. The Eskimos were superior to the Temne on every measure.

Cross-cultural research also is cited as exemplifying the effect of sex sterotypy on spatial skill. Berry (1966) found that the Temne men did significantly better than the women on most measures. But there were no significant differences between the Eskimo men's and women's scores. Sex differences likewise were absent in a study of 9- to 15-year-old Western Eskimo children (98 boys and 69 girls). The point-biserial correlations between sex and performance on an embedded figures test were near zero (MacArthur, 1967). Berry (1966) observed that, unlike Temne women, Eskimo women are not treated as dependent in the society; very loose controls are exercised over women as well as children. Eskimo women thus seem to share in the experiences of men to a greater degree than is the case for Temne women, and since the exigencies of Eskimo culture and environment require spatial discrimination of high order, both Eskimo males and females may be expected to show the trait to a larger absolute degree than the Temne, and to show less sex differentiation as well.

One of the social controls believed to be applied more strictly to females than males, particularly during childhood, is freedom to wander from the home. As the high degree of spatial skill of the wide-traveling Eskimo suggests, wandering and traveling from the home may facilitate spatial skill. Recent research has been interpreted as supporting this possibility. Among the children of the Logoli tribe in Kenya, Munroe and Munroe (1968) observed that girls generally are expected to stay nearby to help adults in their work, such as the care of other children. Because of this, the girls are less free to wander about exploring the countryside. By spot check, the Munroes found that in free time the boys were farther from home than were girls of the same age. In a test of construction of the diagonal – a spatial-conceptual test developed by Olson (1970) – boys were markedly better. On another spatial task – copying of block designs – children who had been more distant from home during the spot checks proved to be significantly more proficient than children who had been less distant. Indeed, only two girls had been farther from home on the average than the age-matched boys, and these two were among the three girls who were better at block building than their male counterparts.

Similar differences have been reported in the life histories of American children who excelled or did poorly on the embedded figures and rod-and-frame tests (Witkin et al., 1962). The children who excelled (more frequently boys than girls) ranged farther from their homes, were less restricted in play areas and activities, were less watched over by their parents, and generally seemed more independent. Witkin et al. (1962) also gave the same tests to the children's mothers and found a modest but significant relation between the mothers' performance and their sons'. Witkin et al. (1962) concluded that the children's "level of psychological differentiation" is enhanced if their parents encourage autonomy in their everyday lives and that mothers foster autonomy in their (male) children according, at least to some degree, to their own level of autonomy.

Evaluation of socialization theory

Societies unquestionably prescribe certain roles and activities for each sex and proscribe others. But whatever other psychological sex differences this socialization experience helps to create, the evidence that it plays the *critical* role in spatial ability is not convincing.

Age of first appearance of sex differences in spatial skills
For example, consider the environmentalist interpretation of the delayed appearance of sex difference in certain spatial tasks. If this delay merely reflects the time needed for culturally prescribed experiential factors to express their effect, then sex differences ought to be *enhanced* with age, since with age more and more sex-typed experiences are amassed. Some evidence bears this out. For example, Wilson (1975) compared 4-, 5-, and 6-year-old boy-girl twins on a maze test and found the boys ahead at each age, but by an increasingly larger degree. But the age range in this study was very narrow; when comparisons are made across a larger age span, the more common finding is that sex differences remain at approximately the same order of magnitude (Book, 1932; Shephard et al., 1962). The study by Shephard et al. (1962) with the Toronto Complex Coordinator is all the more impressive in this regard because of the range of ages tested – from 5 to 70 years. On all three tasks – standard, reverse, and out-of-phase – the sex difference in scores was essentially unchanged across age for the various measures of performance. Furthermore, if prior learning of directional relations accounted for the sex differences, the average percentage of errors would be expected to decrease with age for the standard task and to increase for the reverse and out-of-phase tasks. But all the curves were U-shaped, with between-task differences generally consistent.

It is conceivable, of course, that age-related changes depend on the degree of differentiation of male and female roles in any particular group or culture. Porteus (1965), for instance, reported that male superiority on mazes is enhanced with age only in "primitive" peoples, and not in groups of "higher cultural stand-

Sex differences in spatial ability

ing," and suggested that "the masculine and feminine roles in many features of daily living are more differentiated as one goes down the cultural scale" (p. 122).[7]

Effects of training
As for the trainability of spatial skill, the failures as well as successes must be noted. Faubion et al. (1942) found no differences on the Surface Development Test between young air corps recruits who had just finished a course in drafting and blueprint reading and a comparable group who had not yet had technical training. Myers (1958) found that U.S. Naval Cadets with past training in mechanical drawing did no better on spatial tests than those without training. Ranucci (1952) and Brown (1954) reported that courses in high school solid geometry did not result in improved performance on spatial relations tests. Mitchelmore (1974) found that college students, who over a 4-week period designed, constructed, and sketched models of elementary three-dimensional shapes, afterward showed no improvement on various spatial tests, including one like the Surface Development Test.

Much may depend, of course, on the intensity and quality of instruction. The engineering students described by Churchill et al. (1942), who improved on the Surface Development Test, had much more training than the air corps recruits tested by Faubion et al. (1942) (400 hours versus 40 hours) and probably more than the college students in Mitchelmore's (1974) study. As for geometry, Brinkmann (1966) pointed out that the usual, unsuccessful approach (e.g., Ranucci, 1952; Brown, 1954) has emphasized the development of formal proofs based on geometrical givens and the manipulation of abstract concepts through logical reasoning, and not the actual cognitive operations, such as pattern folding or block manipulation, more likely to be involved in geometrical tasks and used with apparent success by Rovet (1975) in the study with third-graders mentioned earlier.

The same criticism, however, does not seem to apply to other spatial tasks on which sex differences appear. For instance, college women who had failed a standard representation-of-horizontality task were permitted repeatedly to observe a bottle half-filled with red water and tilted at various angles, and then were asked to adjust the "water" level in a second bottle by moving a rotatable disk, half red, half white, representing the water level. Two training methods were used: In one, the subject first made a "predictive adjustment" with the model covered; then the cover was removed from the model, revealing the real waterline, and the subject could readjust the pretend waterline to match the real waterline, if she perceived her adjustment to be in error; finally, the cover on the model was replaced. This multipart procedure, constituting a single trial, was repeated until the subject made 10 successive predictive adjustments within 4° of the horizontal, or until 48 trials had been completed. For the second method, the model was not covered during training, and the subject had to adjust the pretend waterline to match the visible real waterline. These procedures, according to the investigators, were designed to "optimize self-discovery of the concept" (Thomas et al., 1973, p. 173).

With the first training method, only 7 of 30 women reached criterion. With the second method, the subjects improved only slightly and none acquired the principle, judging from their answers to interview questions. In contrast to sophisticated subjects, who invariably state, for example, that "water is always level," these women were more likely to say, "water is level when the bottle is upright, but is inclined when the bottle is tilted" (p. 174).

Why should young college women be less capable than men of seeing the true horizontal level of water in a glass? If the requisite perceptual experience is looking at and noting that liquid levels are always horizontal to gravity (ground) and not to the containers in which the liquids are held, can we reasonably assume such experiences to be more frequent in the lives of males than females? But even if this were so – even if it were found, for instance, that when children are given transparent containers filled with liquid, boys more often than girls spontaneously tilt and inspect the liquid level – girls' less frequent experiences nevertheless should be sufficient for acquisition of so basic a concept. And even if a lack of such experiences underlies the sex differences in adults in this particular task, we nevertheless would expect the concentrated provision of compensatory experiences to bring females up to the male level. The findings of Thomas et al. suggest that such experiences are insufficient.[8]

Cross-cultural studies
Much has been made of the demonstrations of cultural differences in overall spatial skill and the absence of sex differences in some cultures. These demonstrations are dramatic but misleading, for

the absence of sex differences in certain cultures is much more the exception than the rule. The Porteus Maze Test has been given to people from many diverse cultures, including Australids of Central Australia, Kalahari Bushmen of South Africa, the Bhil of Indore State, Central India, the people of Alor, and the Chamorros and Carolinians of Micronesia, as well as Americans and Europeans; in all cases, the male advantage appears (see Porteus, 1965, Chap. 6). Likewise, on the embedded figures and rod-and-frame tests, better performance by males, both children and adults, has been reported for many cultural groups besides the Temne and Americans, including French (Andrieux, 1955; Chateau, 1959), English (Newbigging, 1954; Bennett, 1956), Italian (Korchin, 1962), Dutch (Wit, 1955), and Hong Kong Chinese (Goodnow, 1962).

Of course, it can be argued that this consistency of results merely reflects the consistency of treatment of males and females across national groups and cultures. Porteus (1965) himself acknowledged this possibility. For instance, he remarked, with respect to the Kalahari Bushmen, that on one occasion, men who had been given the maze test asked him for test blanks to take home to their village, "not so that they could apply them to their female relatives but 'to show our women how clever we are.' In other words, the mazes constituted a man's game in which his prowess was to be admired but not copied" (p. 121). Similarly, among the Australids, "a close distinction is drawn between men's and women's 'business,' and any invasion of the other's province is not usually attempted" (p. 121).

The socialization account of sex differences, in its strong form, nonetheless fails basically when comparisons are confined to males, since the differences in male average scores in most cases are *greater* than the sex differences found. For example, on the maze test, Australid males did much better than Bushmen males, but Porteus (1965, pp. 122–123) could find nothing about their environments that reasonably could have accounted for the difference in the same way that Berry (1966) presumed the Eskimo environment to hold advantage over the Temne environment in the promotion of spatial skill.

Parent-child similarities in spatial ability
Some aspects of the family constellations of scores on spatial ability tests also vitiate the social learning hypothesis. Recall that Witkin et al. (1962) found that mothers' scores on rod-and-

frame and embedded figures tests correlated significantly with their sons' scores, and on this basis proposed that mothers fostered autonomy (field independence) in their sons according to their own level of autonomy. Corah (1965) extended the analysis to girls and their fathers. The subjects were 30 boys and 30 girls between 8 and 11 years of age and their biological mothers and fathers (60 families total). The children took the Children's Embedded Figures Test (Karp & Konstadt, 1963), their parents the usual adult test (Witkin, 1950). All subjects also drew a human figure, which was scored on a "sophistication-of-body-concept" scale developed by Witkin et al. (1962) as another measure of field independence or cognitive differentiation. The correlations on the embedded figures test, corrected for age and IQ as measured on a verbal intelligence test (Ammons & Ammons, 1948), were as follows: father-son, 0.18; father-daughter, 0.28; mother-son, 0.31; mother-daughter, 0.02. None of the correlations was statistically significant. For the figure drawing, the only significant ($p < 0.05$) correlation was between mother and son ($r = 0.39$), with all remaining parent-child correlations less than 0.15. Finally, for the combined scores, the only significant correlations were for father-daughter (0.41) and mother-son (0.39). The remaining correlations were 0.25 (father-son) or lower (mother-daughter, zero).

Corah concluded that if level of parental differentiation influences the development of differentiation in the child, the pattern of influence is more complex than the findings of Witkin et al. (1962) imply. Corah suggested that it may be more the different-sex parent who fosters gender identity in the child. This proposal has a certain plausibility and agrees with a report that women who had identified with their fathers did better on rod-and-frame and embedded figures tests than women who had identified with their mothers (Bieri, 1960). On the other hand, adolescent boys whose fathers were absent did worse on these same tasks than boys whose fathers were present in the home (Barclay & Cusumano, 1967). If the mother's behavior vis-à-vis her son is, in fact, more critical than the father's, one would not have expected this outcome.

One can see the complexities of interpretation these studies invite. Presently we shall see that these family constellations of correlations may reflect a process more basic than sex-role identification and the pattern of parent-child social relationships.

Sex differences in play and exploration

Recall, from the study of Witkin et al. (1962), that the children – boys and girls alike – with high scores on the spatial tests generally seemed more independent (ranged farther from home, were less watched over, and so forth) than the children with lower scores. Witkin et al. supposed, as had the Munroes (1968) for their Kenyan subjects, that this was but another example of the negative effect of sex-stereotyped treatment on girls' spatial ability. But these authors assume a direction of causation for what is, of course, only a relationship of correlation. From their findings we may conclude, just as reasonably, that any particular child's natural tendency to behave in certain ways calls out certain behaviors in his parents. A child predisposed to be active and to explore, and who seems more competent than another child, may be harder to confine to quiet games and activities near the home. If differences in the general scope of experience in the environment *are* the critical factor in spatial ability in human beings, these differences could be organized along the biological as well as the sociological dimension of sex. Perhaps boys more than girls naturally involve themselves in experiences critical for the development of spatial ability.

There is, indeed, some evidence for this possibility. A variety of studies indicates sex differences in play and exploration in young children. For example in children as young as 1 year of age, boys maintain less proximity to their mothers than girls do (Goldberg & Lewis, 1969; Messer & Lewis, 1972; Ley & Koepke, 1975). Preschool-age boys also are more restless and physically active (Zazzo & Jullien, 1954; Pederson & Bell, 1970) and cover more ground or are more likely to seek large areas for exploration and play (Otterstaedt, 1962, cited by Garai & Scheinfeld, 1968; Garai, 1970). One of the more impressive studies, by Harper and Sanders (1975), is a report of time-sampled observations of the free play of 65 children between 3 and 5 years of age in a 0.5 acre nursery school complex. At both ages, the boys used more space and consistently spent more time playing outdoors than the girls did.

The quality and object as well as location of play are different as well for boys and girls. For instance, boys more often choose to play with blocks (Honzik, 1951; Farrell, 1957) and, it has been noted for 12-year-olds, more typically build tall structures, while girls prefer to build horizontal structures (Erikson, 1951). Boys also more often pick up and throw things – balls, stones, sticks, snow. Young men are the same, as any count of snowballs thrown or stones skipped on water will reveal. Because boys throw things more often than girls do, and throw in a physically different way at least as early as 4 years of age (Gesell et al., 1940), it is not surprising that boys also throw more accurately. But the male's greater accuracy on the rotary pursuit task, reviewed earlier (Ammons et al., 1955; Noble & Noble, 1972), is harder to account for in terms of practice, strength, or posture.

There is evidence, too, of sex differences in the exploration and manipulation of novel objects. An example is a study by Hutt (1970a, b) of 3- and 4-year-olds' reactions to a novel toy – a metal box on four legs. A lever mounted atop the box could be moved in different directions, causing a bell to ring, a buzzer to sound, or lights to go on. The lights and sounds could be made to work together or independently. Each child was allowed five 10-minute periods with this toy in a room that contained other commercial toys. Significantly more girls than boys failed to approach and investigate the box. These children acted inhibited; they played with the other toys, but did so repetitively and stereotypically. After two or three sessions, however, their responses to the box became more playful, as though the child, who initially was asking, "What does this *object* do?" now asked, "What can *I* do with this object?" That is, the emphasis shifted, as Hutt put it, from inquiry to invention, the child more and more using his imagination. Here another sex difference appeared. Boys used the toy in unconventional ways about four times more frequently than the girls did. None of these sex differences were systematically related to IQ or socioeconomic background.

In general, then, the games and play preferred by preschool boys emphasize exploration of objects, movement, strength, body contact, use of height and downfall, use of the channelization and arrest of motion, and greater range of movement and use of open spaces. Girls' games, on the other hand, are characterized more by choral activity, sociability, verbal behavior, and stasis (Sutton-Smith & Savasta, 1972).

The sex differences found in these studies are difficult to explain simply in terms of different treatment of males and females. For instance, in Harper and Sander's (1975) study of preschool children's use of space, the authors noted that the nursery school staff was composed of women of liberal, egalitarian views, and there was no reason to suspect that the children were pressured to conform to traditional roles. Indeed, girls were enticed to go outdoors. Neither was there any evidence of overt parental pressure on the children to conform to sex-stereotyped behavior nor any indication of more encouragement of outdoor activity for boys than girls. Finally, choice of location was unrelated to play equipment available. For instance, containers such as pots and bowls were used by the boys 73% of the time outdoors compared with 27% for the girls.

One could object to Hutt's (1970a, b) study of 3- and 4-year-olds' reactions to a novel toy on the grounds that the toy was not really neutral but was more masculine than feminine, the implication being that boys would be more motivated in a sex-appropriate task. Of course, motivation and interest can influence performance on any test. But as Hutt noted, it is too easy to label a toy or game as masculine *after* it has been shown to be of greater interest to boys, though this is not obvious beforehand. Hutt also said, of her own study, that of children who did explore, boys and girls approached the toy just as readily and cheerfully, and in the first 10-minute session girls even manipulated the toy slightly more often than boys, on the average.

The evidence, therefore, favors an explanation in terms of endogenous determination of gender-dimorphic behaviors certainly as much as one that emphasizes the prescriptions of the culture. If exploration, movement, and object manipulation are critical in the development of spatial skills, the male's superior spatial ability seems to be at least partly a natural outgrowth of his greater interest in, and aptitude for, such experiences.

GENETIC FACTORS IN SPATIAL ABILITY

In considering environmental explanations of sex differences in spatial ability, we have alluded to the possibility that either or both environmental and neural factors may be genetically (or phyletically) programmed. This possibility deserves further consideration. After all, we no longer can doubt that a substantial component of individual differences in performance on multifactor intelligence tests can be laid to heritable factors, and spatial ability, on evidence from a number of independent studies, seems to be one of the *most* heritable of individual differences. The question is twofold. Is there a genetic basis for sex differences in spatial ability? If so, how can this evidence be linked to the evidence for functional differences between the cerebral hemispheres so as to explain male superiority in spatial ability?

Let us begin by reviewing some of the evidence for a heritability component in spatial skill. A standard technique in behavior-genetic analysis of human intellectual ability is the comparison of twins as a function of zygosity (e.g., Newman et al., 1937). To estimate the heritability of any intellectual factor, the within-pair variance of scores of monozygous twins is compared with that of dizygous twins. The method is based on the assumption that the difference in scores between monozygous twins is the result of environmental effects only, and the difference in scores between dizygous twins is the result of environmental effects in addition to hereditary differences. Vandenberg et al. (1968) made these comparisons with 50 pairs of monozygotic and 25 pairs of dizygotic twin boys and girls on the seven subtests of the Differential Aptitude Test. The comparison of score variances for the spatial subtest for the boys yielded a highly significant heritability ratio – indeed the largest ratio obtained of all seven subtests. But the corresponding ratio for the girls was not quite significant ($p > 0.05$). According to Bock and Vandenberg (1968), this sex difference in the heritability estimate was undoubtedly rooted in the poorer performance of females on tests requiring the visualization of an object in three-dimensional space. (The girls did significantly worse than the boys on the spatial subtest.) Thus on spatial tests, because a large number of girls do not express the trait in sufficient strength to reveal any substantial degree of individual differences, the evidence for heritability is only borderline. By contrast, the trait is expressed fully in boys and shows a more nearly pure heritable variation – purer, indeed, than on any other subtest of the Differential Aptitude Test. Furthermore, for the boys, the heritable variance component was not only large relative to the environmental and error components, but the between-family component was small. The implication is that the spatial visualization test is not particularly sensitive to social class, educational, and other differences usually found between families.

If spatial ability is heritable, then it has been suggested (Goldberg & Meredith, 1975) that it also would be expected to remain relatively constant over age. That is, not only should the shape of the distribution of individual spatial ability scores remain stable over time, but each person's position within the distribution should remain stable as well, as is true for physical characteristics such as stature. This expectation has been confirmed. We earlier reviewed a study by Tuddenham (1970, 1971) of sex differences in several Piagetian tests (e.g., water level, perspectives, house plans) given to children in the early elementary grades. Several years later, when these children were in high school, 76 were retested by Goldberg and Meredith (1975) on four standard spatial tasks, including the cards test from the Primary Mental Abilities Test (Thurstone & Thurstone, 1965), a mental rotations test (Vandenberg's adaptation of Shepard and Metzler's [1971] figures), and a paper formboard and paper-folding test (French et al., 1963).

Two of the Piagetian tasks (water level and house plans), given when the children were younger, correlated highly with all the spatial tests given in high school, and as was true when the children were younger, the boys again did significantly better than the girls. Moreover, correlations of each subject's scores on the Piagetian tests with the later spatial scores disclosed a significant degree of stability over time of individual children's positions in the distribution of group scores.

Genetic basis for sex differences in spatial ability: recessive trait model

Although these findings support the view that spatial ability, like certain physical characteristics, is heritable, the data do not explain why the trait is expressed more strongly in males than in females. At least as early as 1943, O'Connor suggested an answer – that spatial ability is a sex-linked recessive characteristic; that at least one of the genes controlling visuospatial ability is a *recessive* gene carried on the X chromosome and therefore expressed more frequently in males than in females. As Bock and Kolakowski (1973) have noted, no immediate test of O'Connor's hypothesis was made. The obvious way – inspection of the spatial trait in pedigrees – would have yielded genotype classification errors because of test unreliability.

Parent-child correlations

There are alternative methods. The simplest is to correlate the performance of children with their same-sex and different-sex parents. The pattern of these correlations for an X-linked trait is distinguishable from the pattern for autosomal inheritance in which sex of family member is irrelevant (see Mather & Jinks, 1971). To see why, let us review the basic features of transmission genetics as applied to the sex-linked recessive gene hypothesis.

Every normal person has 46 chromosomes arranged in 23 pairs. The twenty-third pair is composed of the sex chromosomes. When both sex chromosomes have the X structure, the sex is female; when one of the chromosomes is Y, the individual is male. The genes borne on the sex chromosomes are sex-linked. Because the mother can endow her child only with an X chromosome, it is the father who determines the sex of the offspring. If the sperm that fertilizes the ovum bears an X chromosome, the result is a female (XX); if it bears a Y chromosome, the result is a male (XY). A sex-linked recessive trait, therefore, can be expressed in females only if it is present on *both* X chromosomes (that is, present in the X contributed by the mother and in the X contributed by the father). But it can be expressed in *any* male because there is no dominant counterpart in the absence of another X chromosome. Thus, if the mother carries the recessive gene for spatial ability on *both* her X chromosomes, then *all* her sons will inevitably express the spatial ability trait. But a daughter's second X chromosome, contributed from her father, may or may not bear the spatial ability gene. If it does not, the daughter will not express the spatial ability trait. We should point out, following Jensen (1975), that this is a model for *enhancement* of spatial ability, and not for the characteristic of spatial ability itself. That is, the model proposes that only the single enhancing gene is recessive with respect to spatial ability. The characteristic itself, in its normal expression, conceivably can be produced by a variety of combinations of genes. Therefore, where reference is made here to the "spatial ability gene," this is shorthand for the single recessive enhancing gene for the trait.

The model is illustrated in Figure 13-1. Shown are the various possible matings that could yield the genotype for spatial ability in male and female offspring, plus those combinations from which the genotype could not be transmitted.

Sex differences in spatial ability 441

The sex-linked recessive trait model thus predicts a higher mother-son than mother-daughter correlation. But similar scores on the spatial visualization test can arise both from the expression of the gene and from the failure of its expression. Therefore, one expects some correlation between the scores of mothers and daughters. Because the transmission by the father of a Y chromo-

Fig. 13-1. Model for sex-linked recessive trait. Probabilities of inheritance of spatial enhancement trait in male and female offspring from all possible parental matches. The figure in each row on the left represents percent of male offspring from each union having genotype ($X_s Y$) for spatial ability. The figure in each row on the right represents percent of female offspring from each union having genotype ($X_s X_s$) for spatial ability.

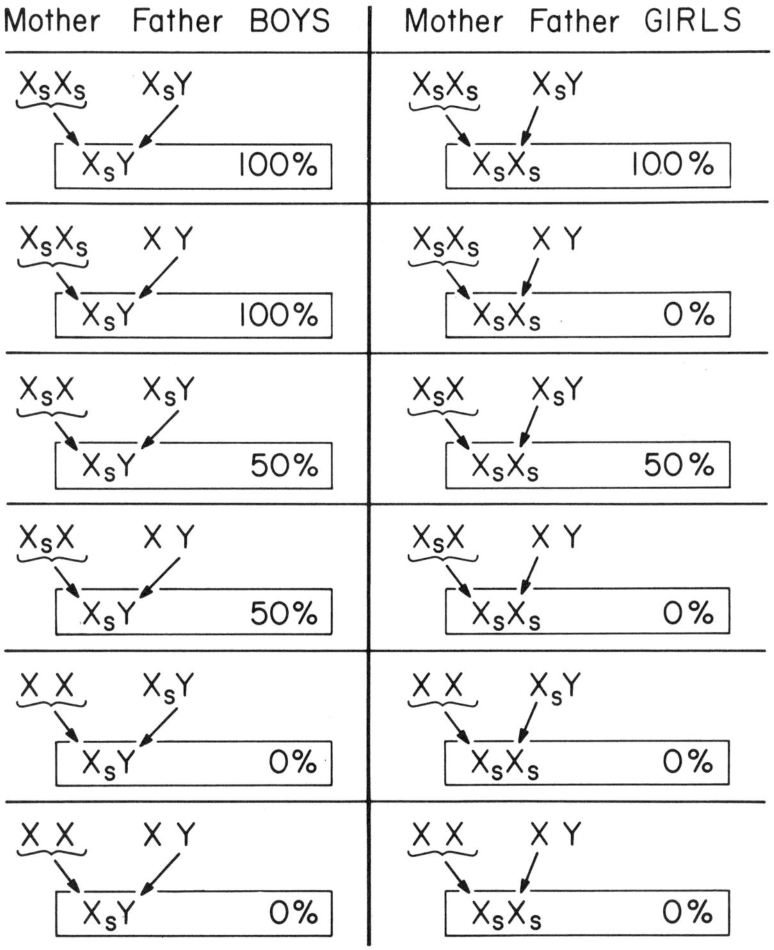

Mother	Father	BOYS	Mother	Father	GIRLS
$X_s X_s$	$X_s Y$	$X_s Y$ 100%	$X_s X_s$	$X_s Y$	$X_s X_s$ 100%
$X_s X_s$	X Y	$X_s Y$ 100%	$X_s X_s$	X Y	$X_s X_s$ 0%
$X_s X$	$X_s Y$	$X_s Y$ 50%	$X_s X$	$X_s Y$	$X_s X_s$ 50%
$X_s X$	X Y	$X_s Y$ 50%	$X_s X$	X Y	$X_s X_s$ 0%
X X	$X_s Y$	$X_s Y$ 0%	X X	$X_s Y$	$X_s X_s$ 0%
X X	X Y	$X_s Y$ 0%	X X	X Y	$X_s X_s$ 0%

some can produce only a male, a sex-linked characteristic can never be transmitted from father to son. (A hemophilic father, therefore, cannot produce a hemophilic son, but can produce a carrier daughter who, in turn, can produce a hemophilic grandson.) Because the father does *not* transmit an X chromosome to his son (doing so would make the son a daughter), the correlation between their visuospatial abilities should be very low or zero. Insofar as the father-son correlation is found to be low, the Y chromosome can be said to lack a significant role in the transmitting of spatial ability.

The results of several investigations of family correlations for spatial ability are summarized in Table 13-1. The first report was by Stafford (1961). Stafford gave an 18-item test of spatial visualization (Identical Blocks Test) to 104 fathers and mothers and to their 58 teenage sons or 70 teenage daughters. The correlation, corrected for attenuation, was fairly high both for mothers and their sons ($r = 0.39$) and for fathers and their daughters ($r = 0.36$). The mother-daughter correlation was much lower ($r = 0.18$); and the father-son correlation was zero.

Similar results have been reported more recently by Hartlage (1970) with the spatial tests of the Differential Aptitude Test

Table 13-1. *Patterns of correlations between parents' and children's scores on various spatial visualization tasks*

Study	Mother-son	Father-daughter	Mother-daughter	Father-son
Stafford (1961) teenagers and parents	0.39[a]	0.36[a]	0.18	0.02
Corah (1965) 8- to 11-year-olds and parents	0.39[a]	0.41[a]	0.00	0.25
Hartlage (1970) teenagers (16–18 years old) and parents	0.39[a]	0.34[a]	0.25	0.18
Bock & Kolakowski (1973) 12-year-olds to teenage, and parents	0.20	0.25	0.12	0.15

[a] Statistically significant, $p < 0.05$.

(Bennett et al., 1959). The subjects were 100 adults and children representing 25 families containing all combinations of father-son, father-daughter, mother-son, and mother-daughter pairs. The age range for the total sample was 16 to 50 years. Like Stafford's results, the fathers' and sons' average scores exceeded the mothers' and daughters'. The correlations were also similar to Stafford's: mother-son, $r = 0.39$, father-daughter, $r = 0.34$. Both correlations were significant ($p < 0.05$). The remaining correlations were lower and nonsignificant: mother-daughter, $r = 0.25$; father-son, $r = 0.18$.

We now can recall the earlier-mentioned study by Corah (1965) of the family correlations for performance on embedded figures and human figure drawing tasks. Like Stafford's and Hartlage's results, Corah's highest and only significant correlations were for the mother-son and father-daughter combinations, a pattern inconsistent with a socialization model but, it can be seen, predicted by the recessive gene model.

The most recent report is by Bock and Kolakowski (1973), who administered the Guilford-Zimmermann Spatial Visualization Test (1953) to 167 families. As in the other studies, the test scores were corrected for regression on age before the correlations were computed. The constellation of correlations again fits the recessive gene model, though less strongly than in earlier studies. The father-daughter and mother-son correlations were 0.25 and 0.20, respectively. The father-son and mother-daughter correlations were 0.15 and 0.12. Bock and Kolakowski also noted, on the basis of their results, together with Hartlage's, that the true father-son correlation may not be exactly zero, possibly because of a small component of autosomally inherited variation, presumably due to autosomally determined general intelligence which, they suggested, must figure to some extent in spatial test performance. Corah's father-son correlation was also above zero, though not statistically significant. Bock and Kolakowski concluded on the basis of the magnitude of familial correlations from their study statistically combined with the Stafford (1961) and Hartlage (1970) data, that approximately 46% of the spatial score variance is attributable to genetic variance whose source is a recessive sex-linked gene with frequency of approximately 0.5 in the population at large.

In contrast to the pattern of correlations predicted by the

sex-linked recessive trait model, no differences in the parent-child correlations according to sex would be expected for autosomal inheritance. For example, the pattern reported for stature is: father-son, 0.51; father-daughter, 0.51; mother-son, 0.49; mother-daughter, 0.51 (Pearson & Lee, 1903; cited in Bock & Kolakowski, 1973).

Distribution of spatial skill
The sex-linked recessive trait model has a second implication. The trait should not be distributed normally in the population. Instead, the proportion of females showing the trait should equal the square of the proportion of males showing the trait; that is, on the average, 25% of the females' scores should exceed the mean male score (Stern, 1960, pp. 218–244). In Figure 13-1, these percentages can be derived by calculating the mean percent of male and female offspring expected to inherit the spatial trait across the six combinations of parental matchings. The score for males is 0.50, for females, 0.25. It was this finding in his own study with the Wiggly Block Test (a three-dimensional visualization test requiring assembly of irregular wavelike pieces into a solid block) that led O'Connor (1943) to suggest the recessive sex-linked trait model in the first place. As noted in the introduction to this chapter, the 20% to 25% figure has been reported in numerous studies with a variety of measures, including the reports described in this section.

The most explicit test of the prediction of nonnormality of distribution of spatial ability in the population is to inspect the distribution of spatial visualization scores separately for males and females (Bock, 1973). Two results in a study by Bock (1967) are consistent with the model. First, the males' scores were distributed bimodally, with an antimode near the 50th percentile. This bimodality may mean that spatial ability within the population is distributed at only two discrete levels rather than continuously – a feature that would be consonant with the picture of trait inheritance influenced by only two different alleles at one gene locus rather than by many alleles at many loci. Second, Bock divided the distribution of all scores at the antimode and found that half the males and 25% of the females excelled – the proportions expected if females must receive the recessive allele from both parents and males must receive it only from their mothers. There was but one inconsistent finding. Bock also noted that all girls who excelled and therefore could be assumed to carry the recessive allele on both X chromosomes should have fathers who excelled too, as their fathers would have had to carry the recessive allele on their one X chromosome for it to be transmitted to their daughters. But there were six females who excelled whose fathers did not excel. Bock suggested that this single discrepant finding may reflect inadequate motivation by these fathers.

The model challenged: evidence from the individual with Turner's syndrome

The family correlation data, together with the size of the discrepancy in proportions of males' and females' spatial scores, appear to constitute strong evidence for a recessive sex-linked model for the inheritance of spatial ability. The model recently has been challenged in an interesting way. Garron (1970) has pointed out an inconsistency between the implications of the model and certain evidence presented in the individual suffering from Turner's syndrome (Turner, 1938), a chromosomal abnormality characterized by short stature in a person with the body morphology of a girl (the victim is nearly always less than 5 ft tall). The victim also bears certain body stigmata, including (though not always in the same individual) webbed neck, webbed fingers and toes, small receding chin, pigmented moles, epicanthal folds (resembling oriental eyes), blue baby and other heart defects, kidney and ureter defects, and hearing loss.

About 80% of individuals with Turner's syndrome are sex-chromatin-*negative* females (the sex-chromatin is missing from the cells when they are test-stained). They therefore have only 45 instead of 46 chromosomes, with the missing chromosome being one of the sex chromosomes and the remaining one being always an X chromosome. Therefore, instead of XX (female), the Turner's syndrome karyotype is X0 (44 + X0). There also are patients in whom the sex-chromatin is present. Sometimes, when the second X chromosome is present, one of its arms is broken (deletion chromosome). Nearly all other individuals with Turner's syndrome show a chromosomal mosaicism in which, for instance, some cells of the body are XX and some X0.

Garron (1970) observed that since both Turner's syndrome females and normal males have but one allele of each gene located on the X chromosome (in contrast to normal females who have two alleles, one on each of their two X chromosomes), one would expect the incidence of expression of sex-linked recessive traits to be comparable in both groups. This expectation, Garron (1970) observed, is borne out with respect to color blindness, a known recessive sex-linked trait found in far less than 1% of normal females but in about 8% of normal males. Administration of the Ishihara (1951) test to 25 individuals with Turner's syndrome disclosed 4 (16%) who gave the typical responses for red-green color blindness (18 to 23 abnormal responses of 24 possible). The remaining subjects made fewer than 4 such errors, except for 4 subjects who made from 5 to 8 errors not unlike the kind that are relatively common among persons with normal color vision (Polani et al., 1956).

If spatial ability, like color blindness, is indeed a recessive characteristic carried on the X chromosome, individuals with Turner's syndrome should express this trait in equal proportion to normal males, that is, significantly *more* frequently than normal *females*. The evidence, however, goes just the other way. The most remarkable psychological characteristic of the syndrome is the existence of a specific cognitive deficit, which Money (1968a) has described as a combination of space-form or visuoconstructional dysgnosia, directional-sense dysgnosia, and mild dyscalculia. Females with Turner's syndrome have less rather than more spatial ability than normal females. This deficit was first described by Shaffer (1962) and confirmed by others (Money, 1964; Buckley, 1971) and is revealed in a particular pattern of scores on the Wechsler Adult Intelligence Scales (WAIS). For the factor structure for the WAIS obtained by Cohen (1957, 1959) with a normal group, performance on perceptual organization (block design + object assembly) and on a factor having to do with numerals and calculation (named freedom from distractibility) was significantly lower than verbal comprehension (information + similarities + comprehension + vocabulary), which was near normal.

The space-form blindness or visuoconstructional dysgnosia associated with Turner's syndrome has been found also on other tests. Evaluation of 10- to 24-year-old patients (mean age 15 years) disclosed deficient performance on the Bender Visual-Motor

Gestalt Test, the Benton Visual Retention Test, the Draw-a-Person Test, the Draw a Floor-Plan exercise, the spatial subtest of the SRA Primary Mental Abilities Test (Money et al., 1965; Money, 1968a), the Road Map Test of Direction Sense (Alexander et al., 1964), Benton & Kemble's (1960) Right-Left Discrimination Battery, and tests requiring imaginary locomotion, rotation, and direction of the body in space (Orientation Test of the Detroit Test of Learning Aptitude) (Alexander & Money, 1966). According to Alexander and Money (1966), the common factor in all the positive findings appeared to be a dysgnosia limited to extrapersonal space and to form perception. Body image appeared to be unaffected.

Case history information was generally consistent with the deficits shown on the psychological tests. Nearly all the patients reported that mathematics was their most difficult school subject, with art frequently named as close behind. Sense of direction also was grossly defective. One 23-year-old, when asked how she would reach her home by automobile, gave incorrect left-right directions. Parents of several patients reported that because their daughters lost their way easily, younger siblings accompanied them on even routine trips. Such difficulties would be normal in very young children just learning their way about the neighborhood. They surely would not be *commonly* expected in older children or young adults.

The space-form dysgnosia, directional-sense dysgnosia, and mild dyscalculia (impairment of ability to deal with numbers and to calculate) are the only deficits to have appeared so far (Money, 1968a, b). There has been no evidence, for instance, of finger agnosia and, most important, no evidence of any language or reading disabilities such as dysgraphia, dyslexia, or aphasia. Indeed, the girl with Turner's syndrome is frequently an excellent reader (Alexander & Money, 1965). Nor is this pattern of cognitive deficits found in other persons with chromosomal abnormalities not involving the sex chromosomes. And it is not characteristic of persons with Klinefelter's syndrome – phenotypic males carrying an extra X chromosome, thus with the XXY karyotype (Money, 1964).

Garron's criticism of the sex-linked recessive hypothesis seems, then, to be well founded.[9] The criticism may be resolvable, however, when certain other characteristics of the individual with

Turner's syndrome are taken into account. The specific pathognomonic symptom of Turner's syndrome is gonadal agenesis or dysgenesis. Exploratory surgery on these persons discloses ovaries that look like streaks instead of being round and plump. They make neither eggs nor female sex hormones. The affected person is sterile and will remain sexually infantile in appearance until treated with female sex hormones, which also induce menstruation. It is known that normal females produce and respond to detectable levels of testosterone (Rosenfield, 1971), and it may be that proficiency in spatial tasks is linked both to the X chromosome and to the presence of testosterone, and that the capacity of the normal monozygous recessive female to express the spatial trait depends on the production of ovarian testosterone above some threshold level.

This hypothesis is made plausible in light of known intellectual deficits in other individuals who, like those with Turner's syndrome, are insensitive to androgen. One such individual is the male with testicular feminizing syndrome. The defect in testicular feminization is a specific end-organ insensitivity to testosterone, and the victim shows a pattern on the WAIS similar to that shown by the individual with Turner's syndrome (Money, 1968a, b; Masica et al., 1969, 1971).

Other supporting evidence has been reported in a study of mine workers in Sierra Leone suffering from the protein deficiency disease kwashiorkor (Dawson, 1967). One of the effects of this disease, which is pandemic in Africa, is liver malfunction that frequently results in an endocrine disturbance involving, in males, gynecomastia (a swelling of the male mammary gland), testicular atrophy, and feminization. These endocrine changes are considered to correspond broadly to those produced by the administration of estrogens, though it is not certain whether the major factor is the inability of the liver to metabolize estrogens or a "perverted conversion" of androgens to estrogens. The disease first appears in the second year, when the child's protein needs are maximal relative to body weight (Trowell et al., 1954). Dawson (1967) found that mine workers who showed the clinical symptom of gynecomastia had significantly lower scores on tests of spatial ability than mine workers not showing the symptom. The afflicted men also had significantly higher verbal and lower numerical scores than normal subjects.

Thus, the model of spatial ability as a recessive sex-linked trait can stand, though now we can see that both genetic and hormonal factors play roles. It is pertinent to note, in this connection, that the directional-sense deficit of individuals with Turner's syndrome is not alleviated by hormone substitution treatment (Money, 1968a, b). Money (1968a, b) has suggested that the syndrome has its origins in embryonic or fetal neurohormonal incidents – presuming there is any foundation to the hormonal explanation in the first place, as now appears likely.

Turner's syndrome and cerebral functioning

This analysis, if reasonable, is not yet complete. The implication of a genetic sex-linked factor in combination with hormonal factors does not itself constitute an explanation of spatial skill. In the particular case of the girl with Turner's syndrome, for example, the demonstrated link between a cytogenetic defect with consequent gonadal agenesis and sex hormone failure and space-form dysgnosia does not by itself *explain* the spatial deficit. It indicates only that the cause is genetic and hormonal rather than environmental, or, to be more precise, that the cause is in the genetic and hormonal environment rather than in the cultural, sociological, or psychological environment. Ultimately, however, there must be certain *cerebral* mechanisms responsible for the gross deficiencies in spatial visualization shown by the individual with Turner's syndrome inasmuch as the cerebral hemispheres are known to subserve the normal expression of spatial skills. What, then, might be the effects of a cytogenetic defect on cerebral functioning, particularly, right hemisphere functioning? Money (1968a) has proposed that the prominence of the space-form deficit, in the absence of other behavioral symptoms (particularly, the absence of aphasia), suggests a deficit or anomaly in functioning of the parietal lobe that is of developmental origin. And since the right parietal lobe is known to subserve space-form abilities of this kind (Critchley, 1953), it is reasonable to guess that the anomaly is more likely to be in the right cerebral hemisphere than in the left. Though Money could not point to any direct evidence of cerebral disease in persons with Turner's syndrome, we can observe certain similarities between the kinds of disturbances of spatial ability shown by these individuals and the spatial deficits shown by patients with known parietal disease.

The classic descriptions are provided by Critchley (1953). For example, Alexander and Money's (1966) descriptions of geographical disorientation or confusion in girls and young women with Turner's syndrome are reminiscent of reports on brain-damaged patients. Zangwill (1951) suggested that the deficit is in topographical memory, that is, in the recognition of familiar places. He described one patient who said that while walking in familiar places he seemed not to know what would be around the next corner. It was as though he were in a strange land. In Critchley's (1953) view, such spatial disorders are more likely to result from damage to the right cerebral hemisphere, with the more critical areas likely to be the territory linking the parietal, occipital, and temporal lobes (Critchley, 1953, p. 355).[10]

NEUROLOGICAL MODELS

The evidence from patients with Turner's syndrome raises the question whether differences in spatial ability between chromosomally *normal* females and males are traceable to differences in those cerebral areas that subserve spatial skills. If spatial ability is a recessive sex-linked trait carried on the X chromosome and programmed hormonally, might one of the differences between normal males and females be in right hemisphere functioning? If so, what is the nature of the difference? Several models have been suggested. This section reviews and evaluates evidence for each. The designation *model* itself suggests a degree of completeness or particularity the authors of the various research studies cited may not have intended. Indeed, the different models, rather than being mutually exclusive, are complementary, often differing primarily in points of emphasis.

Model 1: Earlier right hemisphere lateralization in males

Because the male's superior spatial ability first appears in childhood, albeit less consistently than in later years, it has been suggested that the development of right hemisphere specialization proceeds earlier and faster in males than in females.

At least during infancy, there as yet are no such indications. Three investigations may be cited. Molfese (1973) presented two mechanically produced sounds (a C-major piano chord and a burst of speech noise) to 10 infants ranging in age from 1 week to 10 months. Recordings of evoked potentials from Wernicke's area and from a corresponding area in the right hemisphere disclosed that 9 infants responded to the noise stimulus, and all 10 to the musical chord, with greater right hemisphere auditory evoked response activity. Entus (1975) found the same lateralization effect with a dichotic listening procedure used in combination with a nonnutritive sucking paradigm, where the index of ear differences is the degree of dishabituation of the sucking response after a change of the auditory signal to one ear or the other. The subjects were 48 infants 22 to 140 days old (24 boys and 24 girls). The sounds were musical notes played on various instruments. Of these infants, 79% showed a left-ear superiority for these sounds, proportions consonant with those reported for dichotic listening studies with older children (e.g., chapter 12). Best and Glanville (1976) also used musical stimuli in a study with 12 infants 93 to 178 days old, but substituted heart rate change as the dependent variable. Of the 12 infants, 10 showed left-ear superiority. In none of these three studies were sex differences in strength of response or in lateralization effect reported.

Where sex differences have appeared, it has been with older children. A study by Rudel et al. (1974) offers suggestive evidence. This study was stimulated by a report by Hermelin and O'Connor (1971a,b), so let us first review the earlier study. Hermelin and O'Connor assessed hand differences in braille-reading speed in 16 children, 8 to 10 years old. All were blind from birth; 14 were right-handed, and 2 were apparently bidextrous. For the group as a whole, left-hand scores were significantly faster than right-hand scores in the reading of simple sentences. For some children, the difference was striking: They were fluent with left-hand reading but could produce only gibberish with the right hand.

Hermelin and O'Connor also tested 15 blind adults, 25 to 65 years of age, this time on repeated different orderings of the 26 individual braille letters of the alphabet. The letters were arranged vertically and read from top to bottom to control for possible motoric scanning bias from left to right that might have favored the left hand. Again, the left hand was significantly more accurate.

One might expect that braille characters, being symbols of alphabet letters, would be better discriminated by the *right*, not left hand (i.e., by the left, linguistic, hemisphere). However, dot patterns like those that constitute braille characters are known to be perceived more accurately by sighted subjects when projected to the left visual field, or right cerebral hemisphere (Kimura, 1969). And even though these patterns are not letter symbols for the sighted subjects, the possibility exists, Hermelin and O'Connor have suggested, that for the blind, braille symbols are first encoded as spatial configurations and only then processed as meaningful patterns.

Hermelin and O'Connor's subjects were experienced braille readers and, being right-handed, usually felt letters to be copied with the left hand while copying with the right. The hand asymmetries in braille reading, therefore, may have resulted from this experience. Rudel et al. (1974) addressed this possibility by using sighted children as subjects. Their results suggest that if braille symbols are first encoded as spatial configurations so as to bestow left-hand advantage, the left-hand superiority appears only gradually, and more quickly and reliably for boys than for girls. Their subjects were 80 intellectually normal 7- to 14-year-olds, all right-handed. Each child had to learn six braille letters with one hand, and six different letters with the other hand, in a paired-associates procedure. At age 7 to 8, the boys performed equally well with both hands, but the girl's *right*-hand scores were superior. By age 13 to 14, both boys' and girls' left-hand scores were superior, though the difference was statistically significant only for the boys.

A limitation of the use of braille learning as a test of developmental differences in right hemisphere specialization is that braille reading obviously requires both left hemisphere and right hemisphere modes – perhaps, as Hermelin and O'Connor suggested, at different processing stages. A better test of the hypothesis would use a task that depends to a greater degree on the action of the right hemisphere. The studies by Knox and Kimura (1970), described earlier in this chapter, may meet this requirement. Recall that in dichotic listening studies with 5- to 8-year-olds, and a monaural listening study with 2.5- to 5-year-olds, boys were more accurate than girls in identification of environmental sounds, but, in the dichotic test, did not show an earlier or significantly stronger left-ear advantage.

Still another purer measure of right hemisphere specialization has been described by Witelson (1975). She presented children with complex, unfamiliar, meaningless four- to eight-sided shapes, a different one to each hand and not visible to the child, in a tactual version of the dichotic technique. On each trial, the subject felt the shapes and then identified the forms he had felt by pointing to a visual display of a group of shapes. Witelson suggested that the use of meaningless shapes presented dichotomously would minimize any possible tendency for linguistic mediation. The subjects were 165 boys and 165 girls ranging in age from 3 to 13 years. All the children were normal and right-handed. The 6- to 13-year-olds were allowed to feel the pairs of forms themselves for 10 seconds. For the 3- to 5-year-olds, palpation was passive: The child's fingers were moved across the forms by the experimenter.

Sex differences were clear. For the boys, there was no hand difference for the 3- and 4-year-olds, but a significant left-hand superiority at age 5 and beyond. The girls showed no hand differences until about age 13, when left-hand superiority appeared.

It is hard to explain these sex differences, whether in overall performance on identification of nonverbal sounds (Knox & Kimura, 1970) or in age of first appearance of left-hand superiority in tactual discrimination of shapes, in terms of different experiences. Knox and Kimura (1970) pointed out that most of the sounds used in their experiments were equally common to the experience of both sexes – or seemed to be. And they justifiably were reluctant to credit the boys' superior performance to superior labeling of what was perceived, because, as we shall see later, girls of the age range tested surpass boys in the use of expressive language. Their hypothesis, rather, was that the results may reflect greater lateralization of spatial functions to the right hemisphere in boys than in girls of the same age. Witelson (1975) similarly proposed that, at least for haptic discrimination, the right hemisphere may be specialized for spatial processing as early as age 5 in boys, but in girls not until many years later. Rudel et al. (1974, p. 737) saw their findings as evidence for an earlier and "perhaps superior" pattern of right hemisphere development in boys.

Model 2: Earlier, greater left-hemisphere language lateralization in females and bilateral spatial representation in males

Thus far, we have confined ourselves to speculation about sex differences in right hemisphere functions. According to a different model, sex differences in spatial ability can be understood only as a consequence of sex differences in *left* hemisphere lateralization. Buffery and Gray (1972) argued that "an originally bilateral neural activity mediating linguistic skill progressively lateralizes over the early years to (usually) the left cerebral hemisphere, in which a relatively dormant but structurally predisposed speech perception mechanism is waiting to subserve it" (p. 144). They have further proposed that this structural mechanism is more developed in the female than the male brain of the same age, so the lateralization of language function occurs earlier and progresses more quickly in females than males. The result is female superiority in verbal tasks and male superiority in spatial tasks:

> Linguistic skill, with its need for quick associations and serial ordering, probably demands fast and intricate neural mechanisms. Such mechanisms could benefit from being subserved by specific structures with a clearly lateralized and localized cerebral representation . . . Spatial skill, however, which is usually exercised in a three-dimensional and completely enclosing world, may benefit from a more bilateral cerebral representation. Thus a consequence of the less well lateralized cerebral representation of linguistic skill in the male brain might be a more bilateral cerebral representation of spatial skill than can be achieved in the female brain. [Buffery & Gray, 1972, p. 144]

Anatomical evidence

The question of sex differences aside, the notion that the left hemisphere is *structurally* predisposed to subserve speech is reasonably well supported. In 1968, Geschwind and Levitsky reported the results of an examination of 100 adult brains, obtained at postmortem, and free of significant disease. They divided the hemispheres and then exposed the upper surface of the temporal lobe (supratemporal plane) on each side by cutting into the plane of the Sylvian fissures. There they found that the planum temporale (a portion of the temporal speech cortex in the area behind Heschl's gyrus) was about 9 mm (33%) longer on the left side in 65% of the brains, equal in 24%, and longer on the right in 11%. The handedness of the persons from whom whese brains were taken was not known, but since about 96% of the population (combining left- and right-handers) is left-brained for speech, these cases are likely to have consisted of approximately this proportion of subjects whose left hemisphere was dominant for language.

The order of magnitude of difference found in the planum temporale has been confirmed (Witelson & Pallie, 1973; Wada et al.,

1975), and a comparable asymmetry now is known to exist as well in the speech area of the frontal lobe (Wada et al., 1975).[11]

Whether, as we might suppose on the basis of Buffery and Gray's model, there is greater planum asymmetry in young females than young males is uncertain. In addition to adult brains, Witelson and Pallie (1973) found the same asymmetries in the brains of 11 infants 1 to 21 days old and 3 infants 1 to 3 months old, all of which were free of neurological disease. And Wada et al. (1975) reported comparable asymmetries in fetal brains, in both frontal and temporal areas, as early as the fifth month postconception. About sex differences, the most that can be said is that Wada et al. reported that the frontal area of infant brains was slightly more asymmetrical in females than in males. In Witelson and Pallie's study, in the brains of the youngest infants – 5 boys and 5 girls who had died between 1 and 21 days after birth – the planum was larger, on average, on the left than on the right in both groups, but the difference was statistically significant only for the female brains.[12]

Psychological evidence
Though the anatomical evidence for sex differences is weak at best, the psychological evidence for the first part of Buffery and Gray's proposal – faster language development in females – is substantial.

Sex differences in language development. Girls speak their first words sooner than boys (Mead, 1913; Terman, 1925; Abt et al., 1929; Morley, 1957), develop larger vocabularies, at least through the early years (Doran, 1907; Nelson, 1973); and, in a study of 24- to 50-month-olds, were found to be more likely to use such mature constructions as the full passive, truncated passive, reflexive, conjunction, participle, and subordinate clause (Horgan, 1976).

There also may be sex differences in articulation – in the fine, precise, coordinated movements of the speech apparatus. In spontaneous speech, boys' and girls' articulation skills seem to develop at the same rate, until about 3 to 5 years, when girls start to improve faster than boys (Poole, 1934; Templin, 1953, 1957; Matheny, 1973). Beyond 8 years, reports disagree about whether the girls' lead is maintained (Saylor, 1949; Templin, 1957).

Girls' speech also may be easier to understand: Among 18- and 24-month-olds, McCarthy (1930) reported substantially higher percentages of comprehensible vocalizations for girls than boys; by

36 months, 99% of girls' speech was comprehensible, a level reached by boys only a year later. Slight differences favoring girls also have been reported for 2.5- to 5.5-year-olds (Young, 1941) and 8- to 10-year olds (Eisenberg et al., 1968).

Girls also excel on tests of word fluency – for instance, as rapidly as possible naming words containing a specified letter or things belonging to a given class. On such tasks, significant sex differences have been reported for children between 8 and 18 years of age (Havighurst & Breese, 1947; Herzberg & Lepkin, 1954). There also is some evidence that girls vocalize more than boys in infancy (Gatewood & Weiss, 1930; Moss, 1967; Goldberg & Lewis, 1969) and early childhood (Olson & Koetzle, 1936; Young, 1941; Halverson & Waldrop, 1970), though as many other studies find no sex differences (see Harris, 1977).

Finally, among grade school children, girls generally are better readers than boys (Prescott, 1955; Gates, 1961; Stanford Research Institute, 1972).

Sex differences in time of onset of hemispheric lateralization. If the sex difference in language development first appears in infancy, one might expect the first indications of differences in left hemisphere specialization at this time. Molfese (1973), Entus (1975), and Best and Glanville (1976), in the studies described earlier, also presented speech sounds to their infant subjects. Molfese presented the words *boy* and *dog* and the speech syllables /ba/ and /dae/; Entus and Best and Glanville presented only monosyllables. In all three studies, the great majority of infants showed stronger left hemisphere responsivity, but there was no evidence of sex differences.

Again, as in the case of right hemisphere lateralization, sex differences have appeared only in studies of older children. Buffery (1971) himself presented evidence from two kinds of experiments. An intermodal matching task combined the techniques of tachistoscopic presentation to the binocular visual hemifield with that of dichotic listening. The purpose was to permit comparison of the relative efficiencies of the two hemispheres in boys and girls for comparing a visual stimulus (printed word) with a simultaneously presented auditory stimulus (same or different spoken word). The subjects were 24 boys and 24 girls, all right-handed, and ranging in age from 5 to 8 years. For both boys and girls at all ages, the greatest accuracy was on the split-load condition (each stimulus of a pair presented initially to different hemispheres) in which the *auditory* word was received by the *left* hemisphere for simultaneous comparison with a *visual* word received by the *right*. The most difficult condition was the reversed split load, when both the auditory and visual words were presented initially to the "functionally less appropriate right cerebral hemisphere" (Buffery & Gray, 1972, p. 140). These differences among conditions appeared in girls at 5 years, but in boys only by age 7.

Buffery (1971) also used a "conflict drawing test," which requires the child to draw, simultaneously and with eyes closed, a square with one hand and a circle with the other. Such conditions, Buffery wrote, "encourage the proprioceptive feedback from each hand to be processed initially by the controlling contralateral cerebral hemisphere." The subjects were 80 boys and 80 girls, all right-handed, and ranging in age from 3 to 11 years. The majority of girls of all ages showed a *left*-hand superiority for the drawing of a well-proportioned square, whereas the boys changed from right-hand to left-hand superiority only by 7 years. Furthermore, among the 3- to 7-year-olds, girls showed a greater degree of right-hand preference than boys. And for both boys and girls the degree of nonpreferred left-hand superiority increased with age and with degree of right-hand preference.

Buffery interpreted the results from these experiments as evidence for earlier, stronger language lateralization in girls than in boys. For the conflict drawing task at least, this interpretation is puzzling, because the finding of earlier, greater *left*-hand superiority in girls than boys is inconsistent with the evidence of Model 2 for earlier, greater *right* hemisphere specialization for spatial functions in *males* (e.g., Knox & Kimura, 1970; Witelson, 1975). Buffery's results instead seem to indicate *earlier* right hemisphere specialization in *females*. The resolution of the discrepancy, according to Buffery and Gray (1972), turns on a further implication of their model: If language lateralization occurs earlier and progresses more quickly in the female brain than in that of the male, this leaves the nondominant (usually right) cerebral hemisphere of the female relatively freer than the male's to subserve nonverbal functions in general and the spatial skill of the nonpreferred hand (here the left) in particular. Therefore, girls would be expected to show earlier left-hand superiority on a spatial task. As for Knox and Kimura's findings, Buffery and Gray (1972) allowed that their hypothesis of greater right hemisphere lateralization of spatial functions for males "may be correct insofar as it concerns nonverbal auditory function" (p. 144).[13]

Finally, sex differences have been reported where the measure of language lateralization has been the more familiar dichotic listening test, though the differences are inconsistent and not always in the expected direction. We earlier noted that Knox and Kimura (1970) found similar right-ear advantages for 5- to 8-year-old boys and girls for digit recognition. Kimura previously (1963a) tested 120 right-handed children between 4 and 9 years of age. Boys and girls in each age group showed a right-ear advantage for recognition of spoken digits, and again no indication of an earlier or stronger effect in girls than in boys. The only sex difference was in overall efficiency on the task, with *boys* better at ages 4, 5, and 6.[14]

The children in the earlier study were from high socioeconomic backgrounds and were above average in intelligence. Kimura (1963a) repeated the study with children from similar backgrounds, and with similar results. Later, however (1967), she tested 5- to 8-year-old children from low- to middle-class socioeconomic areas and found a significant and substantial right-ear advantage in all groups except the 5-year-old boys. Of 20 5-year-old boys, 11 had higher right-ear scores, 8 had higher left-ear scores, and 1 child showed no difference between ears. The children in these two sets of studies differed in several ways (intelligence, home background, and verbal ability), any one or several of which may have been critical in the different results found.

Sex differences have been reported also by Pizzamiglio and Cecchini (1971). They presented dichotic pairs of words and digits to 192 Roman children between 5 and 10 years of age and found, among the youngest, a stronger lateralization effect for girls than for boys.

Finally, Bryden et al. (1973) presented natural speech consonant-vowel syllables, but only in the initial stop consonant, to 12 boys and 12 girls in kindergarten, second, fourth, sixth, and eighth grades (age range 6 to 14 years). The girls showed a clear laterality effect by the fourth grade; the boys lagged considerably behind.

Evaluation of Models 1 and 2

The empirical evidence for Model 1 (earlier right hemisphere specialization in males) seems reasonably good, though it remains to be explained why studies of infants (Molfese, 1973; Entus, 1975; Best & Glanville, 1976) yield evidence for right hemisphere specialization, and equally for boys and girls, and in studies with older children the predicted sex difference appears but, puzzlingly, the girl, rather than only showing less lateralization than the boy, sometimes shows none at all. Certain tasks and methods undoubtedly are more sensitive measures than others of the special functions of the right hemisphere. Perhaps as the child moves beyond the simple perceptual skills of infancy to more sophisticated stages, the sex differences begin to be reflected. In other words, the right hemisphere, predisposed to subserve visuospatial skills even at birth, becomes further specialized, or lateralized, through the first several years of life, and it is this further specialization that proceeds more rapidly in males than females.

As for Buffery and Gray's proposal (Model 2), there is evidence that left hemisphere lateralization proceeds faster in females than in males. The exception is Buffery's own conflict drawing test, whose results directly conflict with research indicating earlier right hemisphere specialization in males (Model 1). Buffery and Gray dismissed Knox and Kimura's (1970) findings on the grounds that they are peculiar to nonverbal auditory functions, but this point loses force in light of results obtained by Rudel et al. (1974) and Witelson (1975) for tasks that, like Buffery's, were haptic and somesthetic. Furthermore, the results of Buffery's conflict drawing test have not been found in adults (Clyma & Clarke, 1973).

As was the case for studies of right hemisphere specialization, it is not clear why studies of infants yield evidence for hemispheric specialization for language, and equally for boys and girls, although for older children, sex differences appear, albeit inconsistently, and the age of appearance of laterality effects itself varies for both boys and girls. Perhaps the left hemisphere, disposed to subserve language functions at birth, becomes further specialized, or lateralized, through the first several years of life as the child moves beyond the simpler perception and production skills of infancy to later more complex linguistic stages, and it is this further specialization that proceeds more rapidly in girls than in boys.

Nothing in any of the evidence reviewed here, however, seems to support Buffery and Gray's further proposal that bilateral representation is beneficial to spatial ability, in contradistinction to the advantage of unilateral control for linguistic processes, and that this bilateral control is more characteristic of males than females. There also seems to be a logical weakness in this part of their model, for if bilateral control is advantageous for spatial ability, and unilateral control is advantageous for verbal ability, this ought to imply impairment in language ability of left-handed persons – the group for whom bilateral language representation most reasonably can be inferred. There is no evidence of this.

The evidence for both Models 1 and 2 is restricted to the childhood years, and herein lies a further problem with each. In the case of Buffery and Gray's hypothesis (Model 2), if greater language lateralization in females and bilateral representation of spatial functions in males *could* be accepted as an explanation of male spatial superiority in childhood, then insofar as the same sex difference in spatial skill exists in adulthood, the model by the same reasoning must imply the same sex difference in lateralization in adulthood. But as Harshman and Remington (1975a) have noted, in a discussion of this model, "the additional conclusion of a basic tendency to greater female *adult* lateralization need not follow" from developmental data (emphasis added). Buffery and Gray themselves noted the limitation of developmental data: "Sex differences in children are difficult to interpret when there is an advantage in favor of girls, since this may always be due to their general maturational advance over boys" (1972, p. 131).

The hypothesized earlier right hemisphere specialization for spatial processing for boys than girls (Model 1) seems to be limited in a similar way. For instance, in Witelson's (1975) study of discrimination of dichotomously presented nonsense forms, the girls eventually caught up with the boys so that by age 13, both boys and girls showed similar left-hand superiority on the task. Again, since males' spatial superiority persists through adulthood, something more than developmental lag – this time for females in the case of right hemisphere lateralization – must be involved.

Model 3: Greater lateralization in males

The third model anticipates the problem of exclusive reliance on developmental data and proposes that the male eventually equals and then surpasses the female in degree of left hemisphere lateralization, so that in adulthood, language in females is bilaterally represented. In other words, in adulthood, it is the male brain rather than the female that is further lateralized. Thus, where Buffery and Gray posited bilateral representation for spatial skill in the male brain, from which they drew a beneficial outcome for spatial skill, this new model posits bilateral representation for language functions in the female brain from which it draws a negative outcome for spatial skill.

Relation between cerebral lateralization and spatial ability
Why would bilateral language representation be expected to impede spatial ability? What could lateralization have to do with spatial ability in the first place? The phenomenon of lateralization now is so familiar that it is taken for granted, but consider that other paired internal organs of the body – lungs, kidneys, ovaries, testes – have identical functions, so far as is known, which means that we can survive quite well with but a single member of each pair. The cerebral hemispheres are the exception. Why? And why, under normal circumstances, does lateralization proceed, as it does in nearly all brains, to one hemisphere for language and the other for visuospatial functions, rather than to either side with equal probability? For that matter, why is there lateralization at all? Why do both hemispheres not subserve both linguistic and visuospatial functions equally, just as both kidneys work equally to maintain proper water balance and excrete metabolic wastes, or as both lungs work to remove carbon dioxide from the blood and replace it with oxygen? These questions are about the nature of the difference between left and right hemisphere processes and how the hemispheres are suited for their respective roles.

Nature of hemispheric specialization
Semmes (1968) has offered a provocative characterization of the hemispheres based on a study of the correlates of astereognosis in war veterans who had suffered left- or right-sided brain injuries (Semmes et al., 1960). The patients took a variety of tests, the most revelatory being three cutaneous tests: touch-pressure threshold, two-point discrimination, and point localization. As expected, the losses in sensitivity for the right hand were maximal after lesions of the left sensorimotor region. But for the left hand, the deficits were *not* clearly related to lesions of the right sensorimotor region. In other words, the researchers failed to find significant regional localization in the right hemisphere for these tests. The hemispheres

seemed to be asymmetrical for contralateral representation in somatic sensation. Similar asymmetries were found for motor reactions and for reflex functions, and for ipsilateral as well as contralateral deficits.

Assuming that the hemispheres differ in the principle of representation of such *elementary* functions, Semmes asked whether this difference is the basis for true hemisphere dominance at higher levels. She supposed that the more complex coordinations characteristic of the higher centers might be brought about by convergence of lower-level centers. Sensorimotor integrations involving a set of *similar* functional units presumably would be favored by the anatomical concentration of these units within a small area, that is, by focal representation of elementary functions.

> Where there is a higher concentration of units representing a particular part at one level, the convergence of these units upon those of the next level would bring about a more precise coding of the input and would thus make possible a more finely modulated control of the output. This finer control could be based not only on concentration of similar input elements, but also on an analogous concentration of similar output elements. The development of such precise control of the articulatory apparatus may provide an optimal substrate for speech representation in the left hemisphere. [Semmes, 1968, p. 23]

As for the right hemisphere, Semmes suggested that its diffuse organization may actually be advantageous for spatial abilities.

> The proximity of unlike functional elements in a diffusely-organized hemisphere would be expected to lead to a different type of integration from that characterizing a focally-organized hemisphere: *unlike* units would more frequently converge, and therefore one might predict heteromodal integration to an extent surpassing that possible in a focally-organized hemisphere. [Semmes, 1968, p. 23]

Thus, compared with functions that depend on a high degree of convergence of *like* elements, spatial functions may depend instead on the convergence of *unlike* elements – visual, kinesthetic, vestibular, and others – "combining in such a way as to create through experience a single supramodal space" (Semmes, 1968, p. 24).[15]

This structural characterization is consistent with recent functional descriptions. For example, it has been suggested that there are two distinct modes of coding operations, each specific to a single hemisphere: The left hemisphere operates in a more logical, analytical, computerlike fashion, analyzing stimulus information input sequentially, abstracting out the relevant details to which it attaches verbal labels; the right hemisphere is primarily a synthesizer, more concerned with the overall stimulus configuration, and organizes and processes information in terms of gestalts or wholes (Zangwill, 1960; Levy-Agresti & Sperry, 1968; Bogen, 1969).

Analytical-synthetic incompatibility: basis for cerebral lateralization
There is yet a further implication in Semmes's (1968) characterization that Levy-Agresti and Sperry (1968) have made explicit. They proposed that the left hemisphere would be "inadequate for the more rapid complex syntheses achieved by the [right] hemisphere" (p. 1151). And on this supposition, they proposed, as a possible basis for cerebral lateralization in man, a "basic incompatibility of language functions on the one hand and synthetic perceptual functions on the other" (p. 1151). The consequence of an incompatibility between analytical and synthetic-gestalt perception, they continued, is that during the evolution of the hominids, gestalt perception may have lateralized into the mute hemisphere.[16]

Lateralization thus may serve the useful function of providing separate cerebral loci for the two major, different types of information-coding operations. Unlike the kidneys or lungs, each cerebral hemisphere must be programmed predominantly for one kind of operation or the other. This design, Levy believes, does not provide for either maximum linguistic or maximum spatial function; rather, "given that both abilities are present, it provides for a *joint* maximization" in the majority of people (J. Levy, personal communication). Neither kind of coding operation, of course, is likely to be sufficient for most complex intellectual problems, but cooperation between the hemispheres is made possible through the corpus callosum, which connects them.[17]

Implication of incomplete lateralization
The most important implication of the model for our purposes is that some people may be more lateralized than others. The idea is that during pre- and postnatal development, genetic factors predispose each neural blueprint – language for the left hemisphere, spa-

tial for the right – to seek control of organization, not only for its designated hemisphere but for the other as well.

If the verbal blueprint wins, then the language-dominant hemisphere is fully and appropriately organized for verbal function, but the nondominant hemisphere also is partially organized for verbal functions, so that this hemisphere's organization is, to some extent, misappropriately designed for spatial functions. Such people will manifest perceptual-spatial defects . . . because the neural organization within this hemisphere is incompletely developed to serve spatial functions. [J. Levy, personal communication] [18]

Evidence for model 3

Analytical-synthetic incompatibility

Are left and right hemisphere coding operations incompatible? One source of evidence is clinical studies that relate lesion location to the performance of spatial problems. Removing the left temporal lobe significantly impairs memory for verbal materials but has only negligible effects on nonverbal spatial tasks for which the right hemisphere is specialized, such as maze learning, (Corkin, 1965; Milner, 1965) or memory for faces (Milner, 1968) and nonsense figures (Kimura, 1963b). But on the same spatial tasks, patients who have undergone right temporal lobectomy are significantly deficient, though their verbal ability is unimpaired. Their intact verbal skills thus seem to be inadequate for spatial analysis.

Bilateral language representation and spatial ability

Are nonverbal skills impaired in persons for whom verbal and nonverbal functions are known to be subserved by the same hemisphere? There are two subjects of choice. One is a person who sustained left brain damage early (before the acquisition of speech) and in whom language functions subsequently are known to be lateralized to the right (intact) cerebral hemisphere, so that both verbal and spatial functions are subserved by the same hemisphere. Basser (1962) identified a number of such individuals, but confined his analysis to language ability. Milner (1969), however, studied both verbal and nonverbal skills in such patients and reported evidence consistent with Levy's hypothesis. Right hemisphere speech lateralization, present in some patients with *early* left hemisphere damage, was associated with impaired nonverbal skill. These patients had lower performance than verbal IQ scores on the WAIS.

The second subject of choice is the left-hander. Since left-handers are known to be less well lateralized than right-handers (e.g., Goodglass & Quadfasel, 1954), language in left-handers would tend to be represented bilaterally, thus presumably reducing overall right hemisphere efficiency for spatial analysis. The model therefore predicts worse spatial skill in left-handers than in right-handers or, alternatively, a greater discrepancy between language and nonlanguage skills in left-handers than in right-handers.

As a first test, Levy (1969) compared the scores of 10 left-handed and 15 right-handed men on the verbal and performance scales of the WAIS. The WAIS was chosen on evidence that its two major factor scales – verbal and performance – reflect the operation of the left and right hemispheres, respectively (Reitan, 1955; Arrigoni & DeRenzi, 1964). The prediction was supported: The two handedness groups differed significantly on performance IQ (left, 117; right, 130) but not on verbal IQ (left, 142; right, 138). The results are more striking when expressed as discrepancy scores: For right-handers, the average difference between the verbal and performance scores was only 8 IQ points; for left-handers, 25 points.

As the very high scores indicate, Levy's subjects represented an extreme end of the ability scale (all were graduate science students), though Levy argued that there is no a priori reason to suppose that the difference between left- and right-handers within this select population would not be present in the population at large. Newcombe and Ratcliff (1973), however, failed to find similar differences in just such a population. They examined scores of more than 800 men and women from the general population on a short form of the WAIS. There were no differences either in pattern or absolute level of performance on the verbal and performance scales among 26 pure left-handers, 139 mixed-handers, and 658 right-handers in the sample, where handedness was determined by a seven-item questionnaire. The authors also cited studies of university students (Gibson, in press; Annett, unpublished data) and schoolchildren (Annett, unpublished data) that similarly fail to find the discrepancy in scores predicted by the hypothesis.

The results of four other studies, however, support the hypothesis. Miller (1971) compared 29 right-handed and 23 mixed-handed undergraduates on a test of verbal intelligence and a test requiring the visual manipulation of shapes in both two and three dimensions. Handedness was determined by questionnaire (Annett, 1967). Miller excluded 2 pure left-handers on the assumption that mixed-handers constitute the critical group for the hypothesis. The two groups' mean verbal scores were nearly identical, but on the spatial test, the right-handers were significantly better.

Nebes (1971b) tested two groups of subjects: 20 graduate students and postdoctoral fellows in biology and 32 college freshmen, with an equal number of left- and right-handers in each group. For the spatial task, each subject blindly explored with index finger an arc taken from one of the three sizes of complete circles lying in free view before him, and then had to match the arc with the circle of which it was a part. Nebes previously (1971a) had given this task to right-handed commissurotomized patients and found superior performance by the nonlanguage hemisphere. The result with the normal subjects was significantly poorer performance by the left-handers in both subgroups. Furthermore, the difference was found only on part-whole matching and not when parts were matched to parts or whole circles to whole circles, tasks for which hand differences do not ordinarily appear.

Silverman et al. (1966) found that 10 left-handed undergraduate males did significantly worse than 12 right-handers on a rod-and-frame test, mirror tracing, identification of the sidedness of drawings of body parts, and embedded figures.

Finally, James et al. (1967) compared 119 right-handed and 15 left-handed male college students on a battery of tests, including spatial orientation and speed and flexibility of closure, and found modest but significant differences favoring the right-handers.

On balance, then, Levy's hypothesis finds a reasonable degree of support. The prediction of discrepant verbal-spatial scores in left-handers is confirmed in two studies (Levy, 1969; Miller, 1971), though three others report no differences (Newcombe & Ratcliff, 1973; Gibson, in press; Annett, unpublished data). But the more general prediction of poorer spatial ability in left-handers than right-handers is confirmed in four independent studies, and perhaps had verbal tests as well as spatial tests been given, the predicted greater discrepancy in scores for left-handers also would have been found.

Sex differences in lateralization

With this background, we can go on to ask whether adult females are less lateralized, on average, than males and whether women number more than men among the group for whom the verbal blueprint wins.

Anatomical differences

Let us begin with anatomical data. Our interest here is with the right hemisphere. It is known that the occipital horns of the lateral ventricles are asymmetrical (Penfield, 1925), with the right occipital horn usually smaller or shorter than the left (McRae, 1948). If the occipital horn is smaller, the corresponding part of the cortex will be larger. This asymmetry is associated with handedness. McRae et al. (1968) measured occipital horn length on pneumoencephalograms and ventriculograms of 100 hospitalized neurological and neurosurgical patients. Of the 87 patients who were right-handed, 52 had a shorter right occipital horn, 9 a shorter left horn, and 26 horns of equal length. A shorter right occipital horn (i.e., larger right occipital cortex) thus correlated moderately well with right-handedness. Therefore, the right hemisphere, in right-handed persons, may be slightly larger than the left hemisphere in the posterior regions, the regions most critically involved in visuospatial functions.

If the male is indeed better lateralized than the female, so that his right hemisphere is better specialized for spatial processing, one might expect this asymmetry in the occipital region to be more marked in males than females. McRae et al., unfortunately, did not report separate analyses for their male and female subjects. There may be an indirect indication, however, that such a sex difference exists. Recall from Witelson and Pallie's study (1973) that in the youngest infant brains the appearance of an enlarged planum temporale in the left hemisphere was less marked in the male than in the female brains. Such a difference, if reliable, would be consonant with the known greater physical maturity of the female infant. The critical comparison is in the adult brain – after maturational sex differences no longer can be held responsible. Geschwind and Levitsky (1968) did not report having looked for sex differences in their adult brains, but this comparison has been made by Wada (1972; Wada et al., 1975). In their sample of 100 adult brains, Wada et al. (1975) found that a majority of the brains showed a larger left planum, but in a certain number of cases, the asymmetry was reversed so that the right planum was the larger. Of these brains, there were significantly more female than male.

If the parietal and occipital regions are larger in the right hemisphere, correspondingly reduced size of frontal and temporal areas is implied because there should be a correlation between right occipital horn size and right planum size. (A relation between horn size and planum size is reported to have been found in the case of the *left* hemisphere: Geschwind [1972, p. 194] has mentioned preliminary studies by Sheremata and Geschwind suggesting a high correlation between a longer left occipital horn and a larger left planum.) Thus the enlarged parietal and occipital areas simultaneously bestow on the right hemisphere an advantage for visuospatial functions and a disadvantage for language functions. If, however, a portion of the right temporal cortex is enlarged more frequently among females than males, as the data of Wada et al. suggest, this implies that those posterior regions are smaller in females than in males and that the right hemisphere's effectiveness as a spatial processor is less in females.

Clinical data

The anatomical data should be regarded very conservatively. The absolute number of female brains with reverse planum asymmetry found by Wada et al., though larger than the number of male brains, still was but a small fraction of the total, and the suggested link with right posterior hemisphere asymmetry has no independent confirmation, though Sheremata and Geschwind's findings for the left hemisphere make such a link a reasonable possibility. Nevertheless, the implied sex difference in lateralization is supported by certain clinical findings. Wada (1972) proposed that the language area in the right hemisphere suggested in his anatomical study is a reserve area that figures in determining whether the language center will shift if the left hemisphere is injured during early childhood (while the person is acquiring language). If this reserve language area is small, language functions may not shift or may shift only partially. But if this area is large, language functions will shift to the right, undamaged side, and language will be relatively unaffected (i.e., the person will not be permanently dysphasic). This proposition is a familiar one in neuropsychology, but with respect to left-handers rather than to females (Hécaen & Ajuriaguerra, 1964).

One immediate clinical implication of Wada's proposal, therefore, is that language disturbance after left hemisphere injury would be expected to be less severe and of shorter duration in females than in males, as it is in left-handers. Unfortunately, as Lake and Bryden (1976) have noted, detailed, extensive surveys have not been carried out. But from their own survey they concluded that the number of reported cases of profound aphasia in females is very small, with most of the cases having been reported in detail because they were unusual in some way (e.g., Chesher, 1936; Goodglass & Quadfasel, 1954; Ettlinger et al., 1956; Botez & Crighel, 1971). The implication in these reports, however, is buttressed by accounts of different patterns of deficit in males and females after brain injury.

Lansdell (1961) reported sex differences in the effects of temporal lobe operations on an explanation-of-proverbs test. After an operation on the language-dominant side, women's proverb scores were unaffected, but men's scores dropped. Lansdell (1962) later tested 22 patients on the Graves Design Judgment Test before and after unilateral temporal lobe surgery for the relief of temporal lobe epilepsy. The mean number of days between tests was 17 (range, 13 to 31). An additional 43 patients were tested for the first time 1 to 8 years after such surgery. The age range at surgery for the total of 64 patients was very broad: 11 to 57 years, with a mean of 31 years.

The Graves test is designed to measure "certain components of aptitude for the appreciation or production of art structure" (Graves, 1948) and, presumably, was chosen because it has an important spatial component, though doubt has been raised (Kimura, 1969). The subject chooses the design he prefers from two or three designs; 90 choices are required. Choices presumed to reflect inartistic preference are generally symmetrical; artistic choices, less so. Of patients whose operation was on the language-dominant hemisphere, the men's scores for artistic preference rose postoperatively, and the women's scores dropped. Of patients whose operation was on the spatial-dominant hemisphere, the men's scores dropped, and the women's rose. The interaction of sex, side of operation, and direction of change in performance was statistically significant. The operation apparently did not merely accentuate differences already present; separate analysis of the pre- and postoperative scores indicated a significant interaction between sex and side of operation only for the postoperative scores.

In a still later study of patients with either left or right temporal lobectomy, Lansdell (1968a,b) found that males with operations on the right side had the lowest WAIS nonverbal factor scores of all patient groups. There also was a significant correlation between extent of tissue removal on the right side and the nonverbal score for males but not for females. Similar results have been described in patients who have undergone commissurotomy for the control of epilepsy (Bogen, 1969). Male patients, postoperatively, showed impairment on a perceptual test (Street-Gestalt Test) relative to a verbal test (similarities) to a greater degree than did female patients.

Finally, there is a recent study by McGlone and Kertesz (1973). The subjects were 78 neurological patients chosen on the basis of unilateral brain damage and reported right-hand preference. Of these patients, 57 (35 males, 22 females) had left hemisphere lesions; 21 (13 males, 8 females) had right hemisphere lesions. There was no differences in chronological age or in intelligence (on Raven's Colored Progressive Matrices Test) between the groups with left and right brain damage, or between males and females. Linguistic and visuospatial abilities were measured by an aphasia test and items from the Block Design Subtest of the Wechsler-Bellevue Intelligence Scale (1944 edition), respectively. The aphasia battery measured fluency and grammatical correctness of spontaneous speech production, comprehension, repetition, naming, reading, and writing ability. As in previous studies, patients with left hemisphere lesions had significantly lower language scores and significantly higher spatial scores than patients with right hemisphere lesions. There were, furthermore, no main sex differences in either the aphasia or block design scores. This time, there were no sex differences in language deficit after left hemisphere damage, but on block design, males with right hemisphere lesions did tend to be worse ($p < 0.10$) than all other groups.

There also was a significant correlation ($r = 0.63$) between scores on block design and language for the 22 females with left hemisphere lesions, but no such relation for females with right-sided lesions or for males with left- or right-sided lesions. Thus, following injury to the left hemisphere the degree of language impairment predicted visuospatial disability in women, but not in men.

The authors of all these reports came to similar conclusions. Lansdell (1962) wrote that "some physiological mechanism underlying artistic judgment and verbal ability may overlap in the female brain but are in opposite hemispheres in the male" (p. 854). Bogen (1969) inferred that right-hemisphere-type thinking ("appositional" thinking) is less lateralized in females than in males. McGlone and Kertesz (1973) concluded that spatial ability may be more unilaterally represented in the right hemispheres in males than in females.

We should note that not all the clinical evidence is supportive. For example, Kimura (1969, p. 456) cited a personal communication from Hécaen that in the long series of cases presented in his 1962 publication, where sex of the patient was not reported, there was, in fact, *no* differential association for males and females of right hemisphere damage with unilateral spatial agnosia.

One problem with clinical evidence is that months or even years can go by before examination, in which time not only specific functional systems but their interaction may be affected. Thus, Buffery and Gray (1972) argued that Lansdell's (1962, 1968a,b) data may simply reflect "the disruptive influence of the lateralized epileptogenic lesion upon the brain's functional topography prior to temporal lobectomy" (p. 139). Although Lansdell's interpretation of his findings may be true for the epileptic brain, Buffery and Gray asserted that the reverse would hold for the normal brain. Fortunately, data on normal individuals are available, and they appear to support the clinical evidence.

Dichotic listening
One source of evidence is from dichotic listening studies. Harshman and Remington (1975b), through reanalysis of others' data together with their own new research, have marshaled impressive evidence for a stronger right-ear advantage for verbal stimuli in adult males, which they offer in explicit support of model 3. Another major source of evidence comes from work by Bryden. He reported (1966; cited in Lake and Bryden, 1976) right-ear superiority for spoken digits for 67% of left-handed males and 74% of right-handed males compared with only 50% and 57% of left-handed and right-handed females, respectively, among a large sample from the general population. Bryden also tabulated scores for males and females separately from a number of his earlier published and unpublished studies. Among 98 subjects tested on a

free-recall dichotic task with digit pairs, 73.6% of males (11 left- and 42 right-handers) showed right-ear superiority, compared with only 62.2% of females (3 left- and 42 right-handers) (cited in Lake & Bryden, 1976). Lake and Bryden (1976) corroborated this sex difference in a sample of 144 undergraduate subjects tested with pairs of consonant-vowel syllables.

Electrophysiological activity
Some electrophysiological data also suggest sex differences in lateralization. A comparison of left and right hemisphere activity recorded by electroencephalography in males during verbal and musical tasks (Herron, 1974) and in females during a speech task (Johnson, 1973) disclosed a higher incidence of lateralization among the males (Johnson & Herron, unpublished data; cited in Harshman & Remington, 1975a). Harshman and Remington observed in a discussion of these results that because different electrophysiological techniques were used for the two groups, the findings are only suggestive, though the electrophysiological data were highly correlated with performance on a dichotic listening test that confirmed the sex difference.

Harshman and Remington also cited a study in progress (Brown, Marsh, and Smith) that is indicating greater male lateralization of differences in the processing of nouns and verbs. The comparison is of averaged evoked potential within each hemisphere across two stimulus conditions: the word *fire* in "ready, aim, fire," and the same word *fire* in the sentence "sit by the fire." The difference in shape of the auditory evoked response is greater in the left hemisphere than in the right, but this asymmetry is much more pronounced in male than female subjects.

Visual field effects
Evidence of weaker language lateralization in females appears also in tachistoscopic measures. Males reportedly show stronger right-field superiority for verbal stimuli (Ehrlichman, 1971; cited in Harshman & Remington, 1975a). The same conclusion is implied, indirectly, when *spatial* stimuli are tachistoscopically projected: Males show greater, more consistent left-field effects (Kimura, 1969, 1973; McGlone & Davidson, 1973). McGlone and Davidson's study deserves special mention because it was designed to test directly the bilateral representation hypothesis as applied to

sex differences in spatial skill. Two experiments were done. The subjects in the first study were 48 secondary school students (16-year-olds) and 68 university students (20-year-olds), including an equal number of left- and right-handers in each group matched for sex, age, school, and grade. The spatial tests used were the WAIS Block Design (Weschsler, 1955) and Thurstone's Primary Mental Abilities Spatial Relations Test (SRA, 1958), which requires mental rotation of two-dimensional figures to match corresponding figures drawn in different orientations. All subjects were tested on dichotic word recognition as a measure of hemispheric localization for language.

On the Spatial Relations Test, males performed significantly better than females. There also was a significant interaction between ear superiority and sex: Females with higher left-ear scores (presumptive evidence of right hemisphere language representation) were worse on the Spatial Relations Test than any other group. Finally, there was a significant interaction between ear superiority and hand preference: Relatively more left-handers with higher left-ear scores fell below the median on the Spatial Relations Test. This effect was especially marked in the case of left-handed females (with higher left-ear scores). Finally, females, and left-handers with left-ear superiority on the dichotic test, had the lowest scores on the Block Design Test.

These results corroborate the usually obtained male superiority on spatial tasks, but they also indicate that the females with the lowest scores were those who had higher left-ear scores on the dichotic listening task. Finally, there was a complex handedness effect: Not all left-handers did worse than right-handers; rather, primarily those left-handers with higher left-ear scores. Thus, the worst performers on the spatial test were left-handed females with higher left-ear scores for dichotic word recognition. The results offer, however, only suggestive evidence in support of Levy's hypothesis because, as the authors pointed out, it must be shown that those subjects for whom language was localized in the right hemisphere were also right-hemisphere-dominant for spatial skill.

In a second study, they therefore retested 79 of the 99 subjects from Experiment 1. A tachistoscopic test of dot enumeration was used to determine which hemisphere subserved nonverbal functions. Previous studies (Kimura, 1963b; Warrington & James, 1967) had established that patients with right hemisphere lesions did worse on such a task than patients with left-sided lesions. And normal subjects are known to show left-field superiority for nonverbal targets presented tachistoscopically and right-field superiority for verbally identifiable targets (Kimura, 1966). McGlone and Davidson, therefore, reexamined these subjects' performance on the Primary Mental Abilities Spatial Relations Test (from Experiment 1) in light of their scores on dot enumeration as well as their prior performance on the dichotic listening task. The assumption was that if lower spatial scores reflected the coexistence of language and spatial modes within the right hemisphere, subjects showing right-ear advantage on the dichotic task combined with right-field superiority on the dot enumeration task (the presumptive index of language and visuospatial mediation by the same hemisphere) should have significantly lower spatial scores than subjects whose dichotic and visual field scores reflected different hemispheric representation of cognitive functions. The distribution of the 66 subjects who could be so classified was in a direction consistent with the bilateral representation model, though the differences were not significant.

Evaluation

On balance, the third model – greater lateralization in males – seems to have the broadest degree of support. The second and third models, as already noted, limit themselves to sex differences in childhood but cannot explain their persistence through adulthood when maturational sex differences no longer exist. Model 1 is excepted if we assume that the earlier right hemisphere lateralization in males represents not just a maturational lead but a genuine sex difference in right hemisphere specialization that goes beyond the childhood years. (In this case, it still remains to be explained why, on haptic discrimination tasks, girls eventually do show the same left-hand superiority boys show.) But if the first two models are limited for their restriction to the childhood years, the third model is limited for not taking these years into account. If, indeed, language lateralization has proceeded further in males than females by adulthood, the sex differences in language lateralization in childhood may well influence adult intellectual performance. Girls, because of their earlier linguistic development, may begin a course of *intellectual* development in which, compared with boys, language plays the larger role. This possibility finds support in studies of the relation between intellectual development in infancy and later life.

Cameron et al. (1967) reanalyzed scores on intelligence tests that had been administered to 35 male and 39 female subjects when they were between 5 and 13 months of age and again during the ages 6 to 26 years. Several of the test items on the infant scales were concerned with vocalization, including the vocalization of eagerness and displeasure, making vocal interjections, saying *da-da* or the equivalent, saying two words, and using expressive jargon. Although it is known that school-age intelligence cannot be predicted from total test scores until much later than 13 months, scores from this particular cluster of measures during infancy were related to intelligence at ages 6 through 26 years, but only for girls. Specifically, females with high Stanford-Binet IQs between 6 and 26 years, but especially between adolescence and adulthood, were found to have had high vocalization scores in infancy. Females with low IQs during this period had low vocalization scores in infancy. For boys, there was no relation between later IQ and vocalization scores in infancy.

This finding has been confirmed in a study of 76 children by Moore (1967). Each child was assigned a speech quotient as a measure of spontaneous babbling at 6 months and use of words at 18 months, and then was tested for general intelligence at 3, 5, and 8 years of age. The language scores during infancy were highly predictive of later measures of *general* intelligence in girls, but not in boys. Moore concluded that intellectual development in girls takes place primarily through linguistic channels. In boys, by contrast, *nonverbal* skills play a relatively more prominent role.

If Moore is right, we can begin to understand why, for instance, girls, but not boys, show positive correlations between level of artistic interest and competence on verbal reasoning tests (Bennett et al., 1959), and why the verbal and performance scores on the WISC are more highly correlated for kindergarten girls than boys (Fagin-Dubin, 1974). And perhaps we can understand what surprised Porteus when, in his first normative maze study (1918), he compared the children's performance on the maze test with their scores on the new and, in Porteus's view, too heavily verbal Binet test. The girls' scores showed closer agreement between the two estimates of intelligence: On the maze test, 79% of the girls tested within 1 year of their Binet mental age, compared with 67% of the boys.

The girl's early lead in language skills may dispose her along one intellectual course, and the boy's lag, though it eventually is overcome, may dispose him along a different path inasmuch as it would create an enforced longer period of time during which his primary way of encoding information from his environment would be nonlinguistic. Hutt (1970a, p. 102) put it this way: "When boys are still concerned with active exploration of their environment and with perceptual and motor skills, the girls are becoming increasingly adroit in their verbal and social functions." If we further suppose that boys are actually accelerated in the lateralization of right hemisphere functioning and are intrinsically more interested in and more skillful in mechanical and spatial relationships, then the circumstances are created for sex differences, even among adults, in the degree of linguistic involvement in life activities.

An alternative view: sex differences in preferred mode of cognitive analysis

It is conceivable that in addition to sex differences in hemisphere lateralization in adulthood, the different developmental histories of males and females therefore predispose them to the use of different methods of analysis of spatial problems, with females relying more on the less efficient left hemisphere modes, males on right hemisphere modes. In this light, let us reconsider some of the studies cited earlier in support of the model for lateralization difference.

Recall that Lansdell (1962), on the basis of evidence that performance on an aesthetic preference task depended more on the nonlanguage hemisphere in males, and more on the language hemisphere in females, concluded that the language modes are more bilaterally represented in females than in males. Lansdell's findings also could be seen as evidence that females rely, in making aesthetic judgments, more on left hemisphere, language modes, and males rely more nearly purely on right hemisphere modes. Consequently, one would expect to find the relations Lansdell obtained: in females, greater dependence of aesthetic judgment on the integrity of the left hemisphere; in males, greater dependence on the integrity of the right hemisphere.

McGlone and Kertesz's (1973) data can be interpreted in a similar way. Recall that block design scores were significantly correlated with language scores for female patients with left hemisphere lesions, but not for females with right-sided lesions or for males with lesions on either side. They viewed these findings as indicating "distinct sex differences in the cerebral lateralization of spatial functions" thus apparently embracing the model for lateralization difference. The data could fit a preferential processing model just as well. Indeed, the authors later appear to make this shift themselves, suggesting that the results mean that women use left hemisphere systems on tasks that involve spatial information.

The normal developmental and adult studies may be open to a similar interpretation. Rudel et al. (1974), for instance, saw their braille-learning data as indicating earlier right hemisphere development in boys, but they also suggested that children may use language to codify difficult discriminations, and that girls may rely more heavily on such language strategies than boys do.[19]

If any evidence could help arbitrate between a lateralization and a preferential processing model, perhaps it would be a clear demonstration of sex differences in *normal* individuals on, for example, the Wada test of hemispheric localization (see note 3). The lateralization model presumably would predict the following results: If it is true that in females language functions are more nearly bilaterally represented in the cerebral hemispheres, with residual or reserve language areas represented in the right hemisphere, then in right-handers, barbiturizing the left hemisphere would result in less severe dysphasia in females than in males.

The Wada test, however, has its dangers, and is recommended for use by its developers only in clinical situations where the site of language representation *must* be found prior to surgery. And in the clinical studies done thus far, sex differences seem not to have been considered. In the major report on the use of this test (Branch et al., 1964), the only reference to sex is in the initial description of the sample (78 males, 45 females). Even with analyses for sex differences in clinical uses of the test, the meaning of differences in either direction would be unclear in view of the usual broad age range of patients, differences in time of onset of lesion, and the inclusion of only those patients for whom there is prior doubt about the side of representation of speech, as was the case in Branch et al. (1964). Furthermore, one wonders whether the *clinical* criteria ordinarily used to infer hemisphere lateralization for language functions are sufficient to differentiate degrees of lateralization between the sexes, whether in normal or clinic populations.

For normal individuals, a better test might be continuous monitoring of hemisphere activity while subjects are mentally working on spatial problems to see whether proportionately more activity is confined to the right hemisphere among men than among woman. Such studies have been done (some are cited earlier), but the critical measure is determination of the relation, for both sexes, between the extent and confinement of activity to the right cerebral hemisphere and actual speed and accuracy of problem solving. A potentially revealing test would use men and women who, independently, have been shown to have good and poor spatial ability. If the men did better than the women, and if, moreover, among the women with poor spatial skill, mental activity were confined to the right hemisphere to as great an extent as among the men, this would suggest that the difference in spatial skill is traceable to sex differences in the efficiency of right hemisphere functioning, and not to the intrusion of left hemisphere modes antagonistic to, or incompatible with, spatial processing.

The Shepard and Metzler (1971) study, mentioned at the beginning of this chapter, suggests still another test. They timed adults while they judged whether pairs of objects represented in drawings were of the same or different shapes. The degree of rotation between the pair varied on each trial from 20° to 180° by 20° intervals. The time required to determine shape identity proved to be a linearly increasing function of the angular difference in the portrayed orientations of the objects. The slope of the obtained function corresponded to an average rate of mental rotation of approximately 60° per second. The value of this demonstration for the question at hand is that the mental operation that the results imply is taking place is almost paradigmatic of the kind of spatial coding operation for which the right hemisphere is specialized and on which males generally are superior to females. Consequently, a comparison of men's and women's reaction time functions might help explicate the nature of the difference between them.

Some evidence is available. Metzler and Shepard (1974) have described two experiments with college-age subjects – a total of 8 males and 8 females, and an equal number of left- and right-handed of each sex. In both experiments, the slopes of the reaction time functions were, on average, slightly greater for the left-handed subjects, though the magnitude of differences was not statistically significant. The same differences appeared for the comparison between sexes. In both experiments, the slopes were somewhat higher on average for women, and in one of the studies, the 4 females were among the 5 subjects whose reaction time functions showed the highest intercepts as well as the steepest slopes. And 3 of these women had the highest error scores and most variable reaction times (Metzler & Shepard, 1974, p. 188). What is critical, however, is that for every subject, the overall linearity and coincidence of functions for the depth and picture-plane pairs were found. In other words, the women and the left-handers, but the left-handed women in particular, were absolutely slower than the men in rate of mental rotation for all pairs of shapes. Furthermore, although both men and women took longer the greater the difference in orientation between the two shapes (i.e., reaction time increased as a linear function of the angular difference in rotation), the women's times increased relatively more than the men's as the difference in orientation of the two shapes increased (steeper slope).

These results suggest that the difference between the poorer and better subjects was one of degree, not of kind. Had the women's (and left-handers') greater slowness and lower accuracy stemmed from a different, nonrotational mental operation, such as a language-guided code, their time–angular distance functions presumably would have been nonlinear.

Only a few subjects were tested, however, and more importantly, all were selected from a quite homogeneous population to meet a minimum high criterion of performance on two paper-and-pencil pretests of spatial ability: for both tests, completion of 90% of the items, with less than 10% error on the completed items (Metzler & Shepard, 1974, p. 153). The poorer subjects, therefore, must have been well within the range of good to superior skill on spatial visualization and mental rotation. The procedure should be repeated with more representative subjects. The authors also mention that in another study, this one using 30 high-school students, the girls had substantially higher reaction times and error rates than the boys. No other details are reported.

There is the further point that this particular mental rotation task may be unsolvable by linguistic, analytical means. Other spatial tests, even rotational tasks like Piaget and Inhelder's (1956) three mountain task, may be solvable in less purely right hemisphere ways. Factor-analytic studies of children's performance support this possibility. A study by Mellone (1944) is a good example. Mellone gave 14 picture tests to 7-year-old school children, including maze drawing, picking out the mirror image of given drawings, completing various patterns of crosses and rings, and block counting. The children also took standard achievement tests in reading and arithmetic.

The maze test proved so difficult for the girls that it had to be omitted in the analysis of their scores. But factor analyses of the remaining scores disclosed remarkable sex differences in the patterning of results. The first factor, for both boys and girls, was a general factor present in all the picture tests. A second factor – school learning – was defined largely by the reading and arithmetic achievement tests and probably involved both verbal and numerical components, but, on evidence from the picture tests, more verbal for girls, and more number for boys. For the boys and girls, different picture tests loaded significantly on the school learning factor: for girls, those tests requiring the understanding and recall of precise verbal instructions; for boys, completing patterns of crosses and rings and block counting. Finally, for the boys, a third factor appeared, absent from reading and arithmetic scores but with high loadings on maze drawing and six other picture tests. Mellone called this a space factor, involving the ability to manipulate given material through visualization alone. This factor was *entirely* missing from the girls' scores. This could not have been simply because the maze test was deleted from their test battery, because the six other tests that had contributed to the boys' space factor ought to have been sufficient to appear as a space factor in the girls' scores without being absorbed in either the general factor or one of the specific factors. The boys were superior to the girls on some of these other tests (e.g., on block counting and mazes), but not on others. Mellone therefore concluded that even if girls lack the visual-manipulative ability possessed by boys, they "can yet succeed in these tests by the exercise of their general intellectual ability. Indeed it seems clear that the factorial composition of any test, while it must, if our analyses have any objective factual basis, remain invariant for the same or similar groups of subjects, might yet vary considerably with groups dissimilar in sex, age, or environment" (Mellone, 1944, p. 14).

All these considerations suggest that, depending on the type of spatial problem and level of individual skill, sex differences in both degree of language processing and lateralization can underlie sex differences in spatial ability. If females are *neurologically* disposed to be better in linguistic than visuospatial skills, and (because they are maturationally more advanced) to acquire language earlier than males, it is reasonable that they ultimately come to depend on linguistic modes more than males do and in more situations. Definitive separation and independent assessment of the lateralization and preferential processing accounts may be impossible, for the tendency to exploit those skills to which one is genetically predisposed must begin at the very outset. Such dependency can work to the female's disadvantage in problems for which linguistic modes are less efficient.[20]

POSSIBLE EFFECTS OF SEX HORMONES ON BRAIN SPECIALIZATION AND NERVOUS SYSTEM ACTIVITY

If differences in neurological specialization at least partly underlie sex differences in spatial ability, then how do such differences actually come about? What sets off one brain and nervous system more than another to a better subserver of spatial ability? We already have reviewed evidence strongly implicating the sex hormones. Presumably, then, the male and female sex hormones, respectively, are instrumental in setting into operation the spatial and verbal blueprints for organization of the cerebral hemispheres. Presumably, too, these different patterns of cerebral organization originate in embryonic or fetal neurohormonal events. If so, it is understandable that hormonal substitutes given individuals with Turner's syndrome later in life fail to alleviate their deficits in direction sense (Money, 1968a). The space-form dysgnosia of the girl with Turner's syndrome thus might be traceable to the effects of early hormone insufficiencies on right cerebral hemisphere functioning and organization.

Little is known, however, about how the sex hormones actually influence nervous system activity. An account by Broverman et al.

(1968) has gained some attention. These investigators conceptualized cognitive functioning to be the result of an interplay between two competing systems: the adrenergic nervous system, which has a mobilizing function that prepares it for activation and thereby presumably facilitates performance of simple perceptual-motor tasks; and the cholinergic system, which functions to promote protection, conservation, and relaxation or inhibition of activity and thereby contributes to the cognitive ability to delay initial responses to obvious stimulus attributes in favor of responses to less obvious stimulus relationships. These latter are the requirements, Broverman et al. argued, commonly found in such spatial tasks as mazes, embedded figures, and rod-and-frame tests.

Males, therefore, excel on spatial tasks because males' androgen steroids are presumed to produce a balance of biochemical factors favoring the cholinergic type of neural functioning. Females tend to be adrenergic, and do better than males on tasks that require "rapid, skillful, repetition, articulation, or coordination of 'lightweight,' overlearned responses" (Broverman et al., 1968, p. 25). The authors cited, as examples, speed of color naming (Staples, 1932; Stroop, 1935); the Digit Symbol Subtest of the WISC and WAIS and other tests requiring rapid perception of details and frequent shifts of attention (Schneidler & Paterson, 1942; Paterson & Andrew, 1946; Norman, 1953; Miele, 1958; Gainer, 1962); and fine manual dexterity (Gesell et al., 1940; McNemar, 1942; Tiffin & Asher, 1948). Girls and women excel in all.[21]

The theory has been attacked (Singer & Montgomery, 1969) and vigorously defended (Broverman et al., 1969). Unfortunately, most of the direct supporting evidence is from animals. What evidence there is for human beings, however, suggests that there *is* a relationship between sex hormone level and spatial skill, but it is highly complex and different for males and females. Broverman et al. (1964; Broverman & Klaiber, 1969) determined the relation between androgenicity and performance on spatial tasks (actually, a measure of the contrast between tests loaded on a spatial factor and tests loaded on a fluent production factor, e.g., rapid repetition of overlearned skills) in adolescent boys and young men, in whom androgenicity was indexed by appearance of body and genital hair and other somatic features. The relation was inverse: The more androgenized (in the adolescent group, earlier maturing) males were relatively worse on spatial tests; the less androgenized males, relatively better. Peterson confirmed this relation for males (1973; cited in Bock, 1973), but found the opposite relation in females; that is, the contrast was related directly to androgenicity. On this basis, Peterson proposed that the relation between the cognitive contrast and the androgen/estrogen ratio is curvilinear such that at intermediate levels the androgen/estrogen ratio is most favorable to high spatial ability and least favorable to fluent production. The less androgenized male is a better visualizer and is less fluent than the more androgenized male; the more androgenized female is a better visualizer and is less fluent than the more estrogenized female. The good visualizer of either sex, therefore, is less sexually differentiated than the fluent producer (Bock, 1973, p. 451). That spatial ability requires some *minimum* level of androgen – as we earlier saw could be adduced from the spatial deficits characteristic of Turner's syndrome in females and the testicular feminization syndrome in males – is consistent with the proposed curvilinear relation between androgenicity and spatial ability.

What remains to be explained is the link relating androgenicity and sexual differentiation with the evidence reviewed earlier for a sex-linked major gene influence on spatial ability. Bock (1973) has suggested that the connection may lie in the observation that the timing of bone ossification, which is known to be influenced by steroid hormone levels, appears to be partly under the control of a sex-linked gene (Garn et al., 1969). The implication is that the degree of *within-sex* sexual differentiation is influenced by a sex-linked gene, and this differentiation in turn influences spatial ability and fluent production. As a test, Peterson (1973) calculated correlations of somatic androgyny measures for all sibling pairs in a sample of children and obtained values conforming closely to those predicted by a sex-linkage hypothesis.

This model implies that the finding of Broverman et al. (1968) of better spatial skill in late-maturing (i.e., less androgenized) boys than in early-maturing boys may reflect the operation of a sex-linked gene controlling the timing of release of androgen rather than the expression of spatial skill directly. Perhaps the X-linked recessive trait for enhancement of spatial skill is actually evidence for sex linkage of androgenicity, spatial skill itself being merely an expression of the sex-limiting effect of androgenicity.

EVOLUTIONARY SELECTION FOR MALE SPATIAL SUPERIORITY

Inasmuch as spatial ability is expressed more frequently in males, and inasmuch as this expression appears to be at least partly under genetic and hormonal control, a last and larger question may be posed: What evolutionary processes are responsible? What is the species advantage of this behavioral sexual dimorphism? The answer is sheer speculation, with most speculators offering variations on the theme of man as hunter and territory marker and woman as bearer and nurse of children. Masica et al. (1969) suggested that sex differences in spatial ability are remnants of "specialized evolutionary adaptations, reflecting the sharper visual-perceptual skills employed by the male in territoriality and mating." Lee and DeVore (1968) emphasized the importance of the adoption of a hunting-gathering way of life that, because of the female's different biological role, would have necessitated some division of labor. The female's food gathering, therefore, would be closer to the vicinity of the camp, the male's farther ranging. That these differences continue to appear even among contemporary hunter-gatherers such as the Ainu (Watanabe, 1968) is seen as supporting this line of reasoning. Because males appear to range farther than females even in societies that gather their food at the local supermarket, the ancestral lights may still be flickering.

Another essential of evolution is thought to have been walking erect, freeing the hands for increased manipulation and use of tools and weapons. Leakey (1961) has speculated that prehistoric man (i.e., males) used bone to split flakes from pieces of flint to make stone tools and weapons. It is reasonable to suppose that spatial visualization and an ability to anticipate the results of a given blow would have high survival value in such practices and, therefore, would be selected for the practitioners.

CONCLUSIONS

I began this chapter by noting that I did not wish to set experiential, genetic, and neurological explanations *against* each other as though we merely had to choose among them. Our job, instead, is to try to understand the nature of the interactions between the genetic material, the environment, and the neurological mechanisms culminating in individual variation in spatial ability. These interactions, we now surely can see, are extremely complex. Is spatial ability a heritable trait? The studies of the concordance of ability in monozygotic and dizygotic twins provide strong confirming evidence. And the patterning of family correlations of spatial skill in combination with the proportion of males typically found to exceed females on tests of spatial skill is consistent with the view that spatial ability is a recessive sex-linked characteristic. Finally, research on persons with Turner's syndrome as well as normal individuals indicates that the actual trigger may be hormonal, though much work remains before the nature of the role of the sex hormones is understood.

The implication of a hormonal-genetic sex-linked factor indicates only that the cause is genetic rather than environmental. Ultimately, neurological mechanisms must be identified. The suggested involvement of right-sided parietal-temporal-occipital dysfunction in the case of the girl with Turner's syndrome raises the question whether differences in right hemisphere specialization underlie sex differences in spatial ability in normal individuals. We must be very cautious in these speculations, and extremely careful in our choice of words. If right hemisphere dysfunction is the operative factor in the space-form dysgnosia associated with Turner's syndrome, we do not want to suggest dysfunction in the case of the normal female – at most only less efficiency. After all, Turner's syndrome is not associated with hard evidence of right hemisphere disease. The deficit is presumed functional and would be presumed functional in the case of the normal female a fortiori.

To any biological explanation of human *psychological* variability, one sometimes hears the objection that a single exception – a single society in which the usual behavioral differences are diminished or even reversed – proves that environmental-cultural factors play the vastly more significant, even the only, role. The reported absence of sex differences in spatial skill in Eskimos has been cited in this regard (Berry, 1966). But this environmentalist argument remains logically suspect, for there is nothing in a biological explanation of sex differences in spatial ability or any other behavior to suggest that the expression of biologically rooted differences cannot be suppressed in certain circumstances. The improvement of a particular skill, for example, through a training procedure like Brinkman's (1966), does not necessarily imply that a lack of similar experiences was responsible for the deficient expression of that skill in the first place. We must avoid the fallacy of the evidence from cure. And the point works both ways: We must be equally cautious in interpreting instances in which remedial measures fail. Thus, the failure of Thomas et al. (1973) to teach the concept of horizontality to college women does not necessarily mean that a lack of certain experiences could *not* have underlain their subjects' deficient performance. But on balance, given the near universality of male spatial superiority, one begins to suspect that the *fundamental* reasons may be at least as reasonably sought in phylogenesis as in ontogenesis. One's attention is drawn to biological factors – to hormonal and neurological mechanisms – as the possible *predisposing* conditions.

Environment and learning are still critical, not so much as influences imposed from without as generated from within the individual. We can speculate, therefore, that the genetic-hormonal factors that create male and female children also predispose the operation of modes of both cognitive and physical activity that tend to enlarge and widen initial differences. The boy more naturally involves himself in experiences that sharpen spatial skills; the girl involves herself more in experiences that strengthen interpersonal skills. The environments into which identifiably male and female children are born – the attitudes and expectations of parents, the biological and social roles prescribed, the very structure of human societies – undoubtedly act to strengthen these differences in activity still more. Must it be so? In some ways it must. If, in the play of infancy and childhood, movement, strength, dominance, object manipulation, and exploration of space are critical for the nurturing of spatial ability, then children who find such play more attractive will practice it more often and more intensely. And boys, it appears, have naturally greater appetites for such play than girls do. It is hard to stimulate such interests in a child who lacks them, as parents and teachers know.

In sum, spatial ability is not biologically conferred the way color blindness or blond hair are conferred. Color blindness and hair color are, respectively, a cognitive deficit and a physical characteristic which, when the genotype is present, are expressed whether the environment is accepting or hostile for such traits. The expression of spatial ability is far likelier to depend on an appropriate environment. If quality of early play, exploration, and involvement in one's

Sex differences in spatial ability 487

environment are critical, circumstances must be created that provide the opportunity for such experiences. There must be a visual, tactual, and auditory environment for the child to inspect. But what the child does in that environment is not so easily controlled. We are talking, then, about basic modes of temperament and cognitive style that probably remain fairly stable throughout life. In this respect, the experiential underpinnings of spatial ability are rooted in our biology.

These speculations define new problems for developmental research, such as the correlating of play patterns in infancy and early childhood to later spatial skill. Analysis of Turner's syndrome suggests that for normal children such an investigation could be most revealing. Girls with Turner's syndrome have been described as the "epitome of femininity," as "very maternal in their play and child-care interests from infancy onward" (Money, 1968b, p. 20; Money et al., 1968). One wonders whether these children, in their earliest play behaviors, involved themselves much less than normal children in experiences that might nurture spatial skill.

With normal persons, the evidence seems reasonably strong that there are sex differences in right hemisphere specialization for visuospatial functioning, but the possibility that males and females in addition differ in their manner of solving spatial problems should not be discounted. If intellectual development takes place through linguistic channels to a greater degree in girls than in boys, as several studies suggest, and if, as much research suggests, girls acquire language earlier than boys, the circumstances are set for sex differences even in adulthood in the involvement of language in life activities, including spatial problems. And language modes *usually* are not suitable for spatial analysis.

The qualification "usually" is critical: Mellone's factor-analytic study (1944) of children's problem solving, described earlier, is but one indication that at least some spatial problems *can* be solved in nonspatial ways – in the case of the girls in Mellone's study, apparently through verbal means. This possibility is suggested also in Berry's (1966) study of Eskimos. The salience of spatial location is well reflected in Eskimo language, which possesses an intricate system of words called *localizers* that aid in the location of objects in space. Since the localizers form an integral part of the word, their use – thus the spatial distinction itself – is obligatory for all Eskimos.

Some may object to the argument that spatial problems can be solved by nonspatial means. They may assert that a spatial problem by definition is *unsolvable* by other means, that problems whose solutions require the mental rotation of images or the synthesizing or integrating of information from different spatial locations, such as the various coordination-of-perspectives or conservation problems described earlier, can be solved only by synthetic modes represented predominantly in the diffusely organized right cerebral hemisphere. They may further assert that if the same spatial problems can be solved analytically, through following rote language rules and operations, the problems are not purely spatial, but must contain analytical as well as synthetic elements. An example of this reasoning is in research with clinical populations. In one instance, the investigators (De Renzi et al., 1969) failed to find differences between patients with left hemisphere damage and patients with right hemisphere damage on tests of visual and tactile memory for position. The conclusion, therefore, was that the tests might have involved both a spatial and a language factor. "The first is likely to be deranged to a higher degree by injuries to the right hemisphere, while the second is obviously represented in the left hemisphere only. Thus, different mechanisms might be responsible for the quantitatively not dissimilar failure of the two hemispheric groups"(p. 282). Redefining a test on the basis of the test results is logically questionable, though in this particular instance, the authors' interpretation nevertheless may be correct. On the other hand, the patients with right hemisphere damage could have compensated by attacking the spatial problems in new ways.

As Williams (1970) has observed, the disruption of specific cerebral areas often causes the disruption of specific mental skills, but it does not have to be so: "The dysphasic patients can still communicate, the amnesic one uses a notebook or some other form of mnemonic to organize his daily life. Even the patient with visuoperceptual disorders can compensate for his loss by developing new cues to orientate himself by" (p. 124). Williams offered a convincing illustration of her point: " 'One can always use geometry,' said a world-famous mathematician, as he put the pieces of the WAIS Object Assembly together soon after having lost most of his right cerebral hemisphere as a result of a traffic accident" (p. 124). The rules of geometry are logical rules, expressible in linguistic form.

Semmes et al. (1963) described a similar reliance by brain-injured patients on a route-tracing task. The subject had to walk through paths represented on maps – cardboard plaques depicting nine circular red spots on the floor of the testing room. The path was drawn in black lines, with a circle at the starting point and an arrowhead at the end. Several patients seemed to use a verbal method of translating the coordinates of the map to the coordinates of the room when these systems were not spatially coincident:

> They named each of the nine spots on the floor by its position (northeast dot, south central dot, etc.), and likewise gave a similar name to each spot on the map. At each choice point they consulted the map, seemingly with regard only to its coordinate system, and announced, for example, "Now I must go to the southwest dot." [Semmes et al., 1963, p. 762]

The authors made the sensible observation that difficulty in spatial orientation might be masked in patients whose language functions are preserved and who use verbal mediators.

The message for basic neuropsychological research is clear. In research with clinical populations, for example, we must consider how spatial problems might have been solved before brain injury. Individuals who, prior to right hemisphere injury, depended to a large degree on language operations (whose sense of direction, for example, depended more on the rote learning of landmarks and street names than on cognitive "maps") would be less impaired in spatial tasks than individuals who, before injury, had used primarily spatial modes. So far as sex differences are concerned, the greater likelihood is that women more than men would make up the former group. But sex differences notwithstanding, any neuropsychological analysis of spatial ability must not presuppose that there exist only a few means by which spatial problems can be *successfully* analyzed. Once alternative analytical solutions are identified, we can go further and put left hemisphere skills to use. An interesting possibility arises once again with Turner's syndrome. In light of their relatively unimpaired and sometimes even excellent linguistic ability, perhaps these individuals' space-form dysgnosia can be partly overcome by proper use of their linguistic ability in spatial tasks for which alternative analytical solutions are available.

We also must begin much more thorough evaluations of the relationship between time and side of brain injury and performance in different *kinds* of spatial tasks. Kohn and Dennis (1974) have made a most promising start. They compared hemidecorticate infantile hemiplegics on a variety of spatial tests, some requiring simple recognition (Ghent overlapping figures, Street completion test), others requiring complex directional skills (visually guided route finding, road map test). Both groups performed nearly perfectly on the recognition tests, but on the more difficult tasks, the left hemidecorticates were significantly superior to the right hemidecorticates. The authors, therefore, concluded that perinatal cerebral pathology, regardless of its laterality, need not prevent mediation of basic, simple, visuospatial abilities in the less affected hemisphere. Either hemisphere can provide a substrate for at least some of the functions based on analyses of spatial components.

Spatial ability and cognitive style

I want to end this chapter with a caution. Certain spatial tasks – the rod-and-frame and embedded-figures tests in particular – are usually called tests of field independence and field dependence. Sex differerences on these tasks nearly always are cited as evidence that females are more field dependent, global thinking, or cognitively undifferentiated, and that males are more field independent, analytical, or cognitively differentiated in *general* cognitive or mentative style. Such interpretations are groundless and have been avoided throughout this chapter. As Sherman (1967) has pointed out, because the most popularly used tests of field dependence – the rod-and-frame and embedded figures tests–are visuospatial tasks, the male's superior performance is understandable and should be *narrowly* interpreted, not generalized into an all-encompassing statement about cognitive style. Other measures of field dependence which do not have a spatial component (e.g., rotator-match brightness constancy, paper-square-match brightness constancy, body steadiness) have not yielded sex differences (Witkin et al., 1954). Sherman (1967) has neatly expressed the underlying fallacy of the field dependence interpretation. She wrote that sex differences on such tests might be explainable "without any reference to field, without any need to infer a passive approach to the field, globality, or lack of analytical skill. The fallacy involved is similar to concluding that women are more analytical than men based on findings of superior female ability to decontextualize the red and green figures on the Ishihara Color Blindness Test" (p. 292).

Sherman's point is underscored when the assumption that males are analytical, females global (or synthetic), in general cognitive style is carried to its logical conclusion. As we have seen, the kinds of abilities or skills subserved by the right cerebral hemisphere have been characterized as diffuse, synthetic, or appositional; if females are global in general cognitive style, they should be *better*, not worse, than males in spatial tasks. In other words, the conclusion that women are less analytical in cognitive style, and the proferring of such a conclusion as an explanation of their poorer performance on embedded figures, rod-and-frame, and a host of other spatial tasks, are directly contrary to accepted characterizations of the nature of spatial tasks and spatial ability, and the nature of hemispheric specialization.

A similar point may be made with respect to sex differences on mazes. We have characterized the maze test as a spatial or directional sense test. But Porteus's aim was to evaluate planning capacity and foresight, to measure an individual's ability "to carry out in proper sequence and prescribed fashion the various steps to be taken in the achievement of a goal" (1965, p. 6). That he chose to do this by setting for his subjects the task of finding their way out of a printed labyrinth is not being criticized: It seems to be an excellent measure. But the fact that males do better than females on mazes surely does not mean that males have more planning capacity and foresight than females in any general sense. Rather, it means that where such attributes are measured with a spatial test, males will be favored.

The scope of our inquiry into the sources and nature of sex differences in spatial ability, therefore, has been justly narrow. We have sought to understand a sex difference in a fairly specifiable set of abilities that are not to be equated with thinking style, cognitive style, or foresight, and whose relationship to these more general attributes remains to be determined.

We conclude on a familiar note: Our review leaves us with far more questions than answers. My hope at least is that these are better questions than had been asked before.

NOTES

1 Differences on two other tasks, however, were less clear. On a stick rotations task, the examiner and patient each had four black-tipped

wooden sticks. The examiner made patterns with his sticks that the patient had to copy exactly when seated beside and then opposite the examiner. On a pool reflections test, a single geometrical shape was presented above a row of five identical shapes, only one of which, because of the position of an internal feature, was identical to the standard in 180° rotation. The patient had to select the comparison figure that was identical, after 180° rotation, to the standard.

2 The dichotic listening technique, devised by Broadbent (1954), usually involves simultaneous presentation of one target to one ear and a different target to the other. Several such pairs of targets are delivered in sequence during a trial, and the subject then is asked to report what he has heard, in whatever order he likes. Broadbent wanted to study the limits of immediate memory on two separated channels, in particular to test the possibility that listeners can hear two spatially separated signals successively rather than simultaneously. He found that correct responses to binaural lists of spoken digits were reported in such an order that all digits presented to one ear were reported before any presented to the other ear.

The value of this technique for the analysis of hemispheric specialization was discovered by Kimura (1961). She presented digits as the target sounds in a test with brain-damaged patients and found that patients with left hemisphere damage in the temporal region reported fewer digits correctly than did patients with right temporal damage. This was expected. But, fortuitously, she also found that patients, regardless of the region damaged, reported digits heard with the right ear more accurately than those heard with the left. People without brain damage did the same. As the left and right ears do not have different basic capacities to detect sound, this right-ear superiority presumably is related to how the ears are connected to the brain. Though the auditory system is less crossed, anatomically speaking, than the visual or tactual and motor systems, the crossed auditory connections are still stronger than the ipsilateral connections. The left hemisphere "prefers" to listen through the right, contralateral ear, and vice versa. The experiments most frequently cited as providing the neurophysiological basis for this generalization are studies by Tunturi (1946) with cats and by Rosenzweig (1951) with dogs. Rosenzweig concluded that at the auditory cortex of each cerebral hemisphere each ear is represented by a population of cortical units. The population representing the contralateral ear is larger than the population representing the ipsilateral ear. The two populations overlap, so that some units belong to both populations.

Because the left hemisphere is specialized for speech perception, it is predictable that speech sounds presented to the right ear would be perceived better. And vocal nonspeech sounds should show a left-ear superiority, suggesting that they are better processed by the right hemisphere.

3 This test was devised by Wada (1949, 1951; Wada & Kirikae, 1949) originally to study the mechanism of the spread of epileptic discharge between the cerebral hemispheres. The technique involves the injection of a barbiturate, sodium amytal, into the common carotid artery, which is distributed through the upper neck, the larynx, pharynx, tongue, face, most of the brain, the eyes, and appendages. This injection was found to induce a temporary loss of function in the cerebral hemisphere ipsilateral to the side of injection, including dysphasia when the artery supplying the hemisphere dominant for language was injected. The following description is based on Wada and Rasmussen (1960). Typically, the right and left sides of the patient are injected on different days. The injection is made with the patient counting, with his forearms raised and his fingers either moving or gripping the examiner's hands. The knees are drawn up so that the feet are resting on the bed close to the buttocks. As the injection is completed, the contralateral arm and leg become flaccid and slump to the bed. The ipsilateral arm and leg remain in the air, and voluntary movements can be carried out on this side as soon as an initial few seconds of confusion are over. The patient usually hesitates or stops counting near the end of the injection, but if the artery serving the nonlanguage hemisphere has been injected, the patient resumes on request within 5 to 20 seconds, and then can name objects accurately while the contralateral hemiplegia (a paralysis confined to the opposite side of the body) is complete.

If the artery for the speech-dominant hemisphere has been injected, the patient is unable to continue counting while the contralateral hemiplegia is complete. On command, he can carry out voluntary movements with the ipsilateral hand and leg after the brief period of confusion has passed. This ability demonstrates that he is cooperating and that his lack of speech is not due to disturbances of consciousness or cooperation. Usually there is dysphasia for 1 to 3 minutes, after which normal speech returns.

The procedure is useful, in the surgical treatment of focal epilepsy and other cerebral diseases, for the determination of lateralization of speech function, particularly in left-handed and bidextrous patients and in right-handed patients for whom there is doubt as to the cerebral hemisphere dominant for speech.

4 The poignancy of Ravel's illness is reflected in these biographical passages:
Often in the past when he had been questioned about his work he had replied, "It is all finished – only setting it down on paper remains" – and then perhaps months would go by before he did the actual writing. Therefore, when they [Ravel's friends] inquired about the progress of Jeanne d'Arc [an epic musical poem he

wanted to write] and he replied, "I have it all in my head," they did not realize that now it was literally impossible for him to write it out [Goss, 1940, pp. 249-250]. He used to say: "I've still so much to say, so many ideas in my head." [Myers, 1960, p. 86]

Like his musical sense, Ravel's sense of direction also seemed unimpaired:

> The faithful Hélène Jourdan-Morhange accompanied him on the long walks he liked to take, roaming through the Rambouillet woods together. The violinist was astonished at how accurately Ravel knew every path, every clearing, the habitat of hundreds of plants and trees. [Stuckenschmidt, 1968, p. 244]
>
> His "bump of locality" was, surprising enough, good, and he was able to pilot Leyritz [the reference is to a tour of Spain with Léon Leyritz, in February 1935] to all kinds of out-of-the-way places usually missed by tourists. [Demuth, 1947, pp. 43-44]

5 This "spatial" quality seems to be most pronounced in the music of certain contemporary composers such as Edgar Varèse, Luciano Berio, and George Crumb. Indeed, Varèse has said as much of one of his compositions:

> Intégrales furent conçues pour une projection spatiale . . . Pour mieux me faire comprendre – car l'oeil est plus rapide et plus discipliné que l'oreille – transférons cette conception dans le domaine visuel et considérons la projection changeante d'une figure géométrique sur un plan, avec la figure et le plan qui tous deux se meuvent dans l'espace, mais chacun avec ses propres vitesses, changeantes et variées, de translation et de rotation. La forme instantanée de la projection est déterminée par l'orientation relative entre la figure et le plan à ce moment. [Anhalt, 1961, pp. 34-35]

6 The analysis of possible sex differences in certain aspects of musical skill, and the linking of any such differences to spatial ability, necessarily await further analysis of hemispheric functioning and music perception. The demonstrations reported here are only a beginning. Studies that attempt precise specification of the dimensions of difference between sounds recognized better by each hemisphere (e.g., Darwin, 1969; Gordon, 1970) and that take musical skill into consideration (Bever & Chiarello, 1974; Gordon, 1975) foretell many complexities.

7 Still another answer to the social learning interpretation of the delayed appearance of sex differences in spatial skill is that the boy's slower maturation, until and through adolescence (Tuddenham & Snyder, 1954; Bayley, 1956), may be obscuring any real differences in spatial skill favoring boys at any particular age. (One sometimes hears the phrase, "maturing of spatial ability.") Girls, on average, begin puberty about 16 months earlier than boys (10.6 years versus 11.8 years) and

reach maximum physical growth rate 38 months earlier than boys, with the difference decreasing slowly thereafter (Nicolson & Hanley, 1953). The problem with this answer is that the age period 10 to 11 years would be the time of the boy's greatest maturational lag, but as we already have seen, in many cases, boys' spatial superiority is first clearly expressed at this very age (e.g., the road map test).

8 It remains to be seen whether that much smaller proportion of college men who do not know the principle will also fail to profit from such experiences. The study of Thomas et al. (1973) included 8 such men among 62 unselected male subjects, but they were not included in the training group.

9 Money (1968a,b), however, reported finding no evidence in girls with Turner's syndrome of either color blindness or color agnosia, contrary to the report of Polani et al. (1956), cited in the text, of a significantly higher incidence of red-green color blindness in individuals with Turner's syndrome than in normal females. The question deserves further study in light of its important implications for the recessive trait hypothesis of spatial ability.

10 Money has offered as still another indication of parietal lobe involvement in the individual with Turner's syndrome certain reported similarities of temperament between the girl with Turner's syndrome and the patient with right parietal damage. He cited Langworthy's (1964) description of a woman with right parietal lobe damage who seemed less alert and more passive, placid, and indifferent to things that previously aroused her attention and emotion (Hampson et al., 1955; see also Shaffer, 1962). The girls with Turner's syndrome likewise appear placed and stolid – resigned, as Money put it, "to the special demands which life imposes on them by reason of the dwarfism, pubertal failure before treatment, sterility, and other physical disabilities. Their tolerance of these indignities is in marked contrast to the emotional disturbance that rather frequently characterizes the mothers" (Money, 1968a, p. 29).

It would be helpful, too, to know whether individuals with Turner's syndrome showed other spatial deficits likely to be associated with parietal disease, such as an impaired ability to recognize incomplete figures (e.g., Kinsbourne, 1966; Warrington & James, 1967; Lansdell, 1968b).

11 In fact, it had been known for some years that the planum temporale was larger on the left than on the right side (Kakeshita, 1925; Von Economo and Horn, 1930; Pfeiffer, 1936; cited in Geschwind, 1974; Wada et al., 1975).

12 Buffery and Gray (1972) themselves referred to two other anatomical differences cited by Lansdell (1961) in support of the view that the female brain is neuroanatomically earlier and better equipped than the male brain for language. One is Conel's (1963) investigation of 8 brains

from 4-year-old children. Lansdell reported that there are no consistent differences among the hemispheres until sex is taken into consideration. Then, "two noteworthy differences emerge." First, Conel's Table IX shows that in 4 of 5 female brains, the amount of myelination is greater in the left hand subarea than in the corresponding area on the right, and in the 3 male brains the difference is reversed. Second, in Table X the number of exogenous fibers in layer I of both the primary-motor and primary-sensory areas was greater on the right in the 4 female brains but greater on the left in 2 of the 3 male brains. Lansdell acknowledged that neither set of differences is statistically significant, but he conjectured whether the anatomical differences could be related to the finding (Ghent, 1961) that side differences in tactual thresholds on the thumbs of young children are not the same for boys and girls. (Ghent had found in 6-year-old girls typical adult patterns of greater sensitivity on the nondominant thumb, but boys did not show this pattern until 11 years of age. Ghent concluded that some aspects of development of the two hemispheres might proceed differently for boys and girls.)

There may indeed be the connection Lansdell has suggested, though the likelihood seems small on Conel's evidence. Lansdell apparently confined his examination to the 4-year-old brains. Inspection of Conel's tables for the brains of younger and older children fails to reveal comparable sex differences, significant or not.

Lansdell also cited an investigation by Matsubara (1960-1961) of variations in cerebral venous drainage that Lansdell characterized as suggesting that "the right vein of Trolard is larger than the left in girls, but not in boys" (p. 550). Since, Lansdell noted, "this is often the major vein in the hemisphere opposite to that used in speech [Di Chiro, 1962], is it possible that the differences in venous drainage are related to the superiority of girls over boys in certain verbal skills?" (p. 550).

A closer reading of Matsubara's report is not encouraging for such surmise. Matsubara measured variations in drainage of the superficial veins of normal and pathologic phlebograms (sphygmographic record of a venous pulse, i.e., a record of the movements of, or force of, the pulse of the veins of the body) of 132 persons with neurological disease and 98 "normal" persons. Of the 132 persons with neurological disease, 66 had lesions and 66 had such diseases as vascular malformations, residual syndrome of cerebral hemorrhage, and inflammation. The "normal" phlebograms were from patients visiting a neurological clinic with some complaints but in whom neurological disease was ruled out by angiography (analysis of cerebral blood pressure through roentgenography) and other neurological examinations. These patients instead had psychogenic diseases, epilepsy, migraine, and posttraumatic encephalopathy. The roentgenographic records indicated variations of dominant drainage in 35 normal persons and in 96 patients with neurological disease. In the majority of instances, the dominant drainage was through the vein of Trolard in normal persons and through the vein of Labbé or the middle cerebral vein in the patients.

The subjects ranged in age from 9 to 59 years. The majority of variations in dominance, however, appeared in the right hemisphere of subjects in the age range 10 to 29 years, suggesting, according to Matsubara, that "the differences in cortical venous return in both hemispheres have some relations with the differences in function of the two hemispheres in these age groups" (p. 88).

It is hard to see, however, what *variations* in superficial venous drainage in "normal" persons of a restricted age range and in patients with neurological disease could imply about sex differences in verbal ability in normal persons. Matsubara, for that matter, does not report *sex* differences in variation of drainage – only that females were more common in the group showing variations.

13 As still further evidence of earlier language lateralization in girls than boys, Buffery and Gray cited a study by Taylor (1969) of the relation between chronological age and the incidence of epileptic lesion in each cerebral hemisphere. From an examination of medical records, Taylor (1969) concluded that before about 2 years of age, the left cerebral hemisphere was more at risk than the right; afterward, this differential risk was reversed. Taylor interpreted both the hemisphere difference and the age crossover as indications that the lesion tends to occur in the functionally less active cerebral hemisphere. Therefore, the left hemisphere was more at risk in early years and the right hemisphere less at risk, because before the establishment of language at about 2 years of age, the left and right hemispheres were, respectively, less and more active.

Taylor proposed that the *right* hemisphere is more active earlier than the left hemisphere, because the earliest childhood years are involved in learning the visuospatial environment – skills that, presumably, require greater right hemisphere participation. (See Harris, 1975, for a review of psychological data.) But *after* language becomes established (and therefore is presumed to become lateralized in the left hemisphere), it is not clear what the left hemisphere is presumed to be more active *than*. More active than it was before, certainly. But Taylor's hypothesis apparently turns on the assumption that it is now functionally more active than the right hemisphere so that older human beings (the transition age presumably being about 2 years) are more verbal than they are visuospatial, in some quantitative, measurable sense; that is, the left hemisphere is more involved in day-to-day activities than the right hemisphere.

As for sex differences, Taylor supposed that if language lateralization occurs faster in girls than boys, the girls' left hemisphere would be at risk for a briefer time than the boys'. This is, the right hemisphere in boys should remain functionally more active for a longer period of time than in girls. The left lobe in boys, therefore, would be expected to be at greater risk than the right lobe during the early years, for a longer period of time. Taylor concluded that the records support this expectation. This is questionable. The distributions presented show a smooth decline in risk over the first 2 years for males, but a sharp decline for females occurring mainly in the second year. But the distributions were not converted into percentiles, as they should have been because different numbers of boys and girls were used. With conversion, the sex differences are less dramatic, the point of maximum difference being about 10%; and the Kolmogorov-Smirnov goodness-of-fit test for independent samples, with alpha set at 0.05, shows that the two distributions fall short of being significantly different from one another. There appears to be no statistical basis for Taylor's characterization of the distributions of males and females as representing different populations. Still, Taylor's report is potentially very valuable, for if the trend shown were obtained repeatedly in independent samples, this would suggest a genuine difference in period of risk. Such a demonstration would have important implications for our understanding of sex differences in brain pathology.

14 Kimura (1963a) also tested 25 left-handed children (11 girls and 14 boys) and for them there was a suggestion of a sex difference in the ear asymmetry effect. Of the 11 girls, 10 were more efficient on the right ear (the normal effect), and 1 showed no difference between ears. But of the 14 left-handed boys, only 9 were more efficient on the right ear, and 5 were more efficient on the left ear. This finding, Kimura noted, is typical of subjects with speech represented in the right hemisphere (Kimura, 1961). Kimura also mentioned a study by Taylor (1962) that may be pertinent: Taylor presented digits dichotically to children with reading disorders and found atypical speech lateralization in the boys but not in the girls. Kimura concluded, from this first study, that though there is no evidence of sex differences in normal right-handed children, there may be differences in children who show "other signs of irregular cerebral organization." Kimura's supposition, therefore, is that children with reading disabilities may be such cerebrally disorganized children.

15 It must be noted, however, that Semmes's view that tactile functions are represented focally in the left hemisphere, diffusely in the right, has not been consistently supported. Benton (1972) reviewed a number of the negative studies. For example, severe somatosensory defect has been found to be associated only with lesions of the postcentral gyrus of *either* hemisphere (Corkin et al., 1964); and no differences have been found between patients with left hemisphere and right hemisphere lesions in the effect of intrahemispheric locus of lesion on frequency of tactile deficits (Carmon & Benton, 1969).

Still another conclusion that would seem to follow from Semmes's hypothesis is that tactile deficits should be more closely interrelated on the right than on the left hand. Vaughan and Costa (1962) found a substantial correlation of 0.52 between pressure sensitivity and two-point discrimination for the contralateral (right) hand in patients with left hemisphere disease, but an even higher correlation of 0.87 for the contralateral (left) hand of patients with right hemisphere disease.

Finally, Carmon and Benton (1969) failed to find substantial differences in the strength of relation among performances on various tactile tasks in patients with left and right hemisphere lesions. The correlations among levels of tactile sensitivity on both the contralateral and ipsilateral hands were not higher in patients with right hemisphere lesions.

It is not clear why these results failed to confirm those of Semmes et al. (1960). But until these discrepancies are resolved by further analysis, the characterizations of the structural organization of the left and right hemispheres as focal and diffuse, respectively, probably should be regarded as only a hypothesis – at least in the case of tactile perception.

16 There is a similar proposal in Semmes's (1968) paper. She has suggested that if focal representation is the basis of left hemisphere dominance for language, then qualities that particularly distinguish animals from man – capacity for language and symbolic thought – are "ultimately based on the phylogenetic trend toward increased localization of function and ... the left hemisphere has proceeded farther than the right in this direction" (p. 23).

17 What is missing, or only implicit, in this as well as all other models of lateralization discussed so far is an explication of the purely mechanical reason why lateralization (i.e., brain and nervous system asymmetry) is critical for spatial perception. An explicit and compelling analysis has been outlined by Corballis and Beale (1970). In a paper on the relation between structural features of the body and nervous system and left-right confusion, they pointed out that:

> A bilaterally symmetrical machine [a machine that would be unchanged by mirror reflection] *must* make mirror-image responses to stimuli which are mirror images about its plane of symmetry. [One reason is that] mirror reflection leaves the machine unaltered but converts any given stimulus and its response to their mirror image. Thus the mirror representation shows that the same machine would make the mirror-image response to the mirror-image stimulus ... This means that a bilaterally symmetrical machine could not tell left from right. It could not discriminate mirror-image stimuli because it could not respond to them by giving differential responses which were *not* mirror images of one another. Neither could it differentiate left from right responses, because it must make symmetrical responses to symmetrical stimuli, which

means that it could not make left or right responses in the absence of left-right information. [Corballis & Beale, 1970, p. 452]

The same analysis applies to a bilaterally symmetrical organism. For purely mechanical reasons, it could not tell left from right. It could respond directionally only if directions were already represented in the stimulus, and the directional (i.e., left-right) information could be only simply transmitted, or copied, from stimuli to responses.

If the development toward asymmetry can be seen as an evolutionary principle, then, as Tschirgi (1958) has suggested, the crucial difference between species that cannot make mirror-image discriminations or can make them with only the greatest difficulty (e.g., the octopus; Sutherland, 1960) and species that can discriminate somewhat more easily (man) is in the difference in the extent to which the brain of each species develops asymmetrical functions. Tschirgi traced the development of the organism from hypothesized, archaic states of complete (i.e., spherical) symmetry, through progressively more asymmetrical stages (radial to bilateral to complete asymmetry), showing that with each increase in asymmetry, more can be known about the surface of the body and its environment. He thus presumed that man is approaching that final stage of asymmetry "because he can distinguish right from left, but he frequently makes mistakes" (Tschirgi, 1958; cited in Scheibel & Scheibel, 1962, p. 30).

18 We should emphasize here that the implication of the model is *not* necessarily that some people are less lateralized than others in left hemisphere functioning, but that in some people more than others, language functions are represented in the right hemisphere in addition to their primary representation in the left hemisphere. Furthermore, the kind of right hemisphere language representation presumably meant here is not the emotional or appositional language that, as Jackson suggested over a century ago (1876), is bilaterally represented, but propositional speech as well.

19 These proposals about preferential processing as an alternative or supplement to Levy's lateralization model are offered specifically for the case of females and not left-handers, the original subjects of Levy's hypothesis. Thus, if left-handers show spatial defects, I would not want to attribute this to interhemispheric competition (preferential language processing), because we can be reasonably sure that the brains of right-handed and left-handed persons are *not* identically organized.

20 It is possible to suggest a particular cerebral mechanism whereby the participation of left hemisphere modes in the solution of spatial problems would be set into operation in those individuals disposed, by past experience, to greater reliance on language strategies in spatial analysis. This mechanism is suggested in a model outlined by Kinsbourne (1970)

for the analysis of laterality effects in perception. For the visual system, for instance, the visual field differences typically found (better left-field discrimination for visuospatial targets, better right-field discrimination for linguistic targets) are usually explained as direct manifestations of the linguistic and visuospatial specialization of the left and right hemispheres, respectively. Kinsbourne has suggested, instead, that the effect is an indirect product of postexposure conjugate lateral movements of the eyes in a direction contralateral to the hemisphere activated in solution of the problem – to the left when a spatial target has been presented, to the right for linguistic problems (cf. Kinsbourne, 1972; Gur et al., 1975). Linguistic material is recognized better in the right visual field, and nonlinguistic material better in the left, because the subject's attention is biased to the appropriate side in each case.

If females are more disposed than males to the use of language modes in spatial problem solving, this model could account for the weaker lateralization effects reported in females in recognition of tachistoscopically projected spatial stimuli. It is not clear, though, how the model would account for weaker right-ear effects for females for linguistic stimuli in dichotic listening. Instead, as Harshman and Remington (1975b) have noted, the female's greater reliance on left hemisphere systems would be expected to cause either no change or a more reliable, larger right-ear advantage. Considering also the electrophysiological and anatomical data, as well as still other behavioral data not cited in the present chapter, these authors, therefore, argue for the primacy of a physiological interpretation, on grounds of parsimony if nothing else.

21 The formulation of Broverman et al. (1968) is similar to one outlined nearly 40 years earlier by Book (1932). The following summary is based on Nash's review (1970, pp. 190–191). Book cited research indicating that women do markedly and consistently better than men in tasks requiring quick response and rapid perception, and males are superior in tasks requiring attention to remain relatively fixed (e.g., Woolley, 1910, 1914). Book argued that such sex differences might be the product of basic neurophysiological differences, specifically those differences influencing the rate of conductivity of nerve tissue. Because males characteristically have longer refractory periods, statistically more of rapidly repeated stimuli fall into the refractory period than in previous periods. Consequently, men's adaptations are slower. Women, however, have a shorter refractory period and a greater number of nerve impulses per unit time. They, therefore, are able to respond quickly to more details and to adapt more quickly to changing test situations. Males are "postural"-type receptors; females are "phasic" types (Book borrowed the terms from Adrian, 1928). The postural type is slow to adapt, has a longer excitatory process, and produces a series of impulses such as are

characteristic of sustained response to stimulation. In the phasic type, by contrast, there is quick adaptation and shorter excitation, with many rapid discrete responses producing quickly fluctuating movements.

Book also suggested that women tend to do better in tasks that are divided into discrete parts, and men do better in tasks that are continuous. Women are the quick, perceptual type; men are slow-moving explorers. On these grounds, Book explained the sex differences found in her study, reviewed earlier in this chapter, that men excelled on mazes and block counting – tasks requiring, she suggested, a more sustained, logical, and analytical approach but not requiring attention to detail or to quick change of stimuli. The women in her study did better on tasks such as number checking that required attention to detail and quick adaptation to changing stimuli.

Nash (1970) has prudently cautioned that Book's "jump from functions like speed of response to logical, analytical modes of thinking seems rather a far one" (p. 191), though he noted that her theory at least contains the idea of two kinds of nervous systems with different modes of functioning – a view that Broverman et al. (1968) restated with the additional proposal about the role of the sex hormones.

REFERENCES

Abt, I. A., Adler, H. A., & Bartelme, P. 1929. The relationship between the onset of speech and intelligence. *J.A.M.A. 93*:1351–1355.

Adrian, E. D. 1928. *The Basis of Sensation: The Action of the Sense Organs.* London: Christopher.

Alajouanine, T. 1948. Aphasia and artistic realization. *Brain 71*:228–241.

Alexander, D., & Money, J. 1965. Reading ability, object constancy, and Turner's syndrome. *Percept. Mot. Skills 20*:981–984.

  1966. Turner's syndrome and Gerstmann's syndrome: Neuropsychologic comparisons. *Neuropsychologia 4*:265–273.

Alexander, D., Walker, H. T., Jr., & Money, J. 1964. Studies in direction sense. 1. Turner's syndrome. *Arch. Gen. Psychiatry 10*:337–339.

Ammons, R. B., Alprin, S. I., & Ammons, C. H. 1955. Rotary pursuit performance as related to sex and age of pre-adult subjects. *J. Exp. Psychol. 49*:127–133.

Ammons, R. B., & Ammons, H. S. 1948. *Full-Range Picture Vocabulary Test.* Missoula, Mont.: Psychological Test Specialists.

Anastasi, A., & Foley, J. P. 1953. *Differential Psychology.* New York: Macmillan.

Andrieux, C. 1955. Contribution à l'étude des différences entre hommes et femmes dans la perception spatiale. *Année Psychol. 55*:41–60.

Anhalt, I. 1961. Review. *Can. Music J. 5*:34–39.

Annett, M. 1967. The binomial distribution of right, mixed, and left handedness. *Q. J. Exp. Psychol. 19:*327–333.

Archer, E. J. 1958. Effect of distribution of practice on a component skill of rotary pursuit tracking. *J. Exp. Psychol. 56:*427–436.

Arrigoni, G., & De Renzi, E. 1964. Constructional apraxia and hemispheric locus of lesion. *Cortex 1:*170–197.

Bakan, P., & Putnam, W. 1974. Right-left discrimination and brain lateralization: Sex differences. *Arch. Neurol. 30:*334–335.

Barakat, M. K. 1951. A factorial study of mathematical abilities. *Br. J. Psychol. (Statist. Sect.) 4:*137–156.

Barclay, A., & Cusumano, D. R. 1967. Father absence, cross-sex identity, and field-dependent behavior in male adolescents. *Child Dev. 38:*243–250.

Barnes, R. H., Cunnold, S. R., Zimmerman, R. R., Simmons, H., MacLeod, R. B., & Krook, L. 1966. Influence of nutritional deprivations in early life on learning behavior of rats as measured by performance in a water maze. *J. Nutr. 89:*399–410.

Barrett, R. J., & Ray, O. S. 1970. Behavior in the open field, Lashley III maze, shuttle-box, and Sidman avoidance as a function of strain, sex, and age. *Dev. Psychol. 3:*73–77.

Basser, L. S. 1962. Hemiplegia of early onset and the faculty of speech with special reference to the effects of hemispherectomy. *Brain 85:*427–460.

Batalla, M. 1943. The maze behavior of children as an example of summative learning. *J. Genet. Psychol. 63:*199–211.

Bayley, N. 1935. The development of motor abilities during the first three years. *Monogr. Soc. Res. Child Dev. 1:*1–26.

1956. Growth curves of height and weight by age for boys and girls, scaled according to physical maturity. *J. Pediatr. 48:*187–194.

Baylor, G. W., Jr. 1971. A treatise on the mind's eye: An empirical investigation of visual mental imagery. Doctoral dissertation, Carnegie-Mellon University.

Bennett, D. H. 1956. Perception of the upright in relation to body image. *J. Ment. Sci. 102:*487–506.

Bennett, G. K., Seashore, H. G., & Wesman, A. G. 1959. *Differential Aptitude Tests*, 3rd ed. New York: Psychological Corporation.

Benton, A. L. 1972. Hemispheric cerebral dominance and somesthesis. In M. Hammer, K. Salzinger, & S. Sutton (eds.). *Psychopathology*, pp. 227–242. New York: Wiley.

Benton, A. L., & Kemble, J. D. 1960. Right-left orientation and reading disability. *Psychiatr. Neurol. (Basel) 139:*49.

Berry, J. W. 1966. Temne and Eskimo perceptual skills. *Int. J. Psychol. 1:*207–229.

Best, C. T., & Glanville, B. B. 1976. A cardiac measure of cerebral asymmetries in infant perception of speech and nonspeech. Paper presented at meetings of the Midwestern Psychological Association, Chicago.

Bettis, N. C. 1974. An assessment of the geographic knowledge and understanding of fifth grade students in Michigan. Doctoral dissertation, Michigan State University.

Bever, T. G., & Chiarello, R. J. 1974. Cerebral dominance in musicians and nonmusicians. *Science 185*:587-589.

Bieri, J. 1960. Parental identification, acceptance of authority, and within-sex differences in cognitive behavior. *J. Abnorm. Soc. Psychol. 60*:76-79.

Bieri, J., Bradburn, W., & Galinsky, M. 1958. Sex differences in perceptual behavior. *J. Pers. 26*:1-12.

Bigelow, G. 1971. Field dependence-field independence in 5- to 10-year-old children. *J. Educ. Res. 64*:397-400.

Blade, M., & Watson, W.S. 1955. Increase in spatial visualization test scores during engineering study. *Psychol. Monogr. 69* (Whole No. 397).

Bock, R. D. 1967. A family study of spatial visualizing ability. *Am. Psychol. 22*:571 (abstract).

1973. Word and image: Sources of the verbal and spatial factors in mental test scores. *Psychometrika 38*:437-457.

Bock, R. D., & Kolakowski, D. 1973. Further evidence of sex-linked major-gene influence on human spatial visualizing ability. *Am. J. Hum. Genet. 25*:1-14.

Bock, R. D., & Vandenberg, S. G. 1968. Components of heritable variation in mental test scores. In S. G. Vandenberg (ed.). *Progress in Human Behavior Genetics*, pp. 233-260. Baltimore: Johns Hopkins Press.

Bogen, J. E. 1969. The other side of the brain 2. An appositional mind. *Bull. Los Angeles Neurol. Soc. 34*:135-162.

Bogen, J. E., & Gordon, H. W. 1971. Musical tests for functional lateralization with intracarotid amobarbital. *Nature 230*:524.

Bogo, N., Winget, C., & Gleser, G. C. 1970. Ego defenses and perceptual styles. *Percept. Mot. Skills 30*:599-604.

Book, H. M. 1932. A psychophysiological analysis of sex differences. *J. Soc. Psychol. 3*:434-461.

Botez, M. I., & Crighel, E. 1971. Partial disconnexion syndrome in an ambidextrous patient. *Brain 94*:487-494.

Brainerd, C. J. 1971. The development of the proportionality scheme in children and adolescents. *Dev. Psychol. 5*:469-476.

Branch, C., Milner, B., & Rasmussen, T. 1964. Intracarotid sodium amytal for the lateralization of cerebral speech dominance. *J. Neurosurg. 21*: 399-405.

Brinkmann, E. H. 1966. Programmed instruction as a technique for improving spatial visualization. *J. Appl. Psychol. 50*:179-184.

Broadbent, D. E. 1954. The role of auditory localization in attention and memory span. *J. Exp. Psychol. 47*:191-196.

Broverman, D. M., Broverman, I. K., Vogel, W., Palmer, R. D., & Klaiber, E. L. 1964. The automatization cognitive style and physical development. *Child Dev. 35*:1343-1359.

Broverman, D. M., & Klaiber, E. L. 1969. Negative relationships between abilities. *Psychometrika 34*:5-20.

Broverman, D. M., Klaiber, E. L., Kobayashi, Y., & Vogel, W. 1968. Roles of activation and inhibition in sex differences in cognitive abilities. *Psychol. Rev. 75*:23-50.

Broverman, D. M., Klaiber, E. L., Kobayashi, Y., & Vogel, W. 1969. Reply to "comment" by Singer and Montgomery on "roles of activation and inhibition in sex differences in cognitive abilities." *Psychol. Rev. 76*: 328-331.

Brown, F. R. 1954. The effect of an experimental course in geometry on ability to visualize in three dimensions. Doctoral dissertation, University of Illinois.

Bryden, M. P. 1966. Results of testing at the 1965 Canadian National Exhibition. 1. Dichotic listening. Unpublished manuscript, University of Waterloo.

Bryden, M. P., Allard, F., & Scarpino, F. 1973. The development of language lateralization and speech perception. Unpublished manuscript, University of Waterloo.

Buckley, F. 1971. Preliminary report on intelligence quotient scores of patients with Turner's syndrome: A replication study. *Br. J. Psychiatry, 119*:513-514.

Buffery, A. W. H. 1971. Sex differences in the development of hemispheric asymmetry of function in the human brain. *Brain 31*:364-365 (abstract).

Buffery, A. W. H., & Gray, J. A. 1972. Sex differences in the development of perceptual and linguistic skills. In C. Ounsted & D. C. Taylor (eds.). *Gender Differences: Their Ontogeny and Significance.* London: Churchill.

Butters, N., & Barton, M. 1970. Effect of parietal lobe damage on the performance of reversible operations in space. *Neuropsychologia 8*:205-214.

Buxton, C. E., & Grant, D. A. 1939. Retroaction and gains in motor learning. 2. Sex differences, and a further analysis of gains. *J. Exp. Psychol. 25*: 198-208.

Byrne, R. 1975. *New York Times*, July 29, p. 26; August 12, p. 32.

Cameron, J., Livson, N., & Bayley, N. 1967. Infant vocalizations and their relationship to mature intelligence. *Science 157*:331-333.

Canavan, D. 1969. Field dependence in children as a function of grade, sex, and ethnic group membership. Paper presented at meetings of the American Psychological Association, Washington, D.C.

Carmon, A., & Benton, A. L. 1969. Tactile perception of direction and number in patients with unilateral cerebral disease. *Neurology (Minneap.) 19*:525-532.

Chase, W. G., & Simon, H. A. 1973. The mind's eye in chess. In W. G. Chase (ed.). *Visual Information Processing,* pp. 215-281. New York: Academic Press.

Chateau, J. 1959. Le test de structuration spaciale. *Le Travail Humain 22:* 281–297.
Chesher, E. C. 1936. Some observations concerning the relation of handedness to the language mechanism. *Bull. Neurol. Inst. N.Y. 4:*556–562.
Churchill, B. D., Curtis, J. M., Coombs, C. H., & Hassell, T. W. 1942. Effect of engineer school training on the Surface Development Test. *Educ. Psychol. Meas. 2:*279–280.
Clyma, E. A., & Clarke, P. R. F. 1973. Which hand draws a better square? *Bull. Br. Psychol. Soc. 26:*141 (abstract).
Cohen, J. 1957. A factor-analytically based rationale for the Wechsler Adult Intelligence Scale. *J. Consult. Psychol. 21:*451–457.
  1959. The factorial structure of the WISC at ages 7-6, 10-6, and 13-6. *J. Consult. Psychol. 23:*285–289.
Coltheart, M., Hull, E., & Slater, D. 1975. Sex differences in imagery and reading. *Nature 253* (5491):437–440.
Conel, J. L. 1963. *The Postnatal Development of the Human Cerebral Cortex*, Vol. VII, *The Cortex of the Four-Year-Old Child*. Cambridge: Harvard University Press.
Corah, N. L. 1965. Differentiation in children and their parents. *J. Pers. 33:* 300–308.
Corballis, M. C., & Beale, I. L. 1970. Bilateral symmetry and behavior. *Psychol. Rev. 77:*451–464.
Corkin, S. 1965. Tactually-guided maze learning in man: Effects of unilateral cortical excisions and bilateral hippocampal lesions. *Neuropsychologia 3:*339–351.
Corkin, S., Milner, B., & Rasmussen, T. 1964. Effects of different cortical excisions on sensory thresholds in man. *Trans. Am. Neurol. Assoc. 89:*112.
Cowley, J. J., & Griesel, R. D. 1963. The development of second-generation low-protein rats. *J. Genet. Psychol. 103:*233–242.
Critchley, M. 1953. *The Parietal Lobes*. New York: Hafner. Republished in 1971, facsimile of 1953 edition.
Culver, C. 1969. Test of right-left discrimination. *Percept. Mot. Skills. 29:* 863–867.
Darwin, C. J. 1969. Auditory perception and cerebral dominance. Doctoral dissertation, Cambridge University.
Dawson, J. L. M. 1967. Cultural and physiological influences upon spatial-perceptual processes in West Africa. 2. *Int. J. Psychol. 2:*171–185.
  1972. Effects of sex hormones on cognitive style in rats and men. *Behav. Genet. 2:*21–42.
de Groot, A. 1965. *Thought and Choice in Chess*. The Hague: Mouton.
Demuth, N. 1947. *Ravel*. London: Dent.
DeOreo, K. D., & Wade, M. G. 1971. Dynamic and static balancing ability of pre-school children. *J. Mot. Behav. 3:*326–335.

De Renzi, E., Faglioni, P., & Scotti, G. 1969. Impairment of memory for position following brain damage. *Cortex* 5:274-284.

Di Chiro, G. 1962. Angiographic patterns of cerebral convexity veins and superficial dural sinuses. *Am. J. Roentgenol. Radium Ther. Nucl. Med.* 87:308-321.

Doran, E.W. 1907. A study of vocabularies. *Pedagogical Seminary* 14:401-438.

Drowatzky, J. N., & Zuccato, F. C. 1967. Inter-relationships between selected measures of static and dynamic balance. *Res. Q.* 38:509-510.

Ehrlichman, H. I. 1971. Hemispheric functioning and individual differences in cognitive abilities. Doctoral dissertation, New School for Social Research. *Dissertation Abstracts International* 33:2319B, (University Microfilms No. 72-27, 869).

Eisenberg, L., Berlin, C. I., Dill, A., & Frank, S. 1968. Class and race effects on the intelligibility of monosyllables. *Child Dev.* 39:1077-1089.

Entus, A. K. 1975. Hemispheric asymmetry in processing of dichotically presented speech and nonspeech sounds by infants. Paper presented to Society for Research in Child Development, Denver.

Erikson, E. H. 1951. Sex differences in the play configurations of preadolescents. *Am. J. Orthopsychiatry,* 21:667-692.

Ettlinger, G., Jackson, C. V., & Zangwill, O. 1956. Cerebral dominance in sinistrals. *Brain* 79:569-588.

Fagin-Dubin, L. 1974. Lateral dominance and development of cerebral specialization. *Cortex* 10:69-74.

Farrell, M. 1957. Sex differences in block play in early childhood education. *J. Educ. Res.* 51:279-284.

Faubion, R. W., Cleveland, E. A., & Hassell, T. W. 1942. The influence of training on mechanical aptitude test-scores. *Educ. Psychol. Meas.* 2:91-94.

Fiebert, M. 1967. Cognitive styles in the deaf. *Percept. Mot. Skills* 24:319-329.

Flanagan, J. C., Dailey, J. T., Shaycoft, M. F., Gorham, W. A., Orr, D. B., Goldberg, I., & Neyman, C. A., Jr. 1961. *Counselor's Technical Manual for Interpreting Test Scores (Project Talent).* Palo Alto, Cal. Cited in E. E. Maccoby & C. N. Jacklin. *The Psychology of Sex Differences.* Stanford, Cal.: Stanford University Press, 1974.

French, J. W. 1951. The description of aptitude and achievement tests in terms of rotated factors. *Psychometric Monogr.* No. 5.

French J. W., Ekstrom, R. B., & Price, L. A. 1963. *Manual for Kit of Reference Tests for Cognitive Factors.* Princeton: Educational Testing Service.

Gainer, W. L. 1962. The ability of the WISC subtests to discriminate between boys and girls of average intelligence. *Cal. J. Educ. Res.* 13:9-16.

Garai, J. E. 1970. Sex differences in mental health. *Genet. Psychol. Monogr.* *81*:123-142.

Garai, J. E., & Scheinfeld, A. 1968. Sex differences in mental and behavioral traits. *Genet. Psychol. Monogr.* 77:169-299.

Garn, S. M., Rohman, C. G., & Hertzog, K. P. 1969. Apparent influence of the X chromosome on timing of 73 ossification centers. *Am. J. Phys. Anthropol. 30:*123-138.

Garron, D. C. 1970. Sex-linked, recessive inheritance of spatial and numerical abilities and Turner's syndrome. *Psychol. Rev.* 77:147-152.

Gästrin, J. 1940. *Det intelligenta Lärandets problem.* Abo: Abo akademi.

Gates, A. I. 1961. Sex differences in reading ability. *Elementary School J. 61:* 431-434.

Gatewood, M. C., & Weiss, A. P. 1930. Race and sex differences in newborn infants. *J. Genet. Psychol. 38:*31-49.

Gelman, R., & Weinberg, D. H. 1972. The relationship between liquid conservation and compensation. *Child Dev. 43:*371-383.

Geschwind, N. 1972. Cerebral dominance and anatomic asymmetry. *N. Engl. J. Med. 287:*194-195.

   1974. The anatomical basis of hemispheric differentiation. In S. J. Dimond & J. G. Beaumont (eds.). *Hemispheric Function in the Human Brain.* New York: Wiley.

Geschwind, N., & Levitsky, W. 1968. Left/right asymmetries in temporal speech region. *Science 161:*186-187.

Gesell, A., et al. 1940. *The First Five Years of Life.* New York: Harper & Row.

Ghent, L. 1961. Developmental changes in tactual thresholds on dominant and nondominant sides. *J. Comp. Physiol. Psychol. 54:*670-673.

Goldberg, J., & Meredith, W. 1975. A longitudinal study of spatial ability. *Behav. Genet. 5:*127-135.

Goldberg, S., & Lewis, M. 1969. Play behavior in the year-old infant: Early sex differences. *Child Dev. 40:*21-32.

Goldschmid, M. L. 1967. Different types of conservation and nonconservation and their relation to age, sex, IQ, MA, and vocabulary. *Child Dev. 38:*1229-1246.

Goldstein, A. G., & Chance, J. E. 1965. Effects of practice on sex-related differences in performance on Embedded Figures. *Psychonomic Sci. 3:*361-362.

Goodenough, D.R., & Eagle, C. J. 1963. A modification of the embedded-figures test for use with young children. *J. Genet. Psychol. 103:*67-74.

Goodenough, F. L., & Smart, R. C. 1935. Inter-relationship of motor abilities in young children. *Child Dev. 6:*141-153.

Goodglass, H., & Quadfasel, F. A. 1954. Language laterality in left-handed aphasics. *Brain* 77:521-548.

Goodnow, R. 1962. Cited in Witkin, H. A., et al. *Phychological Differentiation*, p. 214. New York: Wiley, 1962.

Gordon, H. W. 1970. Hemispheric asymmetries in the perception of musical chords. *Cortex* 6:387-398.

1975. Hemispheric asymmetry and musical performance. *Science* 189:68-69.

Goss, M. 1940. *Bolero: The life of Maurice Ravel*. New York: Holt.

Graves, A. J. 1972. Attainment of conservation of mass, weight, and volume in minimally educated adults. *Dev. Psychol.* 7:223.

Graves, M. 1948. *Design Judgment Test: Manual*. New York: Psychological Corporation.

Graves, M. F., & Koziol, S. 1971. Noun plural development in primary grade children. *Child Dev.* 42:1165-1173.

Gross, F. 1959. The role of set in perception of the upright. *J. Pers.* 27: 95-103.

Gruen, G. E., & Vore, D. A. 1972. Development of conservation in normal and retarded children. *Dev. Psychol.* 6:146-157.

Guilford, J. P. 1947. *Printed Classification Tests, AAF Report No. 5*. Washington, D. C.: GPO.

Guilford, J. P., & Zimmerman, W. F. 1953. *The Guilford-Zimmerman Aptitude Survey. IV. Spatial Visualization Form B*. Beverly Hills, Cal.: Sheridan Supply.

Gur, R. E., Gur, R. C., & Harris, L. J. 1975. Cerebral activation, as measured by subject's lateral eye movements, is influenced by experimenter location. *Neuropsychologia* 13:35-44.

Halverson, C. F., & Waldrop, M. F. 1970. Maternal behavior toward own and other preschool children: The problem of "ownness." *Child Dev.* 41: 839-845.

Hampson, J. L., Hampson, J. G., & Money, J. 1955. The syndrome of gonadal agenesis (ovarian agenesis) and male chromosomal pattern in girls and women. *Bull. Johns Hopkins Hosp.* 97:207-226.

Hanson, M. R. 1965. Motor performance testing of elementary school age children. Doctoral dissertation, University of Washington.

Harper, L. V., & Sanders, K. M. 1975. Preschool children's use of space: Sex differences in outdoor play. *Dev. Psychol.* 11:1, 119.

Harris, L. J. 1972. Discrimination of left and right, and development of the logic of relations. *Merrill-Palmer Q.* 18:307-320.

1975. Neurophysiological factors in the development of spatial skills. In J. Eliot & N. J. Salkind (eds.). *Children's Spatial Development*, pp. 5-56. Springfield, Ill.: Thomas.

1977. Sex differences in the growth and use of language. In E. Donelson & J. Gullahorn (eds.). *Woman: A Psychological Perspective*, pp. 79-94. New York: Wiley.

Harris, L. J., & Allen, T. W. 1971. The effects of stimulus alignment on children's performance in a conservation-of-length problem. *Psychonomic Sci. 23:*137–139.

Harris, L. J., Hanley, C., & Best, C. T. 1975. Conservation of horizontality: Sex differences in fifth-graders and college students. Paper presented at Biennial Meetings of the Society for Research in Child Development, Denver.

Harshman, R. A., & Remington, R. 1975a. Sex, language, and the brain. 1. A review of the literature on adult sex differences in lateralization. Mimeographed. 30 pp.

1975b. Sex, language, and the brain. 2. Adult sex differences in lateralization of dichotic verbal stimuli. Mimeographed.

Hartlage, L. C. 1970. Sex-linked inheritance of spatial ability. *Percept. Mot. Skills 31:*610.

Havighurst, R. J., & Breese, F. F. 1947. Relation between ability and social status in a midwestern community. 3. Primary mental abilities. *J. Educ. Psychol. 38:*241–247.

Hécaen, H. 1962. Clinical symptomatology in right and left hemispheric lesions. In V. B. Mountcastle (ed.). *Interhemispheric Relations and Cerebral Dominance.* pp 215–263. Baltimore: Johns Hopkins Press.

Hécaen, H., & Ajuriaguerra, J. de 1964. *Left-handedness: Manual Superiority and Cerebral Dominance.* New York: Grune & Stratton.

Hermelin, B., & O'Connor, N. 1971a. Right and left handed reading of Braille. *Nature 231:*470.

1971b. Functional asymmetry in the reading of Braille. *Neuropsychologia 9:*431–435.

Herron J. 1974. EEG alpha asymmetry and dichotic listening in stutterers. Paper presented at the Annual Convention of the Society for Psychophysiological Research, Salt Lake City.

Herzberg, F., & Lepkin, M. A. 1954. A study of sex differences on the Primary Mental Abilities Test. *Educ. Psychol. Meas. 14:*687–689.

Honzik, M. P. 1951. Sex differences in the occurrence of materials in the play constructions of pre-adolescents. *Child Dev. 22:*15–35.

Hooper, F. H. 1969. Piaget's conservation tasks: The logical and developmental priority of identity conservation. *J. Exp. Child Psychol. 8:*234–249.

Horgan, D. 1976. Sex differences in language development. Paper presented at meetings of the Midwestern Psychological Association, Chicago.

Hubbert, H. B. 1915. The effect of age on habit formation in the albino rat. *Behav. Monogr. 2:*1–55.

Hutt, C. 1970a. Specific and diversive exploration. In H. Reese & L. Lipsitt (eds.). *Advances in Child Development and Behavior,* Vol. V. New York: Academic Press.

1970b. Curiosity in young children. *Sci. J. 6:*68–72.

Ishihara, S. 1951. *Tests for Colour Blindness*, 10th rev. ed. London: Lewis.
Jackson, J. H. 1876. Case of large cerebral tumour without optic neuritis and with left hemiplegia and imperception. *R. Lond. Ophthamol. Hosp. Rep. 8:*434–444.
James, W. E., Mefferd, R. B., & Wieland, B. 1967. Repetitive psychometric measures: handedness and performance. *Percept. Mot. Skills 25:*209–212.
Jenkins, L. M. 1930. A comparative study of motor achievements of children of five, six, and seven years of age. *Teachers College (Columbia University) Contrib. Educ.* No. 414.
Jensen, A. R. 1975. A theoretical note on sex linkage and race differences in spatial visualization ability. *Behav. Genet. 5:*151–164.
Johnson, P. O. 1973. Electrophysiological correlates of overt language behavior in humans. Master's thesis, California State University, San Francisco.
Kagan, J. 1964. Acquisition and significance of sex typing and sex role identity. In M. L. Hoffman & L. W. Hoffman (eds.). *Review of Child Development Research*, Vol. I, pp. 137–167. New York: Russell Sage.
Kakeshita, T. 1925. Zur Anatomies der Operkularen temporal Region (Vergleichende untersuchungen der rechten und linken Seite). *Arb. Neurol. Inst. Wein 27:*292.
Karp, S. A., & Konstadt, N. L. 1963. *Children's Embedded Figures Test.* New York: Cognitive Tests.
Kato, N. 1965. A fundamental study of rod-frame test. *Jap. Psychol. Res. 7:*61–68.
Keogh, B. K. 1971. Pattern copying under three conditions of an expanded spatial field. *Dev. Psychol. 4:*25–31.
Keogh, B. K., & Ryan, S. R. 1971. Use of three measures and field organization with young children. *Percept. Mot. Skills 33:*466.
Kimura, D. 1961. Some effects of temporal lobe damage on auditory perception. *Can. J. Psychol. 15:*156–165.
　1963a. Speech lateralization in young children as determined by an auditory test. *J. Comp. Physiol. Psychol. 56:*899–902.
　1963b. Right temporal lobe damage. *Arch. Neurol. 8:*264–271.
　1964. Left-right differences in the perception of melodies. *Q. J. Exp. Psychol. 16:*355–358.
　1966. Dual functional asymmetry of the brain in visual perception. *Neuropsychologia 4:*275–285.
　1967. Functional asymmetry of the brain in dichotic listening. *Cortex 3:* 163–178.
　1969. Spatial localization in left and right visual fields. *Can. J. Psychol. 23:* 445–458.
　1973. The asymmetry of the human brain. *Sci. Am. 228:*70–78.

Kinsbourne, M. 1966. Limitations in visual capacity due to cerebral lesions. Paper presented to 18th International Congress on Psychology, Moscow.
    1970. The cerebral basis of lateral asymmetries in attention. *Acta Psychol.* *55:*193-201.
    1972. Eye and head turning indicates cerebral lateralization. *Science 176:* 539-541.
Knox, C., & Kimura, D. 1970. Cerebral processing of non-verbal sounds in boys and girls. *Neuropsychologia 8:*227-238.
Kohn, B., & Dennis, M. 1974. Selective impairment of visuo-spatial abilities in infantile hemiplegics after right cerebral hemidecortication. *Neuropsychologia 12:*505-512.
Korchin, S. 1962. Cited in Witkin, H. A., et al. *Psychological Differentiation,* p. 214. New York: Wiley, 1962.
Krechevsky, I. 1933. Hereditary nature of "hypotheses." *J. Comp. Psychol. 16:*99-116.
Lake, D. A., & Bryden, M. P. 1976. Handedness and sex differences in hemispheric asymmetry. *Brain and Language 8:*266-282.
Langhorne, M. C. 1948. The effects of maze rotation on learning. *J. Gen. Psychol. 38:*191-205.
Langworthy, O. 1964. Only half aware: A review. *Am. J. Psychiatry 121:* 116-122.
Lansdell, H. 1961. The effect of neurosurgery on a test of proverbs. *Am. Psychol. 16:*448 (abstract).
    1962. A sex difference in effect of temporal lobe neurosurgery on design preference. *Nature 194:*852-854.
    1968a. The use of factor scores from the Wechsher-Bellevue Scale of Intelligence in assessing patients with temporal lobe removals. *Cortex 4:*257-268.
    1968b. Effect of extent of temporal lobe ablations on two lateralized deficits. *Physiol. and Behav. 3:*271-273.
Leakey, L. S. B. 1961. *The Progress and Evolution of Man in Africa.* London: Oxford University Press.
Lee, R. B., & DeVore, I. (eds.). 1968. *Man the Hunter.* Chicago: Aldine.
Levy, J. 1969. Possible basis for the evolution of lateral specialization of the human brain. *Nature 224:*614-615.
Levy-Agresti, J., & Sperry, R. W. 1968. Differential perceptual capacities in major and minor hemispheres. *Proc. Natl. Acad. Sci. U.S.A. 61:*1151.
Ley, R. G., & Koepke, J. E. 1975. Sex and age differences in the departures of young children from their mothers. Paper presented at Biennial Meetings of the Society for Research in Child Development, Denver.
Liben, L. S. 1973. Operative understanding of horizontality and its relation to long-term memory. Paper presented at Biennial Meetings of the Society for Research in Child Development, Philadelphia.

Long, A. B., & Looft, W. R. 1972. The development of directionality in children: Ages six through twelve. *Dev. Psychol.* 6:375-380.

Lord, F. E. 1941. A study of spatial orientation of children. *J. Educ. Res. 34:* 481-505.

Luria, A. R., Tsvetkova, L. S., & Futer, D. S. 1965. Aphasia in a composer. *J. Neurol. Sci.* 2:288-292.

MacArthur, R. 1967. Sex differences in field dependence for the Eskimo. *Int. J. Psychol.* 2:139-140.

Masica, D. N., Money, J., & Ehrhardt, A. A. 1971. Fetal feminization and female gender identity in the testicular feminizing syndrome of androgen insensitivity. *Arch. Sex. Behav.* 1:131-142.

Masica, D. N., Money, J., Ehrhardt, A. A., & Lewis, V. G. 1969. Fetal sex hormones and cognitive patterns: Studies in the testicular feminizing syndrome of androgen insensitivity. *Johns Hopkins Med. J. 123:*105-114.

Matheny, A. P., Jr. 1973. Heredity and environmental components of competency of children's articulation. Paper presented at Biennial Meetings of the Society for Research in Child Development, Philadelphia.

Mather, K., & Jinks, J. L. 1971. *Biometrical Genetics: The Study of Continuous Variation*, 2nd ed. London: Chapman & Hall.

Matsubara, T. 1960-1961. An observation on cerebral phlebograms with special reference to the changes in the superficial veins. *Nagoya J. Med. Sci.* 23:86-94.

Mattson, M. L. 1933. The relation between the habit to be acquired and the form of the learning curve in young children. *Genet. Psychol. Monogr.* 13:299-398.

McCarthy, D. 1930. *The Language Development of the Preschool Child.* Institute of Child Welfare Monograph Series No. 4. Minneapolis: University of Minnesota Press.

McCaskill, C. L., & Wellman, B. L. 1938. A study of common motor achievement at the preschool ages. *Child Dev.* 9:141-150.

McDermott, V. 1972. A conceptual musical space. *J. Aesthetics 30:*489-494.

McGinnis, E. 1929. The acquisition and interference of motor habits in young children. *Genet. Psychol. Monogr.* 6:209-311.

McGlone, J., & Davidson, W. 1973. The relation between cerebral speech laterality and spatial ability with special reference to sex and hand preference. *Neuropsychologia 11:*105-113.

McGlone, J., & Kertesz, A. 1973. Sex differences in cerebral processing of visuospatial tasks. *Cortex* 9:313-320.

McNemar, Q. 1942. *The Revision of the Stanford-Binet Scale: An Analysis of the Standardization Data.* Boston: Houghton-Mifflin.

McNemar, Q., & Stone, C. P. 1932. The sex difference in rats on three learning tasks. *J. Comp. Psychol. 14:*171-180.

McRae, D. L. 1948. Focal epilepsy: Correlation of the pathological and radiological findings. *Radiology* 50:439-457.

McRae, D. L., Branch, C. L., & Milner, B. 1968. The occipital horns and cerebral dominance. *Neurology* 18:95-98.

Mead, C. D. 1913. The age of walking and talking in relation to general intelligence. *Pedagogical Seminary* 20:460-484.

Mellone, M. A. 1944. A factorial study of picture tests for young children. *Br. J. Psychol.* 35:9-16.

Messer, S. B., & Lewis, M. 1972. Social class and sex differences in the attachment and play behavior of the one-year-old infant. *Merrill-Palmer Q.*, 18:295-306.

Metzler, J., & Shepard, R. N. 1974. Rotation of tri-dimensional objects. In R. L. Solso (ed.). *Theories in Cognitive Psychology: The Loyola Symposium*, pp. 147-201. New York: Wiley.

Miele, J. A. 1958. Sex differences in intelligence: The relationship of sex to intelligence as measured by the Wechsler Adult Intelligence Scale and the Wechsler Intelligence Scale for Children. *Dissertation Abstracts* 18: 2213.

Miller, E. 1971. Handedness and the patterns of human ability. *Br. J. Psychol.* 62:111-112.

Miller, G. A. 1956. The magical number seven, plus or minus two: Some limits on our capacity for processing information. *Psychol. Rev.* 63:81-97.

Milner, B. 1962. Laterality effects in audition. In V. B. Mountcastle (ed.). *Interhemispheric Relations and Cerebral Dominance*. Baltimore: Johns Hopkins Press.

  1965. Visually guided maze learning in man: Effects of bilateral hippocampal, bilateral frontal, and unilateral cerebral lesions. *Neuropsychologia* 3:317-338.

  1968. Visual recognition and recall after right temporal-lobe excision in man. *Neuropsychologia* 6:191-210.

  1969. Residual intellectual and memory deficits after head injury. In E. Walker, W. Caveness, & M. Critchley (eds.). *The Late Effects of Head Injury*. Springfield, Ill.: Thomas.

Mitchelmore, M. C. 1974. The perceptual development of Jamaican students, with special reference to visualization and drawings of three-dimensional geometrical figures and the effects of spatial training. Doctoral dissertation, Ohio State University.

Molfese, D. L. 1973. Cerebral asymmetry in infants, children, and adults: Auditory evoked responses to speech and noise stimuli. Doctoral dissertation, Pennsylvania State University.

Money, J. 1964. Two cytogenetic syndromes – psychologic comparisons. 1. Intelligence and specific-factor quotients. *J. Psychiatr. Res.* 2:223-231.

  1968a. Cognitive deficits in Turner's syndrome. In S. G. Vandenberg (ed.).

Progress in Human Behavior Genetics, pp. 27-30. Baltimore: Johns Hopkins Press.

1968b. *Sex Errors of the Body*. Baltimore: Johns Hopkins Press.

Money, J., Alexander, D., & Walker, H. T., Jr. 1965. *A Standardized Road-Map Test of Direction Sense*. Baltimore: Johns Hopkins Press.

Money, J., Ehrhardt, A. A., & Masica, D. N. 1968. Fetal feminization induced by androgen insensitivity in the testicular feminizing syndrome: Effect on marriage and maternalism. *Johns Hopkins Med. J. 123:*160-167.

Money, J., Walker, H. T., Jr., & Alexander, D. 1965. Development of direction sense and three syndromes of impairment. *Slow Learning Child: Aust. J. Educ. Backward Child.* Cited in J. Money, D. Alexander, & H. T. Walker, Jr. *A Standardized Road-Map Test of Direction Sense*. Baltimore: Johns Hopkins Press.

Moore, T. 1967. Language and intelligence: A longitudinal study of the first eight years. 1. Patterns of development in boys and girls. *Hum. Dev. 10:* 88-106.

Morf, M. E., Kavanaugh, R. D., & McConville, M. 1971. Intratest and sex differences on a portable Rod-and-Frame Test. *Percept. Mot. Skills 32:* 727-733.

Morley, M. E. 1957. *The Development and Disorders of Speech in Childhood*. London: Livingstone.

Morris, B. B. 1971. Effects of angle, sex, and cue on adults' perception of the horizontal. *Percept. Mot. Skills 32:*827-830.

Moss, H. 1967. Sex, age, and state as determinants of mother-infant interaction. *Merrill-Palmer Q. 13:*19-36.

Munroe, R. L., & Munroe, R. H. 1968. Space and numbers: Some ecological factors in culture and behavior. Paper presented at the workshop of the Makerere Institute of Social Research, New York. Cited in Olson, D. R. *Cognitive Development: The Child's Acquisition of Diagonality*. New York: Academic Press, 1970.

Myers, C. T. 1958. *The Effect of Training in Mechanical Drawing on Spatial Relations Test-Scores as Predictors of Engineering Drawing Grades*. Research and Development Report N. 58-4. Princeton: Educational Testing Service.

Myers, R. H. 1960. *Ravel: Life and Works*. London: Duckworth.

Nash, J. 1970. *Developmental Psychology: A Psychobiological Approach*. Englewood Cliffs, N.J.: Prentice-Hall.

Nebes, R. D. 1971a. Superiority of the minor hemisphere in commissurotomized man for the perception of part-whole relations. *Cortex 7:*333-349.

1971b. Handedness and the perception of part-whole relationship. *Cortex 7:*350-356.

Nelson, K. 1973. Structure and strategy in learning to talk. *Monogr. Soc. Res. Child Dev. 38:*(1, 2).

Newbigging, P. L. 1954. The relationship between reversible perspective and embedded figures. *Can. J. Psychol.* 8:204-208.

Newcombe, F., & Ratcliff, G. 1973. Handedness, speech lateralization, and ability. *Neuropsychologia* 11:399-408.

Newman, H. H., Freeman, F. N., & Holzinger, K. T. 1937. *Twins: A Study of Heredity and Environment.* Chicago: University of Chicago Press.

Nicolson, A. B., & Hanley, C. 1953. Indices of physiological maturity: Derivation and interrelationships. *Child Dev.* 24:3-38.

Noble, C. E. 1969. Race, reality, and experimental psychology. *Perspect. Biol. Med.* 13:10-30.

Noble, C. E., Fuchs, J. E., Robel, D. P., & Chambers, R. W. 1958. Individual vs. social performance on two perceptual-motor tasks. *Percept. Mot. Skills* 8:131-134.

Noble, C. E., & Noble, C. S. 1972. Pursuit tracking skill with separate and combined visual and auditory feedback. *J. Mot. Behav.* 4:195-205.

Norman, R. D. 1953. Sex differences and other aspects of young superior adult performance on the Wechsler Bellevue. *J. Consult. Psychol.* 17:411-418.

O'Connor, J. 1943. *Structural Visualization.* Boston: Human Engineering Laboratory.

Okonji, M. O. 1969. The differential effects of rural and urban upbringing on the development of cognitive styles. *Int. J. Psychol.* 4:293-305.

Olson, D. R. 1970. *Cognitive Development: The Child's Acquisition of Diagonality.* New York: Academic Press.

Olson, W. C., & Koetzle, V. S. 1936. Amount and rate of talking of young children. *J. Exp. Educ.* 5:175-179.

Otterstaedt, H. 1962. Untersuchungen über den Spielraum von Vorortkindern einer mittleren Stadt. (Investigations concerning the play area of suburban children in a medium-sized city). *Psychol. Rundschau* 13:275-287.

Paterson, D. G., & Andrew, D. M. 1946. *Manual for the Minnesota Vocational Test for Clerical Workers.* New York: Psychological Corporation.

Pearson, K., & Lee, A. 1903. Laws of inheritance in man. *Biometrika* 2:357-462.

Pederson, F. A., & Bell, R. Q. 1970. Sex differences in preschool children without histories of complications of pregnancy and delivery. *Dev. Psychol.* 3:10-15.

Penfield, W. G. 1925. Cerebral pneumography: Its dangers and uses. *Arch. Neurol. Psychiatry (Chicago)* 13:580-591.

Peterson, A. C. 1973. The relationship of androgenicity in males and females to spatial ability and fluent production. Doctoral dissertation, University of Chicago.

Pfeiffer, R. A. 1936. Pathologie des Hoerstrahlung und der corticalen Hoersphaere. In O. Bumke & O. Foerster (eds.). *Handbuch der Neurologie,* Vol. 6, pp. 533-626. Berlin: Springer.

Piaget, J., & Inhelder, B. 1956. *The Child's Conception of Space.* New York: Humanities Press. paperback edition: New York: Norton, 1967.
Pizzamiglio, L., & Cecchini, M. 1971. Development of the hemispheric dominance in children from 5 to 10 years of age and their relations with the development of cognitive processes. *Brain Res. 31:*363 (abstract).
Polani, P. E., Lessof, M. H., & Bishop, P. M. F. 1956. Color-blindness in "ovarian agenesis" (gonadal dysplasia). *Lancet 2:*118-120.
Poole, I. 1934. Genetic development of articulation of consonant sounds in speech. *Elementary English Rev. 11:*159-161.
Porteus, S. D. 1918. The measurement of intelligence: 653 children examined by the Binet and Porteus tests. *J. Educ. Psychol. 9:*13-31.
    1965. *Porteus Maze Test: Fifty Years' Application.* Palo Alto, Cal.: Pacific Books.
Pratoomraj, S., & Johnson, R. C. 1966. Kinds of questions and types of conservation tasks as related to children's conservation responses. *Child Dev. 37:*343-353.
Prescott, G. A. 1955. Sex differences in Metropolitan Readiness Test results. *J. Educ. Res. 48:*605-610.
Ranucci, E. R. 1952. The effect of the study of solid geometry on certain aspects of space perception abilities. Doctoral dissertation, Teachers College, Columbia University.
Rebelsky, F. 1964. Adults' perception of the horizontal. *Percept. Mot. Skills 19:*371-374.
Reitan, R. M. 1955. Certain differential effects of left and right cerebral lesions in human adults. *J. Comp. Physiol. Psychol. 48:*474-481.
Reshevsky, S. 1948. *Reshevsky on Chess.* New York: Chess Review.
Rosenfield, R. L. 1971. Plasma testosterone-binding globulin and indices of the concentration of unbound plasma androgens in normal and hirsute subjects. *J. Clin. Endocrinol. 32:*717-728.
Rosenzweig, M. R. 1951. Representations of the two ears at the auditory cortex. *Am. J. Physiol. 167:*147-158.
Rosenzweig, M. R., Krech, D., & Bennett, E. L. 1960. A search for relations between brain chemistry and behavior. *Psychol. Bull. 57:*476-492.
Rovet, J. 1975. Mediate and direct experience in the development of children's spatial skills. Paper presented at Biennial Meetings of the Society for Research in Child Development, Denver.
Rudel, R. G., Denckla, M. B., & Spalten, E. 1974. The functional asymmetry of Braille letter learning in normal sighted children. *Neurology 24:*733-738.
Saad, L. G., & Storer, W. O. 1960. *Understanding in Mathematics.* Edinburgh: Oliver & Boyd.
Saarni, C. I. 1973. Piagetian operations and field independence as factors in children's problem-solving performance. *Child Dev. 44:*338-345.

Sadownikova-Koltzova, M. P. 1926. Genetic analysis of temperament of rats. *J. Exp. Zool. 45:*301-318.

Saylor, H. D. 1949. The effect of maturation upon defective articulation in grades seven through twelve. *J. Speech Hear. Disord. 14:*202-207.

Scheibel, A. B., & Scheibel, M. E. 1962. Discussion. In V. B. Mountcastle (ed.). *Interhemispheric Relations and Cerebral Dominance*, pp. 26-30. Baltimore: Johns Hopkins Press.

Schneidler, G. R., & Paterson, D. G. 1942. Sex differences in clerical aptitude. *J. Educ. Psychol. 33:*303-309.

Schwartz, D. W., & Karp, S. A. 1967. Field dependence in a geriatric population. *Percept. Mot. Skills 24:*495-504.

Schwartz, G. E., Davidson, R. J., Maer, F., & Bromfield, E. 1973. Patterns of hemispheric dominance in musical, emotional, verbal, and spatial tasks. Paper read to the Society for Psychophysiological Research, Galveston, Tex.

Seils, L. G. 1951. The relationship between measures of physical growth and gross motor performance of primary grade school children. *Res. Q. 22:* 244-260.

Semmes, J. 1968. Hemispheric specialization: A possible clue to mechanisms. *Neuropsychologia 6:*11-26.

Semmes, J., Weinstein, S., Ghent, L., & Teuber, H.-L. 1960. *Somatosensory Changes after Penetrating Brain Wounds.* Cambridge: Harvard University Press.

1963. Correlates of impaired orientation in personal and extrapersonal space. *Brain 86:*747-772.

Shaffer, J. W. 1962. A specific cognitive deficit observed in gonadal aplasia (Turner's syndrome). *J. Clin. Psychol. 18:*403-406.

Shankweiler, D. 1966. Effects of temporal lobe damage on perception of dichotically presented melodies. *J. Comp. Physiol. Psychol. 62:*115-119.

Shepard, R. N. & Metzler, J. 1971. Mental rotation of three-dimensional objects. *Science 171:*701-703.

Shepard, A. H., Abbey, D. S., & Humphries, M. 1962. Age and sex in relation to perceptual motor performance on several control-display relations on the TCC. *Percept. Mot. Skills* (Monograph Supplement 1-V14) *14:*103-118.

Sherman, J. A. 1967. Problem of sex differences in space perception and aspects of individual functioning. *Psychol. Rev. 74:*290-299.

Siegvald, H. 1944. *Experimentella Undersokningar rörande Intellektuella Könsdifferenser*, Vols. I, II. Lund: Akademisk avhandling.

Silverman, A., Adevai, G., & McGough, W. 1966. Some relationships between handedness and perception. *J. Psychosom. Res. 10:*151-158.

Silverman, J., Buchsbaum, M., & Stierlin, H. 1973. Sex differences in perceptual differentiation and stimulus intensity control. *J. Pers. Soc. Psychol. 25:*309-318.

Singer, G., & Montgomery, R. B. 1969. Comment on "Roles of activation and inhibition in sex differences in cognitive abilities." *Psychol. Rev.* 76: 325-327.

Smith, I. M. 1960. The validity of tests of spatial ability as predictors of success on technical courses. *Br. J. Educ. Psychol.* 30:138-145.

1964. *Spatial Ability: Its Educational and Social Significance.* San Diego, Cal: Knapp.

[The] *SRA Primary Mental Abilities Test: Spatial Relations Test, Ages 11-17.* Chicago: Science Research Associates, 1958.

Stafford, R. E. 1961. Sex differences in spatial visualization as evidence of sex-linked inheritance. *Percept. Mot. Skills* 13:428.

Stanford Research Institute. 1972. Follow-through pupil tests, parent interviews, and teacher questionnaires. Appendix C. Cited in E. E. Maccoby & C. N. Jacklin. *The Psychology of Sex Differences*, p. 83. Stanford, Cal: Stanford University Press, 1974.

Staples, R. 1932. The responses of infants to color. *J. Exp. Psychol.* 15:119-141.

Stern, C. 1960. *Human Genetics.* San Francisco: Freeman.

Stroop, J. R. 1935. Studies of interference in serial verbal reactions. *J. Exp. Psychol.* 18:643-672.

Stuckenschmidt, H. H. 1968. *Maurice Ravel: Variations on His Life and Work.* Philadelphia: Chilton. Translated from German by S. R. Rosenbaum; original German publication, 1966.

Sutherland, N. S. 1960. Visual discrimination of shape by *Octopus*: mirror images. *Br. J. Psychol.* 51:9-18.

Sutton-Smith, B., & Savasta, M. 1972. Sex differences in play and power. Paper presented at annual meeting of the Eastern Psychological Association, Boston.

Tanner, J. M., & Inhelder, B. (eds.). 1958. *Discussions on Child Development.* New York: International Universities Press.

Taylor, D. C. 1969. Differential rates of cerebral maturation between sexes and between hemispheres. *Lancet* 2:140-142.

Taylor, L. B. 1962. Perception of digits presented to right and left ears in children with reading difficulties. Paper presented at meeting of Canadian Psychological Association, Hamilton, Ontario.

Templin, M. C. 1953. Norms on a screening test of articulation for ages three through eight. *J. Speech Hear. Disord.* 18:323-331.

1957. *Certain Language Skills in Children.* Institute of Child Welfare Monograph No. 26. Minneapolis: University of Minnesota Press.

Terman, L. M., et al. 1925. *Genetic Studies of Genius*, Vol. 1, *Mental and Physical Traits of 1000 Gifted Children.* Stanford, Cal: Stanford University Press.

Thomas, H. 1971. The development of water-level representation. Paper presented at Biennial Meetings of the Society for Research in Child Development, Minneapolis.

Thomas, H., & Hummel, D. D. 1972. Paper presented at Regional Meeting of the Society for Research in Child Development, Williamsburg, Va.

Thomas, H., Jamison, W., & Hummel, D. D. 1973. Observation is insufficient for discovering that the surface of still water is invariantly horizontal. *Science 181*:173-174.

Thurstone, L. L. 1938. *Primary Mental Abilities.* Psychometric Monographs, No. 1. Chicago: University of Chicago Press.

1950. *Some Primary Abilities in Visual Thinking.* Report No. 59. Chicago: University of Chicago Psychometric Laboratory.

Thurstone, L. L., & Thurstone, T. G. 1965. *Primary Mental Abilities, Technical Report.* Chicago: Science Research Associates.

Tiffin, J., & Asher, E. J. 1948. The Purdue pegboard: Norms and studies of reliability and validity. *J. Appl. Psychol. 32:*234-247.

Tomlin, M. I., & Stone, C. P. 1933. Sex differences in learning abilities of albino rats. *J. Comp. Psychol. 16:*207-229.

Trowell, H. C., Davies, J. N. P., & Dean, R. F. A. 1954. *Kwashiorkor.* London: Arnold.

Tryon, R. C. 1931. Studies in individual differences in maze ability. 2. The determination of individual differences by age, weight, sex, and pigmentation. *J. Comp. Psychol. 12:*1-22.

Tschirgi, R. D. 1958. Spatial perception and central nervous system symmetry. *Arq. Neuro-psiquiatria 16:*364-366.

Tuddenham, R. D. 1970. A Piagetian test of cognitive development. In W. B. Dockrell (ed.). *On Intelligence: The Toronto Symposium on Intelligence, 1969.* London: Methuen.

1971. Theoretical regularities and individual idiosyncrasies. In D. R. Green, M. P. Ford, & G. B. Flamer (eds.). *Measurement and Piaget.* New York: McGraw-Hill.

Tuddenham, R. D., & Snyder, M. M. 1954. Physical growth of California boys and girls from birth to 18 years. *Univ. Calif. Publ. Child Dev. 1*(2).

Tunturi, A. R. 1946. A study on the pathway from the medial geniculate body to the acoustic cortex in the dog. *Am. J. Physiol. 147:*311-319.

Turner, H. H. 1938. A syndrome of infantilism, congenital webbed neck and cubitus valgus. *Endocrinology 23:*566-574.

Vandenberg, S. G., Stafford, R. E., & Brown, A. 1968. The Louisville twin study. In S. G. Vandenberg (ed.). *Progress in Human Behavior Genetics,* pp. 153-204. Baltimore: Johns Hopkins Press.

Vaughan, H. G., & Costa, L. D. 1962. Performance of patients with lateralized cerebral lesions. 2. Sensory and motor tests. *J. Nerv. Ment. Dis. 134:* 237-243.

Von Economo, C., & Horn, L. 1930. Ueber Windingsrelief, Masse und Rindenarchitektonik der supratemporalflache, ihre individuellen und ihre seitenunterschiede. *Z. Gesamte Neurol. Psychiatr. 130:*687-757.

Wada, J. 1949. A new method for the determination of the side of cerebral speech dominance: A preliminary report on the intracarotid injection of sodium amytal in man. *Igaku to Seibutsugaku 14:*221-222.

1951. An experimental study on the neural mechanism of the spread of epileptic impulse. *Folia Psychiatr. Neurol. Jap. 4:*289-301.

1972. Quoted in *Univ. British Columbia Rep. 18*(5):8, 11.

Wada, J., & Kirikae, T. 1949. Neurological contribution to induced unilateral paralysis of human cerebral hemisphere: Special emphasis on experimentally induced aphasia. *Hokkaido Igaku Zasshi (Acta Med. Hokkaido) 24:* 1-10.

Wada, J., & Rasmussen, T. 1960. Intracarotid injection of sodium amytal for the lateralization of cerebral speech dominance. *J. Neurosurg. 17:*266-282.

Wada, J. A., Clark, R., & Hamm, A. 1975. Asymmetry of temporal and frontal speech zones in 100 adult and 100 infant brains. *Arch. Neurol. 32:* 239-246.

Warrington, E., & James, M. 1967. Disorders of visual perception in patients with localized cerebral lesions. *Neuropsychologia 5:*253-266.

Watanabe, H. 1968. Subsistence and ecology of northern food gatherers with special reference to the Ainu. In R. L. Lee & I. DeVore (eds.). *Man the Hunter*, pp. 69-77. Chicago: Aldine.

Wechsler, D. 1955. *Manual for the Wechsler Adult Intelligence Scale.* New York: Psychological Corp.

Williams, M. 1970. *Brain Damage and the Mind.* Harmondsworth, Middlesex: Penguin Books.

Wilson, R. S. 1975. Twins: patterns of cognitive development as measured on the Wechsler Preschool and Primary Scale of Intelligence. *Dev. Psychol. 11:*126-134.

Wit, O. C. 1955. Sex differences in perception. Master's thesis, University of Utrecht. Cited in Witkin, H. A., et al. *Psychological Differentiation.* New York: Wiley, 1962.

Witelson, S. F. 1975. Age and sex differences in the development of right hemispheric specialization for spatial processing as reflected in a dichotomous tactual stimulation task. Paper presented to Society for Research in Child Development, Denver.

Witelson, S. F., & Pallie, W. 1973. Left hemisphere specialization for language in the newborn: Neuroanatomical evidence of asymmetry. *Brain Res. 96:*641-646.

Witkin, H. A. 1950. Individual differences in ease of perception of embedded figures. *J. Pers. 19:*1-15.

Witkin, H. A., Dyk, R. B., Faterson, G. E., Goodenough, D. R., & Karp, S. A. 1962. *Psychological Differentiation.* New York: Wiley.

Witkin, H. A., Goodenough, D. R., & Karp, S. A. 1967. Stability of cognitive style from childhood to young adulthood. *J. Pers. Soc. Psychol.* 7:291-300.

Witkin, H. A., Lewis, H. B., Hertzman, M., Machover, K., Meissner, P. B., & Wapner, S. 1954. *Personality Through Perception.* New York: Harper & Row.

Woolley, H. T. 1910. A review of the recent literature in the psychology of sex. *Psychol. Bull.* 7:335-342.

1914. General review and summaries: The psychology of sex. *Psychol. Bull.* 11:353-379.

Young, F. M. 1941. An analysis of certain variables in a developmental study of language. *Genet. Psychol. Monogr.* 23:3-141.

Zangwill, O. L. 1951. Discussion on parietal lobe syndromes. *Proc. R. Soc. Med.* 44:343-346.

1960. *Cerebral Dominance and Its Relation to Psychological Function.* Springfield, Ill.: Thomas.

Zazzo, R., & Jullien, C. 1954. Contribution à la psychologie différentielle des sexes au niveau pré-scolaire. *Enfance* 7:12-23.

14
Human handedness

ROBERT E. HICKS AND MARCEL KINSBOURNE

Asymmetrical hand preference is a species-specific characteristic of humans. Surveys from the United States, Britain, Western Europe, Japan, the Solomon Islands, and various nonliterate cultures of Africa and Asia indicate that approximately 90% of the people are right-handed (Chamberlain, 1928; Komai & Fukoda, 1934; Rife, 1940; Hécaen & Ajuriaguerra, 1964; Verhaegen & Ntumba, 1964; Dawson, 1972; Annett, 1973; Rhoads & Damon, 1973).

Indirect evidence suggests that humans have been right-handed since prehistoric times. Dennis (1958) found that the people represented in the paintings in Egyptian tombs were usually engaged in right-handed activities. Wilson (1885) reported that most paleolithic tools, weapons, and ornaments were apparently made with the right hand. Uhrbrock (1973) concluded that many prehistoric handprints came from right-handers. Perhaps the most intriguing claim comes from Dart (1949), who concluded that *Australopithecus* was probably right-handed, based on examination of fossilized fractured skulls.

Forelimb preference in nonhuman mammals seems fundamentally different from human handedness (Hicks & Kinsbourne, 1976a). Although most species exhibit more or less stable forepaw preferences, there appear to be no deviations from symmetrical preference between subjects for any species except humans. In addition, hand preference in humans is correlated with cerebral specialization for various functions. Nonhuman mammals have not been demonstrated to possess cerebral specialization in any manner similar to humans; that is, no double dissociations have been reported in nonhuman mammals (Hamilton, 1976; Webster, 1976).

This chapter reviews evidence concerning three aspects of handedness: its correlation with other dimensions of laterality, its causes, and its measurement and definition.

HANDEDNESS AND LATERALITY

Language

Neuropathology

The earliest evidence concerning cerebral specialization came from studies of language disorder (e.g., dysphasia) following unilateral trauma. This was also the case with respect to handedness. The attempt was to correlate handedness, side of disorder, and incidence of dysphasia.

Zangwill (1967) has reviewed reports on unilateral head trauma of several varieties. There is some variability in the series reviewed, but it is correct to conclude that dysphasia in a right-hander following unilateral trauma to the right hemisphere is very rare (less than 2%). Thus, approximately 98% of the right-handed dysphasics have left hemisphere lesions. In left-handers, however, dysphasia following a right hemisphere lesion is not uncommon (approximately one-third of the patients). Since 1967, similar reports have appeared (Gloning et al., 1969; Luria, 1969; Roberts, 1969; Hécaen & Sauget, 1971).

More recently, administration of unilateral electroconvulsive therapy (ECT) has afforded the opportunity for studying transient hemisphere disability. Left-sided application of ECT in right-handers leads to much greater impairment on a verbal task a few minutes later than if the ECT is applied to the right side (Fleminger et al., 1970; Pratt et al., 1971; Pratt & Warrington, 1972). The results are more variable in left-handers. Typically, patients are asked to name objects from verbal descriptions. Pratt et al. (1971) found that left-sided ECT produced greater impairment on this task than right-sided ECT in 11 of 12 right-handers, whereas 8 of 12 left-handers suffered greater left-sided impairment, and the remaining 4, greater right-sided impairment. Warrington and Pratt (1973) found 7 of 30 (23%) left-handers were more dysphasic after right-sided ECT, compared with 1 of 52 (2%) right-handers (Pratt & Warrington, 1972). No asymmetry was found in 2 of the 30 (7%) left-handers, whereas all 52 right-handers showed asymmetry.

Another procedure that allows the study of temporary hemisphere disability is the unilateral injection of sodium amytal into the carotid artery. This is always followed by a (temporary) contralateral sensory and motor loss (Milner et al., 1964). Milner et al. (1964) reported that 5 of 48 (10.4%) right-handers suffered verbal difficulty (e.g., in naming and counting) following right-sided injection. The remaining 43 patients were verbally impaired following left-sided injection. Of the 44 left-handers they reported on, 28 had left hemisphere dominance for language (64%), 7 were judged to have bilateral language representation (16%), and 9 (20%) had right hemisphere language dominance. Subsequently, Rasmussen and Milner (1975) reported on 140 right-handers and 112 left-handers. None of the right-handers had bilateral speech representation, 96% had left hemisphere dominance for speech, and 4% had right hemisphere dominance for speech. Of the left-handers, 70% had left hemisphere, 15% right hemisphere, and 15% bilateral speech representation.

Dichotic listening and tachistoscopic presentation
The literature reviewed in this section deals for the most part with normal populations. In a dichotic listening study, competing auditory stimuli are simultaneously delivered to the two ears. Typically, right-handers recall significantly more verbal stimuli from the right ear than from the left ear, whereas left-handers as a group show smaller differences between the ears (Satz et al., 1965, 1967; Curry, 1967; Curry & Rutherford, 1967; Zurif & Bryden, 1969; Bryden, 1970, 1973, 1975; Knox & Boone, 1970; Dee, 1971; Orlando, 1972; Hines & Satz, 1974; Shankweiler & Studdert-Kennedy, 1975).

According to several studies, the difference between ear reports correlates significantly with the difference between the hands on proficiency or preference measures (Satz et al., 1967; Orlando, 1972; Bryden, 1975; Shankweiler & Studdert-Kennedy, 1975). Similarly, Knox and Boone (1970) reported that it was primarily the strongly left-handed subjects who showed no difference between the ears or a left-ear superiority. Dee (1971) reported the opposite. However, he also reported a greater right-ear superiority for strongly left-handed subjects than for strongly right-handed subjects, so his results are somewhat suspect.

Another procedure for studying language laterality in normal subjects is to present verbal stimuli to a single visual half-field

(VHF), or to present separate stimuli to each VHF, and ask the subject to identify the material. Usually, right-handed subjects identify stimuli presented to the right VHF (left cerebral hemisphere) better than stimuli presented to the left VHF (right cerebral hemisphere). When left-handers are included, an interaction between handedness and VHF is usually obtained (Bryden, 1964, 1965, 1973; Orbach, 1967; Zurif & Bryden, 1969; Hines & Satz, 1971, 1974; McKeever & Gill, 1972; McKeever et al., 1973), with left-handers showing less difference between VHF scores.

Several investigators have used the Posner and Mitchell (1967) name versus physical identity task in laterality studies. A binary manual response is used to indicate whether two stimuli match physically (e.g., *A-A* or *a-a* match, but *A-B* or *A-a* do not match) or nominally (e.g., *A-A*, *a-a*, or *A-a* all match because they have the same name code). Right-handers perform name matches faster when the display is presented to the right VHF than to the left VHF, whereas no difference or a faster left-VHF reaction is obtained for physical matches (Cohen, 1972; Geffen et al., 1972). Left-handers show no significant difference between the two VHF reaction times for either type of task (Cohen, 1972).

Qualitative differences
The previous sections review evidence that left-handers (as a group) are more likely to have language representation in the right hemisphere. There are also several reports of qualitative differences in lateralization in left- and right-handers. Hécaen and Piercy (1956) studied patients with paroxysmal dysphasia. They concluded that left-handers vary more than right-handers, not only in which hemisphere is dominant for language, but also in the diffuseness of language representation within a single hemisphere.

Rasmussen and Milner (1975) reported a surprising dissociation between naming and serial language tasks in left-handers. Of the left-handers with bilateral speech representation, 57% had associated naming and serial processes; in 43% these processes were dissociated (i.e., naming and sequential verbalization were represented in different hemispheres). In 67% of the instances of dissociation, the more marked deficit in serial performance was during right hemisphere depression, and the more severe deficit in naming performance was during left hemisphere depression.

In a similar vein, Marcie (1972), studying writing disorders in

47 left-handers with unilateral lesions, claimed that writing disorders are more independent of speech and reading disorders in left- than in right-handed patients. Left-handers may, with a lesion restricted to one hemisphere, demonstrate signs usually associated with lesions of both the left and right hemispheres of right-handers (Roberts, 1969; Hécaen & Sauget, 1971).

In tests with normal subjects, Hines and Satz (1974) found that difference scores between the ears in dichotic listening and between visual half-fields were significantly correlated in right-handers, but not in left-handers. This indicates a dissociation between language by eye and ear in left-handers, but not in right-handers. Zurif and Bryden (1969) and Bryden (1973) did not find the two difference scores to be significantly correlated, but there are several methodological difficulties in both studies (see Hines & Satz, 1974, for a critique of Zurif & Bryden, 1969).

In summary, the evidence reviewed indicates that left-handers are more variable than right-handers concerning which hemisphere is superior for language functions and in regard to consistency of lateralization for language within and between the hemispheres.

Performance

Many of the studies reviewed in this section deal with language, but not as directly as the studies in previous sections.

Kinsbourne and Cook (1971) reported that when right-handed subjects balance a dowel rod while repeating sentences, balancing time decreases (compared with a nonspeech control condition) for the right hand, but not the left hand. They interpreted the right-hand verbalization decrement as due to the need for the left cerebral hemisphere simultaneously to program the right-sided motor control and the speech. Hicks (1975) replicated this result, and found that increased phonetic difficulty of the phrases (verbalized) produced an increased decrement in right-handed balancing, whereas left-handed balancing was unchanged. But in left-handers, concurrent verbalization shortened balancing duration with both hands. With a finger-sequencing task, Hicks et al. (1975) found that concurrent verbal rehearsal interfered with right-handed more than left-handed performance in right-handers. Lomas and Kimura (1976) obtained similar results on a similar task, and they found that speaking interfered with both hands in left-handers.

Kinsbourne (1972) theorized that eye and head turning to a given direction indicates activation of the contralateral cerebral hemisphere. Congruent with this idea, when right-handers solve verbal questions they more often turn head and eyes to the right, but while solving musical or spatial problems they more often turn eyes and head to the left (Kinsbourne, 1972; Kocel et al., 1972). Left-handers showed no overall directional bias to either type of question, with some subjects looking left and some right to each type of question (Kinsbourne, 1972). Of additional interest, more than 80% of the left-handers demonstrated a directional bias consistent across all questions (i.e., the leftward lookers looked left to verbal and spatial questions, and vice versa).

Kimura (1973a) found that speaking in right-handers is accompanied by more free motions of the hands than are silent activities. This increase is much larger in the right hand than in the left. The association of right-hand free movements (i.e., movements not bringing the limb into contact with anything) with speech did not seem to be due to a diffuse motor facilitation, since self-touching movements (e.g., ear, glasses) occurred equally often with either hand. Kimura (1973a) concluded that "activation of the speech system in the left hemisphere is associated with concomitant activation of certain other motor systems in the hemisphere" (p. 49). Kimura (1973b) replicated her earlier results with right-handers; she also found that left-handers make more left-handed than right-handed free movements while speaking, but make just as many right-handed movements as right-handers. She concluded that because there is "a connection between speaking and gesturing, it is reasonable to suppose that where speech is not unilaterally organized, gesturing should be manifested less unilaterally" (p. 54).

Efron (1963a,b) studied perception of temporal order in left- and right-handers. In right-handers, he concluded that this function is mediated primarily by the left cerebral hemisphere. As an example of the results supporting this conclusion, for the LVF stimulus to be judged as simultaneous with a RVF stimulus, the former must precede the latter by about 5 msec. He interpreted this as representing the time needed for callosal transfer of information. Left-handers as a group were not lateralized on this function (Efron, 1963a,b). Carmon and Nachshon (1971) found that temporal order perception was more impaired by lesions in the left than in the right cerebral hemisphere in right-handed patients.

Murray and Hagan (1973) studied pain tolerance of the limbs. They found that right-handers had much greater pain tolerance in the right hand and foot than in the left hand and foot. The asymmetry was in the same direction for left-handers, but much less pronounced. Similarly, Varney and Benton (1975) found that in identifying the direction of the linear tactile stimulation applied to the palm, the left palm was superior to the right palm in right-handers. This was not true of left-handers.

Using photographs of human faces as stimuli, Gilbert and Bakan (1973) found a handedness VHF bias, with right-handers strongly biased toward the left VHF and left-handers showing no significant VHF bias. Facial recognition is more dependent on the right cerebral hemisphere in right-handers (e.g., Benton & Van Allen, 1968; Rizzolatti et al., 1971).

A rare instance in which left-handers are more strongly lateralized than right-handers has been reported by Froeschels (1961). Tongue clicking is usually done to the side of the mouth ipsilateral to the preferred hand, and this relationship is more consistent for left-handers.

Perhaps the most intriguing interaction of performance asymmetry with handedness was reported by Olson and Laxar. They used a word-picture verification task using the terms *right* and *left*. Right-handers process *right* faster and more accurately than *left* (Olson & Laxar, 1973), but there is no difference in either speed or accuracy between the two terms in left-handers (Olson & Laxar, 1974).

With the exception of tongue clicking, studies of lateralized performance effects have demonstrated that right-handers are more lateralized than are left-handers.

Nonperformance

McRae et al. (1968) found a difference in the size of the occipital horns. In right-handers with asymmetrical occipital horns, the left occipital horn was larger five times more often than the right. In left-handers who showed asymmetry, the right and left occipital horns were equally often larger.

Hochberg and LeMay (1975) found that in 67% of 123 right-handers, the angle formed by the vessels leaving the posterior end of the Sylvian fissure was greater on the right than on the left by more than 10°, the mean (right-left) difference being 23.5°. In only 21.4% of 38 left-handers was the right Sylvian point angle larger than the left by more than 10°; the mean (right-left) difference, 6.6°. LeMay and Culebras (1972) obtained similar results, and in addition, found that the typical (right-hander) hemisphere differences in the Sylvian fissures are present in fetal life.

Carmon and Gombos (1970) found that the systolic pressure in the ophthalmic artery was higher on the right side in most right-handers and higher on the left side in most left-handers. However, in left-handers the pressures were much more variable. In the group of right-handers, 73% had higher pressure on the right. In the left-handed group, 40% had higher pressure on the left, and 27% had higher pressure on the right. Grouping by relative manual preference made the relationship much clearer, with those subjects having symmetrical hand preference also having symmetrical ophthalmic pressures. Carmon et al. (1972) found that the blood volume was also greater in the hemisphere ipsilateral to the preferred hand in both left- and right-handers.

Galin and Ornstein (1972) reported that during a verbal task, the alpha rhythm over the right side of the head increased relative to the left side. The opposite occurred during a spatial task. Their subjects were right-handed. Butler and Glass (1972) found the alpha rhythm more suppressed on the left during mental arithmetic in right-handers, but in left-handers the alpha rhythm was suppressed on both sides.

In right-handers, about 1 second prior to executing a voluntary unilateral hand movement, a slow, negative potential shift can be measured (maximally) over the contralateral motor cortex (Vaughn et al., 1968; McAdam & Seales, 1969). Kutas and Donchin (1974) had both left- and right-handed subjects squeeze a dynamometer. Replicating earlier work, they found that the premovement "readiness" potentials were larger over the hemisphere contralateral to the responding hand in right-handers. This was true only for the right hand in left-handers. Left-handers showed bilaterally symmetrical potentials for left-handed responses.

Evoked potentials to nonverbal stimuli are usually of greater amplitude over the hemisphere ipsilateral to the preferred hand. This is another of those rare cases where left-handers are apparently more lateralized than right-handers, that is, the interhemisphere difference is larger in left-handers (Pfefferbaum & Buschsbaum, 1971; Gott & Boyarsky, 1972).

Eye and foot preferences have been studied in relation to handedness. Annett (1974) reported that the preferred hand correlated with both the preferred eye and foot.

The relationship between preferred hand and eye appears difficult to replicate, however. Two recent papers have reported no relationship between (any of) a battery of eyedness tests and handedness (Gronwall & Sampson, 1971; Coren & Kaplan, 1973). Both papers also contained thorough literature reviews of eyedness measures and their reported relationship to handedness. Humphis (1969) and Sampson and Horrocks (1967) also reported no relationship of handedness and eyedness. Friedlander (1971) found ocular preference (for the right eye) related to right-handedness, but not to left-handedness (i.e., half the left-handers preferred either eye).

Several studies have reported that the fingerprints and palmprints of right-handers are more asymmetrical between hands than those of left-handers (e.g., Newman, 1934; Rife, 1955).

With the exceptions of tongue clicking and evoked potentials, left-handers appear less lateralized on all the measures reviewed.

ORIGINS OF HANDEDNESS

Genetic

Collins (1970) concluded "that offspring resemble their parents in handedness according to an unspecified environmental mechanism" (p. 132). He based this claim on some controversial (see Nagylaki & Levy, 1973) analyses of previously published data on human twins and on his earlier (Collins, 1968, 1969) demonstration that three generations of selection for pawedness in mice failed to change the original 50:50 ratio of left and right preference. Collins (1970), and in his later paper (1975), argued strongly for the appropriateness of mice as "a useful analogue, if not homologue, to human hand preference." The validity of Collins's assertion is very questionable. Several important distinctions between human handedness and nonhuman forelimb preference are made earlier in this chapter.

Provins (1967) similarly acknowledged the familial handedness correlation, but admitting the ambiguity of such evidence, nevertheless suggested that differential training underlies handedness. The only evidence offered in support of his suggestion is that the potential skill level of the nonpreferred hand is as high as that of the preferred hand on some tasks.

Differential within-pair concordance of monozygotic and dizygotic twins seems unsuited for studying the heritability of handedness. The principle reasons are the complex embryology of monozygotic twins (e.g., situs inversus of ectodermal structures; Corner, 1955) and the considerably greater frequency of left-handedness in both monozygotic and dyzygotic twins compared with the general population (e.g., Nagylaki & Levy, 1973).

It has been frequently demonstrated that left-handedness is familial (Ramaley, 1913; Chamberlain, 1928; Rife, 1940; Falek, 1955; Annett, 1973, 1974; Bakan et al., 1973). However, a familial association may result from environmental influences confounded with family mating combinations. Annett (1973) presented data incompatible with this hypothesis: 84% of left-handed children have two right-handed parents, 72% of the children of left-handed mothers are right-handed, and about 50% of the children of two left-handed parents are right-handed.

In a more direct test of the modeling hypothesis, Hicks and Kinsbourne (1976b) found that hand preference of college students correlated significantly with the writing hands of their biological parents (replicating previous family studies), but not their stepparents. Stepparents' handedness had no effect on the handedness of their stepchildren, and this lack of effect was invariant with how long the stepparents and children had lived together. Hicks and Kinsbourne's (1976b) data are inconsistent with a prenatal environmental hypothesis because of a significant effect of the (biological) father's handedness on the child's handedness.

In addition to the cross-fostering study of Hicks and Kinsbourne (1976b), a substantial amount of incidental evidence also favors a hypothesis of genetic influence on handedness.

Familial left-handedness affects other laterality phenomena. Right-handers with left-handed relatives have a higher incidence of recovery from dysphasia (Subirana, 1958, 1964; Luria, 1969) and a higher incidence of crossed (i.e., resulting from right hemisphere lesions) dysphasia (Ettlinger et al., 1956) than right-handers without left-handed relatives. The observed frequencies were more similar to those of left-handers than to those of right-handers. In addition, several investigators have found a reduced superiority of the right VHF for visually presented verbal materials in normal right-handers having left-handed relatives (Hines & Satz, 1971;

ing up small collars to place on the pins. The task requires accurate coordination of finger and hand movements, as well as control of arm movements. Each subject performed the task with each hand. Among the right-handers, 72% showed strong right-hand superiority, 16% showed only slight right-hand superiority, and 12% showed either no difference between hands or left-hand superiority. The left-handers were more variable: 55% showed left-hand superiority, 30% showed no difference between the hands, and 15% showed right-hand superiority. The investigators obtained similar results on tests using scissors. Similar results have been reported elsewhere (e.g., McKeever et al., 1973; Raczkowski et al., 1974).

Several studies have reported smaller performance differences between the hands of left-handers than of right-handers, with degree of preference even more carefully matched (McGlone, 1970; Provins & Cunliffe, 1972). Annett (1970), however, reported a linear relationship between relative preference and relative manual skill at peg moving. Lake and Bryden (1976) administered a handedness questionnaire and measured performance of each hand on a visually guided repetitive manual aiming task. Relative preference correlated 0.78 with relative proficiency on the aiming task. However, for right-handers alone, the correlation was 0.44; for left-handers, the correlation was 0.00.

In summary, left-handers (compared with right-handers) demonstrate larger between-subject and within-subject variability on measures of manual preference. Similarly, left-handers have greater between-subject variability on relative manual proficiency.

Some of the inconsistency in the relationship between preference and performance measures of handedness may be due to several problems in the latter. There are at least 10 independent dimensions of unimanual performance (Fleishman, 1972). Which of these dimensions to sample is an initial problem. In addition, there are a number of task factors that can change the distribution of performance asymmetries. The relationship of performance asymmetries to preference asymmetry (and other dimensions of laterality) thus changes or even reverses in direction. The implications for a performance-defined criterion of handedness should be obvious.

Steingrueber (1975) reported that relative manual proficiency is a function of task complexity. His two tasks, tapping on squares and dotting, each had three levels of complexity. He found that manual asymmetry decreased with decreasing task complexity and the variance likewise decreased; that is, the difference between the hands moved toward zero, and the dispersion about the mean decreased with decreasing complexity. On virtually any criterion, then, the proportion of subjects classified as left-handed would change with task complexity.

Hellebrandt and Houtz (1950) reported that intermanual differences on an ergographical task increased with increased fatigue. Thus, continuous practice may amplify the distribution of asymmetry. The converse has been argued for distributed practice (Provins, 1967): As practice with each hand increases, the difference between the hands decreases. And sometimes relative manual proficiency is invariant with practice (Hicks, 1975, Experiment 3).

Practice may also have more subtle effects on relative proficiency distribution. Relative manual proficiency measures, of course, require that each hand perform the task at least once. Well-done studies counterbalance order of administration across hands. Counterbalancing assumes symmetrical transfer. There is little assurance that bimanual transfer is symmetrical. Briggs and Brogden (1953) found greater transfer of linear pursuit movements from the left to the right hand than vice versa in right-handers. Ammons (1958) found an asymmetry of bilateral transfer of rotary pursuit performance that varied with distribution of practice. With inverted-reversed printing, both skill (Hicks, 1974) and work decrement (Meier & French, 1965; Hicks, 1972) transfer more in the left-to-right than the right-to-left direction in right-handers. Relative manual proficiency may thus depend on which hand is tested first and on the intertrial interval.

Stimulus-response and response-response compatibility can affect the difference in performance between the hands. Annett and Sheridan (1973) found that the left hand is faster in an incompatible than in a compatible condition in right-handers. The opposite was true for the right hand. Sheridan (1973) did not find a compatibility effect on the left hand, but the right hand did worse in an incompatible than a compatible condition in right-handers. Also, he found no difference between the hands in the incompatible condition. Hicks et al. (1976) obtained comparable results with a keyboard task. Sheridan (1973) reported that requiring increased precision of movement (i.e., Fitts's index of difficulty) produced increased right-hand superiority. It is obvious that the shape of the asymmetry distribution may well depend on the compatibility and precision required. It is also likely that preferred handedness would interact with compatibility (i.e., have opposite effects on the hands of left-handers), but as only right-handers were tested in these studies, this possibility cannot be assessed.

The interhand difference may depend on directional properties of the task. Reed and Smith (1961) found that the preferred direction of horizontal movement is away from the body with each hand, for both left- and right-handers. Similarly, Brown et al. (1948), Downey (1932), and Shimrat (1973) found that horizontal movements away from the body are more skillful than horizonatal movements toward the body. A given manual task then, may change intermanual differences differentially for left- and right-handers depending on the direction of the necessary horizontal movements.

The type of movement is also a critical variable. Nakamura and Saito (1974) found that supination of the right arm is faster than supination of the left arm in right-handers. In these same subjects, flexion of the left arm is faster than flexion of the right arm. Opposite effects were found for left-handers. Similar results were reported by Nakamura et al. (1975). Kimura and Vanderwolf (1970) reported that the left hand is superior on a finger-flexion task for both left- and right-handers. Flowers (1975) reported that the preferred hand is faster on a visual aiming task, but not on a ballistic movement task.

SUMMARY

It does not seem possible to specify the proficiency dimensions underlying hand preference at this time. The difference between performance scores of the two hands appears to be greatly influenced by task factors previously discussed. It is probably correct to conclude that on most manual tasks, the majority of right-handers, at least, are more skillful with the preferred hand. Handedness seems more variable in left-handers than in right-handers.

Handedness is related to most other laterality dimensions. Virtually any lateralized dimension in right-handers appears more symmetrical in left-handers; the mean relative asymmetry score is closer to zero in left-handers. In addition, left-handers appear to be more heterogeneous than right-handers on most lateralized dimensions; the between-subject variance is larger in left- than right-handed populations.

Laterality per se appears to be more homogeneous in right-handers than in left-handers: Different indexes of laterality are more likely to intercorrelate with each other in groups of right-handers than in groups of left-handers.

There is evidence that genetic factors influence handedness, although it is not clear that an adequate genetic model is available. Whereas there is no good evidence that unselected populations of left-handers are neurologically impaired, left-handers do tend to be overrepresented in most poplations with neurological disease.

REFERENCES

Ammons, R. B. 1958. Le mouvement. In G. S. Seward & J. P. Seward (eds.). *Current Psychological Issues: Essays in Honor of Robert S. Woodworth.* New York: Holt.

Annett, J., & Sheridan, M. R. 1973. Effects of S-R and R-R compatibility on bimanual movement time. *Q. J. Exp. Psychol. 25*:247-252.

Annett, M. 1964. A model of the inheritance of handedness and cerebral dominance. *Nature 204*:59-60.

—— 1967. The binomial distribution of right, mixed and left handedness. *Q. J. Exp. Psychol. 19*:327-333.

—— 1970. The growth of manual preference and speed. *Br. J. Psychol. 61*:545-558.

—— 1972. The distribution of manual asymmetry. *Br. J. Psychol. 63*:343-358.

—— 1973. Handedness in families. *Ann. Hum. Genet. 37*:93-105.

—— 1974. Handedness in the children of two left-handed parents. *Br. J. Psychol. 65*:129-131.

Annett, M., & Turner, A. 1974. Laterality and the growth of intellectual abilities. *Br. J. Educ. Psychol. 44*:37-46.

Applebee, A. N. 1971. Research in reading retardation: Two critical problems. *J. Child Psychol. Psychiatry 12*:91-113.

Bakan, P., Dibb, G., & Reed, P. 1973. Handedness and birth stress. *Neuropsychologia 11*:363-366.

Bakwin, H. 1950. Psychiatric aspects of pediatrics: Lateral dominance, right- and left-handedness. *J. Pediatr. 36*:385-391.

Benton, A. L. 1962. Clinical symptomatology in right and left hemisphere lesions. In V. B. Mountcastle (ed.). *Interhemispheric Relations and Cerebral Dominance.* Baltimore: Johns Hopkins Press.

Benton, A. L., Meyers, R., & Polder, G. J. 1962. Some aspects of handedness. *Psychiatr. Neurol. (Basel) 144*:321-337.

Benton, A. L., & Van Allen, M. W. 1968. Impairment in facial recognition in patients with cerebral disease. *Cortex 4*:344-358.

Berman, A. 1971. The problem of assessing cerebral dominance and its relationship to intelligence. *Cortex 7*:372-386.

Bolin, B. J. 1953. Left handedness and stuttering as signs diagnostic of epileptics. *J. Ment. Sci. 99*:483-488.
Brain, W. R. 1945. Speech and handedness. *Lancet 249*:837-841.
Briggs, G. E., & Brogden, W. J. 1953. Bilateral aspects of the trigonometric relationship of precision and angle of linear pursuit-movements. *Am. J. Psychol. 66*:472-478.
Briggs, G. G., & Nebes, R. D. 1975. Patterns of hand preference in a student population. *Cortex 11*:230-238.
Briggs, G. G., Nebes, R. D., & Kinsbourne, M. 1976. Intellectual differences in relation to personal and family handedness. *Q. J. Exp. Psychol. 28*: 591-602.
Brown, J. S., Knauft, E. B., & Rosenbaum, G. 1948. The accuracy of positioning movements as a function of their direction and extent. *Am. J. Psychol. 61*:167-182.
Bryden, M. P. 1964. Tachistoscopic recognition and cerebral dominance. *Percept. Mot. Skills 19*:686.
   1965. Tachistoscopic recognition, handedness, and cerebral dominance. *Neuropsychologia 3*:1-8.
   1970. Laterality effects in dichotic listening: Relations with handedness and reading ability in children. *Neuropsychologia 8*:443-450.
   1973. Perceptual asymmetry in vision: Relation to handedness, eyedness, and speech lateralization. *Cortex 9*:418-432.
   1975. Speech lateralization in families: A preliminary study using dichotic listening. *Brain Language 2*:201-211.
Burt, C. 1950. *The Backward Child*, 3rd ed. London: University of London Press.
Butler, S. R., & Glass, A. 1974. Asymmetries in the electroencephalogram associated with cerebral dominance. *Electroencephalogr. Clin. Neurophysiol. 36*:481-491.
Caplan, P., & Kinsbourne, M. 1976. Baby drops the rattle: Asymmetry of duration of grasp by infants. *Child Dev. 47*:532-534.
Carmon, A., & Gombos, G. H. 1970. A physiological vascular correlate of hand-preference: Possible implications with respect to hemispheric cerebral dominance. *Neuropsychologia 8*:119-128.
Carmon, A., Harishanu, Y., Lowinger, E., & Lavy, S. 1972. Asymmetries in hemispheric blood volume and cerebral dominance. *Behav. Biol. 7*:853-859.
Carmon, A., & Nachshon, I. 1971. Effect of unilateral brain damage on perception of temporal order. *Cortex 7*:410-418.
Chakrabarti, J., & Barker, D. G. 1966. Lateral dominance and reading ability. *Percept. Mot. Skills 22*:881-882.
Chamberlain, H. D. 1928. The inheritance of left-handedness. *J. Hered. 19*: 557-559.
Cohen, G. 1972. Hemispheric differences in a letter classification task. *Percept. Psychophysics 11*:139-142.

Collins, R. L. 1968. On the inheritance of handedness. 1. Laterality in inbred mice. *J. Hered.* 59:9–12.

—— 1969. On the inheritance of handedness. 2. Selection for sinistrality in mice. *J. Hered.* 60:117–119.

—— 1970. The sound of one paw clapping: An inquiry into the origin of left-handedness. In G. Lindzey & D. Thiessen (eds.). *Contributions to Behavior-Genetic Analysis: The Mouse as a Prototype.* New York: Appleton-Century-Crofts.

—— 1975. When left-handed mice live in right-handed worlds. *Science 187:* 181–184.

Coren, S., & Kaplan, C. P. 1973. Patterns of ocular dominance. *Am. J. Optom.* 50:283–292.

Corner, G. W. 1955. The observed embryology of human single-ovum twins and other multiple births. *Am. J. Obstet. Gynecol.* 70:933–951.

Crovitz, H. F., & Zener, K. 1962. A group-test for assessing hand-and-eye dominance. *Am. J. Psychol.* 75:271–276.

Crowell, D. H., Jones, R. H., Kapunai, L. E., & Nakawaga, J. K. 1973. Unilateral cortical activity in newborns: An early index of cerebral dominance? *Science 180:*205–208.

Curry, F. K. W. 1967. A comparison of left-handed and right-handed subjects on verbal and non-verbal dichotic listening tasks. *Cortex 3:*343–352.

Curry, F. K. W., & Rutherford, D. R. 1967. Recognition and recall of dichotically presented verbal stimuli by right- and left-handed persons. *Neuropsychologia 5:*119–126.

Dart, R. A. 1949. The predatory implement technique of Australopithecus. *Am. J. Phys. Anthropol.* 7:1–38.

Dawson, J. L. M. 1972. Temne-Arunta hand-eye dominance and cognitive style. *Int. J. Psychol.* 7:219–233.

Dee, H. L. 1971. Auditory asymmetry and strength of manual preference. *Cortex* 7:236–245.

Dennis, W. 1958. Early graphic evidence of dextrality in man. *Percept. Mot. Skills* 8:147–149.

Doll, E. A. 1933. Psychological significance of cerebral birth lesions. *Am. J. Psychol.* 45:444–452.

Downey, J. E. 1932. Back-slanted writing and sinistral tendencies. *J. Educ. Psychol.* 23:277–286.

Efron, R. 1963a. The effect of handedness on the perception of simultaneity and temporal order. *Brain 86:*261–284.

—— 1963b. The effect of stimulus intensity on the perception of simultaneity in right- and left-handed subjects. *Brain 86:*285–294.

Ettlinger, G., Jackson, C. V., & Zangwill, O. L. 1956. Cerebral dominance in sinistrals. *Brain* 79:569–588.

Falek, A. 1955. Handedness: A family study. *Am. J. Hum. Genet.* 11:52–62.

Fleishman, E. A. 1972. On the relation between abilities, learning, and human performance. *Am. Psychol.* 27:1017–1032.

Fleminger, J. J., de L. Horne, D. J., & Nott, P. 1970. Unilateral electroconvulsive therapy and cerebral dominance: Effect of right- and left-sided electrode placement on verbal memory. *J. Neurol. Neurosurg. Psychiatry* 33:408–411.

Flick, G. L. 1966. Sinistrality revisited: A perceptual-motor approach. *Child Dev.* 37:613–622.

Flowers, K. 1975. Handedness and controlled movement. *Br. J. Psychol.* 66:39–52.

Friedlander, W. J. 1971. Some aspects of eyedness. *Cortex* 7:357–371.

Froeschels, E. 1961. Is handedness organic or functional in nature? *Am. J. Psychother.* 15:101–105.

Galin, D., & Ornstein, R. E. 1972. Lateral specialization of cognitive mode: An EEG study. *Psychophysiology* 9:412–418.

Gardiner, M. F., & Walter, D. O. 1976. Evidence of hemispheric specialization from infant EEG. In S. Harnad, R. W. Doty, L. Goldstein, J. Paynes, & G. Kranther (eds.). *Lateralization in the Nervous System.* New York: Academic Press.

Geffen, G., Bradshaw, J. L., & Nettleton, N. C. 1972. Hemispheric asymmetry: Verbal and spatial encoding of visual stimuli. *J. Exp. Psychol.* 95:25–31.

Geschwind, N., & Levitsky, W. 1968. Human brain: Left-right asymmetries in temporal speech region. *Science* 161:186–187.

Gesell, A. 1938. The tonic neck reflex in the human infant. *J. Pediatr.* 13:455–464.

Gesell, A., & Ames, L. B. 1947. The development of handedness. *J. Genet. Psychol.* 70:155–175.

Gesell, A., & Halverson, H. M. 1942. The daily maturation of infant behavior: A cinema study of postures, movements and laterality. *J. Genet. Psychol.* 61:3–32.

Gilbert, C., & Bakan, P. 1973. Visual asymmetry in perception of faces. *Neuropsychologia* 11:355–362.

Gloning, I., Gloning, K., Haub, G., & Quatember, R. 1969. Comparison of verbal behavior in right handed and non-right handed patients with anatomically verified lesion of one hemisphere. *Cortex* 5:41–52.

Gordon, H. 1920. Left-handedness and mirror writing especially among defective children. *Brain* 43:313–368.

Gott, P. S., & Boyarsky, L. L. 1972. The relation of cerebral dominance and handedness to visual evoked potentials. *J. Neurobiol.* 3:65–67.

Gronwall, D. M. A., & Sampson, H. 1971. Ocular dominance: A test of two hypotheses. *Br. J. Psychol.* 62:175–185.

Hamilton, C. R. 1976. Perceptual and mnemonic lateralization in monkeys?

In S. Harnad, R. W. Doty, L. Goldstein, J. Jaynes, & G. Krauther (eds.) *Lateralization in the Nervous System.* New York: Academic Press.

Hartlage, L. C., & Green, J. B. 1971. EEG differences in children's reading, spelling and arithmetic abilities. *Percept. Mot. Skills 32*:133-134.

Hécaen, H., & Ajuriaguerra, J. de. 1964. *Left-handedness: Manual Superiority and Cerebral Dominance.* New York: Grune & Stratton.

Hécaen, H., & Piercy, M. 1956. Paroxysmal dysphasia and the problem of cerebral dominance. *J. Neurol. Neurosurg. Psychiatry 19*:194-201.

Hécaen, H., & Sauget, J. 1971. Cerebral dominance in left-handed subjects. *Cortex 7*:19-48.

Hellebrandt, F. A., & Houtz, S. J. 1950. Ergographic study of hand dominance. *Am. J. Phys. Anthropol. 8*:225-236.

Hicks, R. E. 1972. Bilateral reminiscence in inverted-reversed printing: Effects of interpolated activity and printing-hand sequence. *J. Mot. Behav. 4*: 241-250.

  1974. Asymmetry of bilateral transfer. *Am. J. Psychol. 87*:667-674.

  1975. Intrahemispheric response competition between vocal and unimanual performance in normal adult human males. *J. Comp. Physiol. Psychol. 89*:50-60.

Hicks, R. E., & Barton, A. K. 1975. A note on left-handedness and severity of mental retardation. *J. Genet. Psychol. 127*:323-324.

Hicks, R. E., Bradshaw, G. J., Kinsbourne, M., and Feigin, D. S. 1977. Vocal-manual trade-offs in hemispheric sharing of performance control in normal adult humans. *J. Mot. Behav.* In press.

Hicks, R. E., & Kinsbourne, M. 1976a. On the genesis of human handedness: A review. *J. Mot. Behav. 8*:257-266.

  1976b. Human handedness: A partial cross-fostering study. *Science 192*: 908-910.

Hicks, R. E., Provenzano, F. J., & Rybstein, E. D. 1975. Generalized and lateralized effects of concurrent verbal rehearsal upon performance of sequential movements of the fingers by the left and right hands. *Acta Psychol. 39*:119-130.

Higgenbottam, J. A. 1973. Relationships between sets of lateral preference measures. *Cortex 9*:402-409.

Hines, D., & Satz, P. 1971. Superiority of right visual half-fields in right-handers for recall of digits presented at varying rates. *Neuropsychologia 9*:21-25.

  1974. Cross-modal asymmetries in perception related to asymmetry in cerebral function. *Neuropsychologia 12*:239-247.

Hochberg, F. H., & LeMay, M. 1975. Arteriographic correlates of handedness. *Neurology 25*:218-222.

Humphis, D. 1969. The measurement of sensory ocular dominance and its relation to personality. *Am. J. Optom. 46*:603-615.

Humphrey, M. E. 1951. Consistency of hand usage. *Br. J. Educ. Psychol.* 21:214–225.

James, W. E., Mefferd, R. B., Jr., & Wieland, B. 1967. Repetitive psychometric measures: Handedness and performance. *Percept. Mot. Skills* 25:209–212.

Kimura, D. 1973a. Manual activity in right-handers associated with speaking. *Neuropsychologia* 11:45–50.

1973b. Manual activity during speaking: 2. Left-handers. *Neuropsychologia* 11:51–55.

Kimura, D., & Vanderwolf, C. H. 1970. The relation between hand preference and the performance of individual finger movements by left and right hands. *Brain* 93:769–774.

Kinsbourne, M. 1972. Eye and head turning indicates cerebral lateralization. *Science* 176:539–541.

Kinsbourne, M., & Cook, J. 1971. Generalized and lateralized effects of concurrent verbalization on a unimanual skill. *Q. J. Exp. Psychol.* 23:341–345.

Knox, A. W., & Boone, D. R. 1970. Auditory laterality and tested handedness. *Cortex* 7:164–173.

Kocel, K., Galin, D., Ornstein, R., & Merrin, E. L. 1972. Lateral eye movement and cognitive mode. *Psychonomic Sci.* 27:223–224.

Komai, T., & Fukoka, G. 1934. A study on the frequency of left-handedness and left-footedness among Japanese school children. *Hum. Biol.* 6:33–42.

Kutas, M., & Donchin, E. 1974. Studies of squeezing: Handedness, responding hand, response force, and asymmetry of readiness potential. *Science* 186:545–548.

Lake, D. A., & Bryden, M. P. 1976. Handedness and sex differences in hemispheric asymmetry. *Brain Language* 3:266–282.

LeMay, M., & Culebras, A. 1972. Human brain: Morphologic differences in the hemispheres demonstrable by carotid arteriography. *N. Engl. J. Med.* 287:168–170.

Levy, J. 1969. Possible basis for the evolution of lateral specialization of the human brain. *Nature* 224:614–615.

Levy, J., & Nagylaki, T. 1972. A model for the genetics of handedness. *Genetics* 72:117–128.

Lomas, J., & Kimura, D. 1976. Intrahemispheric interaction between speaking and sequential manual activity. *Neuropsychologia* 14:23–33.

Luria, A. L. 1969. *Traumatic Aphasia.* The Hague: Mouton.

Marcie, P. 1972. Writing disorders in 47 left-handed patients with unilateral cerebral lesions. *Int. J. Ment. Health* 3:30–37.

McAdam, D. W., & Seales, D. M. 1969. Bereitschaftspotential enhancement with increased level of motivation. *Electroenceophalogr. Clin. Neurophysiol.* 27:73–75.

McGlone, J. 1970. Hand preference and the performance of sequential movements of the fingers by the left and right hands. Research Bulletin No. 153, University of Western Ontario, London, Ontario, Canada.

McKeever, W. F., & Gill, K. M. 1972. Visual half-field differences in masking effects for sequential letter stimuli in the right and left handed. *Neuropsychologia* 10:111-117.

McKeever, W. F., Van Deventer, A. D., & Suberti, M. 1973. Avowed, assessed, and familial handedness and differential hemispheric processing of brief sequential and non-sequential visual stimuli. *Neuropsychologia* 11:235-238.

McMeekan, E. R. L., & Lishman, W. A. 1975. Retest reliabilities and interrelationship of the Annett Hand Preference Questionnaire and the Edinburgh Handedness Inventory. *Br. J. Psychol.* 66:53-59.

McRae, D. L., Branch, C. L., & Milner, B. 1968. The occipital horns and cerebral dominance. *Neurology* 18:95-98.

Meier, M. G., & French, L. A. 1965. Lateralized deficits in complex visual discrimination and bilateral transfer of reminiscence following unilateral temporal lobectomy. *Neuropsychologia* 3:261-272.

Miller, E. 1971. Handedness and the pattern of human ability. *Br. J. Psychol.* 62:111-112.

Milner, B., Branch, C., & Rasmussen, T. 1964. Observations on cerebral dominance. In A. V. S. de Rueck & M. O'Conner (eds.). *Ciba Foundations Symposium on Disorders of Language.* London: Churchill.

Molfese, D. L. 1973. Cerebral asymmetry in infants, children and adults: Auditory evoked responses to speech and musical stimuli. *J. Acoust. Soc. Am.* 53:363-373.

Molfese, D. L., Freeman, R. B., Jr., & Palermo, D. 1975. The ontogeny of brain lateralization for speech and nonspeech stimuli. *Brain Language* 2:356-368.

Morley, M. E. 1965. *The Development and Disorders of Speech in Childhood*, 2nd ed. Baltimore: Williams & Wilkins.

Murray, F. S., & Hagan, B. C. 1973. Pain threshold and tolerance of hands and feet. *J. Comp. Physiol. Psychol.* 84:639-643.

Nagylaki, T., & Levy, J. 1973. "The Sound of One Paw Clapping" is not sound. *Behav. Genet.* 3:279-292.

Nakamura, R., & Saito, H. 1974. Preferred hand and reaction time in different movement patterns. *Percept. Mot. Skills* 39:1275-1281.

Nakamura, R., Taniguchi, R., & Oshima, Y. 1975. Synchronization error in bilateral simultaneous flexion of elbows. *Percept. Mot. Skills* 40:527-532.

Nebes, R. D. 1971. Handedness and the perception of part-whole relationship. *Cortex* 7:350-356.

Nebes, R. D., & Briggs, G. C. 1974. Handedness and the retention of visual material. *Cortex* 10:209-214.

Newcombe, F., & Ratcliffe, G. 1973. Handedness, speech lateralization and ability. *Neuropsychologia 11*:399-407.

Newman, H. H. 1934. Dermatoglyphics and the problem of handedness. *Am. J. Anat. 55*:277-322.

Olson, G. M., & Laxar, K. 1973. Asymmetries in processing the terms "right" and "left." *J. Exp. Psychol. 100*:284-290.

1974. Processing the terms *right* and *left*: A note on left-handers. *J. Exp. Psychol. 102*:1135-1137.

Orbach, J. 1967. Differential recognition of Hebrew and English words in right and left visual fields as a function of cerebral dominance and reading habits. *Neuropsychologia 5*:127-134.

Orlando, C. P. 1972. Measures of handedness as indicators of language lateralization. *Bull. Orton Soc. 22*:14-26.

Pfefferbaum, A., & Buchsbaum, M. 1971. Handedness and cortical hemisphere effects in sine wave stimulated evoked responses. *Neuropsychologia 9*: 237-240.

Posner, M. I., & Mitchell, R. F. 1967. Chronometric analysis of classification. *Psychol. Rev. 74*:392-409.

Pratt, R. T. C., & Warrington, E. K. 1972. The assessment of cerebral dominance with unilateral ECT. *Br. J. Psychiatry 121*:327-328.

Pratt, R. T. C., Warrington, E. K., & Halliday, A. M. 1971. Unilateral ECT as a test for cerebral dominance, with a strategy for treating left-handers. *Br. J. Psychiatry 119*:78-83.

Provins, K. A. 1967. Motor skills, handedness, and behaviour. *Aust. J. Psychol. 19*:137-150.

Provins, K. A., & Cunliffe, P. 1972. Motor performance tests of handedness and motivation. *Percept. Mot. Skills 35*:143-150.

Raczkowski, D., Kalat, J. W., & Nebes, R. 1974. Reliability and validity of some handedness questionnaire items. *Neuropsychologia 12*:43-47.

Ramaley, F. 1913. Inheritance of left-handedness. *Am. Naturalist 47*:730-738.

Rasmussen, K. J., & Milner, B. 1975. Clinical and surgical studies of the cerebral speech areas in man. In K. J. Zulch, O. Creutzfeldt, & G. Galbraith (eds.). *Otfrid Foerster Symposium on Cerebral Localization*. Heidelberg: Springer.

Reed, G. F., & Smith, A. C. 1961. Laterality and directional preferences in a simple perceptual-motor task. *Q. J. Exp. Psychol. 13*:122-124.

Rhoads, J. G., & Damon, A. 1973. Some genetic traits in Solomon Island populations. *Am. J. Phys. Anthropol. 39*:179-183.

Rife, D. C. 1940. Handedness, with special reference to twins. *Genetics 25*: 178-186.

1955. Hand prints and handedness. *Am. J. Hum. Genet. 7*:170-179.

Rizzolatti, G., Umiltà, C., & Berlucchi, G. 1971. Opposite superiorities of the right and left cerebral hemispheres in discriminative reaction time to physiognomical and alphabetical material. *Brain* 94:431–442.

Roberts, J., & Engle, A. 1974. *Family Background, Early Development, and Intelligence of Children 6–11 Years*. National Center for Health Statistics. Data from the National Health Survey, Series 11, No. 142. DHEW Publication No. (HRA) 75-1624. Washington, D.C.: GPO.

Roberts, L. 1969. Aphasia, apraxia and agnosia in abnormal states of cerebral dominance. In P. J. Vinken & G. W. Bruyn (eds.). *Handbook of Clinical Neurology*, Vol. 4. Amsterdam: North-Holland.

Sampson, H., & Horrocks, J. B. 1967. Binocular rivalry and immediate memory. *Q. J. Exp. Psychol.* 19:224–231.

Satz, P. 1972. Pathological left-handedness: An explanatory model. *Cortex* 8:121–135.

1973. Left-handedness and early brain insult: An explanation. *Neuropsychologia* 11:115–117.

Satz, P., Achenbach, K., & Fennell, E. 1967. Correlations between assessed manual laterality and predicted speech laterality in a normal population. *Neuropsychologia* 5:295–310.

Satz, P., Achenbach, K., Patishall, E., & Fennell, E. 1965. Ear asymmetry and handedness in dichotic listening. *Cortex* 1:377–396.

Satz, P., Fennell, E., & Jones, M. B. 1969. Comments on: A model of the inheritance of handedness and cerebral dominance. *Neuropsychologia* 7:101–103.

Shankweiler, D., & Studdert-Kennedy, M. 1975. A continuum for speech perception? *Brain Language*, 2:212–225.

Sheridan, M. R. 1973. Effects of S-R compatibility and task difficulty on unimanual movement time. *J. Mot. Behav.* 5:199–205.

Shimrat, N. 1973. The impact of laterality and cultural background on the development of writing skills. *Neuropsychologia* 11:239–242.

Silverman, A. J., Adevai, G., & McGough, E. W. 1966. Some relationships between handedness and perception. *J. Psychosom. Res.* 10:151–158.

Siqueland, E. R., & Lipsitt, L. P. 1966. Conditioned head turning in human newborns. *J. Exp. Child Psychol.* 4:356–377.

Steingrueber, H. J. 1975. Handedness as a function of test complexity. *Percept. Mot. Skills* 40:263–266.

Subirana, A. 1958. The prognosis in aphasia in relation to cerebral dominance and handedness. *Brain* 81:415–425.

1964. The relationship between handedness and cerebral dominance. *Int. J. Neurol.* 4:215–234.

Turkewitz, G., & Creighton, S. 1974. Changes in lateral differentiation of head posture in the human neonate. *Dev. Psychol.* 8:85–89.

Turkewitz, G., Gordon, B. W., & Birch, M. G. 1968. Head turning in the human neonate: Effect of prandial condition and lateral preference. *J. Comp. Physiol. Psychol.* 59:189–192.

Uhrbrock, R. S. 1973. Laterality in art. *J. Aesthetics Art Criticism* 32:27–35.
Varney, N. R., & Benton, A. L. 1975. Tactile perception of direction in relation to handedness and familial handedness. *Neuropsychologia* 13:449–454.
Vaughn, H. G., Jr., Costa, L. D., & Ritter, W. 1968. Topography of the human motor potential. *Electroencephalogr. Clin. Neurophysiol.* 25:1–10.
Verhaegen, P., & Ntumba, A. 1964. Note on the frequency of left-handedness in African children. *J. Educ. Psychol.* 55:89–90.
Vernon, M. D. 1957. *Backwardness in Reading.* London: Cambridge University Press.
Wada, J. A., Clark, R., & Hamm, A. 1975. Cerebral hemispheric asymmetry in humans: Cortical speech zones in 100 adult and 100 infant brains. *Arch. Neurol.* 32:239–246.
Warrington, E. K., & Pratt, R. T. C. 1973. Language laterality in left-handers assessed by unilateral ECT. *Neuropsychologia* 11:423–428.
Webster, W. G. 1976. Hemispheric asymmetry in cats. In S. Harnad, R. W. Doty, L. Goldstein, J. Paynes, & G. Krauther (eds.). *Lateralization in the Nervous System.* New York: Academic Press.
Wilson, D. 1885. Palaeolithic dexterity. *R. Soc. Can. Proc. Trans. III* 2:119–133.
Wilson, M. O., & Dolon, L. B. 1931. Handedness and ability. *Am. J. Psychol.* 43:261–268.
Witelson, S. F., & Pallie, W. 1973. Left hemisphere specialization for language in the newborn: Neuroanatomical evidence for asymmetry. *Brain* 96:641–646.
Wold, R. M. 1968. Dominance – fact or phantasy: Its significance to learning disabilities. *J. Am. Optom. Assoc.* 39:908–916.
Wussler, M., & Barclay, A. 1970. Cerebral dominance, psycholinguistic skills and reading disability. *Percept. Mot. Skills* 31:419–425.
Zangwill, O. L. 1960. *Cerebral Dominance and Its Relation to Psychological Function.* London: Oliver & Boyd.
  1962. Handedness and dominance. In J. Money (ed.). *Reading Disabilities.* Baltimore: Johns Hopkins Press.
  1967. Speech and the minor hemisphere. *Acta Neurol. Psychiatr. Belg.* 67:1013–1020.
Zurif, E. B., & Bryden, M. P. 1969. Familial handedness and left-right differences in auditory and visual perception. *Neuropsychologia* 7:179–188.
Zurif, E. B., & Carson, G. 1970. Dyslexia in relation to cerebral dominance and temporal analysis. *Neuropsychologia* 8:351–361.

PART V

BIOLOGICAL ORIGINS: PERSPECTIVES

15
Evolution of language in relation to lateral action

MARCEL KINSBOURNE

Human language is centrally represented in the cerebral cortex and, to some extent, the upper part of the brain stem. At these levels it is left-lateralized in its representation for more than 95% of people. I shall present evidence that language as a function is not purely emergent, but evolved in relation to central programs for motor behavior, and that the organization of its central representation reflects that fact. In this view, language develops in relation to right-sided action and, particularly, rightward orienting. The reason for this is the principle of double decussation, applicable to vertebrates in general, in whom unilateral input is transmitted to the opposite side of the brain and instructions for unilateral action are also contralaterally programmed.

I shall argue that language represents an elaboration, extension, and abstraction of sensorimotor function and, specifically, of the basic sequence of approach behaviors: lateral (rightward) orientation, locomotion, grasp, and consummation. I shall marshal evidence in support of several new theoretical models: (1) to account for the evolution of decussation in vertebrates; (2) to rationalize the relationship of central language and central motor representations; (3) to account for the asymmetrical, rather than bisymmetrical, organization of language; and (4) to suggest how that lateralization develops and by what mechanism deviations from it, in non-right-handers, might occur. Each of these models is offered for its heuristic value. For each, the evidence is, at present, strictly circumstantial; but each generates further testable predictions.

EVOLUTION OF DECUSSATION DURING INVERTEBRATE-VERTEBRATE TRANSITION

Those invertebrates that are bisymmetrically organized with respect to both their bodies and their nervous systems (Bilateria) all show a predominantly ipsilateral mode of control. Individual intercalated neurons are quite common in certain invertebrate species, but do not infringe the general rule that each half of the central nervous system is primarily in control of the musculature on the same side of the body. Those invertebrate species that have specialized sense organs show a similar ipsilateral pattern of central sensory representation. The predominantly contralateral mode of control that evolves with the appearance of decussation has attracted a number of post hoc attempts at explanation that rely upon various hypothetical adaptive advantages of this type of organization (Braitenberg, 1965; Sarnat & Netsky, 1974). Although such attempts arguably succeed in demonstrating adaptive advantage for a particular real or hypothetical vertebrate precursor, they do not explain why such a radical departure from previous nervous organization should be perpetuated through multitudinous species for whom no such adaptive purpose would be met. It would be more convincing if the postulated adaptive advantage were clearly relevant to most, if not all, of the creatures whose nervous systems conformed to the principle in question. It does not seem possible to arrive at such an adaptive advantage with respect to decussation itself, but I hope to show that there is another design characteristic of the vertebrate nervous system that quite plausibly has major adaptive advantage for all members of the phylum and that decussation is a consequence of the anatomical rearrangements necessitated by this other development in evolution (Kinsbourne, 1976).

The chordate phylum, almost all the species in which are, in addition, vertebrate, derives its name from the presence of a relatively rigid structure running along the animal's dorsal longitudinal axis, the notochord or the vertebral column. Invertebrates, by definition, do not have this structure. In invertebrates, as in vertebrates, the nervous system may show an anterior specialization meriting for some the description of head ganglion and for others of brain; as in vertebrates, this is dorsally located in relation to the more ventral oropharynx. But in invertebrates, the posterior extension of this structure sweeps sharply ventrally, then courses longitudinally along the ventral surface of the animal toward the posterior end. Thus, a change from ventral to dorsal position of the main length of the neuraxis laid the foundation for, and perhaps necessitated, the evolution of the notochord. But whatever the adaptive purpose of the relocation or be there some other as yet unknown purpose, it is reasonable to accept the ventral-to-dorsal displacement of the neuraxis as a fundamental innovation at the invertebrate-vertebrate transition. How did this change come about?

My attempt to answer this question will be aided by taking into account yet another radical deviation of vertebrate from invertebrate anatomical design. This affects an ostensibly quite unrelated parameter, the change in which is also mysterious with respect to its conceivable adaptive purpose. There is a change in the direction of flow of blood within the main longitudinal vessels. In invertebrates the blood that flows toward the posterior end of the body does so ventrally, and the blood that is returned anteriorly courses through dorsally located channels. Any heartlike structure that evolves is dorsally located. In vertebrates the exact opposite is the case. The main axis of posterior flow runs in the dorsal part of the body, and the main axis of return to the anterior end runs ventrally. The heart is ventral.

Thus, I must deal with three major changes in evolution: the dorsal positioning of the neuraxis, decussation within that structure, and a reversal of main direction of blood flow. One need not, however, suppose each one of them was brought about in response to a different adaptive pressure. Instead, I shall argue that the rearrangement involved in one of these changes of necessity carried the other two in its wake.

How can one most economically swing a ventral neuraxis to a dorsal position? This can be done in one step by a 180° rotation of the organism upon itself. This rotation would occur at about the segmental level at which the invertebrate neuraxis sweeps ventrally away from the dorsal brain or head ganglion; that is, at a point posterior to the oropharynx. Thus, the anterior end of the body can be regarded as unperturbed, but the rest of the body swings around so that the neuraxis travels dorsally the full length of the organism. That same rotation of the body would, of necessity, reverse the direction of blood flow and swing the heart into its typically vertebrate posterior location. The 180° twist would have one more consequence: a twist of the neuraxis at the point of rotation, giving rise to decussation. Decussation is thus explained (as well as the reversal of direction of blood flow), not on the basis of some hypothetical advantage bestowed by that reorganization, but as a mechanical consequence of the bodily rearrangement by which dorsal position of the neuraxis was obtained.

The occurrence of 180° rotation is not without precedent in ontogeny. It is found in the male of some examples of the Diptera, in whom the last two or three segments are thus rotated, presumably to help bring the copulatory apparatus to its position in the adult. In snails (Prosobranchia), a very elaborate process of 180° rotation during ontogeny has been well recorded. It would, of course, be a most conclusive finding if the postulated somatic twist could actually be found in the ontogeny of some protochordate organism, and for all we know it will be. However, there is a consensus to the effect that ontogeny does not necessarily recapitulate phylogeny (de Beer, 1951), and certainly one would not require that all phylogenetic rearrangements be demonstrable from an ontogenetic perspective. A simple rider attached to existing genetic instructions might implement the proposed mechanical change without need for a prerotatory stage in the development of the species.

There is, however, one other aspect of the model that yields testable predictions. The model ties together the three major design characteristics of vertebrates listed as resulting from a single event. This generates an all-or-none prediction for the incidence of these three characteristics. Any species should either have all three – dorsal neuraxis, decussation, and vertebrate pattern of blood flow – or have none of these. Intermediate stages in which one or two follow the vertebrate mold, whereas the others do not, are inadmissible. A search for intermediate situations of this type has so far been fruitless. No invertebrate so far studied, with one exception to be mentioned below, is credited with any of these characteristics, and the simplest chordates documented (Cephalochordata and Urochordata) already incorporate all three. The one interesting potential test case, which might indeed have even wider implications for notions of invertebrate-vertebrate transition, is the beard worm or pogonophore (Ivanov, 1963).

This simple marine worm seems an unpromising vertebrate ancestor in that it is sessile rather than free moving and totally lacks the customary digestive system. However, it has been recorded as possessing two characteristics otherwise found only among the chordates: a posterior neuraxis and the vertebrate pattern of direction of blood flow. According to the model proposed here, this could be a transitional case and disconfirm the hypothesis if there is no decussation in this organism's neuraxis, or confirm it if there is. The matter remains to be studied. Indeed, if there is decussation, then the pogonophores might become viable candidates for resemblance to the original protochordate. The absence of a digestive system might not be an insuperable obstacle insofar as it is recognized that the absence is a regressive phenomenon rather than an indication of failure to evolve such a system in the pogonophore ancestor.

SYNERGY OF SPEECH AND BODY MOVEMENT

Given that vertebrate control centers do lie contralateral to the relevant musculature, I shall now marshal the evidence that language in humans relates to right-sided and specifically rightward motion. I shall mention the experimental demonstrations of synergistic effects between language and right-sided movement and then speculate on how the relationship might have come about.

In an extensive series of experiments with right-handed subjects, we have been able to show that both overt and covert verbal activity is associated with a rightward shift of attention or orienting response. For instance, while people are pondering verbal problems, they tend to swing head and eyes to the right (Kinsbourne, 1972). In various experimental situations, the rightward orienting may variously entail turning of the whole body to the right, turning head and eyes only, or, if the head is held in place, the eyes only. It may even be restricted to a submotor rightward shift of visual or auditory attention (Kinsbourne, 1970, 1973, 1975). In none of these cases does the rightward orienting serve any obvious adaptive purpose. Rather, it is a manifestation of the basic wiring of the human nervous system, presumably reflecting an ancient blueprint upon which further elaboration has occurred during the development of the human brain (Kinsbourne, 1974). It is, of course, possible to overcome or even reverse that rightward attentional shift while speaking, but apparently only at the price of some proportion of the mental capacity available for the main task (Kinsbourne, 1975). The sum total of these demonstrations entitles us to entertain models that relate the origins of language to lateral orienting behavior.

In another series of experiments, we have been able to show a similar relationship between overt or covert vocal activity and motor performance on the right side of the body. Briefly, if the limbs on the two sides of the body are concurrently active in an unrelated manner, concurrent vocal activity is more effective and less characterized by error if it conforms to the overall program of the right-sided limbs. If the right-sided limbs are employed in action unrelated to that of the vocal apparatus, there is much mutual interference; if the left-sided limbs are active in the exact same way, the interference is slight or absent. For instance, skilled right-handed performance (dowel balancing, finger tapping, sequential finger movements) is more embarrassed by concurrent vocalization than is the same performance when carried out by the left hand (Kinsbourne & Cook, 1971; Hicks, 1975; Hicks et al., 1975). This series of experiments illustrates the presumably biologically preprogrammed linkage between language functions and skilled movement of the opposite side of the body.

The archaic nature of the relationship between language and right-sided action is further supported by the ontogeny of the interaction. It is a characteristic of synergisms in general that they are most marked at the youngest testable age and diminish as the nervous system matures. If the relationship between speaking and right-sided orienting and moving is properly described as synergic, it too should conform to this developmental pattern. With respect to two relevant paradigms, dichotic listening and voice-finger tapping interference, Kinsbourne & Hiscock (1977) were able to document just such a decrease of asymmetrical effect from age 3 to age 11 years in normal children.

We emerge with a model of language origins in relation to the right hand and right-sided action. We would suppose that protoman first made utterances that were coincident with and driven by the same rhythm as the movement in question. As this skill further evolved, the utterances became internalized and detached from overt action, so that they became capable of assuming a signaling role in the absence of their referents. A vast degree of further elaboration and transformation would then ultimately result in the highly intricate language systems prevalent in most human societies. However, even in the most sophisticated speaker, verbal activity is not completely free of corollary somatic movement produced on an involuntary basis.

It is usually not possible to validate evolutionary speculations by direct experimental tests. It then becomes tempting to scrutinize relevant phases in ontogeny in the hope of finding telltale phenomena that could be regarded as recapitulating the phylogeny in question. This is legitimate as long as it is realized that such recapitulation is quite variable in its incidence and that no conclusions can be drawn from a negative outcome to a search through ontogeny. In the event, ontogenetic analogs to our hypothesized phylogenetic sequence are not hard to find in the human infant. Prominent within the very limited behavioral repertoire of the newborn infant is the lateral orienting synergism (tonic neck response). The human infant orients spontaneously and initially silently to the environment, shifting his point of regard and focus of attention to one side or the other. When he enters the babbling phase, a significant association between vocal and orienting response is evident. The child babbles as he points. When the production of meaningless speech sounds becomes punctuated by intelligible words, these are at first usually the names of objects, and the child always points while naming and does not name without pointing. Only after further maturation can nominalization be detached from overt bodily orienting.

These observations (which await quantitative verification) are not intended to suggest that early speech is exclusively nested within the orienting synergism. Babbling of distinctive quality can also be heard in conjunction with the motor sequences that are sequelae of the orienting response – locomotion, grasp, manipulation. It remains to be established whether action words are initially of necessity nested in corresponding movement sequences, only later to be abstracted therefrom. But for the present purposes it suffices to say that ontogeny at least reassures us about the validity of the notion that language evolved in a motor context – specifically the approach sequence initiated by the orienting response and terminated by some consummatory act.

ORIGIN OF ASYMMETRICAL CENTRAL REPRESENTATION OF LANGUAGE

Given that language evolved in relation to centrally located motor programs, why is the customary representation of language lateralized to one hemisphere, rather than related to motor mechanisms bilaterally? As with matters related to invertebrate-vertebrate transition, here too a strenuous attempt to enlist a post hoc adaptive advantage for lateralization has been made (Levy, 1969). Fortunately, the notion that lateralization is advantageous can be "tested" directly. This is because a significant minority of humans diverge to a variable extent from the left-lateralized norm. These are the non-right-handers, almost half of whom are regarded as other than left-lateralized for language. This minority should be demonstrably inferior on adaptively important tasks.

Much debate has been triggered by the recurrent observation that there is a higher than customary incidence of left-handedness among certain cognitively disabled subject groups, and indeed, it is true that the more profoundly mentally retarded a child is, the more likely he is to be non-right-handed (Hicks & Barton, 1975). But, scattered claims to the contrary, it is now overwhelmingly obvious from large-scale studies that within the general population neither left-handers nor ambidextrous individuals are at any measurable disadvantage (see Chapter 14). Thus, whereas brain damage may induce deviation from left lateralization of language, the preponderance of left-handers – who are so for genetic reasons (Hicks & Kinsbourne, 1976) – are not in the least cognitively embarrassed by their deviant cerebral organization. Therefore, we need a different explanation for left lateralization of language.

When obvious explanations fail, it is appropriate to review the implicit premises. The assumption here is that bisymmetry is the norm, and deviations from that norm must have had adaptive justificaton. But bisymmetry is not an inexorable design characteristic of vertebrate anatomy. Many organs are grossly asymmetrical in shape and location, and symmetrical structures in many species exhibit numerous minor asymmetries, ranging from the pectoral fins of fish to the temporal planum of the human brain (see discussion in Kinsbourne, 1974). Symmetry is most exact with respect to the somatic musculature and its central control mechanisms. This is not surprising, as contingencies that call for rapid response are equally likely to emanate from either side of extrapersonal space, so that reaction to either side needs to be equally efficient. Only at the stage of consummation, when the object has been grasped and is being manipulated, does the need for rapid action, and thus for bisymmetry, relax. Once trapped, the object can be transferred to one side, if that happens to be the more skilled with respect to the intended manipulation. Thus, there is adaptive sense in asymmetry of manual dexterity for unimanual tasks and in lateralization of control centers for skilled action sequences, which, in fact, exist (e.g., Denny-Brown, 1950).

Whereas language originated in a motor context, it is not in itself deployed toward specific points in ambient space. Words are not aimed at exact locations, and therefore language needs no bisymmetry of representation. If we accept that it is bisymmetry which develops in response to an adaptive imperative, we can easily explain why language does not conform to that same pattern. Its organization was not influenced in this way; there was no specific need for bisymmetry, and so no need for bisymmetrical language representation to evolve. Therefore, like any other structure not forced into the bisymmetrical mold, language-related cerebral cortex conformed to the less constrained asymmetrical state, which in humans happens to be left-sided in a species-specific manner.

This discussion does not preclude the possibility that different patterns of lateralization have predictable implications for cognitive mechanisms. Indeed, one such is suggested in the final section.

POSSIBLE ROLE FOR BRAIN STEM LANGUAGE MECHANISM

The existence of left-lateralized thalamic representation of language processes is now beyond question (Ojemann & Van Buren, 1968; Ojemann et al., 1971). The role of this neural mechanism remains obscure. I shall suggest that the thalamic facility is not merely redundant with more elaborate cerebral language mechanisms, but rather subserves a unique function in the adoption of verbal mental set.

Two ingredients are essential for specific cognitive processing to occur: the availability of the appropriate processor and its activation by a selector mechanism. Behaviorally, the action of the selector is manifested in the individual's adoption of a particular problem-solving mode. When the processor is switched in, it begins to function; and while it remains switched in, it continues to function.

Selective activation of the processor is thus viewed as necessary at least for the more profound aspects of verbal processing. Now, suppose the selector is (1) active in the absence of processor or (2) not selectively tuned to its processor.

The former case may occur in the very young infant, in whom most cerebral neural networks are not yet functional.

It has been clearly established that infants can perform differentially in relation to different speech sounds (Eimas et al., 1971). It has, more recently, been shown that reverse asymmetry of evoked potential amplitude obtains in preverbal infants, depending on their exposure to speech or nonspeech stimuli (Molfese, 1973). Another report claims shorter latency control of nonnutritive sucking by right-sided than left-sided change in the phonemic category of auditory input, the reverse being the case for musical input (Entus, 1975). If these reports prove replicable, they need not be construed as compelling belief in a hitherto unsuspected functioning cerebral processor for language in very young infants. More simply, these findings could be attributable to lateralized ascending activation by an already functional thalamic selector for the adoption of verbal set.

The notion that a cognitive set may be adopted earlier in infancy than the stage at which it first can be used for adaptive processing is quite consistent with some well-known aspects of infants' behavior. After all, much of infants' play lends itself to just such an interpretation. Play is clearly not adaptive in a straightforward goal-oriented fashion. The notion that it represents practice is quite implausible: Children persist in a particular play activity well after they cease to show any further change in the rate or accuracy of performance. Play appears to foreshadow subsequent adaptive functioning. The infant manipulates in ways that lead to no tangible result, but will later do so, when he has learned which manipulation is best applied to which object. He vocalizes during action (babbles) before he has learned which particular phonemic sequence conventionally signals which event. We see this as the activation of the superficial (input and output) stages of what will become the complete cognitive processing sequence. The central processor is not yet in action.

If a lateralized selector selectively activates the left hemisphere in verbal circumstances, the left lateralization of language is understandable without recourse to assumptions about preprogrammed lateral differences in cerebral organization (Semmes, 1968). It also becomes understandable how, in cases of early left cerebral destruction, the right hemisphere might assume a language role. However, it would, except in a special circumstance, to be discussed below, not have the full benefit of selective activation during the adoption of verbal set. If the claim of Kohn and Dennis (1974) that the substituting hemisphere is less adept at the most difficult language task than the hemisphere originally destined for the language role proves replicable in the context of larger series of cases, then a mechanism springs to mind. The compensating hemisphere is less able to be activated in intensely selective fashion.

The explanatory potential of the selector model is finally illustrated by the account it gives of a particular well-documented characteristic of at least some non-right-handers: their accelerated rate of recovery from aphasia. A recent discovery about the nature of recovery from aphasia provides the groundwork for a possible explanation. This is the discovery that in severe aphasia the control of verbal utterances is likely to have been assumed by the undamaged right hemisphere (Kinsbourne, 1971; Czopf, 1972).

Let us revert to our account of the mechanism by which language processing might lateralize during injury: the action of a selector with lateralized projection. Suppose that the individual who is poorly lateralized as regards manual performance and manual dexterity (the non-right-hander) is likely also to be poorly lateralized with respect to the projection of his language selector. Two consequences might accrue. First, he might find it difficult to implement an intensely selective verbal mental set, leading perhaps to some as yet undetected insufficiency with respect to particularly taxing verbal problem-solving skills. Second, and relevant to this discussion, is the expectation that, if the selector projects diffusely, then wherever language migrates to after brain damage it might still remain within the range of its selector and thus adventitiously benefit from a degree of supportive activation denied the individual with a well-lateralized selection facility.

CONCLUSION

In overview, I have in this discussion assumed that the persisting failure to explain a series of crucially important cognitive phenomena relative to language lateralization results from the unwitting adoption of questionable premises. In suggesting four new models, we have questioned four current but unvalidated assumptions: that decussation is (or was) of adaptive advantage; that language differs qualitatively from sensorimotor function; that bisymmetry is a preordained vertebrate base state, deviations from which need further explanation; and that the language-specialized hemisphere is so for structural reasons. Questioning assumptions does not, of course, disprove them. But by proposing possible alternatives we open up these areas to scientific investigation.

REFERENCES

Braitenberg, V. 1965. Taxis, kinesis, and decussation. In N. Weiner & J. P. Schacter. *Cybernetics of the Nervous System*. New York: Elsevier.

Czopf, J. 1972. The role of the non-dominant hemisphere in the restitution of speech in aphasia. *Arch. Psychiatr. Nervenkr. 216*:162–171.

deBeer, G. R. 1951. *Embryos and Ancestors*. Oxford: Clarendon Press.

Denny-Brown, D. 1950. Disintegration of motor function resulting from cerebral lesions. *J. Nerv. Ment. Dis. 112*:1–45.

Eimas, D., Siqueland, E., Jusczyk, P., & Vigorito, J. 1971. Speech perception in infants. *Science 171*:303–306.

Entus, A. 1975. *Hemispheric Asymmetry in Processing of Dichotically Presented Speech and Non-Speech Sounds by Infants*. Society for Research in Child Development, Denver.

Hicks, R. E. 1975. Intrahemispheric response competition between vocal and unimanual performance in normal adult human males. *J. Comp. Physiol. Psychol. 98*:50–60.

Hicks, R. E., & Barton, A. K. 1975. A note on left-handedness and severity of mental retardation. *J. Genet. Psychol. 127*:323–324.

Hicks, R. E., & Kinsbourne, M. 1976. Human handedness: A cross-fostering study. *Science 92*:908–910.

Hicks, R. E., Provenzano, F. J., & Rybstein, E. 1975. Generalized and lateralized effects of concurrent verbal rehearsal upon performance of sequential movements of the fingers of the left and right hands. *Acta Psychol. 39*:119–130.

Ivanov, A. V. 1963. *Pogonophora*. New York: Consultants Bureau.

Kinsbourne, M. 1970. The cerebral basis of lateral asymmetries in attention. *Acta Psychol. 33*:193–201.

1971. The minor cerebral hemisphere as a source of aphasic speech. *Arch. Neurol. 25*:302–306.

1972. Eye and head turning indicate cerebral lateralization. *Science 176*: 539–541.

1973. The control of attention by interaction between the cerebral hemispheres. In S. Kornblum (ed.). *Attention and Performance*, Vol. IV. New York: Academic Press.

1974. Mechanisms of hemispheric interaction in man. In M. Kinsbourne & W. L. Smith (eds.). *Hemispheric Disconnection and Cerebral Function*. Springfield, Ill.: Thomas.

1975. The mechanism of hemispheric control of the lateral gradient of attention. In R. Rabbitt & S. Dornic (eds.). *Attention & Performance*, Vol. V. New York: Academic Press.

1976. Discussion on lateralization: Invertebrate to vertebrate transition. *Proceedings of the Conference on Evolution and Lateralization of the Brain*. New York: New York Academy of Sciences.

Kinsbourne, M., & Cook, J. 1971. Generalized and lateralized effect on concurrent verbalization on a unimanual skill. *Q. J. Exp. Psychol.* 23:341-345.

Kinsbourne, M., & Hiscock, M. 1977. Does cerebral dominance develop? In S. Segalowitz & F. A. Gruber (eds.). *Language Development & Neurological Theory*. New York: Academic Press.

Kohn, B., and Dennis, M. 1974. Patterns of hemispheric specialization after hemidecortication for infantile hemiplegia. M. Kinsbourne & W. L. Smith (eds.). *Hemispheric Disconnection and Cerebral Function*. Springfield, Ill.: Thomas.

Levy, J. 1969. Possible basis for the evolution of lateral specialization of the human brain. *Nature* 224:614-615.

Molfese, D. 1973. Cerebral asymmetry in infants, children and adults: Auditory evoked responses to speech and musical stimuli. *J. Acoust. Soc.* 53:363A.

Ojemann, G., & Van Buren, J. M. 1968. Anomia from pulvinar and subcortical parietal stimulation. *Brain* 91:99-116.

Ojemann, G., Blick, K., & Ward, A. 1971. Improvement and disturbance of short term verbal memory with human ventrolateral thalamic stimulation. *Brain* 94:225-240.

Sarnat, H. B., & Netsky, M. G. 1974. *Evolution of the Nervous System*. New York: Oxford University Press.

Semmes, J. 1968. Hemispheric specialization: A possible clue to mechanism. *Neuropsychologia* 6:11-26.

Name index

Abt, L. A., 455
Ach, N., 275
Adrian, E. D., 294, 299, 501
Aird, R. B., 298
Ajuriaguerra, J., 54, 330, 374
Akelaitis, A. J., 366
Akert, K., 333
Alajouanine, T., 423
Albert, M., 87
Alexander, D., 447, 450
Ames, L. B., 374, 533
Ammons, C. H., 269
Ammons, R. B., 269, 413, 434, 436
Anastasi, A., 426
Andersen, A. L., 66
Andersen, P., 294
Anderson, I., 153, 161, 179, 187, 227, 249
Anderson, J. B., 221
Andersson, A., 294
Andrew, D. M., 482
Andrews, G., 108
Andrieux, C., 407, 433
Anhalt, I., 494
Annett, M., 8, 465, 466, 523, 531, 532, 534, 535, 537, 538
Antonitis, J. J., 295
Applebee, A. N., 534
Apter, J. T., 333
Arnett, M., 312
Arrigoni, G., 75, 76, 331, 465
Asanuma, H., 360
Asher, E. J., 482
Atkinson, R. C., 243, 249
Aulhorn, O., 180
Ayres, J. J. B., 149, 153, 154, 160, 196, 221, 248

Bakan, P., 412, 529, 532, 534, 535
Bakwin, H., 534
Barakat, M. K., 408
Barclay, A., 434, 534
Barlow, H. B., 379
Barlow, J. S., 298
Barnes, R. J., 411, 416
Bartlett, F. C., 381
Barton, M. I., 69, 77, 168, 169, 249, 417, 534, 560
Basser, L. S., 104, 330, 374, 464
Battalla, M., 410
Battersby, W. S., 54
Bayley, N., 412, 494
Baylor, G. W., 409
Beale, I. L., 379, 499, 500
Beaubaton, D., 378
Bechtoldt, H. P., 87
Bell, S. K., 306, 435
Bender, M. B., 58
Bennett, D. H., 407, 433
Bennett, G. K., 115, 407, 433, 443
Benson, D. F., 69, 77
Benton, A. L., 49, 75, 77, 78, 81, 87, 114, 447, 498, 529, 536
Berger, C. I., 181
Berkhout, J., 298
Berlucchi, G., 276, 281, 282, 283, 285, 298
Berman, A., 535
Bernstein, N., 360, 370
Berry, J. W., 428, 433, 485, 487
Bertelson, P., 269
Best, C. T., 451, 456, 459
Bettis, N. C., 415
Bever, T. G., 392, 369, 494

Bieri, J., 407, 434
Bigelow, G., 405
Bigum, H. B., 301
Bingley, T., 19
Bishop, A., 356
Bisiach, E., 165
Black, P., 363
Blackwell, H. R., 208
Blade, M., 426
Blakemore, C., 402
Blyth, K. W., 269
Bock, R. D., 405, 438, 439, 443, 444, 445, 483
Bodian, D., 20, 41
Bogen, J. E., 422, 463, 470, 471
Bogen, N., 87, 102, 117, 121, 132, 365, 366
Bogo, N., 415
Bolin, B. J., 534
Book, H. M., 407, 410, 430, 501
Boone, D. R., 525
Boring, E. G., 275
Bose, R. C., 59
Bossom, J., 361, 363
Botez, M. I., 469
Bower, T. G. R., 193
Boyarsky, L. L., 302, 530
Boyd, R., 20
Boynton, R. M., 239
Bradshaw, J. L., 281, 282
Brain, W. R., 536
Braine, L. G., 223, 233, 243
Brainerd, C. J., 418
Braitenberg, V., 554
Branch, C., 102, 103, 108, 447
Brandt, H. F., 221
Brazier, M. A. B., 298
Breese, F. F., 456
Bremer, F., 288
Bresson, F., 378
Briggs, G. G., 268, 270, 271, 535, 538
Brinkmann, E. H., 431
Brinkmann, J., 377, 427, 431, 485
Broadbent, D. E., 193, 243, 250, 251, 492
Broca, P., 20, 99, 102
Brodal, A., 359
Broerse, A. C., 155
Brookshire, K. M., 373
Broverman, D. M., 481, 482, 483
Brown, F. R., 19, 58, 431, 472
Brown, J. S., 539

Bruce, R., 311
Bruder, G. E., 154, 158, 213, 214
Bruner, J. S., 374
Bryden, M. P., 143, 149, 150, 153, 162, 164, 165, 166, 167, 168, 169, 170, 172, 173, 174, 175, 177, 187, 189, 190, 196, 197, 212, 217, 220, 225, 243, 244, 248, 250, 393, 394, 396, 397, 399, 400, 458, 469, 471, 472, 525, 526, 527, 533, 537
Buchsbaum, M., 300, 302, 303, 304, 530
Buchwald, J., 359
Buckley, F., 446
Buffery, A. W. H., 453, 454, 455, 456, 457, 459, 460, 461, 471, 478, 495, 497
Burke, R. S., 224
Burkland, C. W., 109
Burt, C., 534
Butler, C. R., 374
Butler, S., 104, 106, 310, 530
Butters, N., 69, 417
Buxton, C. E., 413
Byrne, R., 408

Cameron, J., 475
Camp, D. S., 148, 159, 166, 203, 207, 230, 231, 235, 251
Campbell, D. T., 143, 158, 160
Caplan, P. J., 8, 533
Carmon, A., 86, 87, 96, 282, 285, 498, 499, 528, 530, 535
Carr, H. A., 180
Carson, C., 393, 534
Casby, J. U., 298
Catlin, J., 275, 277
Cattell, M., 274
Cechini, M., 458
Cernacek, J., 267
Chaikin, J. D., 185, 227
Chakrabarti, J., 534
Chamberlain, H. D., 523, 532
Chambers, R. A., 336, 361
Chartock, H. E., 297
Chase, W. G., 409
Chateau, J., 433
Chen, L. K., 180
Chesher, E. C., 19
Chiarello, R. J., 494
Chomsky, N., 381
Chorover, S. L., 217
Churchill, B. D., 427

Name index

Clark, W. C., 158, 199, 206
Clark, W. H., 248
Clarke, P. R. F., 459
Clyma, E. A., 459
Cohen, G., 280
Cohen, L., 268, 446, 526
Cohn, R., 298
Cole, J., 21, 40
Coles, G. R., 152
Collins, R. L., 531
Colonna, A., 66
Coltheart, M., 407
Conel, J. L., 495, 496
Cook, J., 11, 270
Cook, T. W., 269, 527, 558
Cooper, C. F., 402
Corah, N. L., 405, 407, 426, 434, 443
Corballis, M., 379, 499, 500
Corballis, M. C., 219
Coren, S., 531
Corkin, S., 69, 73, 464, 498
Corner, G. W., 532
Cornil, L., 294
Corwin, T., 239
Costa, L. D., 54, 66, 499
Cowley, J. J., 411
Craik, K. J. W., 381
Cress, R. H., 21
Crighel, E., 469
Critchley, M., 449, 450
Cropley, A. J., 220
Crosland, H. R., 153, 161, 163, 176, 179, 193, 215, 227
Crowell, D. H., 299, 533
Crovitz, H. F., 149, 150, 154, 164, 165, 172, 185, 189, 206, 212, 216, 217, 232, 235, 238, 250, 536
Culebras, A., 530, 533
Culver, C. M., 302, 412
Cunliffe, P., 295, 537
Curry, F. K. W., 525
Cusumano, D. R., 434
Czopf, J., 563

Dalby, T., 274
Dallenbach, K. M., 153, 158, 160, 179, 224, 226, 233
Damon, A., 523
Dart, R. A., 523
Darwin, C. J., 494
Daves, W., 150, 238, 250
Davidson, W., 472, 474

Davis, R. C., 267
Dawson, J. L. M., 411, 448, 523
Dax, M., 26
deBeer, G. R., 556
Dee, H. L., 74, 77, 78, 525
Deese, J., 251
de Groot, A., 409
Deininger, R. L., 286
De Laguna, G. A., 337, 374
Delse, F. C., 310
Demuth, N., 494
Dennis, W., 18, 489, 523, 562
Denny-Brown, D. L., 336, 361, 560
De Oreo, K. D., 413
de Renzi, E., 51, 54, 61, 63, 66, 67, 70, 71, 72, 73, 75, 76, 77, 78, 82, 87, 92, 118, 121, 331, 465, 488
Derks, P. L., 214, 258, 261
Descantes, R., 380
Devine, J. V., 40
De Vore, I., 484
Diamond, S. P., 58
Di Chiro, G., 20, 496
Dick, A. E., 186
Dixon, N. F., 281
Doerries, L. E., 246
Doll, E. A., 534
Dolon, L. B., 534
Donchin, E., 530
Donders, F. C., 275, 276
Doran, E. W., 455
Dornbush, R. L., 164, 227, 250
Downer, J. L. C., 351, 363
Downey, J. E., 539
Drowatzky, J. N., 412
Duensing, F., 78
Dustman, R. E., 302
Dyer, D. W., 160, 178, 183, 199, 221, 230, 248

Eagle, C. J., 426
Eason, R. G., 302
Ebner, F. F., 373
Eccles, J. C., 120, 130
Efron, R. L., 86, 171, 218, 528
Egeth, H., 153
Eguchi, S., 402
Ehrlichman, H. I., 472
Eimas, D., 561
Eimas, P. D., 401, 402
Eisenberg, L., 456
Eldred, E., 359

Elithorn, A., 66
Ellis, N. R., 249
Entus, A. K., 450, 456, 459, 562
Erikson, E. H., 435
Ettlinger, G., 157, 165, 362, 373, 469, 532
Ewert, J. P., 333, 335

Fabersek, V., 238
Fagin-Dubin, L., 475
Faglioni, P., 54, 58, 61, 66, 67, 70, 71, 72, 75, 76, 77, 78
Falek, A., 532
Farrell, M., 435
Faubion, R. W., 431
Fedio, P., 126, 303, 304
Feigenbaum, E. A., 247
Feinberg, I., 20
Fiebert, M., 406, 415
Filbey, R. A., 275, 277, 278, 279, 280, 281
Filion, R. D. L., 187, 199, 206, 221, 228, 229, 230, 250
Finkel, M. E., 161, 164, 168, 222, 245, 248
Fiske, D. W., 143, 160
Fitts, M. P., 252, 286
Fitzgerald, K. E., 212, 234
Flament, F., 374
Flanagan, J. C., 407
Fleishman, E. A., 537
Fleming, D. E., 304
Fleminger, J. J., 524
Flick, G. L., 535
Flowers, K., 378
Folb, S., 399
Foote, W. E., 155, 158
Forgays, D. G., 155, 163, 175, 177
Freeburne, C. M., 153, 161
Freedman, N. L., 299
Freedman, S. J., 241
French, J. W., 87, 405, 538
Friedlander, W. J., 302, 531
Friedman, S. M., 155, 161, 172, 185, 221, 248, 252
Fudin, R., 161, 178
Fukoda, G., 523

Gainer, W. L., 482
Gainotti, G., 54, 77, 87, 128
Galin, D., 296, 530
Garai, J. E., 435
Garfinkle, M., 22

Garn, S. M., 483
Garner, W. R., 142, 155, 206, 248
Garoiner, M. F., 533
Garoutte, B., 298
Garron, D. C., 445, 446
Gastaut, H., 294
Gastrin, J., 408
Gates, A. I., 456
Gatewood, M. C., 456
Gaze, R. M., 334
Gazzaniga, M. S., 100, 102, 104, 105, 109, 110, 112, 113, 114, 117, 128, 275, 277, 278, 279, 280, 281, 282, 285, 310, 331, 351, 363, 365, 366, 367
Geffen, G., 280, 399, 526
Geffner, D. S., 393, 401
Gelman, R., 418
Gerjuoy, I. R., 226
Geschwind, N., 3, 306, 331, 365, 366, 454, 467, 468, 495
Gesell, A., 374, 413, 436, 482
Ghent, L., 24, 496
Giannitrapani, D., 298
Gibson, J. J., 465, 466
Gilbert, C., 529
Gilden, L., 39
Gilinsky, A. S., 143
Gill, K. M., 281, 283, 285, 526, 533
Glanville, A. D., 153, 179, 226, 232, 295, 451, 456, 459
Glanzer, M., 157, 196, 199, 248
Glass, A., 310
Glees, P., 20, 21, 40
Gloning, I., 524
Goff, W. R., 39
Goldberg, J., 435, 438, 439, 456
Goldberg, S., 435, 456
Goldman, R. D., 153, 161
Goldschmid, M. K., 418
Goldstein, A. G., 407
Gombos, G. H., 530
Goodenough, D. R., 405, 412, 426
Goodenough, F. L., 412
Goodglass, H., 19, 40, 168, 465, 469
Goodnow, R., 407, 433
Gordon, H. W., 111, 114, 121, 128, 422, 494
Goss, M., 494
Gott, P. S., 102, 105, 302
Gottschalk, J., 220
Gottwald, R. L., 248

Name index

Graham, J. T., 308
Grant, D. A., 413
Graves, A. J., 406, 415, 418, 469
Graves, M. F., 415
Gray, J. A., 453, 454, 455, 457, 459, 460, 461, 471, 495, 497
Green, J. B., 24, 298
Greenberg, H. J., 308
Griesel, R. D., 411
Gronwall, D. M. A., 531
Gropper, B. A., 248
Gross, F., 415
Gruen, G. E., 418
Guilford, J. P., 405
Gur, R. E., 501

Haber, R. N., 153, 157, 158
Hagan, B. C., 529
Haldane, E. S., 380
Halperin, X., 96
Halstead, W. C., 26
Halverson, H. H., 374
Hamilton, C. L., 363
Hamilton, C. R., 363, 379, 523
Hampson, J. L., 495
Hanley, C., 495
Hanson, M. R., 413
Harcum, E. R., 146, 148, 149, 153, 154, 155, 156, 157, 158, 159, 160, 161, 162, 164, 165, 166, 167, 169, 178, 180, 182, 183, 184, 187, 188, 189, 190, 191, 192, 193, 194, 195, 199, 200, 203, 206, 207, 208, 209, 210, 212, 213, 216, 221, 222, 223, 224, 226, 227, 228, 229, 230, 231, 232, 233, 234, 235, 241, 245, 246, 247, 248, 250, 251, 253
Hariton, B., 39
Harper, L. V., 435, 536
Harris, C. S., 153, 241, 412, 418, 419, 456
Harris, L. J., 412
Harshman, R. A., 460
Hartlage, L. C., 534
Hartman, R. R., 155, 227
Haslerud, G. M., 206, 221
Havens, L. L., 155, 158
Havighurst, R. J., 456
Hayashi, T., 175
Hays, W. L., 161, 162
Head, H., 33
Hebb, D. O., 26, 141, 244, 246, 249
Hecaen, H., 54, 87, 330, 331, 374, 379, 523, 524, 526, 527, 533

Heilbrun, A. B., 66
Hein, A. V., 241
Held, R., 241
Hellebrandt, F. A., 538
Helmholtz, H., 274, 275, 288
Hermelin, B., 378, 451, 452
Heron, W., 144, 162, 163, 164, 165, 166, 167, 172, 173, 175, 177, 178, 179, 191, 193, 198, 206, 225, 233, 241, 245, 246
Herron, J., 282, 472
Hershenson, M., 158, 159, 226
Herzberg, F., 456
Hess, W. R., 333, 335
Hicks, R. E., 270, 271, 527, 532, 533, 534, 538, 558, 560
Higgenbottam, J. A., 533
Hill, D., 294
Hill, J. H., 402
Hillier, W. F., 103, 109
Hillyard, S. A., 105, 112, 113
Hines, D., 214, 525, 526, 527, 532, 533
Hirata, K., 165, 170, 172, 174, 214, 217, 227
Hirsch, H. V. B., 402
Hiscock, M., 11, 271, 558
Hochberg, F. H., 529
Hochberg, I., 393, 401
Holke, F. A., 321
Holland, B. F., 216
Holmes, G., 33
Hommes, O. R., 128
Honzik, M. P., 435
Hooper, F. H., 418
Horgan, D., 455
Horowitz, L. M., 155
Horrocks, J. B., 531
Horst, E. von, 320
Houtz, S. J., 538
Hubbert, H. B., 410
Hubbs, C. L., 6
Hubel, D. H., 374, 379, 402
Huling, M. D., 278
Humphis, D., 531
Humphrey, M. E., 19, 40, 536
Humphrey, N. K., 336
Hutt, C., 436, 437, 476
Hyman, L. H., 5

Ingle, D., 378
Inglis, J., 393, 394, 397, 400
Inhelder, B., 418, 420, 425, 480

Ishihara, S., 490
Ivanov, A. V., 556

Jackson, J. H., 19
Jacobson, M., 234
James, W. E., 160, 186, 466, 473, 495, 535
Jarvik, M. E., 24
Jasper, H. H., 237, 242, 292
Jastak, J., 295
Jeeves, M. A., 281
Jenkins, J. J., 211, 413
Jenkins, L. M., 413
Jensen, A. R., 440
Jinks, J. L., 440
Johnson, L., 418, 472
Johnson, P. O., 472
Jones, R. K., 108, 164, 165, 171, 178, 218, 222, 234, 245, 253
Jullien, C., 435

Kakeshita, T., 495
Kaplan, E., 365, 531
Karp, S. A., 405, 407, 415, 426, 434
Kato, N., 415
Kaufman, E. L., 208
Kay, H., 250
Kemble, J. D., 447
Kennard, M., 364
Kennedy, F., 19
Keogh, B. K., 406, 411, 415, 416
Kephart, N. C., 217
Kimura, D., 3, 7, 19, 92, 96, 121, 170, 189, 330, 374, 378, 392, 397, 398, 399, 420, 422, 452, 453, 457, 458, 459, 464, 469, 471, 472, 473, 474, 492, 498, 527, 528, 539
Kinsbourne, M., 3, 4, 5, 6, 7, 8, 10, 11, 103, 122, 131, 160, 170, 186, 268, 270, 271, 277, 278, 280, 284, 287, 309, 400, 495, 500, 501, 527, 528, 533, 557, 558, 560, 563
Kirikae, T., 493
Kirk, S. A., 126
Kirssin, J. E., 160, 224, 226
Klaiber, E. L., 482
Kleist, K., 74
Klemmer, E. T., 154, 166, 167
Knehr, C. A., 216
Knox, A. W., 392, 397, 420, 452, 453, 457, 458, 459, 525
Kocel, K., 528
Koepke, J. E., 435

Koestler, A., 211
Koestzle, V. S., 456
Kogan, J., 425
Kohn, B., 489, 562
Kolakowski, D., 439, 443, 444
Kolers, P. A., 242
Komai, T., 523
Konstadt, N. L., 405, 426, 434
Kooi, K. A., 300
Korchin, S., 433
Koziol, S., 406, 415
Krashen, S. D., 96
Krechevsky, I., 415
Kreuter, C., 11, 271
Krieg, W. J. S., 267, 268
Krise, B. M., 242
Kumar, S., 124, 127
Kutas, M., 530
Kuypers, H. G. J. M., 330, 359, 360, 361, 377

L'Abate, L., 228
LaBerge, D. L., 251
LaGrone, C. W., 149, 163, 177, 208, 216, 241
Lake, D. A., 469, 471, 472
Langhorne, M. C., 411
Langworthy, O., 495
Lansdell, H., 118, 469, 470, 471, 476, 495, 496
Lansing, R. W., 299
Lashley, K. S., 26, 243, 360, 361
Lawrence, D. H., 152, 251, 360, 361
Leakey, L. S. B., 484
Ledlow, A., 279, 280, 281, 288, 289
Lee, R. B., 444, 484
LeMay, M., 529, 530, 533
Lenneberg, E. H., 374, 402
Lepkin, M. A., 456
Levin, H., 155, 242
Levitsky, W., 306, 331, 454, 467, 533
Levy, J., 106, 111, 114, 115, 121, 123, 124, 132, 376, 378, 379, 464, 465, 466, 473, 500, 532, 534, 535, 560
Levy-Agresti, J., 366
Lewis, E. G., 301, 456
Ley, R. G., 435
Liben, L. S., 419
Liepman, H., 366
Lindsley, D. B., 298
Ling, A. H., 393, 396
Lippman, H. S., 374
Lipscomb, D. B., 217

Name index

Lipsitt, L. P., 7
Lishman, W. A., 129, 536
Liske, F., 298
Loeb, J., 382
Lomas, J., 527
Long, A. B., 412
Lord, F. E., 415
Low, M. D., 312
Luessenhop, A. J., 108
Luria, A. L., 423, 524, 532

McAdam, D. W., 268, 311, 530
MacArthur, R., 428
McCarthy, D., 455
McCarthy, J. J., 126, 455
McCaskill, C. L., 412
McCulloch, W. S., 267
McDermott, V., 424
McDonnell, P. M., 378
McFie, J., 77, 242, 331
McGinnis, E., 410
McGlone, J., 470, 471, 472, 474, 476, 537
Mach, E., 363
MacKay, D. M., 129
McKee, G., 296
McKeever, W. F., 275, 278, 279, 280, 281, 283, 285, 526, 533, 535, 537
McKenna, V. V., 232
McKinney, J. P., 218
McMeekan, E. R. L., 536
McMurray, J., 11, 271
McNemar, Q., 410, 482
McRae, D. L., 467, 529
Mandes, E. J., 150, 182, 185, 250
Marchbanks, G., 155, 242
Marcie, P., 526
Mark, R. F., 363
Markowitz, H., 170
Marsh, G. R., 309, 472
Marshall, A. J., 212, 234
Martin, J. I., 302
Masani, P. A., 243
Masica, D. N., 484
Massa, R. J., 165, 184, 185, 190, 191, 213, 227, 238
Matheny, A. P., 455
Mather, K., 440
Mathewson, J. W., 197
Matsubara, T., 496, 497
Matsumija, Y., 307
Matthews, B. M. C., 294
Mattson, M. L., 410

Mead, C. D., 455
Meier, M. G., 87, 538
Mellone, M. A., 480, 481, 487
Melton, A. Q., 246
Melville, J. R., 168
Mensh, J. N., 102
Meredith, W., 438, 439
Messer, S. B., 435
Metzler, J., 405, 409, 428, 478, 479
Mewhort, D. J. K., 156, 186, 189, 190, 206, 234
Michon, J. A., 190
Miller, B., 200, 247
Miller, E., 482
Miller, G. A., 155, 198, 409, 465, 466, 535
Milner, B., 69, 73, 106, 111, 126, 127, 330, 331, 367, 396, 464, 525, 526
Milner, P. M., 244, 247
Mishkin, M., 155, 163, 175, 374
Mitchelmore, M. C., 431
Moffett, A., 373
Molfese, D. L., 308, 450, 456, 459, 533, 562
Money, J., 412, 416, 446, 447, 448, 449, 450, 481, 487, 495
Montgomery, R. B., 482
Monty, R. A., 191
Moore, T., 475
Morf, M. E., 415
Morgan, A. H., 296
Morley, M. E., 455, 534
Morrell, J. K., 306
Morris, B. B., 419
Morton, H. B., 362
Moscovitch, M., 275, 277, 281, 282
Moss, H., 456
Motoyoshi, R., 221
Mountcastle, V. P., 271
Muller, G. E., 155
Munroe, R. L., 429, 435
Murray, F. S., 529
Myers, C. T., 431
Myers, R. H., 362, 363, 373, 431

Nachshon, I., 86, 96, 528
Nagylaki, T., 532, 534
Nakamura, R., 539
Napier, J. R., 356
Nash, J., 408
Nash, M. D., 304, 408, 502
Natsoulas, T., 158

Nebes, R. D., 106, 110, 118, 119, 466, 535
Neff, W. D., 96
Neisser, V., 142, 153, 156, 246, 248
Nelson, K., 455
Netsky, M. G., 554
Newbigging, P. L., 407, 433
Newcombe, F., 70, 465, 466
Newman, H. H., 438, 531, 533
Nicolson, A. B., 495
Noble, C. E., 413, 436
Norman, R. D., 482
Norrsell, U., 104, 106
Ntumba, A., 523

O'Conner, J., 444
O'Conner, N., 378, 451, 452
Ogle, K. N., 20
Ogle, S. W., 236, 237, 241
Ojemann, G., 561
Okonji, M. O., 407, 415
Olson, D. R., 429, 456, 529
Olson, G. M., 529
Ommaya, A. K., 361, 363
Orbach, J., 174, 249, 526
Orlando, C. P., 525, 535
Ornstein, R. E., 130, 296, 530
Osaka, R., 165, 170, 172, 174, 214, 217, 227
Otterstaedt, H., 435
Overton, W., 170, 174, 177

Paillard, J., 378
Pallie, I., 454, 455, 467, 533
Panhuysen, L. H. H. M., 128
Papcun, G., 399
Papen, G., 87
Pasik, P., 363, 374
Paterson, D. G., 482
Patterson, A., 331
Pearson, K., 444
Pederson, F. A., 435
Penfield, W., 19, 143, 330, 331, 467
Perkins, D. N., 242
Perria, L., 128
Perriment, A. D., 281, 282
Peterson, A. C., 483
Peterson, G. M., 40, 482, 483
Pfefferbaum, A., 300, 302, 530
Pfeiffer, R. A., 495
Piaget, J., 374, 381, 420, 425, 480
Pierce, J., 156
Piercy, M., 66, 77, 78, 81, 526

Pillsbury, W. B., 226
Pilzecker, A., 155
Pizzamiglio, L., 458
Poffenberger, A. T., 276, 277, 278, 279, 281, 282, 283, 285, 286
Polani, P. E., 495
Pollack, I., 247
Poole, I., 455
Poon, L. W., 305
Popen, R., 87
Porteus, S. D., 430, 433, 475, 491
Posner, M. I., 526
Postman, L., 251
Poulton, E. L., 250
Powell, T. P. S., 271
Pratoomraj, S., 418
Pratt, R. T. C., 524
Preilowsky, B. F. B., 378
Prescott, R. H., 456
Price, R. H., 156, 158
Provins, K. A., 8, 295, 531, 537, 538
Putnam, W., 412

Quadfasel, F. A., 19, 40, 465, 469

Rabbitt, P. M. A., 269
Rabe, A., 146, 166, 182, 184, 185, 188, 194, 195, 199, 200, 207, 209, 210, 221, 226, 248
Rabin, P., 115
Rabinovich, M. S., 393
Raczkowski, D., 536, 537
Rainey, C. A., 149, 165, 166, 168
Ramaley, F., 532
Ramony Cahol, S., 382
Raney, E. T., 237, 242, 294
Ranucci, E. R., 431
Rasmussen, T., 19, 20, 101, 103, 109, 312, 330, 493, 525, 526
Ratcliff, G., 465, 466
Ray, O. S., 411, 416
Rebelsky, F., 419
Reed, G. F., 539
Regan, J. O., 220
Reitan, R. M., 465
Remington, R., 460, 472, 501
Reshevsky, S., 409
Revesman, S., 217
Rhoads, J. G., 523
Richlin, M., 301
Rife, D. C., 523, 531, 532, 534
Rigney, J. W., 40
Rizzolatti, G., 529

Name index

Roberts, J., 524, 527
Roberts, L., 19, 143, 330, 396
Robinson, G. M., 96
Rommelveit, R., 248
Rosadini, G., 129, 143, 396
Rosenfield, R. L., 448
Rosenzweig, M. R., 492
Ross, G. T. R., 380
Ross, S., 221
Rossi, G. R., 129, 143, 396
Rovet, J., 431
Roy, J. N., 59
Rubino, C. A., 87
Rudel, R. G., 451, 452, 453, 459, 477
Russell, D. J., 298
Russell, R. W., 70
Russo, M., 66
Rutherford, M., 525
Rutledge, L. T., 288
Rutschmann, R., 171, 239

Saad, L. G., 408
Sadownikova-Koltzova, M. P., 410
Sait, P. E., 401
Sakata, H., 306, 360
Sampson, H., 215, 218, 531
Sanders, A. F., 232, 435, 436
Sarnat, H. B., 554
Satz, P., 393, 525, 526, 527, 532, 533, 534, 535
Sauget, J., 524, 527, 533
Savasto, M., 436
Saylor, H. D., 455
Scheinfeld, A., 435
Schenkenberg, T., 302
Schiffman, H. R., 149, 164, 165, 189, 206
Schiller, P. H., 356
Schlosberg, H., 251
Schneider, D., 334
Schneider, G. E., 335
Schneidler, G. R., 482
Schneirla, T. C., 4, 6
Schonell, F. J., 242
Schrier, A. M., 331
Schwartz, G. E., 407, 415, 422
Scotti, G., 51, 54, 61, 63, 64, 70, 71, 72
Seales, D. M., 305, 530
Seeger, C. M., 252
Seils, L. G., 412
Semmes, J., 29, 38, 69, 99, 239, 355, 461, 462, 463, 488, 498, 499, 562

Serafetinides, E. A., 126, 129
Sersen, E. A., 22
Seyffarth, G., 318
Shaffer, J. W., 446, 495
Shankweiler, D., 96, 397, 399, 422, 525, 535
Shapiro, G., 30
Sheibel, A. B., 500
Shelburne, S. A., 304, 305
Shepard, R. N., 405, 409, 428
Shephard, A. H., 414, 430, 478, 479
Sheridan, M. R., 538
Sherman, J. A., 490
Sherrington, C. S., 3, 277
Shiffrin, R. M., 247, 249
Shimrat, N., 539
Siegvald, H., 408
Silverman, A., 415, 466, 535
Silverman, J., 415
Simon, H. A., 247, 409
Simon, J. R., 286, 287
Simonsen, J., 19
Singer, G., 482
Siqueland, E. R., 7
Skrzypek, G., 149, 154, 183, 184, 189, 206, 213
Smart, R. C., 412
Smith, A., 102, 109, 117, 121, 125, 127, 128, 408, 472,
Smith, K. U., 366
Smith, N. F., 149, 153, 155, 193, 221, 227, 229, 250, 251
Smyth, V. O., 66, 78, 81
Solomon, D. J., 96
Sommers, R. K., 393, 396, 397, 398, 408
Spehlmann, R., 304
Sperling, G., 153, 243, 246
Sperry, R. W., 100, 102, 104, 105, 106, 109, 110, 115, 116, 124, 128, 129, 132, 331, 362, 363, 365, 366, 367, 368, 373, 375, 376, 377, 378, 463
Spinelli, D. N., 402
Spinnler, H., 51, 58, 118
Spong, P., 215, 218
Sprague, J. M., 335, 378
Stafford, R. E., 442
Staples, R., 482
Steingrueber, H. J., 537
Stern, C., 444
Sternberg, S., 249, 275
Stone, C. P., 411

Storer, W. O., 408
Strauss, H., 294
Stroop, J. R., 482
Stuckenschmidt, H. H., 494
Studdert-Kennedy, M., 397, 399, 525, 535
Subirana, A., 532
Sutherland, N. S., 500
Sutton-Smith, B., 436
Swaan, E. T., 155
Swanson, J. M., 279, 280, 281, 284, 287
Swets, J. A., 158
Sykes, D. H., 393, 394, 397, 400

Takala, M., 211, 218, 239, 243
Tanner, W. P., 158
Tauber, G., 321
Taylor, M. L., 106, 126, 393, 396, 397, 398, 497, 498
Tecce, J. J., 309
Templin, M. C., 455
Teng, E. L., 105
Terman, C. M., 455
Terrace, L. M., 168, 175, 282
Terzian, H., 101, 103, 109, 128
Teuber, H. L., 26, 27, 120, 241, 330
Thomas, H., 299, 419, 432, 485, 495
Thompson, L. W., 309
Thurnam, J., 20
Thurstone, L. L., 118, 120, 405, 427, 439
Tiacci, C., 78, 87
Tiffin, J., 482
Tinker, M. A., 177, 178, 180, 241
Toman, J., 299
Tomlin, M. I., 411
Trask, F. P., 249
Treisman, A. M., 399
Trevarthen, C., 115, 122, 335, 336, 351, 357, 363, 364, 366, 367, 368, 373, 376, 377, 378, 379
Trowell, H. C., 448
Tryon, R. C., 410
Tschirgi, R. D., 500
Tuddenham, R. D., 416, 418, 439, 494
Tunturi, A. R., 492
Turkewitz, G., 7, 374
Turner, H. H., 535
Twitchell, T. E., 374

Uexkull, J. Von, 332
Uhrbrock, R. S., 523

VanAllen, M. W., 529
Van Buren, J. M., 561
Van de Meer, H. C., 171, 240
Vanderberg, S. G., 405, 407, 438
Vanderwolf, C. H., 378
Varney, N. R., 529
Vaughan, H. G., 66, 499
Vaughn, H. G., 530
Vella, E. J., 304
Verhaegen, P., 523
Vernon, M. D., 534
Vetter, R. J., 33
Vignolo, L. A., 82
Von Bonin, G., 20
Von Economo, C., 495
Vore, D. A., 418

Wada, J., 19, 20, 103, 109, 312, 454, 455, 467, 468, 493, 495, 533
Wade, M. G., 413
Wagner, J., 20, 179, 181
Waldrop, M. F., 456
Walker, E. L., 246
Wallace, R. J., 279, 287, 288
Walter, D. O., 298, 533
Warren, J. M., 373
Warrington, E. K., 77, 78, 115, 117, 160, 186, 473, 495, 524
Watanube, H., 484
Watson, W. S., 426
Watts, J. W., 364
Webster, W. G., 523
Wechsler, D., 473
Weinberg, L. K., 126, 418
Weinstein, L., 22, 25, 26, 27, 28, 29, 31, 33, 34, 36, 41, 120, 355
Weiskrantz, L., 336
Weiss, A. P., 456
Weitzman, D. O., 170
Welch, J. C., 268
Wellman, B. L., 412
Wells, C. E., 294
Werner, H., 160
Wertheim, T., 179
Wertheimer, M., 120
Whitaker, H. A., 268, 311
White, M. J., 176, 197, 212, 224, 236, 282, 312, 330
Wiener, M., 170, 174, 177
Wiesel, T. N., 375, 402
Williams, M., 488
Wilson, M. O., 294, 430, 523, 534
Winnick, W. A., 154, 158, 164, 213, 214, 227, 250

Name index

Winters, J. J., 226
Wit, O. C., 433
Witelson, S., 393, 452, 453, 454, 533
Witkin, H. A., 406, 407, 415, 429, 433, 434, 435, 490
Wkye, M., 157, 165
Wold, R. M., 534
Wolff, B. B., 24, 25
Wood, C. C., 307
Woodworth, R. S., 181, 274, 275, 288
Woolley, H. T., 501
Wussler, M., 534
Wyatt, D. F., 158
Wyke, M., 217

Yntema, D. B., 249
Young, F. M., 456
Young, J. Z., 39, 40, 330, 456

Zaidel, D., 116
Zaidel, E., 103, 109, 112, 113, 114, 120, 121, 125, 126, 378, 379
Zangwill, O. L., 19, 40, 77, 143, 330, 331, 374, 450, 463, 524, 534
Zazzo, R., 435
Zener, K., 235, 536
Zuccato, F. C., 412
Zurif, E. B., 143, 174, 220, 393, 401, 525, 526, 527, 533

Subject index

acetylcholine, 21
agenesis, 375
alpha activity, rhythm, 293; hemispheric, 298, 303, 312, 422, 530
alpha blocking, 422
amnesic, 488
amorphosynthesis, 361
amphibian, 330
amputation, 31, 32, 34
androgen, 483, 488
animal sounds, 420
aphasia, 69, 86, 102, 103, 126, 447, 449, 470; crossed, 18, 19, 40; jargon, 87; motor, 87; recovery from, 563; Wernicke, 423
aplasia, 31
approach, 4
apraxia, 117
Army General Classification Test, 26
articulation apraxia, 534
astereognosis, 461
attention, 8, 9
auditory evoked responses, 472
auditory localization, 424
Australopithecus, 523
averaged evoked potentials, 300–12

baboons, 337–58
Balint's syndrome, 54
bassoon, 424
Bender Gestalt, 126, 446
Benton Visual Retention Test, 127, 447
beta activity, 294
Bhil, 433
Bilateria, 554
bilingual subjects, 163
binocular parallax, 347

birth trauma, 534, 535
bisymmetrical, bisymmetry, 3–11
block counting, 407
body schema test, 29
braille letters, symbols, 451, 452
braille reading speed, 451, 477
Broca's area, 268, 311, 312

callosal section, 11, 100, 104–6, 112, 120, 122, 271; and consciousness, 129, 130; and emotion, 127, 128; left hemisphere after, 102; in monkey, 364–78; right hemisphere after, 108; and spatial perception, 115, 124
Carolinians, 433
cats, 330, 335, 372, 373
cell assemblies, 244
cello, 424
cephalochordata, 556
cerebral dominance, 17, 40, 41, 86, 143, 160, 218, 220, 230, 233, 239, 248, 292; and depression, 128; and directional habits, 164, 173, 177, 217; evolution of, 372–4; in reading disability, 242; of stutterers, 108
cerebral palsy, 534
Chamoiros, 433
chess, 408
chiasm, 347, 351, 363, 372, 373
Children's Embedded Figure Test, 426, 434
chimeras, 123, 124, 132
Chordate Plylum, 554
chromosomal mosaicism, 445, 447
chromosomes, 440, 442, 445, 446, 448, 450
closure flexibility, 120

Subject index

closure speed, 118
cognitive style, 176
color agnosia, 495
color blindness, 446, 495
consciousness, 129, 131, 330, 378
conservation of number, distance, volume, area, 417, 418
constructional apraxia, 74
consummatory response, 335
contingent negative variation, 309, 310, 312
core motor system, 359
corpus callosum, 463, 528; anterior commissure section, 347, 378; and EEG, 298, 310, 311; and IHTT, 276, 277, 283–5
cortical evoked potentials, 39
corticorubrospinal projections, 359, 360
corticospinal connections, 358

decentration, 417
decussation, 554–7, 563
Detroit Test of Learning Aptitude, 447
developmental aphasia, 534
dichotic listening, 5, 10, 111, 244, 420, 458, 471, 472, 492, 525, 527; development, 392–401, 558; nonverbal, 96, 422, 451, 452
Differential Aptitude Test, 407, 420, 426, 427, 438, 442
Diptera, 556
directional sense, 415
dowel rod, 11
Draw-a-Floorplan, 447
Draw-a-Person, 447
dual-task interaction, 9
Dutch, 433
dynamic balance, 412
dynamometer, 530
dysarthria, 534
dysgraphia, 447
dyslexia, 447, 535
dysphasia, 477, 488, 493, 524, 532

EEG, 292, 293, 299, 300, 303, 472
electroconvulsive therapy, 524
electromyographic 267, 268
Elithorn's Test, 66
embedded figures, 425; test, 407, 415, 429, 433, 434, 443, 482, 490, 491
English, 433
epilepsy, 534
Eskimos, 428, 429, 485, 487

estrogen, 448, 483
evoked potentials, 530, 531, 562
explanation of proverbs test, 469

feminization, 448
finger agnosia, 447
finger oscillation, 36, 37
fingerprints, 533
fins, 6
fish, 330
Fitt's index of difficulty, 538
flounder, 330
flute, 424
French, 433
frequency analysis, 96
frog, 336
functional distance, functional space, 9, 267–9, 271, 278

geographical ability, 415
geographical disorientation, 450
gesturing, 528
Ghent overlapping figures, 490
gonadal agenesis, 448, 449
Goodenough-Harris Human Figure, 535
Gottschaldt's Test, 66, 67, 406
grasp, 7, 8
Grones Design Judgment Test, 469
Guilford-Zimmerman Spatial Visualization Test, 443
gynecomastia, 448

hamsters, 335
heart, 555
Hebrew, 163, 164, 169, 174, 223, 249, 283
hemispatial neglect, 347, 348, 364, 371
Heschl's gyrus, 454
Hong Kong Chinese, 433
human infant, 356, 374, 378

Identical Blocks Test, 442
IHTT, interhemispheric transfer time, 270, 277–89
Illinois Test of Psycholinguistic Abilities, 120, 126
infantile hemiplegia, 490
invertebrate, 554, 555, 556
invertebrate-vertebrate transition, 555, 556
Ishihara Color Blindness Test, 446, 490
Italian, 433

Japanese words, 165, 170, 172, 174, 175, 217, 221

Kalahari Bushmen, 433
Klinefelter's syndrome, 447
Kohs Blocks, 428
kwashiorkor, 448

Landolt rings, 165, 170, 172, 216
lateral gaze, 5
lateral inhibition, 181
Laterality-Discrimination Test, 412
left-right discrimination, 412
Logoli, 429

mammals, 336, 378
manual dexterity test, 536
map reading, 411
mastectomy, 34
maze performance, 410
memory disorders, 71
memory for designs test, 127
memory for position tests, 71, 72, 73
menstruation, 448
mental retardates, 534, 535
mental rotation, 425, 439
metacontrol, 378
Morrisby Shapes, 428
music composition, 421
music performance, 421

neonates, 374
neuraxis, 555-7
nonnutritive sucking, 451, 562
notochord, 554, 555

optokinetic nystagmus, 363
orientation, orienting, 4, 5
oropharynx, 554, 555

pain sensitivity, 24, 25
palmprints, 533
paper-folding test, 439
paper formboard, 439
paraplegia, 31, 34
paroxysmal dysphasia, 526
pathological left-handedness, 534, 535
paw preference, 7, 21, 40, 531
pectoral fins, 560
personal orientation, 29
phantom perception, 31
phi test, 237
photic driving, 299, 533
Piagetian Tests, 416, 418, 439
pineal gland, 329
planum temporale, 454, 455, 468, 533, 560
plasticity, 19

pogonophore, 556, 557
point localization, 22
Porteus Maze Test, 433
precision motor system, 359
pressure sensitivity threshold, 21, 22, 24, 31, 34, 38
Primary Mental Abilities Test, 439, 447
prosimians, 356
Prosobranchia, 556
psychosurgery, 26
pyramidal, 360

rats, 335, 415
Raven's Colored Progressive Matrices Test, 61, 66, 116, 470
reading disability, 534
response-response compatibility, 538
rhesus monkeys, 377
Richtfeld, 330
Right-Left Discrimination Battery, 447
road map test, 411, 412, 447, 490, 495
rod-and-frame tests, 414, 415, 429, 433, 434, 482, 490, 491
rotary pursuit task, 269, 413, 436
roughness discrimination, 28, 29

Seashore Test, 422
sex chromatin, 445
sex-role stereotypes, 425, 428, 435, 437
shuttlebox, 416
situs inversus, 532
size discrimination, judgment, 27, 28, 31
sodium amytal test, 19, 20, 100-8, 109, 121, 126, 128, 129, 312, 422, 493, 525
sophistication of body concept scale, 434
spatial memory, 69, 70, 71, 92
spatial perception, 61, 65, 69, 71, 92
speech, 18, 19, 41
stammering, 534
Stanford-Binet I.Q., 475
stepparents, 532
stock brainedness, 19
Street Test, 470, 490
stutterers, 108
Surface Developmental Test, 427, 431
Sylvian fissure, 533
syntax, 113

tactile memory, 126
tactile shape discrimination test, 63
tactual mazes, 411
tactual sensitivity, 22, 30
tectobulbar, 359
Temne, 428, 429

Subject index

temporal lobectomy, 29, 30
testicular atrophy, 448
testicular feminizing syndrome, 448
testosterone, 448
thalamic, 561, 562
three mountain task, 480
Thurstone's Primary Mental Abilities Spatial Relations Test, 473, 474
toad, 336
tongue clicking, 529, 531
tonic neck response, 559
topographical memory, 450
Toronto Complex Coordinator, 413, 414, 430
Turner's syndrome, 445–50, 481–9; color blindness, 495
turns: 485, 531, 532
two-point discrimination, 22, 28, 31, 38

Umwelt, 330
unilateral neglect, spatial agnosia, 8, 54, 58, 60, 82, 363, 471
Urochordata, 556

vertebral column, 554
vertebrate, 356, 378, 555–7, 560, 563
visual hemi-inattention, 54, 61, 82
visual mazes, 411
visual memory, 73

Wada, 477
WAIS (Wechsler Adult Scale of Intelligence), 470; block design, 66, 473; object assembly, 66, 482; similarities, 125, 488; testicular feminizing syndrome, 448; in Turner's syndrome, 446
water level test, 418, 419
Wechsler-Bellevue Intelligence Scale, 470
Wechsler Memory Scale, 127
Wernicke's area, 305, 308, 423, 451
Wiggly Block Test, 444
WISC, 475, 482, 535
withdrawal, 4
word fluency, 456

X-linked trait, 440, 442, 445